Monad to Man

Monad to Man

The Concept of Progress in Evolutionary Biology

Michael Ruse

HARVARD UNIVERSITY PRESS
Cambridge, Massachusetts
London, England
1996

Printed in the United States of America

Library of Congress Cataloging-in-Publication Data

Ruse, Michael.
Monad to man : the concept of progress in evolutionary biology / Michael Ruse.
p. cm.
Includes bibliographical references and index.
ISBN 0-674-58220-9 (alk. paper)
1. Evolution (Biology)—Philosophy 2. Evolution (Biology)—History. I. Title.
QH360.5.R87 1996
575—dc20 96-18951

For my children

Nigel

Rebekah

Emily

Oliver

Edward

Contents

Acknowledgments

I owe this book to three people. First, to F. H. Legg, my history teacher at Bootham School York back in 1955. If some of his enthusiasm for the Victorians comes through in my pages, then I have started to repay my debt to him. Second, to Robert M. Young, under whose direction I spent a year in the Wellcome Unit at Cambridge University in the early 1970s. I agree with few of his conclusions and he agrees with none of mine, but I still think that his is the most exciting mind ever to have turned to the Darwinian Revolution. He—and Martin Rudwick and Roy Porter, both then also at Cambridge—taught me that you simply cannot think about science divorced from the social context. The third debt I owe is to Edward O. Wilson, at Harvard University. He was important in two ways. First, because he urged me to attempt at least once to write a really big book. In one sense, I have certainly done that! Second, because in the course of writing a paper with him on ethics I came to see that, although we were (and are) close friends and committed Darwinian evolutionists, his vision of the field is simply out of focus with the way that I see things. I think I now know wherein lies our difference.

Several institutions welcomed me while I was working on the manuscript. In England, these were the Department of History and Philosophy of Science, Wolfson College, and Pembroke College, all of the University of Cambridge: I am grateful especially to Michael Redhead, Mary Hesse, and Nicholas Davies. In France, welcome came from the Ecole d'Hautes Etudes in Paris and the Laboratoire de Paleontologie Vertebré in Montpellier: I am grateful especially to Jacques Michaux and Jean Gayon. My home university of Guelph has been very tolerant of my absences. I am much obligated to David Murray, Carole Stewart, and Brian Calvert. I am also deeply in the debt of my several typists and assistants: Gail McGinnis, Linda Jenkins, Moira Howes, and David Castle.

The staffs of the archives and libraries I consulted were always friendly and helpful, and the same is true of the scientists I interviewed. Both private and public funds supported my labors, and I am especially grateful to the Isaak Walton Killam Memorial Fund and the (Canadian) Social Sciences and Humanities Research Council.

Many people read part or all of my text. If you had seen the earlier versions, you would know that this really was an act of true friendship. The list includes: Peter Bowler, Jean Gayon, Jon Hodge, David Hull, Jim Lennox, Dan McShea, Ernst Mayr, Greg Mitman, Ron Rainger, Marc Swetlitz, and Polly Winsor. My closest intellectual friends and sternest critics have been John Beatty and Robert J. Richards. Bob's reaction, on reading the manuscript, was to throw it across the room. I have striven to ensure that he will continue to feel that way.

My editors at Harvard, Howard Boyer, Michael Fisher, and Kate Schmit, have been terrific, and the same is true over and over again of my wife, Lizzie. We have now done with rental homes and at long last she can spend the summer tending her own beloved garden. Finally, a word about my dedication. I do not know if this is the best book I shall ever write; but I do know that it is the one which has taken the most time, demanded the most effort and caused the most stress, and it is the one for which there is the biggest gap between the original bright idea and the finished product. It is therefore peculiarly appropriate that it be dedicated to my children.

Monad to Man

Introduction

Le Jardin des Plantes lies just a few hundred meters upriver from Notre Dame, on the left bank of the Seine, in Paris. Dominating the grounds is one of the most beautiful buildings in a city of beautiful buildings. La Grande Galerie du Muséum National d'Histoire Naturelle was opened in 1889 (as was the Eiffel Tower) to commemorate the hundredth anniversary of the beginning of the French Revolution.

Alas, in our time, for many years the building stood empty, in a seemingly endless state of repair and renovation. Finally, in the spring of 1994, the doors were opened and the public was once again invited in. The wait was justified. Entering the vast hall, one is simply overwhelmed by a huge display celebrating the diversity of life. Adam and Eve never dreamed of this! Yet, although it commands the visitor's attention at first, in a way this flamboyant exhibition of the curator's skill is but a filler. The real message and purpose of the museum is to be found in the side galleries, tier upon tier of them around the walls.

It is the goddess of *evolution* to whom the museum is dedicated—the story of life from its first beginnings, and of the causal mechanisms which fuel the way forward. From displays using the most simple of graphics to those relying on all of the tricks of high technology, you are guided on a trail from life's earliest forms to our own species, *Homo sapiens,* and treated to glimpses of what perhaps lies beyond. Memorably, in what is surely the greatest triumph in the hundred-year history of the Entente

Cordiale, pride of place is given to he who is known as the "father" of evolution, the Englishman Charles Robert Darwin (1809–1882).

Evolution: Triumphant or Troubled?

There is no real surprise that the museum is structured as it is. Evolution is one of the ideas of our age. What child of the playground is ignorant of the dinosaurs? Has not seen them in pictures, has not collected them in plastic, has not eaten them as pasta? Who could be a serious reader of the newspapers and be unaware of the fabulous hominid finds in Africa—near-complete skeletons showing that our ancestors rose up on two legs, before their brains exploded in size? And who has not bought, for themselves or a friend or a relative, one of those best sellers explaining so vividly some aspect of development in time? *Ever Since Darwin* by the American paleontologist Stephen Jay Gould, or *The Blind Watchmaker* by the British sociobiologist Richard Dawkins?

Nor is this fascination with evolution merely a phenomenon of the popular realm. Historians of science are producing scholarly editions of key texts, most especially Darwin's private notebooks and all of his correspondence, both the letters to and the letters from him (Barrett et al. 1987; Darwin 1985–). Philosophers like myself have discovered evolution, to such an extent that we have to be reminded that there is more to biology than evolution and more to science than biology (Callebaut 1994). The same is true in other fields, both the social sciences and the humanities. Recently it is the students of speech and rhetoric who have turned to the subject, happily deconstructing the metaphors and other stylistic tricks in the evolutionists' texts (Myers 1990; Selzer 1993).

Most importantly, there are the scientists themselves. Where before we had departments of zoology and botany, now we find departments of evolutionary biology (often linked with ecology) cutting right across traditional divides. Associated with this "new" discipline, there are places now for evolutionists to display their labors. Not only do the generalist publications like *Science* and *Nature* carry much on evolution, but there are specialist outlets also. In 1994, for instance, the journal *Evolution* appeared in six issues, in 2,066 pages in all. There were a total of 171 articles, authored by 358 people, on organisms from fruit flies to salamanders, from white-tailed deer to milkweeds.

Yet, all is not well. Evolution may be one of the dominant ideas of our time. It is also one of the more troubled. Stay with the scientists for a moment. The Russian-born American evolutionist Theodosius Dobzhansky (1900–1975) used to boast that nothing in biology makes sense except in the light of evolution. Perhaps so. But few who have fought over undergraduate biology curricula can be unaware of how difficult it can be to insert evolution explicitly into the program. There is always another class in biochemistry which is thought absolutely essential for entry into graduate or medical school. It is hard to quantify these sorts of things, but a quick survey of *Peterson's Guide to Graduate Programs in Biological and Agricultural Sciences* is very suggestive. The 1994 edition advertises about 325 programs in molecular biology and only 45 in evolution—nearly an order of magnitude of difference. Overall in the *Guide,* cell and molecular biology get over three hundred pages whereas ecology, environmental biology, and evolutionary biology get barely fifty. This is hardly a stellar performance.

Listen to paleontologist Anthony Hallam talking about perceptions at his home university (Birmingham) of England's most distinguished evolutionist:

> John Maynard Smith, whom I have a great admiration for, was invited to give the Huxley lecture. I'm on the Huxley committee, which selects these things, and I pushed for him. But my biological professor colleagues were saying—"Ah yes, Maynard Smith. Isn't he old hat? Couldn't we get a molecular biologist like Alec Jeffries?" Well, in fact, we got Alec Jeffries, the year after! It isn't either/or really, but there's no question about the fact that my biological colleagues—in this university and I think they speak for many biologists—don't see the sort of thing that Maynard Smith does as too pertinent to the mainstream of biology and what really turns them on. (Interview with author, spring 1991)

Others, physiologists and those whom Hallam dismisses as "cell crunchers," may belittle evolutionary theorizing and its practitioners, but it is not as if full-time evolutionists always inspire much confidence in themselves and their ideas. One of the most unpleasant disputes in recent science occurred when the Harvard entomologist Edward O. Wilson published *Sociobiology: The New Synthesis*—a magisterial overview of applications of Darwinian theory to animal and human social behavior. One might have expected the social scientists to have felt threatened and to have reacted nastily. Much more of a shock was the fact that Wilson's leading critics were his fellow department members, the population geneticist (and student of Dobzhansky) Richard C. Lewontin and the aforementioned Stephen Jay Gould. If they are troubled, how should we feel? (See Allen et al. 1976; Segerstrale 1985.)

Outside the narrow confines of science the controversy rises, not only about theories but about the very idea of evolution itself. The Creationist debates of the 1920s have had to be fought all over again in the 1980s, as American fundamentalists, convinced of the literal truth of all of the Bible, have insisted that the early chapters of Genesis give an origins story far superior to that of evolution (Montague 1984; Ruse 1988). It is easy and tempting to dismiss these people as fanatics, and many of them are, but they have influence—especially with boards of education in the United States. Moreover, recently they have found more sophisticated supporters, among lawyers for instance, and even more so among philosophers (Johnson 1991; Plantinga 1991).

Not that the philosophical community taken as a whole has ever been entirely enamored with evolution and its implications. Most Anglo-Saxon philosophy of this century—continental philosophy for that matter also—would go unaltered if the biblical six days of creation were indeed shown true. The influential Ludwig Wittgenstein, for example, was contemptuous in his dismissal of the significance of evolution (Wittgenstein 1923, 4.1122). Nor was his great rival, the late Karl Popper, all that happy on the subject. It is true that he did try, very hard, to use and internalize evolutionism. But, notoriously, Popper characterized Darwin's thinking as less than genuine science and more a "metaphysical research programme" (Popper 1972). He qualified this until the day he died; but, somehow, the stain would not scrub out.

Paris or Potemkin Village? Which is the better metaphor for evolution's success as a revolutionary idea? This is a challenge, and, for all that we philosophers may contribute to the problem, as students of the theory of knowledge, this is a challenge for us. What is the nature of evolutionary theory and what, if anything, is it about the subject that makes it so addictive and yet leaves so many feeling so queasy? This book is one philosopher's attempt to answer these questions. Not through logical analysis of the ideas today. My approach is through history for, as a deeply committed evolutionist, I believe that the answers to the present are to be found in the past. Hence, what I will offer you is a history of evolutionary ideas. But not just a history of evolutionism *per se,* for mine is a philosophical history of science. Not a history of philosophy nor yet a history of philosophy in science, although there is some of that. Rather, I write a history which is used to *understand* the present. For this is what philosophy is all about.[1]

Professional Science/Mature Science

In turning to history, as does the evolutionist, I reveal to you my philosophical allegiance. I am not one for *a priori* theorizing in a time-frozen vacuum. I am following the scientists themselves. I am a *naturalist,* meaning that since science is the best kind of knowledge that we have, as a philosopher I take science as a model (Ruse 1995). But this means that if I am to move forward in my task, I must have some guidelines or signposts. Charles Darwin himself expressed matters exactly, grumbling that if you do geology without thought, "a man might as well go into a gravel-pit and count the pebbles and describe the colours." He added: "How odd it is that anyone should not see that all observation must be for or against some view if it is to be of any service!" (Darwin 1985– , 9, 269). I must have some hypothesis (or hypotheses) about evolution and its history which I can test, if I am to achieve my ends.

The easiest way to generate such a hypothesis is by searching first for the ideal. Suppose we were entirely satisfied with evolution. Suppose we all counted it as good science, great science even, and that no-one questioned this status in any way whatsoever. What would this mean exactly? What features would we expect evolution to satisfy then? We can take as a background presupposition that the aim of science is to give us knowledge of, understanding of, the world of experience. Yet, this very presupposition enjoins me to take particular care. Because I am trying to work in the spirit of the scientist, it is not for me to stipulate what makes for science, and especially not what makes for good science. It is not for me to prescribe merit, and it is certainly not for me to make that which we have today, by definition, the best. Instead, I must work descriptively, telling what people seem (or have seemed) to regard as good or satisfactory science, what people judge to be the better kind of science.

Putting matters this way, we might decide to pursue our inquiry by following the lead of *sociology.* You may feel that, as philosophers, such a direction is misguided, beneath our dignity perhaps. Let me say that, if nothing else, I hope this study will force you to change your opinion. Here, without apology, I ask you to remember that when characterizing our sense of pride in our thinking about evolution—and even more when digging at our sense of worry—I spoke of such things as the success of the journal *Evolution* and the failure of evolutionists to occupy the academic ground of molecular biologists. These are things which touch very directly on our judgments of status or worth. Because the journal is doing well, we feel good about evolutionary studies; because departments of molecular biology much outnumber departments of evolution, we have questions.

Simplistically, one might say that good or genuine or top-quality science is that produced by good or genuine or top-quality scientists. But what are the latter? The sociological notion I shall employ in this study is that of *professional* science, something done by professional scientists. What I have in mind here is the science of the person who has made a full commitment to his or her subject, who has achieved the skill and status of a practitioner of the traditional professions—medicine, law, the clergy (Shils 1968; Ben-David 1972; Cardwell 1957). Recognizing that this is an evolving notion, we should anticipate that much of our interest will center precisely on and around nuances in the idea of scientist as professional, and that it is a mistake to expect rigid necessary and sufficient criteria. Nevertheless, one can fairly safely say that today, in speaking of a "professional scientist," one would probably be referring to a person working full-time (or nearly so) at a university or research laboratory. Such a person will have qualifications and training—a doctorate and more years beyond that—and a recognized position in society. And although they may be paid well for what they do, it will be understood that scientists practice their art for the sake of the art, rather than for mere gain. A professional scientist has much in common with the person prized by Socrates in the first book of the *Republic.*

Today's professional scientist has students: many have remarked how the culture of science mirrors the old apprenticeship patterns. He or she belongs to professional organizations—some more prestigious and restricted than others—and a mark of one's status is often one's contribution to the functioning of these groups. The professional scientist publishes in professional outlets—especially in refereed journals, but also in monographs and in multi-authored edited collections. Here, particularly, we have a significant line of demarcation from the nonprofessional world. Within and only within the accepted outlets does one find the esoteric understanding of the true professional.

Closely connected to the notion of a professional scientist is that of a *scientific discipline* (Hull 1988; Kohler 1991). Scientists do not look at everything, indifferently. They break up into certain subject areas—working in different *fields* (or *domains*) of inquiry. Without wanting to preempt future discussion, it seems fair to say that here, at least, things get more specialized as time goes by. The field (and discipline) is narrowed into sub-fields (and sub-disciplines), which may then take on autonomous existence. The discipline is the social group working on or in the field. It is marked especially by its own specialized journals and organizations. Today, the success—the very existence—of a discipline is a function of its success in attracting bright students and adequate grant money or other support. Hence, the building of a discipline often involves the finding of potential consumers, for one's intellectual products and for one's students.

A field will have various ideas or sets of ideas which unite its practitioners. It is important to note that professional competence in one discipline does not necessarily imply professional competence in other disciplines. One is reminded in this context of silly things claimed in the last two decades by senior astronomers about the origins of life. However, it is true that what the sociologists of science call the "Matthew effect" is often evident—people at the top of the field get taken more seriously (and get more credit) than people at the bottom (Merton 1973). One's efforts to achieve discipline status—perhaps moving oneself out from other disciplines or up from nonprofessional science—may get a major boost from the enthusiasm of respected professionals in other fields. If one can get such people to contribute to the new field, all of one's worries may be over.

Let us speak of the aim of professional scientists as being the production of *mature* science. I mean by this term work which is valued and respected, especially by fellow professionals. Clearly professionals may get involved in a new area of science, which almost by definition is not particularly mature; indeed, its *im*maturity may be the attraction. But the intention is to move forward toward maturity. This is not to say that there is no place in professional science for the independent thinker. My point is just that a thinker who persistently goes against or ignores the norms or values of a particular professional community is going to lose respect—be excluded from the journals and the awards and the organizational kudos—and may eventually be pushed to or beyond the limits of accepted professionalism.

A notion like "mature" science starts to take us from the purely sociological to more traditional philosophical questions, to questions about the status of science in itself as good or bad at what it is supposed to be or to do—that is, to questions about its worth as knowledge: *epistemology*. But before turning directly to this, ask first about the flip side to professionalism. It is natural to think of science beyond the borders marked out by professional scientists as being "amateur"; but, since I will look at the past as well as the present, I am loathe to use this term with its connotations of unpaid activity. In England, particularly, with (as we shall see) a history of reluctance to give state support to science, we run the danger of anachronism: illicitly reading the present into the past. I shall rather speak of "nonprofessional" science as *popular science,* in the sense of something accessible to the general public. I think of *Scientific American* as being an exemplar of (the very best) popular science. It is true that, in a way, its contents are very professional—but the articles are intended for any reader with an interest in the subject, not only for the authors' disciplinary peers.

Note, therefore, that I am not using the term *popular science* as such in a pejorative sense. Nor do I want to suggest that there is a hard and fast line between professional science and popular science. What interests me is that a group of people might want to upgrade their science across the boundaries. Theoretically, it seems that there are a number of ways in which one might attempt this, including redrawing the boundary! More tangibly, one might set about organizing the social side of a professional science—journals, organizations, and the like—although, showing that norms or values are bound to be involved here, once this effort is under way certain criteria will be invoked to justify decisions on whose work is to be included and whose is not. The whole point of a profession is that its members have moved beyond the popular realm.

An important sub-branch of what I call "popular science" is *pseudo-science* or *quasi-science* (Hanen et al. 1980). Here we are getting pejorative, for these words refer to things which are generally thought discreditable, bogus even, by professional scientists. In the past, phrenology fitted the bill. Today, Scientology qualifies. Some topics are on the border, perhaps moving one way or the other. Homeopathy strikes me as being in this state. All pseudo-science is popular science. Not all popular science is pseudo-science. The articles in *Scientific American,* although popular, are not pseudo or quasi. You might want to elevate pseudo-science to the same hierarchical level as popular science. I prefer to see the former contained in the latter, not only because of the border cases and because pseudo-science is popular but because the denial of pseudo-science is certainly going to be (non-pseudo) popular science. Someone who takes time out to attack astrology is most surely not addressing professional scientists (exclusively). Of course, the denial of popular science is not necessarily pseudo or quasi. It could just be wrong.

In talking of things being "right" and "wrong" we have gone about as far as we can without being overtly epistemological, so let us drop the pretense and turn directly to these issues. At once we find ourselves in murky waters. Intense discussion in recent years has shown that if we hope for one crisp "criterion of demarcation" between science and non-science or good science and bad science, we shall almost certainly be disappointed. No such divider exists. Notoriously, Popper (1959) claimed that *falsifiability* would do the trick. Good science, genuine science, is that which could in principle be shown false by empirical evidence. Bad science, non-science (Popper does not always keep these quite as separate as one might like), could not even in principle be shown false. I myself like to characterize science in terms of natural *law* (Ruse 1988). Genuine science is that which tries to explain in terms of unbroken regularities and not miracles and so forth. But, while both of these criteria probably get us out of the difficulty of calling Noah's Flood or the Resurrection "scientific," neither really pushes us forward very far. There are all sorts of things which are not obviously falsifiable (like Newton's laws of motion) that we would count scientific, and the converse holds also. Likewise, even if one thinks that the world is law-bound, there is the question of whether and in what sense genuine or good science must appeal to general, universal laws. What is the nature of an "appeal" in this context?

There is no need to despair too readily. There may be no single, unique criterion of demarcation for good, genuine science—what I have decided to call "mature" science—but perhaps there is a cluster of concepts or standards, the satisfaction of more and more of which is what counts. It is far from essential that a poem have rhymes, but it does rate for quite a bit. Thus directed, I am going to introduce the notion of an *epistemic value*. Against the background presumption that our aim is to understand the world of experience, a world of unbroken regularity, these values are tools or standards that we cherish, since "they are presumed to promote the truth-like character of science, its character as the most secure knowledge available to us of the world we seek to understand." Hence, an "epistemic value is one we have reason to believe will, if pursued, help toward the attainment of such knowledge" (McMullin 1983, 18; see also Kuhn 1977, 321–322).

What are the key epistemic values we find exemplified in mature science? One, much discussed in recent years, is that of the aesthetic or conceptual *elegance* which characterizes certain parts of science. Some parts are *simpler*, they have a "ring of truth" that others do not. In the

Copernican revolution, for instance, the inferior/superior planet distinction arose naturally from the new theory, whereas the Ptolemaic system had to introduce all sorts of additional *ad hoc* hypotheses to explain the distinction (Kuhn 1957). In addition to this value, other suggestions include: *predictive accuracy,* the virtue of a science in being able to hit an unknown target with some skill; *internal coherence,* that the components of a theory hang together properly, with no part contradicting other parts; *external consistency,* the obvious worth in having science which does not (*à la* Velikovsky) demand that all of one's well-established prior theories be jettisoned; *unifying power,* the ability to tie together disparate parts under one or a few overarching hypotheses; and *fertility:* "The theory proves able to make novel predictions that were not part of the set of original explananda. More important, the theory proves to have the imaginative resources, functioning here rather as a metaphor might in literature, to enable anomalies to be overcome and new and powerful extensions to be made" (McMullin 1983, 16).

This is not a definitive, official list, and other commentators would slice the pie in different ways. For instance, the nineteenth-century philosopher of science William Whewell (1840) spoke of the virtues of achieving a *Consilience of Inductions,* within which notion he included both unification and fertility and which he then identified with that of simplicity! The point is that there is a cluster of factors like these that do seem important in science. Of course, these values do not exist in splendid isolation. A scientist must go out and examine the world, experiment and so forth, and there is an increasingly large data base of empirical information. But as more and more information is gathered, scientists are more and more able to build systems which exhibit and are controlled by the epistemic values. Their work approaches mature science.

What is the alternative, the opposite? Most obviously one would say "immature science"; but one has to take care not to lump too many different things together under the same (possibly misleading) heading. If one's world picture is not being informed and constrained exclusively by epistemic values, then other values (*non-epistemic,* or what I shall often call *cultural*) may well be at work. Because in science as in the real world one's reach always exceeds one's grasp (less metaphorically, one's theorizing is bound to outstrip the evidence), "presumably all sorts of values can slip in: political, moral, social, religious. The list is as long as the list of possible human goals" (McMullin 1983, 19). Note that this rather suggests (what is surely true) that one's judgment of the status of non-mature science could be a function of the intent of the practitioners. Given two people who accept the early chapters of Genesis, one might judge the one a bad scientist because he insists on a literal reading and on opposing other views informed by epistemic values, and one might judge the other no scientist at all but a good theologian because, taking Genesis metaphorically, he uses the reading for moral and spiritual purposes. (Not "When were animals and plants formed?" but "What are our duties to animals and plants?")

Some philosophers argue that the growth of science to maturity involves the gradual replacement of cultural values by epistemic values. This is an attractive idea. Think, for instance, of the growth of anthropology from the racist, sexist writings of men such as Richard Burton to the work of today's men and women, who are eager to make sure that nothing in anthropology is externally inconsistent with our knowledge of human genetics. It is not so much a question of whether you accept or reject Burton's views on blacks and women as a matter of acknowledging that today we simply know that any such views are negated by modern biology. However, I am unwilling to endorse this position too strongly here for fear of prejudging matters which are at stake. In particular, I am reluctant simply to accept that a necessary condition for scientific maturity is that all the cultural values be expelled. Perhaps so. We shall have to see.[2]

What does seem fair to say is that there is a distinction to be made between non-mature science that, for various reasons, may not be strong on the epistemic values and non-mature science that, again for various reasons, breaks from or denies the epistemic values. There is a difference between the person who writes on modern physics in a book for schoolchildren and who therefore drops all of the mathematics, even though it is mathematics which gives modern physics its predictive power, and the neo-Nazi who persists in claiming that Jews or Gypsies or homosexuals are species apart. There is flat external inconsistency here with modern biology.

Already, I am hinting at connections between the sociological and the epistemological, but first just a brief word about the actual products of science. Traditionally, the ultimate aim of science has been thought that of going beneath the phenomenal surface of experience, to find the powers or forces or *causes* that make things work. Thinking on these is incorporated within the key unit of the scientist, the "theory," as in "Newton's theory of gravitational attraction." Much effort has been devoted to explicating the notion of theory, in particular the extent to which it is (and necessarily must be) an axiomatic system, with initial assumptions or hypotheses and deductively derived consequences or theorems (Hempel 1966). Here, I shall take no stand on the issue, but I will say unequivocally that in the pursuit of mature science, trying to exemplify such values as predictive fertility, scientists do rely very heavily on the methods and findings of the deductive enterprise *par excellence,* mathematics. Theory building often makes heavy demands on formal techniques.

Another matter on which I shall take no stand is whether, as is argued by many of today's philosophers, theories are better thought of not as single monolithic axiom systems but as families of related systems, "models," applicable to limited areas of experience (Giere 1988). I will agree, however, that this is the way that working science often shows itself. I will agree also that there is more to science than just ideal systems of thought—techniques and methods are very important. This is brought out strongly by Thomas Kuhn's (1962) popular notion of a "paradigm," which is rather more than a theory, being a whole way of looking at and explaining and manipulating an area of study.

What of the connections? A good part of this book—a crucial part of this book—concerns the relations between the sociological and the epistemological, and where they do or do not come into focus together. Hence, I do not want to prejudge the issue with simplistic equivalences. But obviously there are major parallels, starting with the already-drawn link between professional science and mature science. You may think that this is simply a matter of definition, for it is true that I have said that mature science is what professional scientists aim to produce. But I intend the claim as a synthetic identity. Professional science has certain features; mature science has certain features; as a matter of empirical fact mature science is what professional science aims to produce. This holds both for the central features professional science aims to exemplify (the epistemic values) and for lesser features and side effects. A major reason for the esoteric nature of so much professional science is that mature science is heavily dependent on mathematics, something notoriously opaque to the outsider.

We likewise see a reasonably nice fit between other divisions made in the two spheres of sociology and epistemology. Non-mature science corresponds to popular science, inasmuch as both are to be considered in the realm of science. The power of epistemic values is relaxed and possibly (probably, since there has to be some burning reason to make people want to do it) cultural values come to the fore—perhaps as part of the content (if, say, one's values are religious), or perhaps in their effect on form (if, say, one's primary intent is to teach or to entertain). There is nothing dishonorable about cultural values *per se,* the very opposite, and likewise there is nothing wrong with popular science *per se,* possibly the very opposite. It is just not professional science. However, as violations of epistemic values are accepted, perhaps in order to maintain certain cultural values, the connection with professional science breaks down. It is here, one presumes, that one finds pseudo-science. That is why Nazi race theory, judged as science, is not just immoral. It is bogus. It does not give a fig for any of the epistemic criteria that real scientists hold dear. It was driven exclusively by cultural values: the supposed worth of Aryans and the hatred of Jews.

Disciplines, with their related fields or domains, are where you find theories or paradigms. It is these latter—supporting them, taking comfort from them, working within them—that bring scientists together. I would not want to say that all scientists within a discipline hold to the very same ideas. That is far from the truth. But inasmuch as a functioning discipline exists, there must be a reasonable amount of shared agreement, a shared language, and respect for difference, if not conformity. And to pick up on the point made earlier about finding (financial) support for disciplines, it is here that we start to see why, even if there is no logical connection, there is nevertheless a causal connection between sociological and epistemological categories. The identity may be synthetic, but it is not fortuitous. When Johann Sebastian Bach wrote and produced his *Passions,* he had a role within and support from his Lutheran church and its members. They wanted his work and they were prepared to pay for it. When science is produced, however, although it may have an internal elegance or beauty, people are usually not prepared to pay that much to contemplate it in itself. Some funds are found for the purest of research; but, other than exceptional cases, as when there is private support, today utilitarian factors—technological implications, national pride, and so forth—lurk not far behind. And here the epistemic values—predictive fertility, for instance—obviously come right to the front. This is the pragmatic sizzle to the theoretical steak.

Hence, without denying that we probably have only part of the story, we can see why the epistemological and the sociological might be expected to coincide. And, conversely with non-professional and non-mature science, for I do want to stress that popular science, including (sometimes especially including) pseudo-science, can be very popular. If people like the values that it promotes, then the funds may well flow in—if not from governments, then from private individuals. I would be prepared to wager that in North America today, there is more money spent on astrology than on astronomy.

Hypotheses about Progress

In this book I shall make much use of the foregoing ideas about professional/mature science, and I will extend and refine them as needed. For now, we have sufficient background against which to work. Yet, we have still no hypothesis (or hypotheses) with which to face and analyze evolutionary thought, past and present. The place to look, though, is obvious. Since we are troubled about evolutionary ideas, our questions should center on their status: as professional science, as mature science. And the key factor here, especially with respect to the failure of evolutionary theory to achieve the status of a mature, professional science, seems to lie in culture. Popular science, including its sub-class of pseudo-science, is impregnated with culture. Our query must therefore be whether there is in evolutionary theory, beginning in the past and coming to the present, evidence which leads us to suspect that culture (let us assume one cultural value) has dominated the field. If so, has this been to the exclusion or at least belittlement of epistemic values? Indeed, to push the point, has a cherished value or idea been the driving force of thinking about evolution, to the exclusion or detriment of its status as mature science? Moreover, do we still live with this problem, in some wise?

Theorizing alone will not suggest any such value. But, relying on the proven heuristic that critics are the best guide, we may readily find a candidate. There is little doubt that a major reason so many people find thoughts of evolution troublesome, in the past and especially with the Creationists today, is that it challenges the place of the Christian religion and gives an alternative account of origins (Peacocke 1986; Numbers 1992). This is not to deny that many (most of today's?) Christians have little trouble accepting evolution, nor that those who have trouble with evolution (Popper, for instance) are not all Christian. But the rivalry does count, and it leads one to ask if it is simply a question of replacement. Is the opposition of evolutionism and religion just a matter of alternatives, or, perhaps more significantly, could there be something inherently offensive to traditional Christians about evolutionism? In our terms, is there

Adam Sedgwick

some cultural value in evolutionism that is deeply threatening and challenging to the values of traditional religion?

All who have studied the subject agree that there is a major applicant for our study, one such value which (irrespective of evolution) deeply perturbs the orthodox religious thinker: the idea of, and the hopes for, human-driven improvement, or *progress* (Wagar 1972). I am not proving this point at the moment, merely stating it. Shortly, I shall present the case that the relationship between Christianity and progress is more complex than one of simple antagonism. For now, simply to underline the plausibility of the suggestion, let me show you a wonderfully rhetorical (and entirely typical) letter, written in 1847 at the height of the railway-building boom, by Adam Sedgwick, Woodwardian professor of Geology at the University of Cambridge, old friend and teacher of Darwin, and devout, evangelically inclined priest in the Church of England:

> What wonderful days we live in! Parsons by the dozen turning blind Papists; . . . men and women talking to one another at 100 miles distance

> by galvanized wires; Old England lighted with burning air; the land cut through by rails till it becomes a great gridiron; men and women doing every day what was once thought no better than a crazed dream, doubling up space and time and putting them in their side-pockets; new planets found as thick as peas; nerves laughed at, and pain driven out of the operating-room; some sleeping comfortably, some cutting jokes while you are lithotomizing them or chopping off their limbs (this I have not seen, but D.V. I hope to see it soon) in short 'tis a strange time we live in! But is there no reverse to this picture? Yes! a sad and sorrowful reverse! our friends are dying around us; famine is stalking round the land; peace is but a calm before a tempest; sin and misery are doing their work of mischief; by God's judgement, the same kind of disease which has destroyed the daily bread of our Irish brethren, may, for aught we can tell, next year consume our daily bread by attacking the grain on which we live. And then what becomes of art and science and civilisation? Gold will not feed us; the heart of man will not beat by steam. (Clark and Hughes 1890, 2, 116)

Sedgwick is almost seduced despite himself; but he pulls back just in time. Put not your faith in man but in God!

The worry here is only of progress, or rather of its illusory gifts before the real promise of religion; but Sedgwick's thought was ever a seamless whole. He may have been sarcastic about the Oxford Movement.[3] He was vitriolic about evolution, and had no hesitation about linking progress with the vile heresy: "I am no believer either in organic or social perfectibility and I believe that all sober experience teaches us that there are conditions both moral and physical which must entail physical and moral pain so long as the world lasts" (letter to Herbert Spencer, on being sent an evolutionary essay, July 29, 1853; Spencer Papers).

Hence, for remember that I am looking now only for ideas and not yet for proofs, it seems worth exploring whether in some way the idea of progress shares a more general history with the idea of evolution. Whether perhaps, in some sense, thanks to progress, thoughts of evolution have functioned as a true rival of Christianity, as what someone like Sedgwick would have regarded as a man-made or secular religion? Admittedly, this is a strong way of putting things, for we have seen that one person's religion may be another person's science. Let me therefore formulate my guiding hypothesis in a relatively non-inflammatory way. Could it be that, down through time, evolution (that is, our ideas about it) has failed to achieve the status of professional, mature science, and that the valuing of progress has been a significant factor in this failure? Is there a battle between the epistemic and the cultural?

My suspicion is that many people would agree that there has been such a battle, but they would rush to add that the epistemic forces have won, decisively (Hesse and Arbib 1986). They would accept that evolution and progress have a shared history but would insist that two key events in the history of evolutionary theorizing have brought about a parting of the ways. First, there was Charles Darwin's contribution, most specifically his key mechanism of natural selection. Especially under its alternative name of the "survival of the fittest," we see that biological change is all very relative, with those that survive being those that survive. There is therefore no absolute scale of winners and losers, and hence no genuine progress in post-Darwinian evolution. Second, there was the coming of an adequate theory of heredity, that associated with the name of Gregor Mendel. In this theory, the "raw blocks" of change, the so-called genetic mutations, are essentially random: not in the sense of being uncaused, but in the sense of not appearing on order to their possessor's needs. Hence, again, directed change is ruled out. Hopes of progress recede. This cultural value, therefore, has no standing in the epistemically admirable evolutionary theorizing of today.[4]

A neat story, but one which must be proved rather than assumed. Without prejudice, I will say simply that, if indeed evolution and progress did once share their history, today a number of possibilities—subsequent or subsidiary hypotheses—present themselves. It may be that, the popular story notwithstanding, progress continues to ride high in evolutionary work today, and this is why people feel discomfort. Or it may be that the popular story is essentially right, that progress is gone and that only epistemic factors operate but that, through ignorance (willful or otherwise), people fail to realize this. A variant here might be that progress remains a central idea in evolutionary thought but that it is no longer a cultural value. Yet another possibility is that progress is gone but that another cultural value has replaced it. Whatever this might be, once again we would have an explanation for why evolutionary thought fails against the highest standards of professional science.

Perhaps none of these suggestions is entirely right and the true story is something else, possibly more complex. This, however, must be a question for the future. For now, we have quite enough hypotheses to carry us forward and into our history. No longer are we forced merely to gather pebbles in the gravel pit. We can test our ideas against the facts, and it is to precisely this task that we turn at once. I shall begin by focusing on the notion of progress in itself. What is the idea, what is its history, has it always taken the same form, who are the critics and why, what would it mean to say that it functions in biology, how would we prove such a claim? After we have considered these matters, we can turn to evolution and its history. We can see if and how the ideas of evolution and of progress have interacted. Right at the end, we can reflect and see how our various hypotheses have fared.

1

Progress and Culture

What do we mean when we speak of *progress?* Consider the following simple statement: "I've never been much of a typist, but I've made a great deal of progress since I got a word-processor." One thing which comes right out at once is the notion of *change.* If we are all standing still, then there is no progress. But, progress is more than just change. There is change as the tide comes in and goes out, but that is hardly progressive. Progress implies that there is change in a certain *direction.* You must be going somewhere to have progress.

We need more than direction alone. Despite years of schooling, when I am faced with real-life situations my French has a tendency to show directional changes, as I forget not merely the subjunctive but the imperfect also. I would hardly speak of this as "progress." Progress, more than anything, implies direction toward an *improved* state. My typing is improved—I am quicker and I make fewer mistakes. My French has a nasty tendency to go downhill, and that is not an improvement. Remembering, therefore, that we chose progress as an example of a value that potentially influences science, our choice seems a happy one. It is almost tautological to link progress with value, although perhaps we might not always want to speak of the process of progress as a value. We may not particularly enjoy change as such, but rather value the end result (Ayala 1982).

However, a note of caution—or perhaps of clarification—should be struck. Suppose we set up some system whereby everybody in society gets adequate health care. Here, talk of value is unproblematic. Yet, consider the actions of some evil man who sets about manufacturing some quite undetectable poison. We might speak of his having made considerable "progress" toward finding the "perfect" substance, in the sense that he has edged closer to the top of the scale of undetectable poison; but, the poisoner apart, no-one else values his end result. Nobody else wants him to succeed. And it is easy to think of other examples where no-one at all values an end.

It is useful at this point to draw a distinction between *value* and *evaluation* (Nagel 1961). Progress that people desire, especially when (by and large) everyone has an interest in the end result, centers on value. Progress against some standard, which may or may not be valued, centers on evaluation. We can think of them as *absolute* progress (or "progress" without qualification) and *comparative* progress. If one arrives at the Heavenly City, one has made absolute progress. If one makes a bigger and better atom bomb, one has made comparative progress. Although value judgments are required in both cases, it is the former which really interests us. Whether and how the latter will arise in our discussion is a question for the future.

The Idea of Progress and Its Antecedents

Let us move from abstract discussion to what most people think about when the talk is of progress—or, adopting the capitalized form favored by J. B. Bury, the most distinguished historian of the subject, when the talk is of *Progress.* (From now on, I shall use this convention for talk of cultural progress, and I shall reserve the uncapitalized "progress" for biological notions.) What do most people have in mind when they are asked about Progress? What is the general conception? Simply, a belief in Progress is the belief in a doctrine about the course of history. It is a belief about change, from the past, to the present, and most probably onwards and upwards into the future.

Bury summed up the concept as follows: "The idea of human Progress then is a theory which involves a synthesis of the past and a prophecy of the future. It is based on an interpretation of history which regards men as slowly advancing—*pedetemtim progredientes*—in a definite and desirable direction, and infers that this progress will continue indefinitely" (Bury 1920, 5). He added that he could not see it as a straight scientific position, or even one made empirically probable from the study of history. It is more of a metaphysical or even theological notion, like thoughts of personal immortality. And what is crucially important about the idea of Progress is that it demands *human* effort. Progress is not simply laid on us—it would in fact be meaningless and fail without our involvement. (The story is in fact more complex, as I discuss later.)

But what are people's actual ideas about Progress and of the causes which drive it? Here, we need to turn to history. And the first thing that we find is that although Progress is a very familiar idea to us, within the history of Western civilization it is a recent notion. The Greeks were, at best, ambiguous toward the concept. They knew—they certainly assumed—that they were better than the savages around them. But a drawn-out, all-consuming metaphysic of Progress was an alien notion, for they had other views on history. On the one side, there was a belief in decline, from a previous Golden Age. On the other side, there was a belief in cycles, of life eternally recurring. But, no real Progress. We see this very clearly in the work of Plato. His views on Progress are part of a general philosophy. He posits the existence of a world of perfection, the world of Ideas or Forms. Things of this world are shadows or reflections of reality. Thus, in principle, we can never achieve the ultimate, here on earth. And that which we do achieve will, eventually, decay and corrupt.[1]

The coming of Christianity likewise proved hostile to the germination of the notion of Progress. Underpinning that outburst of rhetoric from Sedgwick, there are strong theological objections to Progress—at least, there were for St. Augustine. The story of Christianity is the story of the Fall: that is, one of human failing and dropping to a point lower than before. Moreover, the hope of improvement and eventual salvation can never come through human efforts—an essential component of Progress. It is achievable purely through God's grace and goodness. (This is the notion properly called "Providence." I shall speak more to the relationship between Progress and Providence shortly.) If theology needed more justification, there was the march of history. Even as Christianity spread and became a world religion, Western civilization collapsed. As the barbarians invaded Rome, who would dare think in terms of Progress?

However, there were ideas in the ancient world which were to prove crucial to the development of a notion of Progress. One was the idea of a "Chain of Being" or a *scala naturae* (Lovejoy 1936). The belief grew that the things of this world—particularly the living things—form a

Ramon Lull's *Ladder of Ascent and Descent of the Mind* (1305). (From the first printed edition of 1512.)

graduated chain, from the least to the most "advanced." In *De Generatione,* Aristotle (1984a, 2.4.737b7–24, 3.9.758a26–758b6) suggested a ranking from insects producing a grub, through those animals (like fish and birds) which produce an egg, up to those with blood which are viviparous (producing live offspring). In medieval theology, where the idea proved particularly congenial, supposedly the chain started with the most primitive forms of life, and then worked up right through humans, and so on up past the various orders of angels until (at the top) one finds God. One should note that, as is so often the case with ideas like this, although some empirical evidence was offered, the chain in its Christianized form rested primarily on *a priori* foundations. God's perfection, His magnificence, implies that He will create all possible forms (the "principle of plentitude") and hence there must be an upwardly rising chain, else there would be gaps—which might have been filled but which were not.

Another idea or set of ideas which might legitimately be mentioned in the pre-history of Progress is a sub-class of Christian (albeit sometimes condemned as heretical) eschatologies, often of a millennial nature—that is, beliefs that were based on prophecies, usually drawn from the books of Daniel or Revelation, supposing that at some future historical point there will be clashes between good and evil, leading to times of much-improved, even paradisiacal, existence. Influential were the thoughts of the twelfth-century monk, Joachim of Flora, who divided history into three periods: the Age of the Father, stretching from Adam to Jesus; the Age of the Son, from Uzziah to the year A.D. 1260 (some sixty or so years in his future); and the Age of the Spirit, from St. Benedict to the world's end. Apparently, these correspond to times of carnality, mixed pleasure and purity, and monastic celibacy. Hardly all agreed that this last state is paradisiacal, but Joachim did stress a historical move forward, to a supposedly better time. Interestingly, his metaphor was that of the tree of life, the mythical plant from Eden, and at his time a potent symbol in Jewish mystical thought. He gave it a distinctively temporal cast (Coulton 1927; Brett 1931; Cook 1974).

Yet, to mix metaphors, the chain and the tree were straws in the wind. Essentially, Progress is a modern (that is, post-medieval) notion. Moreover, its genesis is hardly a matter of mystery. With the brilliant scientific discoveries of the sixteenth and seventeenth centuries, people developed a new confidence in their abilities—what they had already done, they could continue to do, yet better. Then, following on this belief in the possibilities of scientific advance, connected also with the way that the new science was demanding fundamental rethinkings of theology, a belief in the possibility of ongoing moral and social improvement—in short, a belief in Progress—arose in the eighteenth and nineteenth centuries (Bury 1924; Pollard 1968; Nisbet 1980; Almond et al. 1982).

Progress in the Enlightenment

Starting with the eighteenth century, the century of the Enlightenment, and selecting drastically, let me pull out three themes of Progress. First, we have the *French* radicalism, which exploded, ultimately, in their revolution. Inspired by Descartes, the leading thinkers, known as *philosophes,* pushed reason to explore and challenge every aspect of their conservative society. As always, Voltaire was a seminal and stimulating figure. But more significant was Anne Robert Jacques Turgot, whose vision—in essays, written while he was a student and published in 1750—was one of science and reason, an idea that improved social organization helps civilization to move ever forward. Even disasters contribute to the onward march. Barbarians, for instance, learn from those whom they conquer: "in the midst of their ravages manners are gradually softened, the human mind takes enlightenment . . . and the total mass of the human race . . . marches always although slowly, towards still higher perfection" (Turgot [1750] 1895, 160). Thanks to modern science and technology, "the evils inseparable from revolutions disappear, the good remains, and Humanity perfects itself" (p. 162).

Powerful though Turgot's writing may have been, it was but to set the stage for the definitive voice of Progress: the Marquis de Condorcet. Sanctified by its author's martyrdom in the Terror, the *Sketch for a Historical Picture of the Progress of the Human Mind* ([1795] 1956) was to inspire right down through the nineteenth century. Again we get the emphasis on rationality and on the belief that human powers of reason, working through science, can lead to ever-better material conditions, away from religion and superstition and ignorance and want and toward improved physical comforts and benefits. But, there is much more than material Progress. There is, thanks to reason, moral and social Progress. Vice is a function of ignorance. Once you have knowledge, it is vanquished forever: "No one can doubt that, as preventative medicine improves, and food and housing becomes healthier, as a way of life is established that develops our physical powers by exercise without ruining them by excess, as the two most virulent causes of deterioration, misery and excessive wealth, are eliminated, the average length of human life will be increased and a better health and stronger physical constitution will be ensured" (p. 199). And with all of this, there will be a general *moral* improvement, with people showing a more caring attitude toward all.

Very much at one with Condorcet's approach was a contemporary school of sensationalist philosophy, the *idéologues*. Influenced by John Locke's program of reducing knowledge to experience, its leading members tried to use the advances of the eighteenth century to improve on and complete Locke's work. One who influenced the group, Étienne Bonnot Condillac, was noteworthy for his insistence on the philosophical and psychological significance of language—the acquisition of which is a key mark of human distinctiveness and importance. A leading member, Pierre-Jean-Georges Cabanis, was medically trained, and it was therefore his aim to give a biological basis to human thinking, a task he attempted through a notorious comparison between the belly and the brain—the one for digesting food and the other for digesting impressions. Although what bound the school together were epistemological concerns, as one can readily imagine these led naturally to schemes for individual and social improvement—especially as developed by Claude Adrien Helvétius. By the end of the century, French Progressionism spanned the whole of human interest and experience.

The second strand of eighteenth-century optimism about Progress is to be found in *Britain,* particularly in that remarkable flowering of thought which occurred north of the Border. In major respects, it paralleled (and was derived from) French thought, but it molded itself in ways reflecting the distinctive politico-economic structure and needs of that country. Living in a society already entering into an Industrial Age, the British thinkers knew full well that political economy is a crucial part of our understanding of forward movement and can be neglected only at our peril. It is no surprise, therefore, to find Adam Smith playing a key role. Arguing that humans do (and should) follow their own self-interests, he concluded that we all benefit because nature (God working through His "invisible hand") so orders it that the best possible outcome will ensue when we do so. Even the selfish rich, "without intending it, without knowing it, advance the interest of the society, and afford means to the multiplication of the species" (Smith [1776] 1937, 184–185). One should stress that, unlike the French, Smith wrote from within a society with which basically he was well satisfied. For him, capitalism was the epitome of the proper social order. What he looked for was improved economic fortunes—more goods, more wages, more rents. Hence he was indifferent to Gallic plans for central planning. He (like his God) favored the free play of market forces. He was for the application of *laissez-faire* economics.

Down in England, the emphasis was perhaps less on political economy as such, but there was the same faith that effort would bring us all to a better land. As one would expect, it was often those with a vested interest in society as it then was—conservatives, landowners, members of the Established Church (of England)—who looked for and wanted stability. Not that orthodoxy was always entirely indifferent to Progressionism. For instance, Edmund Law, sometime Master of Peterhouse College Cambridge, Knightsbridge Professor of Moral Philosophy at that University, and for nineteen years Bishop of Carlisle, explicitly temporalized the Great Chain of Being, speaking of "a regular progress, in a growing happiness through all eternity" (Law 1820, 290; quoted in Spadafora 1990, 245). But generally it was those on the outside looking in—dissenters, Unitarians, businessmen, and the like—who endorsed and strove for Progress: the chemist Joseph Priestley, a non-conformist minister, for one instance, and the novelist, philosopher, and freethinker William Godwin, for another. After all, they were the ones wanting change!

Finally, we come to our third strand, that of the *German* idealists. The British were the pragmatists; it was their intent to deal with (what they perceived to be) the realities of human emotions as they translate into economic forces. But then, pragmatism came naturally to them because they were part of a real, united, functioning country. Eighteenth-century Germany was fragmented into virtual city-states. Intellectuals, therefore, were almost pitchforked into focusing on ideals, rather than realities. Their hopes were of Progress, nevertheless. This was certainly true of thinkers like Herder and Immanuel Kant. Where German Progressionism really blossomed, however, was in the work of the great philosopher G. F. W. Hegel, whose thought was planted at the end of the eighteenth century and matured at the beginning of the nineteenth. Building on Kant and his immediate philosophical successors, Fichte and Schelling, Hegel was an idealist—thinking that the world of empirical sense is but part of the truth. Mind or spirit *(Geist)* is that which in some way defines and constitutes reality. World force, not reason or political economy, would have to drive Progress.

Mixed in also was a strong empathy with nature as represented by Romanticism. Naturally, therefore, Hegel was a major figure in the influential *Naturphilosophie* movement—one which stressed not merely idealism but the dynamic, forward-moving, purposive thrust to reality. In this view reality can be comprehended only as an organic whole, for it loses much when broken down into parts (Cunningham and Jardine

1990). A key element in *Naturphilosophie* is the notion of polarity—things present themselves in opposition, and it is out of their tension and unity that new levels of reality, involving new properties, arise. From this notion Hegel devised his powerful philosophical method of dialectic: one starts with a thesis, moves on to the opposite or "contradictory," the antithesis, and then out of the clash a new level arises, the synthesis. At a conceptual level (particularly in the *Phenomenology of Mind*), this led Hegel into a hierarchical vision of reality, as one moves dialectically from the here and now, the sensory given, right up to Absolute Spirit, which in some way Hegel wanted to identify with the God of Christianity. Later, in his university lectures, Hegel made it very clear that at a temporal level the dialectic fuels and propels a Progressive movement through history, as the spirit manifests itself at ever-higher levels. Notoriously, Hegel had trouble separating his ideal from the reality of Prussia, where as professor at Berlin he lived for many years (Taylor 1975).

These developments define the concept of Progress in the eighteenth and into the nineteenth centuries. As we prepare to move on, let me draw our attention to one important point, made pressing by German Progressionists but present already in the British philosophy. Given that Progress contains a *human* factor, it would be nice were one able to keep to the sharp theological distinction between Progress and Providence, between our own acts and *God's* plan and action in the world. With French thinkers, the distinction is possible—the cause of Progress is unaided human effort working through human intellect. But, contrary to Bury's (1920) claim that Progress and Providence are always antithetical concepts, the simple fact is that many have combined the two.

German Progressionism particularly has its roots in Christian idealism. It gives off a strong whiff of teleological inevitability. In Hegel's case, there are those who would put the matter more strongly. British Progressionism was never that spiritual, but the introduction of something like Smith's "invisible hand" clearly mixes the sacred and the secular. In a similar manner, there is strong evidence that that strand of English Progressionism, which owed a major debt to millennialism, often confused Providence and Progress. Richard Price, Presbyterian divine, preaching on "Thy kingdom come" (Matthew 6:10), assured his congregation that the belief "that there is a progressive improvement in human affairs which will terminate in greater degrees of light and virtue and happiness than have been yet known, appears to me highly probable" (Price 1787, 1, 4–10; quoted in Spadafora 1990, 369). The point is not now to swing the other way, pretending there is no distinction between Progress and Providence, but to recognize that the two could blend—whatever orthodox theology might say.

Progressionism Triumphant

The eighteenth century ended with the French Revolution, and with it the end of much of the optimism about Progress. It did not take Napoleon long to sense the threat of the *idéologues,* whom he opposed with scorn and hostility. In Britain the reaction was even more angry, as people watched with horror the violent happenings across the channel. It is little wonder, therefore, that works attacking (at least) simple optimism found much favor. Noteworthy above all in this respect were the writings of the Reverend Thomas Robert Malthus, especially his *Essay on the Principle of Population, as it affects the Future Improvement of Society, with Remarks on the Speculations of Mr. Godwin, M. Condorcet, and other writers* (1798; much expanded in later editions, up to the sixth, 1826). All efforts at improvement and movement forward are doomed, argued Malthus. Indeed, to try to make things better (particularly through any kind of comprehensive state-run scheme) is only to store up much more misery for the future. Food supplies can be improved only at an arithmetic rate. Population numbers strain potentially at a geometric rate. Hence, there are bound to be ongoing "struggles for existence," with the weaker or inadequate being destroyed.

It is true that Malthus saw a rising ladder from savages to (English-speaking) Europeans, but the essential framework was Providential. And the same was true if one turned, for theological backing to Malthusian pessimism, to the writings of Malthus's fellow Anglican clergyman, Archdeacon William Paley (1743–1805). For him the Christian religion was not only the correct one, it was the scientifically sensible one. In his *Evidences of Christianity,* Paley showed at great length that the Jesus of the Gospels—the Jesus of the miracles—compels belief by any reasonable person. Additionally, as Paley showed at even greater length in his writings on natural theology, the world of the Creator is the world of exquisite design. To think that one should, let alone could, improve on His handiwork comes close to blasphemy.

Yet, for all of the shock of the French Revolution, for all of the pessimism of writers like Malthus, the tug of Progress was too strong to be ignored for long. Science had triumphed and continued to do so. The technological payoffs existed and increased. There was just too much success in these and related areas to be ignored. And with them and through them, confidence in general growth and improvement—in morality, in social doctrines, in politics and art and literature and whatever—continued to grow. With so many doubts about orthodox religion and the beliefs of earlier years, Progress became quintessentially a nineteenth-century belief. It gave meaning to life—it offered inspiration—after the collapse of the foundations of the past.

In France, one had first the highly influential writings of Claude Henri Saint-Simon. Much concerned with the finding and justification of the actual *laws* of Progress, he opened the way for his most famous disciple, he with whom the idea of Progress will always be associated: Auguste Comte. The key insights lay in a three-stage law (Comte [1822] 1975, 29). Supposedly, all thought, at least in all of our established branches of understanding, goes through three phases: the theological, the metaphysical, and the positive or scientific. At the most general level, Comte outlined a great theological period, ending around 1400; then a metaphysical period, drawing as he wrote to a close; and, finally, would come a positive epoch (for which Comte was readying the way). Moreover, not only does each science go through three stages, but also does science itself: science began with mathematics and astronomy and progressed via physics and chemistry, and so on upward to biology, sociology, and (ultimately) ethics.

Progress, governed by law—that was the message of the French. Going back to Britain, we had best put the clock forward to around 1830. This was the point of renewal of optimism and of looking eagerly toward the future, the point at which the fears of the French Revolution and the deprivations of the Napoleonic wars were starting to fade. Thomas Carlyle (1831) and the empiricist philosopher John Stuart Mill (1859), although later to go their separate ways, were the twin prophets of change: not so much reflecting an already existent belief in Progress in Britain in the early 1830s, as trying (thanks to shared enthusiasm for Saint-Simon) to turn people's thoughts favorably to Progress (Houghton 1957, 31). A careful philosopher, Mill drew a distinction between laws dealing with underlying causes (like Newton's gravitational law) and empirical laws dealing with the phenomena (like Kepler's laws). General laws of Progress, laws at the sociological level like Comte's, can only be empirical—and thus unsupported they are troublingly open to modification and refutation. What we need to do is dig down into human nature and come up with psychological laws which tell us about causes: "the immediate or derivative laws according to which social states generate one another as society advances—the *axiomata media* of General Sociology" (Mill [1843] 1974, 924).

As he grew older, Mill had increasing doubts about the worth of Comte's three-stage law. You need a capacity for speculative metaphysics, alien to Mill's nature, whole-heartedly to embrace such generalizations. This capacity was certainly not missing from that Victorian who became Britain's chief prophet of Progress, Herbert Spencer (1820–1903). He jumped into the notion with both feet, made its promotion a life-long enthusiasm, wrote extensively (even for a Victorian) on it, and was much responsible for the popularity of the idea in the second half of the nineteenth century. In particular, Spencer drew on German biologists for the fact "settled beyond dispute that organic progress consists in a change from the homogeneous to the heterogeneous"—from the simple to the complex.

> Now, we propose in the first place to show, that this law of organic progress is the law of all progress. Whether it be in the development of the Earth, in the development of Life upon its surface, in the development of Society, of Government, of Manufactures, of Commerce, of Language, Literature, Science, Art, this same evolution of the simple into the complex, through successive differentiations, holds throughout. From the earliest traceable cosmical changes down to the latest results of civilization, we shall find that the transformation of the homogeneous into the heterogeneous, is that in which Progress essentially consists. (Spencer [1857] 1868, 2–3)

Spencer was not the only one turning to Germany (Ashton 1980). Responding to neo-Hegelian currents was a whole school of Anglican historians, the best-known member of which was Thomas Arnold, headmaster of Rugby School and father of Matthew. The Anglicans argued that God's Providence manifests itself in a Progressive realization of spirit, as successive civilizations go through an organic process of growth, maturity, and decay—cycles making an onward-moving chain (Forbes 1952; Bowler 1990b). In his highly controversial *History of the Jews* ([1830] 1909), Henry Hart Milman (later Dean of St. Paul's) wrote that: "Nothing is more curious or more calculated to confirm the veracity of the Old Testament history than the remarkable picture which it presents of the gradual development of human society; the ancestors of the Jews and the Jews themselves pass through every stage of comparative civilization" (p. 25). Milman was adapting a theme, for taking the Jews as an allegory for Providential development has a long tradition in Protestant (especially German mystical) thought. A point to be noted for future discussion is that the parallel was generally drawn between sacred history and the upward path to salvation of the individual soul (Berlin 1993).

Related to the ideas of Arnold and Milman, in the philosophy of secular history, were the thoughts of the champion of consilience, William Whewell: Anglican clergyman, powerful conservative Master of Trinity College Cambridge, scientific polymath, and author of major works on the history and philosophy of science (Fisch and Schaffer 1991). In the 1830s, influenced directly by Kant and indirectly through such British Germanizers as Samuel Taylor Coleridge, Whewell expounded an explicit, developmental vision of science's history. And it was Progressionist, for thanks to significant "Inductive Epochs" he thought that we can struggle from ignorance to attain absolute truth. Moreover, he linked this intellectual struggle to a moral struggle, with a like end. "All things work together for good" (Whewell Papers, R 18 1789, fols. 12–13, 136).

And so back to Germany and to the post-Hegelian scene. The immediate impact of the master's thought was in theology, as people wrestled to bring the dialectic to bear on Christian belief. Yet, it was only after the theological link was firmly broken that the Hegelian philosophy found its fullest scope, most particularly with Karl Marx and his friend and collaborator Friedrich Engels. In their system—the dynamic process of "dialectical materialism"—there is the reaction against Hegel and against those who would criticize him but continue to view history at the level of ideas: "My dialectic method is not only different from the Hegelian, but is its direct opposite. To Hegel, the life-process of the human brain, *i.e.* the process of thinking, which, under the name of 'the Idea,' he even transforms into an independent subject, is the demiurgos of the real world, and the real world is only the external, phenomenal form of 'the Idea.' With me, on the contrary, the ideal is nothing less than the material world reflected by the human mind, and translated into forms of thought" (Marx 1873, 25).

One must turn from idealism to materialism, Marx insisted, but not to a crude materialism. It must be one which emphasizes man as a social animal. It must be one which sees a dynamic between men and their society. "The mode of production of material life conditions the social, political and intellectual life process in general. It is not the consciousness of men that determines their being, but, on the contrary, their social being that determines their consciousness" (Marx [1859] 1977, 389). All of this leads to a Progression forward, as it has done from aristocracy to capitalism and as it will do, first to socialism, and then ultimately to communism.

Progress Falls from Favor

These are the three principal strands of Progressionism. The French supposed an upward climb to a better world—better politically, socially, morally—one where reason dominates emotion (especially religious emotion); later versions of this theory were marked increasingly by various stages through which ideas or society is supposed to progress. The chief causal force is the rational human intellect, and science (including technology) functions both as a model for Progress and as a mechanism generating Progress.

The British tended to put much more emphasis on economic forces. For them, particularly, the free play of market forces was important. In this they reflected the fact that theirs was a rapidly industrializing nation—much more so than most—and so they could see material wealth and benefits flowing from what, in itself, might seem like short-sighted self-interest. For the Progressionists, competition was not merely a fact but a necessary condition.

The Germans were idealists. For them, history has a natural upward flow. In the work of some, the more purely philosophically inclined like Hegel, the role of a kind of world spirit is more explicit. In the work of others, like Marx and Engels, such non-material forces are less evident. Though we must keep in mind that all Progress is teleological (end directed) in some sense, the Germanic version does stress this more than most. Sometimes, indeed, this inevitability, this determinism, is made quite explicit, as for instance in a famous sentence in the introduction to *Das Kapital:* "The country that is more developed industrially only shows, to the less developed, the image of its own future" (Marx 1867, 13).

No country—not even England—has its borders tightly sealed against the influx of foreign ideas. Thoughts on Progress originating in one country were liable to emigrate, eventually to alter thoughts on Progress in another country. Mill was influenced by Comte. Carlyle read and was moved by the Germans, as were the Anglican historians and Whewell (who had, as it happens, also read Comte). In other words, by about (say) 1850, or even earlier, a well-educated young person might reasonably have been expected to know of and to respond to all of the various strands of thought.

Moreover, the plant of Progress was one which took root and throve in foreign climes—in Russia, for instance. And, in the other direction, across the Atlantic, there was much enthusiasm by many in nineteenth-century America for Spencerianism, particularly with its promise of Progress through *laissez-faire* economics (Hofstadter 1959; Russett 1976). It is important to note, however, that many Americans (with Spencer himself) tended to emphasize the Progress of the diligent and talented over the failure of the inadequate. This tied in nicely with German idealism, which also went to America. American pragmatism, despite its can-do, commonsense image, had a strong streak of Romantic optimism, drawn in part at least from the *Naturphilosophen*. For instance, C. S. Peirce owed much to Schelling (Peirce [1892] 1955, 339). For Peirce, nature moves teleologically toward goals, and these are violently opposed to the reductionistic individualism of the *laissez-faire* "Gospel of Greed." Rather: "Progress comes from every individual merging his individuality in sympathy with his neighbors" (Peirce 1955, 364). Related sentiments can be found in the writings of William James, as he worked toward his own distinctive vision of salvation.

Indeed, individual figures apart, a good case can be made for saying that any kind of Progressionism which did not cut itself off too far from its religious roots would find favor in the New World. To the average American, certainly to the average late-nineteenth-century Protestant American, the idea that God does not stand behind the country would have seemed not merely heretical but absurd. Here, certainly, the Progress/Providence dichotomy collapsed. "The progress of the human race is fixed by laws immutable as the nature of God. The fidelity of man may hasten it; the wilfulness of man may retard it, but Divine Providence has decreed its certain issue" (quoted in Wagar 1972, 96).

This was Progress in the eighteenth and nineteenth centuries. And yet, today, many of us cringe before such sentiments. What went wrong? What led to the decline of this faith, this optimism in the powers of reason, or the forces of economy, or the inevitability of world history? A number of factors were germane. Most obviously, as we come into this century, there were the World Wars—particularly, the dreadful carnage of the First. Who could—who would dare—speak of Progress as the blood of millions of young men soaked the Belgian and French fields? But long before the First World War voices were raised against the general popularity of Progress. In Britain, for instance, by the early 1870s there was a growing feeling by many that things were coming unstuck. To refer again to the distinction between comparative Progress and absolute Progress, people in the land of the Industrial Revolution always tended to start with the former and move toward the latter. They were used to making bigger and better machines—and to trying to beat their rivals with respect to price and efficiency and so forth. In itself, comparative Progress does not imply absolute Progress, but the British philosophy was that it will. Build bigger and better steam engines, and eventually in the wash it will all lead to an improved economy and society.

However, if things go wrong with your comparative Progress—the ends are not being achieved, or the wrong ends are being achieved—then your absolute Progress is threatened. If you are simply building a railroad, on speculation, rather than because it is needed, then there is no hope of lasting Progress. Decline will set in. It is true that, even as decline occurs, a hope for reversal may live on, as people redefine their comparative Progressive ends or think of new ways of achieving still-cherished goals. Nevertheless, by the end of the nineteenth century, an increasing number of people thought chimerical even the prospect of regaining the right Progressivist track. The last quarter of the century saw dreadful agricultural and industrial depressions in Britain, made no easier by the threat of competition from other lands. Imperialism was one solution; but, as the Boer War showed, that was not without its problems. Militarism was as much a symptom of decline as sign of Progress. Those with a classical education found themselves drawn increasingly to a Greek philosophy of history, forswearing the guarantee of genuine advance. Nations rise and fall in their cycles. History gives no hope for the yet-to-come (Keohane 1982, 28).

After the First World War, this gloomy outlook became commonplace, in Britain and elsewhere. At best, hopes of Progress could only glimmer, with nothing to spark a real flame. With a Hegelian metaphysical intensity, Oswald Spengler's bleak best-seller, *The Decline of the West* ([1926] 1966), saw humankind locked into eternal cycles of rise and fall, growth and decay. Even natural optimists lost their faith when the heady excitement of the Twenties was ended by the Great Crash, followed by the appalling depression of the Thirties. The hopes of the Weimar Republic soured into the totalitarianism of the Nazi era. Worst of all, the inspiration of the Russian Revolution, and the coming to power of the Bolsheviks, collapsed all too quickly into the repressive Russia of Stalin.

Little wonder, then, that in 1940, as the world was consumed by yet another total conflict, the theologian Reinhold Niebuhr should write that: "History does not move forward without catastrophe, happiness is not guaranteed by the multiplication of physical comforts, social harmony is not easily created by more intelligence, and human nature is not as good or as harmless as had been supposed" (Niebuhr 1940, 188; quoted in Chambers 1958, 211). Such thinking was confirmed at the end of the Second World War with the realization not only of its terrible price—twenty million Russians dead, to start—but of the Holocaust, the greatest systematic crime in history, the deliberate slaughter of six million Jews, by people who had seen the highest flowering of civilization.

Add the Cold War, the atomic bomb, the world population crisis, the non-stop exploitation and pollution of natural resources, and the hopes of a Condorcet or a Carlyle seem a hollow obscenity. Yet, as we prepare to turn now from the history of Progress in itself to its possible influence on the thinking of evolutionists, it is important not to end on a totally negative note. Western intellectuals may revel in despair, but humans are optimists. Not everyone in this century has denied the possibility and hope of Progress. One of the greatest enthusiasts for Progress—a man whose eagerness was unrivaled at any period—was the French Jesuit theologian/paleontologist Pierre Teilhard de Chardin. Mixing science and religion in a synthesis which infuriated Catholic authorities but which delighted and inspired a fanatical band of followers, Teilhard (1955) argued that the whole of life, and matter even, is part of an upward-thrusting developmental process.

One must allow that Teilhard's troubles with his church show the continuing tensions between Progress and Christianity. Although there are many like him who happily embrace both, to this day many others think them contradictions. Among contemporary Christian critics of Progress, Joachim of Flora has become the anti-Christ (Wagar 1972). But what intellectuals may think, whether they be for or against Progress, is one thing. What people in the real world think is quite another. It is certainly the case that the general population thinks brighter times are coming—at least, so various sociological surveys tell us. Progress is a far-from-forgotten hope with the public at large, especially those for whom religion no longer provides an overall world picture. Or for those in a culture like America where there has been a tradition of blending Progress and Providence (Iggers 1982, 31).

Let us keep in mind this qualifying point: for all the naysayers, it is certainly not impossible for people today to believe in Progress. It is all a question of pressing the right buttons and of giving urgent motives and convincing reasons. But for now, this is a thought to be kept for the future. What cannot be denied is that in many influential Western circles, Progress is in eclipse.

Evolution, Cultural Progress, and Biological Progress

The time has come to start turning toward biology, and as we do we face a major question. How, even in theory, can a cultural notion like Progress be connected to biology, specifically to evolutionary biology? To address this question, I will begin with a brief look at the notion of "evolution" itself, stressing three things we normally mean when we speak of (organic) evolution. Although they are not completely separate, it is important to tease them apart. (See Bowler 1975 and Richards 1992 for historical analyses of the idea, and Ruse 1984 and Ayala 1985 for conceptual analyses.)

First, we have evolution as *fact*. By this, I mean simply the idea or claim that all organisms, those living as well as those dead, those represented in the fossil record as well as those forever lost, those big as well as those small, came to exist and to be as they are by a natural (non-miraculous, law-bound) process of reproductive descent from organisms very different from themselves. It might be, though I shall not insist for the mere fact of evolution, that organisms have shared origins. It might be, although again I shall not insist, that the earliest life has its origins in inorganic, non-living materials. The point is that from the beginnings to the ends, there is continuity, with the first being connected to the last by descent.

Second, we have evolution as *path*. By this I mean the actual tracks ("phylogenies") that organisms took through time, from first to last. Did all organisms have one common origin, or were there several starts, and if the latter is true were they all at the same time? Was there gradual change down through the ages, or were there some fairly abrupt switches on occasion? Did, to start to get more particular, some highly atypical events in life's history occur in the Cambrian, or is that perception a function of our own interests, together with a biased fossil record? Did the birds evolve from the reptiles, and if so did they come via the dinosaur branch? When did humans evolve and what were the main events in our past? Are we more closely related to chimpanzees than to gorillas?

Third, we have evolution as *theory*. Here, we deal with mechanisms or causes. What makes evolution go? Is it a force, perhaps something akin to Newton's force of gravitational attraction? Or is this example from physics altogether inappropriate when we turn to biology? Should we be looking for something entirely new? Should we be looking for just one cause, or for as many as are appropriate? I speak of "theory" in this context because it is within the framework of theory that scientists traditionally embed their causal speculations; but, in line with comments made in the Introduction, I make no assumptions here about the nature of theory.

These three categories are not distinct. You can hardly have path or theory without fact, and it is vain to think that thoughts on paths will be separate from thoughts on mechanisms. If, for instance, you think that the path is one of constant (or frequent) splitting, you will have theoretical notions quite different from someone who thinks that most change occurs in unbroken, non-fragmented lines. But, for all of the overlapping and intertwining, the categories do represent different ideas and it is as well to recognize this.

It is certainly as well to recognize this when you are dealing with notions of progress (referring, in this non-capitalized form, to biological progress). For biological progress to be absolute, the endpoint of the sequence must be higher—have greater value—than the beginning point. (For comparative progress, it is enough that the endpoint be closer to some set standard than the beginning point.) And even at this rather abstract level, it is clear that you need to specify in what sense you are speaking of "evolution" before you can make meaningful statements about its proper connection with Progress.

Obviously, at one level—what you might think the most crucial level—if you are talking of evolution as progressive, you are talking of the *path* of evolution as progressive. The later members fit some criterion or standard or value better than the earlier members. But then, not only have you got to specify the standard, and show why it might involve interests, but you have got to go on to answer questions about *causes*. Why does evolution come about, and do the causes of evolution have anything to do with the progressiveness of the path of evolution? *Prima facie* you might think that it is necessary that they do. But while it may be odd if they do not, logically it seems possible that progress be an epiphenomenon of the evolutionary causal process. Or perhaps progress occurs and causally it has nothing to do with biology at all. These are all questions that must be tackled on a case-by-case basis. The point is that they are questions that must be tackled.

Yet, how are the questions to be tackled? Let us dig a little more deeply into this before we turn to history. What exactly is it that we might hope to show? The main point to grasp right now is the most obvious. Whatever else we might find in the world of organisms, or rather in the domain of biology (which is where people express their ideas about organisms), whether this world/domain be evolutionary or otherwise, we shall not find Progress. At least, we shall not find Progress in any direct or central way. The theory or philosophy of Progress is a theory or philosophy about human beings: about their achievements and capacities and hopes for improvement. Biology is all about organisms: plants, animals, fish, dogs, trees, viruses, and (whether last or least) humans. The sphere of Progress is part of the sphere of biology, but only part.

This means that at most we are working with an analogy or a *metaphor*—although "at most" is not necessarily a mark of discredit. Presumably, if the notion of Progress has any relevance to biology, what is happening is something like this. Believers in the notion of Progress are abstracting, not necessarily consciously, certain salient features from this philosophy about human life and applying them to the world of organisms. The world of organisms is in some wise being viewed as if it were the world of intellectual/moral/social humans. Conversely, and not necessarily in a vicious circle, the world of organisms is probably being used as support for the doctrine of Progress. It is felt that there is enough material in the world of organisms—and independent evidence—to make more reasonable claims about the existence and importance of (human) Progress.

But how is the world of organisms viewed as if it were the world of humans? Progress (with a capital *P*) has variously referred to an increase in knowledge (especially scientific knowledge), technological ability, moral and social behavior, economic success, religious sensitivity, and more. Progress is about the sorts of things that humans value. Some of these could be transferred fairly readily to the organic world—to the world of animals, at least. We might look, for instance, to a rise in social behavior. Other features, like scientific understanding, make little sense outside the human realm—and absolutely none at all when we turn from animals to plants. At a point like this the biological progressionist presumably looks for a rise in features on which human achievements depend. One thinks here particularly of intellectual capacity, broadly construed. One thinks (reverting to a point made above) of the kinds of qualities something like intellectual capacity represents, even though it might be the case that these qualities are themselves contingently connected with (human-valued) organic capacities or features. One thinks very obviously of something like complexity, defined in some way or another.

> Plants, again, inasmuch as they are without locomotion, present no great variety in their heterogeneous parts. For, where the functions are but few, few also are the organs required to effect them . . . Animals, however, that not only live but perceive, present a greater multiformity of parts, and this diversity is greater in some animals than in others, being most varied in those to whose share has fallen not mere life but life of high degree. Now such an animal is man. (Aristotle 1984b, *De Partibus,* 1022.656a1–7; quoted by Russell 1916, 16)

It is certainly not necessarily the case that one would think a rise in complexity itself a matter of value, with implications for progress. Indeed, we have seen that a much prized epistemic value is simplicity. But if organic features that we value usually or always involve an increase in complexity, then such an increase might well be taken as a mark of progress.

Yet, what of proof? Suppose that there really is in evolutionary theorizing a metaphoric move from Progress (however defined) to progress (however defined). How could we ever hope to *show* that such a shift has taken place? One can plausibly suggest that a number of levels of evidence are pertinent, of which the following three types are certainly significant. Whether they exhaust the potential evidence is another matter; but, taken together they make (or break) a powerful case.

First, we look for signs that the biological theorizing has in fact been influenced by the idea of Progress—or, rather, we look for signs that the biology is of the form which would obtain were it so influenced. In particular, do we find a thesis about progress in the biological world? Here, some of the qualities just listed—sociality, intelligence, complexity—become important. At this point, it is vital not to buy victory too cheaply or to impute views too readily. Because someone thinks humans are the endpoint of evolution, it does not follow that he or she is a progressionist. After all, since we are still around, humans could hardly help being the endpoint of evolution, in some sense. What we are after are properties which are not merely catalogued or measured but valued, for themselves or because they are associated with other properties. And we are after (the postulation of) causes which lead necessarily to the ever-greater manifestation of valued properties. Then we can start to feel confidence that there are in biology claims about progress.

Second, hoping to find that there is a move from Progress to progress, we look for evidence that the biologist concerned consciously or unconsciously leaned on the idea of Progress, or, conversely, hoped to support the idea indirectly through the organic world. Pertinent here would be explicit claims, sympathy with ideals, reference to appropriate literature, and so forth. Obviously all and any of this is highly supportive of our cause, although one should beware of taking its absence as definitive counter-evidence. Intentionally or unintentionally, a scientist might be most loathe to bring what he or she perceives as extra-scientific factors into the act of theorizing. Conversely, one cannot really assume, without supporting evidence, that what a scientist claims to be doing is what he or she is in fact doing.

Third, least tangibly but in a way most significantly, especially for our overall inquiry, we look to the extent that the biology outstrips the evidence. Now, it has been pointed out that *all* biology—all science—outstrips its evidence, so there is hardly much to comment on in that fact alone. But some instances of outstripping are more equal than other instances. In particular, some outstripping really starts to make you look for the influence of cultural values. If a scientist joins up three points in a straight line rather than a wiggle, that is no cause for comment. Clearly, the epistemic virtue of simplicity is at play here. But if the scientist prefers the wiggle to the straight line, even though logically the choice is defensible, you start to look for other factors at play. There is literally an infinite number of ways to join three points. Why choose this one? You search for outside influences. Likewise, if there are many options other than progress, but a biologist insists on seeing progress, then we can legitimately start to look for outside influences—and Progress is a good place to start.

One final point, and we can turn to history. What is the strongest possible case that we might be building, and what are weaker but still acceptable cases? The strongest case would be that a belief in Progress is a necessary and sufficient condition for a belief in progress: all and only those people who believe in the social doctrine of Progress believe in the biological doctrine of progress. I take it that it is highly unlikely that we would ever get this exact a correspondence between the two. For instance, almost perversely, because someone believed in Progress, given the gap between theory and evidence that person might be led to deny progress. Conversely, someone who believed in progress might be led to deny Progress—for reasons which we might (at some point) uncover.

However, with this said, we can hope for more (we *should* hope for more) than the weakest correlation between Progress and progress, if we are to say anything significant about the importance of cultural values for science. We would expect a belief in Progress generally to lead to a belief in progress, and conversely a lack of belief in Progress to lead to no belief in progress—that possibly a sharp denial of Progress would imply a sharp denial of progress. And, since exceptions often prove rules, we would expect the mismatches to be informative in themselves. For this reason, if for no other, we can in our historical survey legitimately look at believers and non-believers in both Progress and progress. At the basic level, everyone is grist for our mill—which is now about to grind.

2

The Birth of Evolutionism

How does one write a philosophical history of science? Any such history as I attempt must avoid disembodied "themes" or "positions." I want to look at the way in which people's cultural views influence (or do not influence) their scientific views. Almost logically, therefore, I must take people at an individual level and ask how their convictions spilled into their work. However similar the scientific conclusions reached by two different people may be, I have no right to assume *a priori* that cultural premises were shared—or, even if they were shared, that they led in the same way to the same conclusions.

Yet, in dealing with people individually, one risks ending up with a row of biographical beads on a chronological string—with all connections reduced to the trivial or the contingent—or, put another way, a timetable rather than a history. The outcome would be not merely uninteresting but profoundly misleading, for the story I have to tell will prove to have very strong explanatory links between the individual and the general, making full sense of the history of evolutionary theorizing. My real hope is that, as the dots of pure color in a pointillist painting blend into a coherent picture, so will the lives and work of the individual scientists blend into a coherent story. But, fearing that I lack the genius of Seurat, I shall impose a continuity and structure through my overall interest in the question of status. In this chapter, and in those following, I shall integrate the findings by speaking to this issue, and, since they tend to be the easier to approach, I shall focus in detail on sociological notions, namely, professional science and its contraries. I shall take these notions not only in their own right but as avenues to epistemology, that is, to an understanding of the place and role of values, epistemic and cultural, in evolutionary theory.

Evolutionism in France

The Greeks were not evolutionists. Aristotle had important things to say about the living world, and he saw clearly that individual organisms are functionally end-directed or "teleological," tending also to be structurally isomorphic with other organisms—what we now call, following Richard Owen (1846), "homologous." But although both of these facts were to prove crucial in the history of evolutionism, Aristotle never thought of the *scala naturae,* or Great Chain of Being, the ordering of organisms from least to most "perfect," in terms of time (Gotthelf and Lennox 1987). Early Christianity, bringing in the Jewish creation story, was even less concerned with development in the natural world. The clock really does not start to move in that direction until the Scientific Revolution, when an increasing flood of biological discoveries put increasing pressure on the traditional static view of origins. At the same time, the physical sciences started to hint at gradual developmental beginnings. Admittedly, these earliest notions of temporal change referred only to the inorganic world, but obvious analogies lurked close by. In the eighteenth century, for example, the so-called nebular hypothesis was developed. This supposes that solar systems (like ours) condensed out of gaseous masses, nebulae which can be seen dotted throughout the universe.

The idea of organic evolution began to make its appearance around the middle of the eighteenth century (Bowler 1984), with modifications to the Great Chain of Being. Two pioneers were the Swiss naturalist Charles Bonnet (1720–1793) and the French philosopher Jean-Baptiste Robinet (1735–1820). Subscribing to the venerable theory of preexisting germs—that organisms contain within them the sparks of all future generations—Bonnet argued that the germs give rise to organisms in an ordered series along a nigh perfectly ascending scale, the Chain of Being. This was hardly genuine evolution, for supposedly the germs create each organism ready for its own particular place; but it is dynamic, for the germs at the lowest level sprung first into action, and then up the Chain,

IDÉE D'UNE ÉCHELLE
DES ETRES NATURELS

L'HOMME
Ourang-Outang
Singe
QUADRUPÈDES
Écureuil volant
Chauve-souris
Autruche
OISEAUX
Oiseaux aquatiques
Oiseaux amphibies
Poissons volans
POISSONS
Poissons rampans
Anguilles
Serpens d'eau
SERPENS
Limaces
Limaçons
COQUILLAGES
Vers à tuyau
Teignea
INSECTES
Gallinsectes
Tænia, ou Solitaire
Polypes
Orties de mer
Sensitive
PLANTES
Lychens
Moisissures
Champignons, Agariez
Truffes
Coraux et Coralloïdes
Lithophytes
Amianthe
Talcs, Gyse, Sélénites
Ardoises
PIERRES
Pierres figurées
Crystallisations
SELS
Vitriols
MÉTAUX
DEMI-MÉTAUX
SOUFRES
Bitumes
TERRES
Terre pâte
EAU
AIR
FEU
Matières plus subtiles

Bonnet's Chain of Being: "Idée d'une échelle des êtres naturels." (From Bonnet 1745, 1, opposite p. xxxii.)

until finally the germs for humankind started to work. Although likewise no genuine evolutionist, Robinet was even more explicit than Bonnet in his temporalizing Aristotle's ordering. Arguing that the *scala naturae* is really a continuous rope rather than a discontinuous chain, he found it relatively easy to argue that the germs contain within themselves, not identical germs for future generations, but germs of organisms higher up the scale. Given the right conditions and so forth, the germs would produce a gradual unfurling of the whole upward Chain.

As significant figures in eighteenth-century French biology, neither Bonnet nor Robinet nor any other could compare to the author/editor of the fifteen-volume *Histoire naturelle, génerale et particulièr,* Georges-Louis Leclerc, le Comte de Buffon (1707–1788) (Roger 1989, 1993). He too subscribed to the Chain, *prima facie* a temporalized version, but he was no real evolutionist either, and in ways was even further from the idea than Bonnet and Robinet. Although Buffon argued for limited change (significantly labeled "degeneration") from certain basic types—more at what we today would call the generic than the specific level—this was never more than an elaboration on a very static theme, one underpinned by the idea that each type has its own form, or *"moule intérieur"* (Roger 1989, 182). You might get a certain variation, especially thanks to domestication, through the effects of the environment, as among the horse or sheep or dog families, but it could only be limited—it could not produce a horse, say, from a cow—and all change was in theory ever reversible. Furthermore, although Buffon never relinquished the Chain of Being metaphor, he toyed constantly with other physical metaphors, notably nets (and maps) and (for a while) trees (Barsanti 1989). Of course, these alternatives are not necessarily anti-evolutionary, especially the tree, but they do warn against simplistic analyses.

Finally, we arrive at he who has the fullest claim in France to being the first genuine, thorough, organic evolutionist, Jean Baptiste Chevalier de Lamarck (1744–1829). Born into the minor nobility, he enrolled in the army but was soon forced by injury to seek a less active life. This he found in Paris, gaining status among the capital's scientists, as well as the patronage of Buffon, for producing the first definitive French Flora (*Flore françoise,* 1778). Later Lamarck moved from botany to zoology, studying those lowly animals which include the insects and the worms. It was Lamarck himself who was to invent the term *invertebrate* for this hotchpotch group, and in the course of his long career he was to produce a major study of this field—one on which his contemporary reputation was

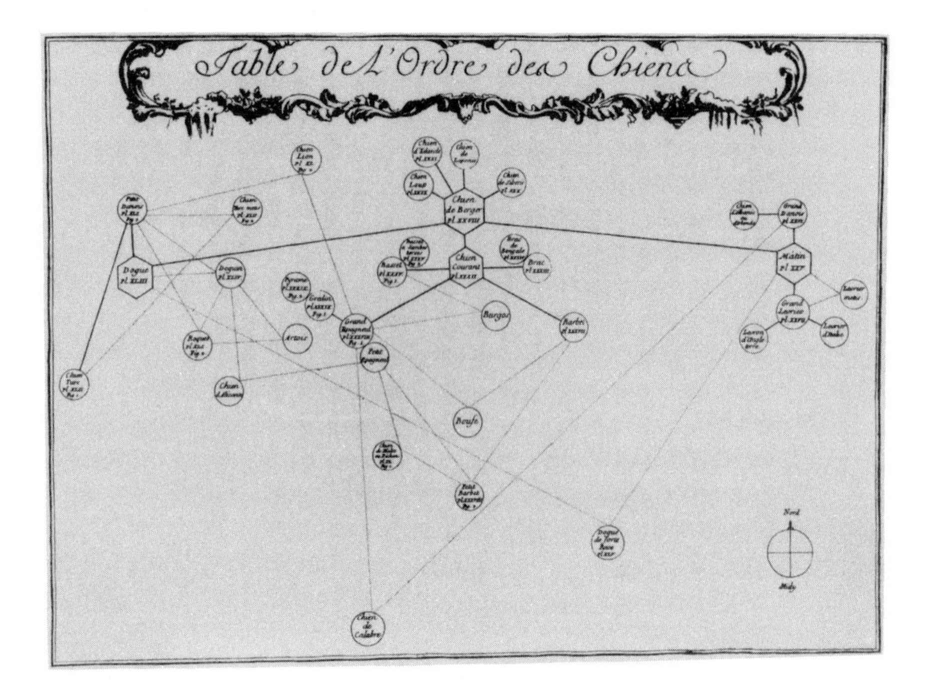

Buffon's map or net of dog relationships: "Le chien avec ses variétés."
(From Buffon 1755, 5, 225.)

to rest (Lamarck 1801a, 1815–1822). Interested in just about every branch of science, the young Lamarck was strongly opposed to evolutionary ideas—apart from anything else, they would play havoc with botanical classification. But, starting cautiously in 1800, Lamarck became more and more sympathetic to beliefs about organic change, and he produced his best-known discussion, the *Philosophie zoologique*, in 1809. (See Burkhardt 1972, 1977; Mayr 1972; Hodge 1971; Barthelemy-Madaule 1982; Sheets-Johnstone 1982; Appel 1987; Jordanova 1976; and Schiller 1971 for an unorthodox view.)

It is important not to approach Lamarck too directly through ideas familiar to us today. To use a term which did not come into general modern use until the second half of the nineteenth century, there is little question that Lamarck was an "evolutionist," unambiguously so. He believed that all organisms living today came gradually, by law-bound processes, from forms widely different. He believed, indeed, that all organisms come from the same kinds of most primitive forms. But here—starting to move now from the notion of evolution as fact to that of evolution as path—the kin-

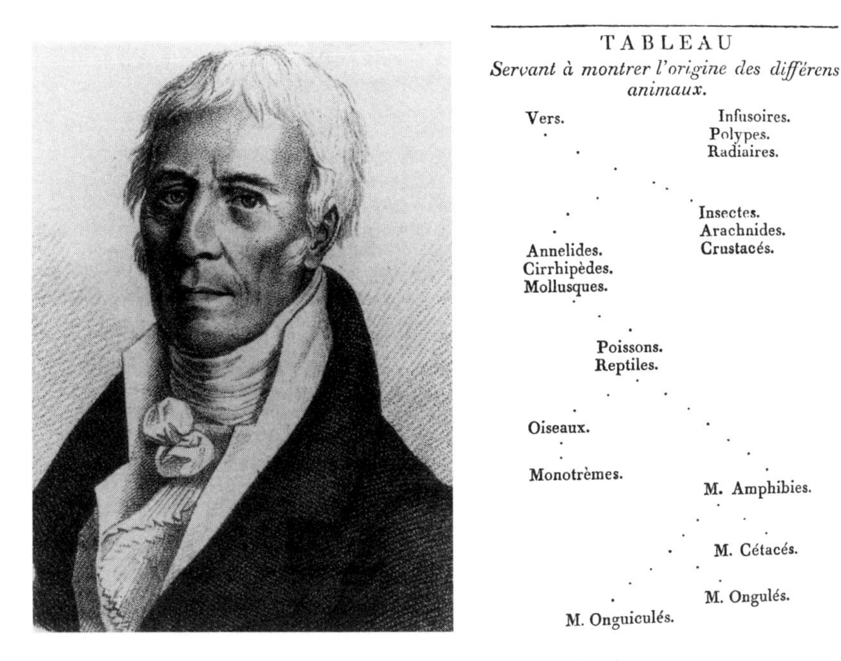

Jean Baptiste de Lamarck and his (inverted) tree of animal life (from the *Philosophie zoologique,* 1809).

ship with modern thought ends. We, influenced by Darwin's theory in the *Origin of Species,* tend to think of evolution as a branching path, with all organisms descended from (in Darwin's words) "one or a few forms." Lamarck, like Bonnet and Robinet, planted his thinking firmly in the Chain of Being tradition, believing evolution to be a climb up a main path, in animals from monad (the most primitive form) to man. "One can therefore truly say that there exists for each kingdom of living beings a unique and gradual series in the range of sizes, corresponding to the known degree of organization, rising, in the kingdom of animals, from the most simple 'animalcules' to the most perfect animals" (Lamarck 1802, 505; in Daudin 1926, 111, n. 1). More than this, Lamarck thought that new forms are being spontaneously created (by the powers of electricity and heat and the like) all of the time. Thus, new organisms are forever getting on the bottom of the Chain—which then carries them up.

In fact, as we move now toward evolution as theory, we find even more complexity in Lamarck, quite apart from the fact that he believed in a separate chain for plants. Lamarck ever denied "vitalism"—the

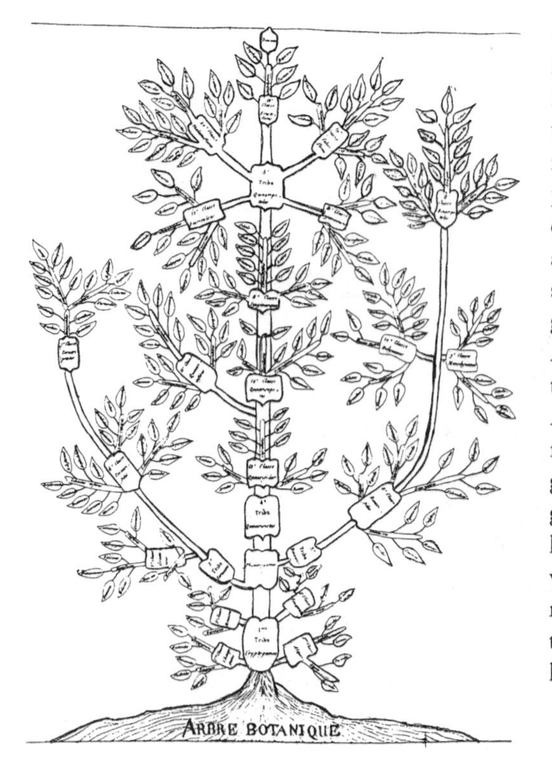

Lamarck was not alone in finding the chain metaphor inadequate as new organisms were discovered which could not be slotted neatly into place. Indeed, the tree was becoming a favored metaphor, as is shown by this purely systematic and atemporal diagram of plant relations—*Arbre botanique*—drawn by the French botanist Augustine Augier (1801, frontispiece). The tree diagram was given a firmly progressionist interpretation, however, suggesting that it was intended to preserve, rather than to downgrade, the image of progress established by the chain metaphor.

notion that there is a kind of force which directs life, a view to be found in Aristotle. He rather thought of himself as a mechanist and spoke of nature as this "blind force" (Lamarck 1820, 35), which for all its results has "however neither goal nor intention, able to do only that which it does, and is itself only a collection of limited causes, and not a particular being" (p. 38). Nevertheless, it is capable of driving the path of organisms upward. At the more immediate level, Lamarck argued that organisms experience "needs" *(besoins)*. These needs, brought about by changes in the environment, then trigger subtle fluids (like electricity) which, circulating in the body, enlarge or develop the appropriate organs. In higher animals, a crucial causal factor is the "inner consciousness" *(sentiment intérieur)*, which makes parts respond and develop: "it is the impetus of all the actions of the individual, that which directs all the movements it can make, and inasmuch as an individual is capable of intelligent thought, is again that alone which guides the actions" (p. 191).

Then, overlaid on top of this upward drive, Lamarck located a secondary mechanism, the one everyone knows of and which now indeed carries his name, "Lamarckism": the inheritance of acquired characteristics. Although this was not in fact an original idea with Lamarck (it occurs in Buffon and others), initially he may have seen it as the primary mechanism of change, and there are some suggestive passages hinting that he may have bound the idea up with an embryological analogy (Corsi 1988, 134). By the time of the *Philosophie zoologique,* all of this seems somewhat downgraded, but the effect was still sufficiently powerful to spoil the beauty of the scale of nature, for Lamarck believed that the inheritance of acquired characters brings about all kinds of side effects, leading various organisms off on tangents. Although his own diagram was inverted, he was on the way to making the tree metaphor become inseparable from the history of evolutionism.

Yet, do not be misled. In Lamarck's mind, all diversions were always tangents, however much they may complicate the overall picture. Essentially, evolution involves an upward climb, from monad to orangutan, and then on to our own species. And this was the position to which he stuck for the rest of his life.

From Culture to Biology

With Lamarck's theory before us, we have reached a natural break. Can we say something about the connection (assuming there is one) between evolution and Progress? It has to be admitted that there is something very suggestive about the timing. For two thousand years there is stasis on the organic origins front—both Progressionism and evolutionism were blocked by Greek thought and Christian theology. Then, just at the point (and in the place) where people start to preach the virtues and inevitability of upward social change, biologists start to talk about the virtues and inevitability of upward organic change. But suggestiveness is not enough. We must dig more deeply, using criteria suggested at the end of the last chapter: a theme of progress, a belief in Progress, and a move beyond the evidence.

Start with Bonnet and Robinet. Since neither was committed to continuous developmental change, one can speak of them only in a limited sense as "progressionists" (if this term be reserved strictly for evolutionists); but, inasmuch as they edged toward such change, the conceptual framework or backbone of their work was one of increasing perfection.

And in the case of the genuine evolutionist Lamarck, we have had evidence already of an upward flow that can only be called "progressive."

> Ascend from the simplest to the most complex; leave from the simplest animalcule and go up along the scale to the animal richest in organization and facilities; conserve everywhere the order of relation in the masses; then you will have hold of the true thread that ties together all of nature's productions, you will have a just idea of her *marche,* and you will be convinced that the simplest of her living productions have successively given rise to all the others. (Lamarck 1802a, 38; quoted by Burkhardt 1977, 141)

If anything, Lamarck's splitting the chain into two emphasized the importance of humans. They could thus be portrayed unambiguously as the endpoint of the animal chain, in any case higher than anything on the plant chain. Moreover, it is intelligence—flagged by complexity—which is the key factor in upward progress. Complexity is not valued in its own right as such, but because it is the sub-stratum for qualities which we humans value: "one observes a kind of gradation in intelligence of animals, as it exists in the increased perfection of their organization" (Lamarck 1802a, 124). Tying in with this belief, Lamarck explicitly divided the animal kingdom into three: *les animaux apathiques,* those without a nervous system; *les animaux sensibles,* the higher invertebrates; and *les animaux intelligents,* which lead up to humankind. It is for the final category that the *sentiment intérieur* is needed.

Buffon provides the contrast. Remember that the Chain of Being was always one among alternatives. Most particularly, there was the (very non-progressionist) net (or map), which became more and more popular. Buffon always viewed the Chain statically, particularly with respect to our own species. We are on top and that is as it always was and had to be. "Everything marks man, especially his appearance, in his superiority above all living beings" (Roger 1993, 536; quoting Buffon, *Histoire naturelle de l'homme: De l'age viril,* 298B). And, forget not that for Buffon change was degeneration. Apparently, this is particularly true in humans, which have seen a sad decline in the standards set by the first (white) men.

Move on to the second point. Have we direct evidence that the idea of Progress was (or was not) influencing people's biological theorizing in an evolutionary direction and that conversely the biology was being used to support a general metaphysical belief in Progress? The early pre- or

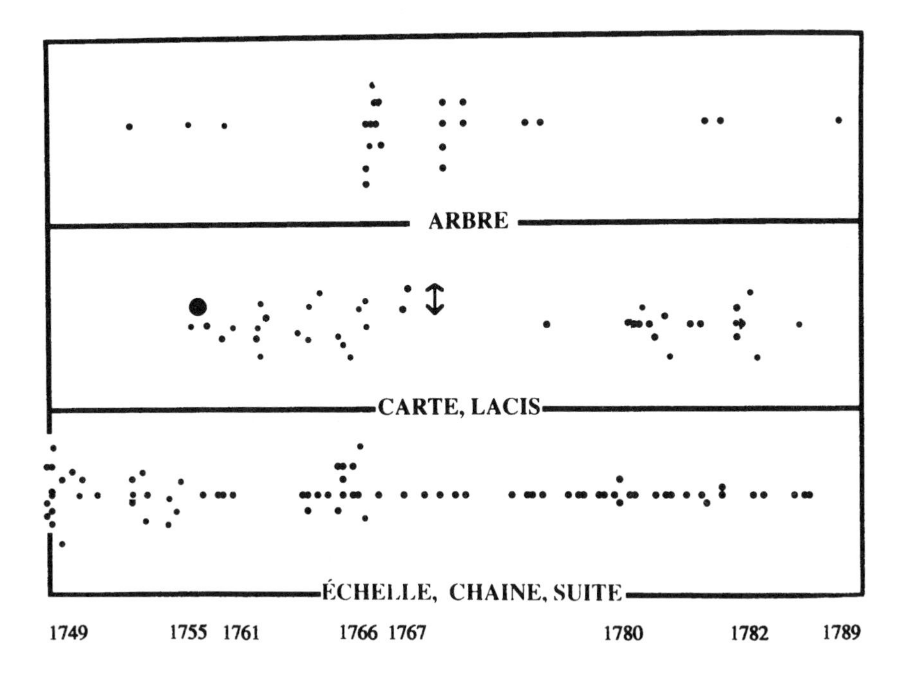

Chronological chart of the three main methods (tree, map, chain) that Buffon used to illustrate animal relationships. Note that the use of the tree metaphor declined with time. (From Barsanti 1992, 290.)

proto-evolutionists were undoubtedly affected by such links. Robinet, for instance, was influenced by Turgot, and explicitly endorsed some kind of Progressionist thesis for humans: "The human mind must be subject to the general law. We cannot see what could arrest the progress of its knowledge, or oppose its development, or stifle the activity of this spirit, all of fire as it is, which has certainly a destiny, since nothing has been made in vain. Its destiny can be nothing other than to exercise imagination, to invent, and to perfect" (*Origin and Progress of Language,* 2d ed., 1, 35; quoted by Lovejoy 1936, 273). Moreover, he allowed that his aim was to promote this view throughout the organic world: "Every being cherishes its own existence and seeks to expand it, and little by little attains the perfection of its species."

Buffon, to the contrary, was very much a figure of the Establishment, a friend of princes and of people with power, and an opponent of the Progressionists like Condillac and Turgot in the various struggles for power and patronage. A compliment returned, as when Condorcet wrote: "I'm taken up with another charlatan, the great Buffon. The more I study him, the more I find him an empty windbag" (Roger 1989, 570; quoting Condorcet 1988, 240–241). How much Buffon's thinking was a direct function of Christian Providentialism is a matter of debate. But, religious or not, he turned his gaze away from natural upward change and took human degeneration—a function of the move to an alien climate and to strange food—as the model for all animal degeneration, including the fact that one never crosses beyond limits: "The most stupid of humans has the wherewithal to drive the most spiritual of animals" (Roger 1993, 537; quoting *De la nature de l'homme,* p. 296).

What of Lamarck? If Buffon looked in one philosophical direction, did Lamarck look in the other? The story of his life provides lots of straw for our bricks. Politically, Lamarck was forward-looking, endorsing moderate change and reform and optimistic about the possibility of Progress in improving social and political conditions. A politic philosophy during the Revolution, given his birth. Certainly his personal life confirms that Lamarck hardly fitted a conventional, conservative mold: by one of his three or four wives he fathered six children, though he married her only on her deathbed. For this reason alone it is well that Lamarck had little empathy for traditional Christian beliefs; he was quite open in his "deism," the philosophy that God stands at a distance, letting all happen according to His (or Its) unbroken law.[1]

Lamarck believed that humans are improvable, especially through the inheritance of acquired characteristics; that ultimately this capacity for positive change aims at social improvement, in the sense of harmony and balance between people, and between society and nature; and that the intellect seems to be the key. Indeed, Lamarck went so far as to suggest that what we need are a few philosopher-kings running the state, with scientists and science occupying a prominent place. It will come as no surprise to learn that Lamarck linked his position with views about the natural superiority of Europeans. Apparently, they have been in existence longer than other races of humans, and they have therefore had more opportunity to move up the scale of perfection (Lamarck 1820). Note the contrast with Buffon, who saw Europeans as the original best stock and others falling away, thanks to alien climates.

Of course, at a point like this, Lamarck's progressionism and his belief in Progress really collapse into one—as they do also when he talks of the human propensity to form groups and the consequent needs which (naturalistically) trigger the evolution of language. Apparently, inferior animals can get by with sighs. We, however, have a great number of diverse needs, multiplied proportionately to our ideas, and hence "it is necessary to use more complex methods to communicate with one's fellows" (Lamarck 1820, 152). Furthermore, intellectual achievements become increasingly complex as civilization advances: "In considering each society with respect to its degree of civilization, one might say that there is a direct proportion between the sophistication of science required for its members' well-being and the needs that they express" (p. 92). Savages are simple folk, with simple needs. We Europeans, on the other hand, . . . !

Lamarck was committed to Progress. Further, this commitment did not come from thin air. He was identified with the *idéologues* and their predecessors and friends, those who supported and preached Progress, specifically Condillac, Helvétius, and Condorcet. Lamarck's beliefs about the genesis of language point to a firm link with the group, given the way that the *idéologues* stressed the significance of speech in the development of reason. Moreover—and here comes the key point—we know that Lamarck himself saw his work in biology as directed by and backing up the philosophical thesis about Progress. A good half of the *Philosophie zoologique* deals with "the physical causes of sentience"—that is, with the development of and reasons behind the nervous system generally and the brain and thought in particular. Explicitly, Lamarck tells us that he sees the non-human world through the lens of the human world, and conversely he thinks the non-human world a key to understanding ourselves. Change in the human world echoes change elsewhere, and vice versa.

Specifically following another *idéologue,* Cabanis, Lamarck located physical and moral attributes of individual human beings in the organization of their social world and argued that using this perspective to view the lower world tells us much about ourselves. Indeed, what he writes could as well have been written by Cabanis—the author most often referred to in the *Philosophie zoologique,* a man who believed not only in spontaneous generation but a limited form of evolution, and one sufficiently committed to the philosophy of Progress that it was he who sheltered Condorcet in the Terror:

> Without doubt, it is possible, by a plan of life, wisely conceived and faithfully followed, to alter the very habits of our constitution to an appreciable degree. It is thus possible to improve the particular nature of

> each individual; and this goal, so worthy of the attention of moralists and philanthropists, requires that all the discoveries of the physiologist and physician be considered. But if we are able usefully to modify each temperament, one at a time, then we can influence, extensively and profoundly, the character of the species, and can produce an effect, systematically and continuously, on succeeding generations. (Cabanis 1802, 434; quoted in Richards 1987, 28–29)

The associationism of Condorcet, Cabanis, and Lamarck—the belief that through experience one can pile up information which can then be transmitted wholesale, whether culturally or biologically—may be traced to Condillac's seminal work of 1754, *Traité des sensations*. In this treatise Condillac tried to meet the challenge of Berkeleian idealism while building on John Locke's attack on innate ideas and arguing that all knowledge could come from sensory experience and habit.

Finally, in our quest to link Progress and evolution, we have the third point: the gap, and its extent, between theory and evidence. I take it that the creative leap needed for belief in the germ theory specifically and the theories of Bonnet and Robinet generally does not need proving or even stressing. Note that I am not saying that it was ridiculous to believe in the germ theory in the middle of the eighteenth century. I am saying that one looks for non-evidential causal factors for such belief—and I take it also that epistemic values do not leap at once to mind. Buffon makes a similar point, in a reverse way. Given that we are dealing with a man who had picked up on and discussed the homologies between vertebrates, and who had admitted fully that we humans have to be classified as animals and that there are significant resemblances (Roger 1989, 536), his views on degeneration and on the in-principle unbridgeable gap between humans and other animals seem strikingly unsupported.

Nor should we think that things were much different for Lamarck, however we today might revere him as a major figure in our history. Whatever the cause of conversion, once he became an evolutionist Lamarck's attitude to the evidence—real or apparent—was positively cavalier. Most obviously, if one were going to build a case for progressionism, at least with respect to evolutionary paths, one would show some interest in the fossil record. This Lamarck singularly failed to do. He postulated progression right up to humankind, and that is it. His only worry was that of fitting all living organisms into the picture. There is a chasm here between belief and evidence and—given the background belief in the Chain of Being—it is Progress which provides the bridge we are seeking.[2]

The three possible levels of evidence concur. The case is made for the connection between the philosophical thesis about Progress and the beginnings of organic evolutionism in France in the second half of the eighteenth century. Particularly for Lamarck, the form of evolutionism he proposed—deeply progressionist—reflects the form of Progressionism the philosophers promoted. The French view of culture posited a kind of upward movement fueled by the development of ideas, as needs arise, and by their subsequent direct transmission. It is precisely the biological equivalent of this view that Lamarck proposed. It is often said today that cultural evolution is "Lamarckian," meaning that it centers on the spread of acquired ideas. Such a truth is hardly contingent. Culture is Lamarckian because Lamarck was cultural.

Erasmus Darwin

Let us cross the channel to Britain and ask about the rise of evolutionism there. North of the Border, James Burnett (1714–1799), Lord Monboddo, balanced his beliefs in a human intellectual development with what he thought was a corresponding degeneration of the body (Greene 1959). Admittedly, he did separate us off from the rest of the organic world, and seemed to see little permanent change in other beings. However, even this idea can be found in some mid-eighteenth-century British writings—significantly, it was floated by one English author in a context which shows that one did not necessarily have to be beyond the pale of orthodoxy in order to entertain proto-evolutionary ideas. William Worthington, Anglican clergyman and sincere Christian, in *An Essay on the Scheme and Conduct, Procedure and Extent of Man's Redemption* (1743), assured his readers that although "the Stomachs of carnivorous Creatures seem to be formed for animal food," it is possible that "this might not have been by any original Constitution of Nature, but at first contracted by Habit, and derived down through their successive Generations" (p. 404). Not, I hasten to add, that Worthington was in any sense a genuine precursor of Lamarck, for his discussion was set in the firm theological context of explaining why there were no meat-eating animals in the Garden of Eden, and why we might again expect the lion to eat straw like the ox. Simply put, what habit has given us, "therefore may be worn off again by Degrees; and the original Form and Tone of their Stomachs be recovered" (p. 404).

Such speculations as these are but premonitions, which pale besides the

Erasmus Darwin

one person who does stand out unambiguously as Britain's first real systematic evolutionist, the provincial physician Erasmus Darwin (1731–1806). Educated at Cambridge and at Edinburgh universities, he settled to practice in the cathedral town of Lichfield. There he married, having three sons, the youngest of whom was Robert, Charles Darwin's father. Some years after his first wife's death, having sired two illegitimate daughters, he became enamored of one of his patients, the young wife of an elderly colonel. This husband conveniently dying, and there being no professional association to frown on such a liaison, the corpulent, somewhat ugly, and much older Darwin promptly married the new widow, moved to Derby, and fathered seven more children, one of whom was to be the mother of the late-nineteenth-century student of human heredity, Francis Galton (McNeil 1987; Harrison 1971; Porter 1989; King-Hele 1981).

Erasmus Darwin's evolutionism is woven through several texts, including poetry, but it is explicit and unambiguous. Most detailed is the discussion in *Zoonomia,* Darwin's major medical treatise, published in 1794. However, the ideas go back at least to 1770, when he had been much impressed by fossil discoveries thrown up by excavations for a tunnel in a new canal. This experience was backed by a number of other factors which seem, collectively, to have tipped him toward evolutionism.

Darwin himself listed, first, the drastic changes which take place in the ontogeny of individuals, being a kind of model for the line, "from the feminine boy to the bearded man, and from the infant girl to the lactescent woman" (Darwin 1794, 500); second, the analogy of "the great changes introduced into various animals by artificial or accidental cultivation," as with horses and dogs and cattle (ibid.); third, such changes as have been effected in groups, as when the docking of dogs' tails has led eventually to heritable changes (p. 502); fourth, the evidence of comparative analogy and from isomorphisms (p. 502); fifth, the changes animals can bring about in their own forms through various desires and wants (p. 503); and sixth, the transitional animals between different classes, as the whales and seals and (especially) frogs, which bridge fish and land animals (p. 505).

Overall, with respect to the path of evolution, Darwin offered a picture similar to that articulated by the French, with the primitive coming first and then working its way up to the final point, man: "mushroom to Monarch," as one commentator has characterized it, picking up on terms used by Darwin (McNeil 1987). Unlike Lamarck, however, Darwin seems to have seen all life (certainly all animal life) as having one shared origin rather than an ongoing sequence of starts. But, what of the causal theory behind evolution? At the most basic level, Darwin ascribed evolutionary change to the inheritance of acquired characters. Drawing on the associationist psychology of the mid-eighteenth-century English thinker David Hartley, who like Condillac had tried to answer Locke's challenge by showing how, through habit and repetition, experience could lead to the development of knowledge, Darwin generalized: "I would apply this ingenious idea to the generation or the production of the embryon, or new animal, which partakes so much of the form and propensities of the parent" (Darwin 1794, 480). Later he supposed also that small particles are given off by the bodily parts of both parents, circulated in the blood, ending finally in the sex organs, from whence they can be combined in reproduction to form the nucleus of an offspring. This theory is a remarkable anticipation of his grandson's theory of "pangenesis." (See the third edition of *Zoonomia,* Darwin 1801, 2, 296–297.)

Backing Darwin's claims about heredity was a wide range of purported supportive instances. Much was fairly anecdotal, but his examples showed that Darwin had more in mind than just physical changes leading to physical changes. Rather, he saw a direct link between the mental and the physical. For instance, in one story which might have been lifted bodily from the pages of the *National Enquirer,* Darwin told of a man who had one child with dark hair and eyes among a brood of fair children. While his wife lay childbearing, the man became sexually obsessed with the darkly colored daughter of one of his tenants. Despite his unsuccessful efforts in this direction—even the offer of money was spurned—he admitted that "the form of this girl dwelt much in his mind for some weeks, and that the next child, which was the dark-eyed young lady above mentioned, was exceedingly like, in both features and colour, to the young woman who refused his addresses" (Darwin 1794, 523–524).

Some of Erasmus Darwin's examples drew on his classical training: "the phalli, which were hung round the necks of the Roman ladies, or worn in their hair, might have effect in producing a greater proportion of male children" (Darwin 1794, 524). Other examples drew on Darwin's great experiences as a busy and successful doctor. He was concerned particularly about the effects of alcohol on the drinker, and the drinker's progeny: "the diseases occasioned by intoxication deform the countenance with leprous eruptions, or the body with tumid viscera, or the joints with knots and distortions" (Darwin 1794, 501). No doubt linked to these observations was his own decision to become a virtual abstainer, an act that was made the more poignant by the fact that the family of his first wife had, he believed, a heritable tendency to drunkenness.

It would be a mistake to leave Darwin's causal speculations merely at the surface level. It is clear that the medical training struck more deeply. The modification of Hartley's associationism was fitted into a four-fold division of organic capacities: irritation (brought on by external bodies), sensation (by pleasure or pain), volition (by desire or aversion), and association, whereby "fibrous motions . . . become so connected by habit, that when one of them is reproduced the others have a tendency to succeed or accompany it" (Darwin 1794, 49). Drawn from various medical sources, these capacities enabled Darwin to show how even the most primitive of organisms would respond to the environment: the first response would be brute reaction but then, through experiences which it found pleasurable or painful, the organism would begin to exert some control over its behavior, which would then lead it to seek out the environment, acting positively to meet its needs, and finally the features produced by the capacities would become part of the basic blueprint of the organism.

Darwin made use of these capacities to show how and why the course of evolution has been as it has been. We find that from the beginnings to the ends of their lives, "all animals undergo perpetual transformations; which are in part produced by their own exertions in consequence of their desires and aversions, of their pleasures and their pains, or of irritations, or of associations; and many of these acquired forms or propensities are transmitted to their posterity." Consequently, "the three great objects of desire, which have changed the forms of many animals by their exertions to gratify them, are those of lust, hunger, and security" (Darwin 1794, 502–503).

This led Darwin straight into a discussion of secondary sexual characteristics, phenomena of which his grandson was to make much. In like manner, Darwin dealt with adaptations for getting food: "The trunk of the elephant is an elongation of the nose for the purpose of pulling down the branches of trees for his food, and for taking up water without bending his knees" (Darwin 1794, 504). Then, with typical rhetorical flourish, having made explicit reference to the evolutionism he found implicit in David Hume's *Dialogues,* Darwin exalted: "What a magnificent idea of the infinite power of *The Great Architect! The Cause of Causes! Parent of Parents! Ens Entium!*" (Darwin 1794, 509).

An Industrialist's View of Evolution?

With respect to cultural views, I will rush past the pre-Darwinian thinkers. Monboddo had some very odd beliefs about Progress—namely that the luxurious society of modern Europeans has made them thoroughly miserable. However, he thought that the root problem was the Fall; that we had to regain our intelligences and that today's unpleasantnesses are a necessary step in that direction; and that we are therefore now on the upward path to Redemption. His beliefs about human change were part and parcel of this funny mix of Progressivism and Providentialism. Likewise with Worthington, one of the English divines who set Progress in a millennial context: "we may observe that all Parts of Nature are endued with a Principle not only to preserve their State, but to advance it, and *that every Thing has a Tendency to its own Perfection*" (Worthington 1743, 223, his italics).

Erasmus Darwin yields a rich harvest bearing on our interest in science and culture, even though, taking a cosmic view of things, he was no P/progressionist. Like the Greeks, he believed in ultimate cycles, with everything developing and then collapsing. However, once we leave the cosmic scale and come down to our world and its history, we find many pertinent thoughts in Darwin's writings. In his very first published words on evolution, Darwin made his biological progressionism an integral part of the picture: "Perhaps all the productions of nature are in their progress to greater perfection? an idea countenanced by the modern discoveries and deductions concerning the progressive formation of the solid parts of the terraqueous globe, and consonant to the dignity of the Creator of all things" (Darwin 1789, 9–10). More significant than this, perhaps, is the fact that the poem on which this passage is a comment is in its entirety a testament to the "pathetic fallacy," namely the viewing of the non-human world as if it were human. Darwin pretends that plants have emotions, especially of the sexual variety, and draws the consequences thereof. Obviously, moving from Progress to progress is a similar move, since one is going from a purported human process to a purported general organic process.

In his later writings the language of progress persists and, more importantly (since "progress" might be just a general term for "change"), in word and idea Darwin makes it very clear that it is upward change for the better, change leading to humans, that he has in mind. This is not concealed, but is the dominant theme: "all nature exists in a state of perpetual improvement . . . the world may still be said to be in its infancy, and continue to improve FOR EVER AND EVER" (Darwin 1801, 2, 318). There is some branching in the path to "perpetual improvement," but ultimately there is progress to humankind.

> Organic Life beneath the shoreless waves
> Was born and nurs'd in Ocean's pearly caves;
> First forms minute, unseen by spheric glass,
> Move on the mud, or pierce the watery mass;
> These, as successive generations bloom,
> New powers acquire, and larger limbs assume;
> Whence countless groups of vegetation spring,
> And breathing realms of fin, and feet, and wing.
> Thus the tall Oak, the giant of the wood,
> Which bears Britannia's thunders on the flood;
> The Whale, unmeasured monster of the main,
> The lordly Lion, monarch of the plain,
> The Eagle soaring in the realms of air,
> Whose eye undazzled drinks the solar glare,
> Imperious man, who rules the bestial crowd,
> Of language, reason, and reflection proud,

With brow erect who scorns this earthy sod,
And styles himself the image of his God;
Arose from rudiments of form and sense,
An embryon point, or microscopic ens!

(Darwin 1803, 1, 295–314)

What is the nature of this progress? What is its defining mark? Are there independent criteria, or is progress simply a matter of becoming more humanlike? The move from simplicity to complexity was certainly an element of Darwin's thinking. There was more than this, however. Often Darwin wrote as though progress consists in ever-greater manifestation and sophistication of the four capacities of irritation, sensation, volition, and association. Again humans come out on top. Additionally, interesting himself in agriculture, Darwin was much impressed by organisms' abilities to fend off attackers and look after themselves. Survival ability in some sense seems to be an important mark of progress. Thanks to "the great Author of all things," organisms "not only increase in size and strength from their embryon state to their maturity, and occasionally cure their accidental diseases, and repair their accidental injuries," they also possess "a power of producing armour to prevent those more violent injuries, which would otherwise destroy them" (Darwin 1800, 350).

In light of his grandson's views on evolution, we might come to think of Erasmus as a precursor. I do not want to make too much of this, but we shall learn that the two men did, at least, share some concerns. The need for defense was very much a function of Malthusian worries about overpopulation. In the *Temple of Nature,* some passages are virtually Malthus in verse.

So human progenies, if unrestrain'd,
By climate friended, and by food sustain'd,
O'er seas and soils, prolific hordes! would spread
Erelong, and deluge their terraqueous bed; But war, and pestilence, disease, and dearth,
Sweep the superfluous myriads from the earth.

(Darwin 1803, 4, 369–374)

What is the answer or consequence to all of this? Darwin's response embodies what seems above all to be the essence of his notion of progress: *happiness.* Although there is necessarily death and destruction, this leads ultimately to change and to a happier state. The way that the world works is that as soon as an organism falls sickly or aged, a more vigorous organism destroys it and, as often as not, eats it. Or feeds the victim to its young. "By this contrivance more pleasurable sensation exists in the world, as the organized matter is taken from a state of less irritability and less sensibility, and converted into a state of greater" (Darwin 1800, 557). As for all utilitarians, the end tends to justify the means.

Move on to our second question. What direct evidence is there that Darwin moved from Progress to progress? The case is as strong. Darwin was a fervent believer in Progress. His religious convictions opened up the way for such a belief, for (in line with comments in passages quoted above) his was a god of non-interference, a god who worked through unbroken law—that is, the god of deism. ("That there exists a superior ENS ENTIUM, which formed these wonderful creatures, is a mathematical demonstration. That HE influences things by a particular providence, is not so evident. The probability, according to my notion, is against it, since general laws seem sufficient for that end"; letter to Thomas Okes, November 23, 1754, in King-Hele 1981.) In this, Darwin was at one with many of his friends, including the famous geologist James Hutton, whose beliefs about the indefiniteness of time may have been an important influence.

Darwin's political commitments pointed toward Progress, too, for he was a radical. Not only did he support the American Revolution, but he was a friend and admirer of Benjamin Franklin ("the greatest statesman of the present"; letter to Benjamin Franklin, May 29, 1787, in King-Hele 1981, 166). Likewise, Darwin responded warmly toward the French Revolution, at least before the Terror. He hoped the French would spread the holy flame of freedom over Europe and typically broke into verse on the subject (McNeil 1987, 65). Most importantly, Darwin's own beliefs were rooted in a commitment to Progress—a social Progress of a kind that would be reflected in his notion of biological progress. He was a founding member of the Lunar Society, a late-eighteenth-century dining club of industrialists and scientists meeting in Birmingham, the heart of the British Midlands and the center of the heavy-industry wing of the Industrial Revolution. His friends and fellow thinkers included Joseph Priestley, the industrialist Matthew Boulton, the engineer James Watt, and the potter Josiah Wedgwood. They were committed to change, to science, to technology, to machines, to an Adam Smith–type of free enterprise, to bigger, better, stronger. And this included Darwin, who for himself entered into the spirit of invention with gusto, and for others hymned their achievements in verse:

> So with strong arm immortal BRINDLEY leads
> His long canals, and parts the velvet meade
>
> (Darwin 1791, 3, 349–350)

His belief was that science is a good thing and that science and technology lead to Progress. But for Darwin, it was more than just a (French-like) notion of intellectual Progress. His Progress was woven right into the social fabric and would, through industry and enterprise, eventually yield a happier society—something which fitted right in with Darwin's (and the rest of the Lunar Society's) utilitarianism.

This secular ethic, brought about by the play of the forces of free enterprise, is of course precisely the human analogue of Darwin's biological progress. Moreover, as we have seen, at the causal level Darwin thought there was a direct link between Progress and progress, because inasmuch as gains are made in the former, through association and the inheritance of acquired characters, they manifest themselves in the latter. Especially for humans, but in rudimentary form for other organisms also, advances in ideas get translated into advances in biology. Indeed, if this is not enough, Darwin himself linked Progress and progress, explicitly: "this idea [that the organic world had a natural origin] is analogous to the improving excellence observable in every part of the creation; . . . such as in the progressive increase of the wisdom and happiness of its inhabitants" (Darwin 1794, 509).

Since the case is so obvious, it is hardly necessary to go on to our third method of assessment: the extent to which Darwin outstrips his evidence. In fairness, without in any sense wanting to say that Darwin sounds particularly modern (certainly not "more modern than his grandson"), we find many points at which today his ideas do match the evidence—in the significance of comparative anatomy, for instance. But, this said, major parts of his theorizing—however we would judge its worth today—had to come from somewhere, and that somewhere was not the real world. Especially in his thoughts about progress, given the then virtually unmapped fossil record on the one hand and his beliefs about the effects of associationism on the other, we know that there had to be some non-evidential source. Once again, as with all else we have discussed, the finger points to Progress.

Thus, overall the case is strong that Erasmus Darwin's evolutionism was progressionist, and that the idea of progress came from his beliefs about Progress. Moreover, the form of Darwin's progress was a function of the form of his Progress.

Lorenz Oken: "Naturphilosoph"

We cross back now to the Continent, to Germany (Russell 1916; Lenoir 1982; Gould 1977a). Here, unlike France and England, there is no one person who stands out in quite the way that does Lamarck or Darwin. This is not to say that there was no interest in evolution or that everyone rejected it. In fact, a case could be made for saying that the interest in evolution and sympathy for development was greater and more universal in Germany than anywhere else. Nor is it to say that there were no full-blown evolutionists. These there surely were, in one degree or another. (Richards 1992 champions Friedrich Tiedemann, Gottfried Treviranus, and Johann Friedrich Meckel.)

The really interesting question is why so many of the German thinkers, living in a culture saturated with ideas about development, including people that later real evolutionists were to look back to as deeply inspirational, did not take that crucial extra step. The suspicion has to be that there is more to the story than simply a missed boat. We must not fall into the trap of judging the past by the present and thinking that these Germans would have been evolutionists if only they had been a little luckier, brighter, braver, etc., etc. To the contrary, the German propensity to produce non-evolutionary developmentalists may have been more than a matter of chance or of individual contingencies. In other words, as with Buffon we may have an opportunity to test ideas about P/progress in a negative way, to look at the question, as it were, from the other side—although I would want to stress that (unlike the example offered by Buffon) we are not now faced with people who are at all ambiguous about upward change.

Whom should we take as our representative? Immanuel Kant could be our choice. A proponent of the nebular hypothesis; recognizing in the *Critique of Judgement* ([1790] 1951) the significance of homology (it "strengthens our suspicions of an actual relationship between [forms] in their production of a common parent . . . from man down to the polype"; pp. 337–338); he was nevertheless typical of those who drew back from evolution at the crucial point (Lovejoy 1959). Turning to the biologist

Johann F. Blumenbach, who argued that there is a life force (*Bildungstrieb,* akin to Newtonian force) which has a certain capacity for variation but which eventually gets exhausted, Kant agreed: "That crude matter should have originally formed itself according to mechanical laws, that life should have sprung from the nature of what is lifeless, that matter should have been able to dispose itself into the form of self-maintaining purposiveness—this he rightly declares to be contradictory to Reason" (Kant 1951, 274).

The great poet Johann Wolfgang von Goethe is another possible choice. Stressing the ground plan or ideal archetype *(Urtyp)* connecting organisms homologously; accepting from Bonnet the Chain of Being; he was the first to see the mammalian skull as but a composite of modified vertebrae. "The lowlier the creature, the more these parts resemble both each other and the whole. The more perfect the creature becomes, the more do its parts become dissimilar" (Goethe 1962, 37, 9; quoted by Wells 1967, 539). Yet he too was no true evolutionist. Nor even was the biologist Carl Friedrich Kielmeyer (1765–1844), though it was he who first articulated the linchpin of much later evolutionism, a three-fold organic parallelism—the kind of isomorphism much cherished by *Naturphilosophen*—between the sequence into which living organisms can be put, the developmental stages of the individual through embryology, and the sequence of forms as they appeared here on earth: "even men and birds are plant-like in their earliest stages of development" (Kielmeyer 1938b, 91–92).

It is true that within (literally) a year or two of the publication of Lamarck's evolutionary speculations, Kielmeyer was endorsing them: "as de Lamarck urges, all those things which are cited as being a case of extinction are equally explicable as [the consequence] of the changed direction of the formative force which ensues upon the transformation of our earth" (Kielmeyer 1938a, 210). But it is clear that Kielmeyer's evolutionism was a limited phenomenon, occurring within certain classes only. Organization does have its bounds, and certain newly appearing organisms might be "original products of the fertile earth." Moreover, as with Lamarck, the contingent changing nature of the real earth was taken as a disruptive element in the developmental process, and for that reason it is certain that Kielmeyer saw the three-fold parallelism ultimately as existing perfectly only in ideal form. In the real world it was broken, incomplete, and with various side lines and branches.

My true representative of the archetype will be he who came to epito-

Lorenz Oken

mize *Naturphilosophie* in biology, one whose life and work were enveloped in controversy, both while he was alive and after his death. I refer to Lorenz Oken (1779–1851), romanticist, embryologist, and for a stormy decade in Jena (1807–1819), students' friend and radical polemicist. His major work, *Lehrbuch der Naturphilosophie,* was published in 1809 (the English translation, *Elements of Physiophilosophy,* in 1847). Written as a series of 3,652 aphorisms, it takes us right through the natural world, from the nature of philosophy to "the blessed condition of Man and of humanity—the Principle of Peace" (p. 665). Rushing past such mundane problems as the nature of God and of Man, barely pausing to deal with side issues like gravity and matter—somewhat earlier Schelling had thought it necessary to caution Oken that "A little diminution of that petulant and sarcastic style of writing, especially about Newton will do no harm" (letter of May 6, 1809; in Schelling 1962–67, 1, 441)—Oken then turned his full attention to the organic world.

There is much that is familiar to those who are students of German biology at this time. For instance, having got him a job at Weimar, Goethe was then rewarded by Oken's claiming the vertebrate theory of the skull for himself: "The number of cephalic vertebrae is 4; namely, nasal, ocular, lingual, and auditory vertebrae" (Oken [1809] 1847, 370). Also, there are versions of the law of parallelism: "The Insect passes through three stages prior to its attaining the adult or perfect condition. It is at first worm, next crab, then a perfect, volant animal with limbs, a Fly" (p. 542). More generally: "During its development the animal passes through all stages of the animal kingdom. The foetus is a representation of all animal classes in time" (p. 491).

But, what of development and evolution? On the surface, the prospects seem favorable. Certainly, there is a naturalistic stance, with organisms being created out of a universal organic substratum that Oken labels "mucus": "Every Organic has issued out of mucus, is naught but mucus under different forms. Every Organic is again soluble into mucus; by which naught else is meant, than that the formed mucus becomes one devoid of form" (p. 185). Moreover, the formation of life—analogous to the views of Lamarck—seems to have come about through the direct action of natural forces, specifically light on seawater.

However, each form—presumably each species—seems to have its own unique starting point. Generation leads only to organisms of the same kind. There certainly seems to be a sequential appearance of organisms on this earth, but each kind started from scratch—or, at least, from mucus.

> 912. The first organic forms, whether plants or animals, emerged from the shallow parts of the sea.
> 913. *Man also is a child of the warm and shallow parts of the sea in the neighbourhood of the land.*
> 914. It is possible, that Man has only originated on one spot, and that indeed the highest mountain in India. It is even possible, that only one favourable moment was granted, in which Men could arise. A definite mixture of water, definite heat of blood, and definite influence of light must concur to his production; and this has probably been the case only in a certain spot and at a certain time. (Oken [1809] 1847, 186)

There is nothing whatsoever here about humans coming from apes or any other kind of organism. Nor is there any hint that Oken was looking this way. Some mucus points—those produced by existent organisms—produce organisms of existent kinds. Some mucus points, those produced fresh from seawater, produce new kinds. Hence, if anything, although in respects Oken's developmentalism is more radical than some (he believed every organism had a natural origin), in other respects he is more conservative (he apparently denied even limited change for existent forms). Even with this most speculative of thinkers, we do not find the German equivalent of Lamarck or Erasmus Darwin. With other *Naturphilosophen,* Oken drew back at the crucial point.

Idealism Taken Seriously

Let us play things straight, analyzing Oken in exactly the same way as we have examined the full-blown evolutionists. First, therefore, there is the matter of the belief in biological progress, though in Oken's case it is not to be interpreted as implying complete continuity (and hence is perhaps not the strictest use of the term). Before him, Kant had definitely edged toward some kind of progressionist interpretation of life's history. Specifically in the case of humans, Europeans seem to come out on top, even though—a consequence of his belief in extraterrestrial beings—humans as such apparently do not win the inter-planetary stakes. More generally, in the *Critique of Judgement,* Kant argued that better adaptedness is the mark of advance. Goethe likewise was an organic progressionist, echoing the Aristotelian criterion of complexity to mark increased value. And with Kielmeyer, there was also a belief in advance—as for anyone who ties their thinking to a law of parallelism, given that human ontogeny involves a development of intelligence, the increased significance of "the power of understanding" seems to be an important mark of advance.

Oken combined all of this and more. His whole system is predicated on an upward climb—in the case of the animals, from the simplest Infusoria up to the most complex or developed Mammalia. Moreover, let there be no doubt about the endpoint: "Animals are only the persistent foetal stages or conditions of man" (Oken [1809] 1847, 492). Man supremely has *freedom* of the body, and through reason and understanding has *freedom* of the mind. Not, however, that we should think all humans exactly equal. There is an ordering of the senses, with the sense of touch at the bottom and the sense of sight at the top (p. 419). It is perhaps predictable that the "Skin-Man is the *Black,* African," whereas the "Eye-Man is the *White,* European" (p. 651, his italics).

Turning next to beliefs in Progress and to possible links between Progress and progress, we find that in Oken's work the positive evidence exists, abundantly. Again before Oken, we know already of Kant's commitment to Progress, and there is certainly no cause to think that he was intellectually schizophrenic, putting his philosophy/social beliefs in one compartment and his biology in another. Likewise with Goethe. We have the all-embracing world picture of the *Naturphilosophen:* "I had not failed to learn from Kant's scientific writings that forces of attraction and repulsion are essential properties of matter, and that, within the latter concept, the two are inseparable; this opened my eyes to the fundamental polarity of all things, which infuses and animates the infinite variety of the phenomenal world" (Goethe 1887–1919, 1, 33, 196).

Progress came as part of the package. If one wanted to make ideas constitutive of the world, in a kind of neo-Platonic way seeing the archetype being progressively revealed in the course of history, and combined bits and pieces of Kantianism (or Platonism) with a belief in the Chain of Being, you end up with something very much like the biology of Goethe—or of Kielmeyer, if your focus is embryological.

In the case of Oken this idealistic philosophizing finds its strongest biological exemplar, for his is virtually the philosophy of (the post-Kantian, pre-Hegelian) Schelling transferred to the organic world. For that philosopher—the idealist's idealist—with the denial of external reality, we are left with the ignorant human mind and blind matter (that is, the unknown eternal mind) in opposition. Toward self-realization, we must and do work up dialectically from the attempts to understand the inorganic world, through the organic world, to total understanding.

> Because there is in our spirit an infinite striving to organize itself, so in the outer world must a general tendency of organization reveal itself. It is so. The world system is a kind of organization, which has formed itself from a common centre. The powers of chemical matter are already beyond the boundaries of the merely mechanical. Even raw materials which separate out of a common medium crystallize out as regular figures. The general formative drive [*Bildungstrieb*] in nature loses itself finally in an infinitude, which even the prepared eye is unable to measure. (Quoted by Morgan 1990, 31, from Schelling in the *Philosophical Journal,* 1897)

Crucial in all of this is a version of the law of parallelism, for in understanding of ourselves, we achieve understanding of reality: "It is certain that whoever could write the history of his own life from its very ground, would have thereby grasped in a brief conspectus the history of the universe" (Oken [1809] 1847, 132). And the end result of all of this is *Freedom*—God, and through It ourselves, transcends the limits, is "not at all conditioned by Objects" (Morgan 1990, 29, quoting Schelling *Of the I as the Principle of Philosophy*), but has full self-autonomy.

All of this is precisely what we have seen in Oken's biology, with the historical climb paralleling the intellectual climb. Even the final act of transcendence is spelt out:

> The universal spirit is *Man.*
>
> In the human race the world has become individual. Man is the entire image or likeness of the world. His language is the spirit of the world. All the functions of animals have attained unto unity, unto self-consciousness, in Man. (Oken [1809] 1847, 662)

I stress that one should not crudely conflate different notions of Progress. British Progress translates into social and cultural improvement, whereas Germanic Progress is an ascent of the spirit, toward self-realization. But certainly the Germanic Progress can incorporate social improvement—inasmuch as one sees an increase in freedom, probably as a function of a unified German state, the idealism of the *Naturphilosophen* incorporates social Progress, even if it is not the essence of change. Given his life-long commitment to personal freedom, we can truly say that in every sense Oken's progressionism was an extension of his Progressionism.

I stress also, however, that in biology as in the social world, there is undoubtedly a teleological flavor to Oken's ideas over and above anything in French and British thought. For Germanic Progressionism, there is an essential human element; but there is a directedness to the unfolding which is uniquely theirs. This is why in biology the law of parallelism, between the development of the individual and the evolution (however understood) of life, was and was to remain so crucially important, both as cause and as effect. In the embryological development of the individual, one sees a kind of internally driven momentum toward a predetermined mature state—not necessarily vitalistic (although many have thought so) but *prima facie* end-directed and (again, as many have thought) end-controlled. It was precisely this presupposed momentum that was so characteristic of Germanic thought about the unfurling of life as a whole.

Remember also the commonplace of German Protestant mystical thought that one can draw a developmental parallel between history (the Jews) and the individual (oneself)—phenomena both end-directed and intensely human. As it happens, he who most vigorously pushed this analogy in the late eighteenth century, Kant's sometime friend and fellow citizen of Königsberg, Johann Georg Hamann, was an implacable foe of the Enlightenment. But he greatly influenced those in the Germanic Progressionist line—Herder and then Goethe and the *Naturphilosophen,* notably Schelling (Berlin 1993).

And so, finally, to our third question. By now, it is almost anti-climactical to go on and ask about the relationship between German biology and the evidence. Or, at least, let us turn the issue on its head and note that it was certainly not a caution in interpreting the empirically given which stopped German thinkers from being evolutionists. The theory of the vertebrate skull was to come under withering fire later in the nineteenth century, and the same is true of laws of parallelism in this century. And for every age, then and now, Oken's speculations about "primary mucus" had at best a metaphorical relationship to the truth. I do not mean to imply that German thinkers, specifically those we have considered and generally the community at large, were indifferent to the empirical world. I do imply that, for or against developmentalism, the facts of the matter were not the decisive factor.

So, what can we say in the final analysis? Inasmuch as the German idealists were evolutionists, it was because they were progressionists. Inasmuch as they were progressionists, it was because they were Progressionists. But why were they, or (at least) why were so many of the crucial figures, not full evolutionists? I have warned against thinking this failure—if that be the right word—was simply a contingent matter of luck or courage or whatever. If courage were needed, the idealists had it in abundance. The point is that "one is not in the business of genealogy, constructing family trees" (Knight 1990, 16). The idealists ultimately were not on the track of evolution.

In the case of individual thinkers, one can pick out good reasons (in the sense of adequate reasons to them) why they would *not* be evolutionists. Kielmeyer was worried about the evidence. Yet, there has to be more to the story than this. And the true answer is surely that just as German Progressionism led them to evolutionism, it just as firmly led them away again. The Germans were idealists. And, notwithstanding the increasingly constitutive nature of their idealism, if idealism is to have any force it has to cut somewhere. Apparently, it cut here. Evolutionism, as a doctrine about the physical, *material* world, was simply not necessary to provide support or to fill out the story—indeed, it had a tendency to mess up the picture with facts. Germanic development—and Germanic Progression—was in essence a function of the world of spirit and not of the world of physical being. This was why the parallel between the individual and the whole was so important, and why physical reality could and must remain at the level of the former. Listen, for instance, to Hegel:

> Nature is to be regarded as a *system of stages*, one arising necessarily from the other and being the proximate truth of the stage from which it results: but it is not generated *naturally* out of the other but only in the inner Idea which constitutes the ground of Nature. *Metamorphosis* pertains only to the Notion as such, since only *its* alteration is development. But in Nature, the Notion is partly only something inward, partly existent only as a living individual: *existent* metamorphosis, therefore, is limited to this individual alone. (Hegel 1970, 20)

The point could not be clearer, and in his students' lecture notes the implications for evolution are drawn explicitly: "If we want to compare the different stages of Nature, it is quite proper to note that, for example, a certain animal has one ventricle and another has two; but we must not then talk of the fact as if we were dealing with parts which had been put together" (Hegel 1970, 21).

The contrast with British Progressionism, in particular—with the Progressionism of Erasmus Darwin, which was about real people in a real nation using real science and real machines to make real Progress—is stark and distinctive. For Darwin, it simply would have been incomprehensible to talk of Progress, or of progress, without intending real physical links. For the German idealists, to talk of Progress, or of progress, was to say nothing of such links—and indeed their introduction would cloud the full picture. Evolutionism in the physical sense was not only not necessary, it was not desirable. That is why they did not go the final mile.

Science as a Profession

My claims are simple. The idea of evolution was the child of the hopes of Progress. Like the parent, it too incorporated the hope of upward climb. And, as the parent appeared in different lights in different countries, so also did the child. In France and England the ascent resulted in a fully continuous move to our own species. In Germany, the thought was father to the child. These conclusions now raise interesting and important questions about status. How are we to regard the speculations (theories, hypotheses) of the evolutionists? More significantly—for what *we* think about Lamarck or Darwin or Oken is not really the point—how was their work regarded by their contemporaries? Did it qualify as professional science? Or did it fall over the divide, on the popular science side? Was it perhaps even down at the quasi-science end of the spectrum?

To answer these questions, we must first raise the issue of science as a profession in the relevant time period, around the end of the eighteenth century. This is a much-discussed topic, for one cannot simply take modern criteria and transpose them back two centuries. Nevertheless, it is not a hopeless or pointless question, for it was around this time that science as a profession was congealing: more so in some countries than in others. Hence, as with the concept of Progress, even though there were points of overlap, it is probably best to deal with the topic on a country by country basis.

France, despite and because of the Revolution, was the first and most vigorous country in offering opportunities for professional scientists, in any sense as we know them (Hahn 1971, 1975; Crosland 1978, 1992). The Paris Académie des Sciences went back to the seventeenth century, and, apart from giving its members acknowledged status and an outlet for publication, as a body of the state it offered them financial support—although the numbers were few (around fifty) and the remuneration not really adequate without supplement. But there were other institutions through which a man could support himself—the Collège de France, an organization for public teaching, for instance. And for life scientists there was the Jardin du Roi, long dominated by Buffon and his followers (Appel 1987).

The Revolution brought changes—the great chemist Lavoisier lost his head, literally, and because it was an instrument of the state the Académie was closed. But the new rulers were hardly less appreciative of science than were the old, especially given the fact that scientists proved an indispensable part of the effort in the wars that were to engage France for a quarter-century (Crosland 1967, 1992; Fox 1984). The Académie was reconstituted as the Institut de France (science being the "First Class"), with state incomes and natty uniforms—leading one English observer to liken a meeting to a gathering of well-behaved butlers. (After the Restoration, in 1814, the Institut went back to its original name.) The Jardin du Roi was reformed as the Muséum d'Histoire Naturelle, with twelve professorships offering support, living quarters, and excellent research facilities—not to mention the loot of Europe, once Napoleon's armies got into gear (Burkhardt 1977). Additionally, there were jobs in the newly formed teaching academies *(Écoles)* and later in the Imperial University. And there were formal and semi-formal groups where people could meet to share interests and exchange ideas. Best known was the grouping of chemists, the Society of Arcueil, formed around Claude Louis Berthollet. It met regularly at his home near Paris, published memoirs, and offered real facilities and advice for promising young men (Crosland 1967).

England had virtually none of the state support offered by France (Musson and Robinson 1960; Schofield 1963). This was a major reason why—the Industrial Revolution notwithstanding—Britain lagged behind France in the profession of science (and in major respects the production of professional science). The Royal Society of London was as old as the Paris Académie but, needing to support its own way, was heavily dependent upon a large body of patrons, whose lack of qualifications and achievements in science was balanced by their willingness to pay hard cash for the privilege of signing themselves "F.R.S." Apparently this was considered a sound investment in the medical profession, for the initials were popularly understood to stand for "Fees Raised Since"! (See McClelland 1985; Stimson 1948.)

Around the turn of the century a number of other organizations/societies were founded, notably the Linnaean Society and (for popular instruction) the Royal Institution. Increasingly, under the long-term, conservative leadership of Sir Joseph Banks, the Royal Society opposed such upstart newcomers. Nor, in England, was much of a balance offered by its two universities, Oxford and Cambridge. Anglican institutions, they offered no science teaching (except mathematics and some theoretical physics), and indeed the professors in various science subjects felt the need neither to know of nor to lecture in their subjects (Engel 1983). Despite important contributions to the *Philosophical Transactions* of the Royal Society, the official picture looks pretty bleak.

But there was another side to the story. Scotland had more and better universities, unencumbered by the religious restrictions, where (since the professors got paid through fees) genuine efforts were made to teach socially useful subjects, including science (Davie 1961). And in England, the non-conformists (Protestant non-Anglicans) started their own academies and supported a network of itinerant teachers (Musson and Robinson 1960). Since non-conformists were much engaged in trade, they were at the forefront of the Industrial Revolution. The things they wanted taught always included science. In addition, the major provincial cities had literary and scientific societies and like groupings. Although it was more exclusive than most, the Lunar Society was an exemplar (Schofield 1963).

In the nineteenth century, Germany was to overtake France as the true home of professional science, thanks particularly to its powerful and extensive state-supported university network. But this development gained momentum only after 1810, with the founding of a university in Berlin, the capital of Prussia. Before that, despite reforms and advances (particularly in Göttingen), support for academic science had been slight (McClelland 1980). Yet, also in this country—more strictly, this grouping of countries bound into a quasi-unit by a common language—through the universities, schools, and various academics, a man of science could make a way. Already noteworthy, even before the shock of Napoleonic invasions, was a German chemical community strong enough to support its own monthly professional journal, *Chemische Annalen* (Hufbauer 1982).

Summing up, it is fair to say of our time period—say by 1809, when Lamarck published the *Philosophie zoologique*—a recognizable idea of a professional scientist was emerging, if it had not already emerged. What then was the science—the professional science—of such a scientist? Leaving the life sciences on one side for the moment, we look at the work of Laplace and Lavoisier, of John Dalton and Thomas Young, of (among the younger generation) Joseph Louis Gay-Lussac and August Jean Fresnel. And, thus directed, the impression one has is that their science was, in recognizably modern terms, "professional." It was careful, it often made (heavy) use of mathematics, it was empirical, showing great and ingenious use of experiment, and it paid detailed attention to measurement and quantification.

Take Gay-Lussac, for instance, worth noting here precisely because he was working and publishing right at our moment of focus (1809). A younger member of the Society of Arcueil, a protégé of Laplace and Berthollet, he was a meticulous experimentalist whose claim to fame was the great accuracy with which he was able to measure physical materials, from an early reconfirmation of Charles's law (about the thermal expansion of gases) to his best-known work on the combination of gases, specifically his law that gases always combine in simple ratios (as oxygen and hydrogen combine in the ratio 1:2). Gay-Lussac was the paradigm of the professional scientist, given his dedication, his social connections, and his recognition, and his work was the paradigm of professional science, given the fanatical care with which he stayed close to the phenomena (Crosland 1978).

We are now starting to link up with the epistemological, so let us make our interest explicit. How well can one translate professional science, as we understand it, into mature science, as we understand it? There was certainly much talk about criteria of good science. The official line, pushed by people like Condillac and almost every British thinker, was that above all one must avoid sweeping hypotheses, especially those involving unknown and unknowable causes. The Scottish common-sense philosopher, Thomas Reid, for instance, begged us to agree with Francis Bacon and Isaac Newton "that every system which pretends to account for the phenomena of Nature by hypotheses or conjecture, is spurious and illegitimate, and serves only to flatter the pride of man with a vain conceit of knowledge which he has not attained" (Reid 1785, 250; quoted in Laudan 1981, 90). Proper practice, apparently, is to collect one "inductive" instance after another, until one builds up one's beliefs in an empirically pure fashion.

As always when people (especially philosophers) are thus talking, however, it is important to look behind the pronouncements at the practices. One of the biggest scientific advances as we come into the nineteenth century was the triumph of the wave theory of light over the previously dominant corpuscular theory. Clearly the new theory was much more than simple inductive enumeration, being about as hypothetical as one could imagine. Moreover, it made direct reference to causes which, by their very nature, could not be sensed directly. As we shall see in the next chapter, this led the philosophers to revise their methodological dicta; but the science did not wait on the philosophy. The really important point was that people like Young and Fresnel were able to score absolutely stunning successes at the epistemic level—most famously, they were able to make devastatingly accurate predictions, whereas the rival theory was simply left standing. Their work compelled people to take it as the epitome of mature (or potentially mature) science. Associated with the high standing science had gained was a sea change in the power of the mathematics being employed—without a switch from geometry to algebra, none of the revolutionary theories would have been possible.

It was not hypotheses as such to which people (and here I refer to people across all our countries) objected, but sweeping conjectures uncontrolled by and indifferent to epistemic norms. Professional scientists were sensitive to pseudo-science and judged it unacceptable precisely because it failed epistemically. Mesmerism, the idea that there is some kind of animal magnetic fluid essential for good health, is a case in point.

Professionals thought its practitioners charlatans, an attitude leading Louis XVI (in 1784) to establish a Royal Commission (chaired by Franklin and including Lavoisier) to investigate. In short order, the hypothesis was dispatched, but not because it was a hypothesis as such, and certainly not because it was causal. Rather, Mesmerism was derided because it failed simple epistemic tests: one had to ignore the way that fluids normally work, one had to downplay the *ad hoc* complexities necessitated by anomalies, and above all one had to accept that prediction worked only when the results were known beforehand. "Since the imagination is a sufficient cause, the supposition of the magnetic fluid is useless" (Franklin et al. 1970, 124). The obvious is simpler.

None of this is to say that there was a fixed methodology and that everyone (of professional standing) at once agreed on the content of good science.[3] In Germany, for instance, there was intense debate when the phlogiston theory was challenged. But, at this time, there is much good sense in the links we have drawn between professionalism and maturity. And these, as I have noted, were underpinned by mutual self-interest. Scientists wanted to do science. Countries and individuals wanted the products of science—the products of mature science—and were prepared to pay professionals accordingly.

Professional Science, Popular Science—Pseudo-Science?

Against this background, what can we say of the evolutionists and of their ideas? Take first the statuses of the principal players. Lamarck is easy to categorize, for if anyone was a professional scientist of the time, it was he. Before the Revolution, he was a member of the Académie des Sciences. Then, when it was reconstituted as the Institut, he became a member, representing botany. He held minor posts in the Jardin du Roi, and when this was transformed into the Muséum d'Histoire Naturelle he grabbed one of the twelve new professorships. One might well argue that at that point he was hardly professionally qualified for the assignment that he got, on worms and insects and the like (his "invertebrates"); but those were troubled times and in the situation he was really as good as anyone (Burkhardt 1977).

Erasmus Darwin does not fit as easily into pigeonholes. As a physician, he was a fully qualified professional, by training, by reputation, by practice. As a scientist, as a life scientist, his status as a professional was more ambiguous. He never held posts in the field, he showed no interest in teaching, and aside from one or two slight papers in the *Philosophical Transactions* he did not go out of his way to build a research record. On the other hand, he did mix with and was respected by professionals, as in the Lunar Society; he had a strong interest in biology broadly construed; and in England (as opposed to France) we must beware of putting too much weight on the way that a man earned his bread. The professional/amateur distinction (thinking now in terms of money) is not very helpful.

In Germany there was significant professionalism, if not always directly in science. Kant, of course, was a philosopher and a very professional one at that. He held a professorial chair in Königsberg. Goethe was no academic, and indeed (his literary work aside) he was a civil servant in the court of Weimar, supervising the mining within the duchy. Kielmeyer and Oken were professional biologists. The former was a full-time academy/university teacher, holding various posts around Germany. As it happens, these include five years as professor of chemistry at Tübingen, but always Kielmeyer's main (and acknowledged) interests were in the biological sciences. Oken likewise held academic appointments at a number of universities. He taught anatomy and embryology and much more, right across the sciences. More than that, he was a leader in science organization, founder in 1822 of the Gesellschaft Deutscher Naturforscher und Aerzte, an ongoing conference for workers in science and medicine, and for thirty years ran his own journal *(Isis)*.

Hence, taken overall, it is probably fair to say that as biologists thinking about evolution our subjects had one foot firmly across the professional scientific divide. Considered collectively, they are more than just an assortment of unqualified outsiders—which, without pretending that one could have made any firm predictions, is surely much as one might have expected. To think creatively about evolution at any level did require some serious knowledge of anatomy, embryology, paleontology, psychology (associationism), and more. Even though not every person was equally interested or competent in every branch, each one had to have a significantly sophisticated scientific background. Yet, if we focus specifically on evolution, we could hardly claim that our subjects had both feet planted across the divide. More categorically, this is not to say that the evolutionism of the day qualified as professional science. Professional science may be produced by professional scientists, but the converse does not necessarily hold. Such is the case here. All of the evidence suggests that evolutionary speculations were regarded very much on the popular side of the divide. Moreover, this was a judgment with a strong epistemological component. Not that this should be a great surprise if you compare what we have seen in this chapter against the work produced by the professional physical scientists of the day. Conspicuous by its absence has been a theory of evolution that included an attempt to quantify, or measure, or experiment, or stay close to the phenomena.

Let us go back over our evolutionists. In respects, Lamarck is the most interesting of them all, for he was—and was regarded by his contemporaries—a real Dr. Jekyll and Mr. Hyde of the professional/popular division. Some of his work was praised by all as precisely that which one expects of a professional scientist focusing on organisms. It was careful, it was descriptive, it was repeatable, it was useful. It was acknowledged as being of the best quality. I refer specifically to Lamarck's taxonomic labors, including (before he became an evolutionist) his work in botany and (after he became an evolutionist) his massive treatise on the invertebrates (Lamarck 1778, 1815–1822). But there was another side to Lamarck, and here he developed a reputation for being very unsound—with good reason, for he was wildly speculative, flew in the face of solid empirical evidence, and predicted in ways that were not simply wrong but embarrassingly so.

In chemistry, Lamarck adopted the pre-Lavoisier theory of four elements—earth, air, water, and fire—and persisted through his career in using them to explain everything, from sound to magnetism, from color to organisms, despite nigh definitive evidence as to the untenability of his initial premises (Burkhardt 1977, 69–71; Lamarck 1794, 1796). In meteorology—where Lamarck really thought he was a founding genius—he speculated on the influence of the moon and made proclamations that truly must be judged pseudo-predictions, the inadequacy of which became apparent to all (including an irate Napoleon) after much state expenditure (Burkhardt 1977, 10; Lamarck 1800–1810). And in geology likewise, Lamarck was much given to wild speculation: again the moon was a key operative factor, this time in swirling water around the globe, thus causing mountains to be built from the precipitates of organic remains within the waters (Burkhardt 1977, 84–87; Lamarck 1801b, 1802b).

Lamarck's evolutionism fell firmly into the speculative branch of his labors and was seen as such by his contemporaries, who (as professionals) responded (for many years) with (at the Institut) professional silence and (at the Collège de France) public scorn. They regarded his evolutionism as being in major respects part and parcel of his ridiculous chemistry and geology and more—which it was. To suppose spontaneous generation was to fly against both conventional biology and conventional chemistry. To suppose that all change comes through needs and environmental pressures is to suppose an ongoing changing earth, as predicted by Lamarck's geology. To suppose . . . Unfortunately to suppose evolution after the fashion of Lamarck was simply not to suppose professional science.

> It is often said that Lamarck's evolutionary theory was rejected in its own day simply because people at the beginning of the nineteenth century were unaccustomed to thinking in evolutionary terms. Lamarck, in other words, was too far ahead of his time to be appreciated. What seems to be more nearly the truth, at least with respect to the French scientific community, is that Lamarck's theory of evolution was rejected not because the idea of organic mutability was virtually unthinkable at the time, but because Lamarck's support of that idea was unconvincing and because, more generally, the kind of speculative venture Lamarck had embarked upon did not correspond with contemporary views of the kind of work a naturalist should be doing. (Burkhardt 1977, 201–202)

Not that Lamarck was particularly repentant. Thinking himself a Newton of the life sciences, he was open in his love of sweeping hypotheses: "Those who would conclude that in the study of nature we must always limit ourselves to amassing facts resemble an architect who would advise always cutting stones, preparing mortar, wood, iron-work, &c. and who would never dare to employ these materials to construct an edifice" (Lamarck 1801b, 3; quoted in Burkhardt 1977, 41).

Erasmus Darwin poses a slightly different problem. It would be wrong to say that he never wrote at a level that would be appreciated and accepted as professional. In his prose writings, however, Darwin veered frequently toward the anecdotal and colorful rather than the carefully descriptive. This applies especially to his evolutionary speculations—remember the Roman ladies—and apparently general contemporary opinion was that Darwin (in prose) was altogether too "philosophical" and not nearly sufficiently empirical and systematic. About his writings in agriculture, one reviewer remarked (in a not-very-complimentary fashion) that he had produced "a work in every page of which some curious hypothesis, some fanciful theory, startles and amuses us"; and another that: "If Doctor Darwin had indulged less in theory, and had enlarged the number of his facts, our satisfaction would have been complete" (McNeil 1987, 174, quoting an anonymous writer in the *Monthly Review,* 1800). Significantly, Darwin seems never to have submitted his evolutionary ideas to the *Philosophical Transactions.*

The case for regarding Darwin's work within the popular realm is strong, even before we turn to the clinching argument, namely his promotion of so many of his ideas through verse. This was just not the way that one did professional science—or preserved one's reputation as a professional physician, for that matter: "I would not have my name affix'd to this work on any account, as I think it would be injurious to me in my medical practise, as it has been to all other physicians who have published poetry" (letter to publisher, about to bring out the first part of the *Botanic Garden*; in King-Hele 1981, 139). It was only when his poetry proved so popular with the public that Darwin dropped his anonymity.

In part, Germany proves in the contrary way the point I am making about early evolutionary speculation having the status of popular science. Kant may not have been a professional scientist, but he set forth the conditions for professional science. If you look again at an above-quoted passage, you will see that he is worrying about how any kind of purposiveness could come about through blind law, without a directing intelligence (Kant 1951, 274). His point was that evolutionary speculations failed these conditions. Kielmeyer, ever the careful professional, agreed with Kant. In part, Germany proves the point the same way as the others do. Full romantic *Naturphilosophie* probably influenced many more serious people than they or others realized or cared to admit, but it had a dreadful reputation among professional scientists—precisely because they were appalled at the idea that one could deduce everything about the world, *a priori* (Knight 1975, 1986). This was just not the way one did physics and chemistry.

Nor were their fears quelled when someone like Oken expressed his beliefs in aphorisms. This was poetry, philosophy, or worse.[4] Justus Liebig, the great German organic chemist of the first half of the nineteenth century, a man who had studied with French chemists, referred to *Naturphilosophie* as the "pestilence, the Black Death, of the nineteenth century" (Reddick 1990, 335). Additionally, Goethe's support of anything pertaining to science was not likely to help its status, given that he had the temerity (stupidity, many would say) to attack Newton's claim that white light is a synthesis of colors—and then followed with his own rival theory, one which by its indifference showed its contempt for mathematical theory. To this day, as he was in that day, Goethe is taken as a paradigm of a poet meddling beyond his ken or ability (Burwick 1986).

The certain conclusion to be drawn—both on the grounds of content and from people's reactions—is that in its early days evolutionary theorizing did not have the status of professional science. The kind of work that the evolutionists were doing was simply not of the epistemic quality to be found in the physical sciences, or (I shall speak more to this point in the next chapter) what was becoming expected in the biological sciences. It was out of tune with the prejudice against sweeping hypotheses, and it did not have overriding merits (as did the wave theory of light) to compel belief, such as predictive fertility and the like.

Moreover, there was certainly no question of discipline building, even if one discounts the fundamental differences imposed by the different visions of Progress. Apart from anything else, whereas (say) a massive job in taxonomy could call for whole squads of ancillary workers—all united by the task at hand—grand speculations just did not demand the organizational structure that characterizes a functioning scientific discipline. Good science gives people work, and that was not offered by the evolutionists. Nor, obviously, were there economic or technological payoffs, of a kind Napoleon (for instance) might have used against the British.

If not professional science, if not (proto-) mature science, then what? Popular science, obviously, and judged thus as popular science, evolutionism was identified with the quasi/pseudo end of the scale. Popular but not quasi-science certainly existed, and was appreciated. The classic examples were produced by Laplace, his *Exposition du système du monde* (1796) and his *Essai philosophique sur les probabilités* (1814); these works, which gave general nontechnical expositions of his ideas on astronomy and probability (respectively), were aimed at the educated public. They are still the great exemplars of popular science. But the evolutionists were not writing in this manner, digesting and simplifying professional work. They were off on their own tracks. At best, they were loose with the epistemic norms. At worst, they were indifferent. Lamarck simply brushed aside counter-evidence; Darwin gathered uncritically; and Oken made it up.

What about P/progress? The conclusions of this chapter, that evolution was so directly and strongly used as a vehicle for thoughts of Progress, support the belief that evolution would be (for proponents as well as opponents) a popular science. Proponents of social or moral Progress might not like the judgment that their work was pseudo-science, but remember that this status does not preclude support from others. And indeed, there is evidence that, the opposition of professionals notwithstanding, there were those who were attracted to evolution precisely because of its P/progressionism. In Lamarck's case, whatever the general disapproval, because of his standing as a professional in other areas, some people were always prepared to dip into his evolutionary works. And of these who read Lamarck, some found values to their liking. The same was true of Oken, and not just in Germany, as we shall learn in the next chapter.

Conversely, in the case of Erasmus Darwin, it was the cultural implications of evolutionism which were seen as the threat. After the French Revolution, anything to do with Progress was anathema. Darwin was attacked savagely by conservatives: a brilliant parody of his poetry, *The Loves of the Triangles* as against *The Loves of the Plants,* quite destroyed his reputation as a poet and a sage (Canning, Frere, and Ellis 1798). He was no longer considered a serious thinker. The point of significance to us is that it was political satirists who did this demolition job, not professional scientists. This was because his work was considered popular science, of a rather disreputable kind. Erasmus Darwin was not out to produce mature science, and his critics knew that.

Evolution failed as professional science, and its failure is mirrored at the epistemological level. Notions like predictive ability and internal coherence were important to the turn-of-the-century professionals, both for their own work and in their criticisms of others. People did not value observation and experimentation and mathematicization in themselves, but rather because such practices did more adequately satisfy epistemic goals. Hence, evolution was judged pseudo-science. Ultimately, however, one senses that the evolutionists just did not care—certainly not enough about status—to make the epistemic values all-deciding. What counted for evolutionists—and what stuck in the craws of opponents—was Progress. All else was dressing. At least, this was the case in those early years.

3

The Nineteenth Century: From Cuvier to Owen

Lamarck's *Philosophie zoologique* was published in 1809, the year of the birth of Charles Darwin. In 1859, exactly one-half century later, Darwin published the *Origin of Species*. In this chapter, I shall cover this period, going as before from country to country—France, Britain, Germany. Especially toward the end, the barriers do start to break down, somewhat, so my ultimate guide is allegiance to ideas rather than nationality.

The Conditions of Existence

The dominant figure in French biology in the first three decades of the nineteenth century was Georges Cuvier (1769–1832). Rightfully acclaimed as the founder of modern comparative anatomy, he was born of modest (although bourgeois) background, in Montbéliard (annexed by France only in 1793), and was educated at the Karlsschule in Stuttgart. This was essentially a training college for potential civil servants, but on graduation Cuvier received no appointment. Therefore, he acted as tutor to a noble family living in Normandy during the worst excesses of the French Revolution, and arrived finally in Paris in 1795. He lived in that city for the rest of his life, and by its end had so far risen in the social scale that he had been created a baron. This was a matter of great personal satisfaction, for Cuvier was much given to outward displays of pomp, rank, and appropriate deference (Coleman 1964; Outram 1984; Appel 1987).

Cuvier's perspective on the organic world, exhibited above all in his great *Le règne animal,* came courtesy of the lens of Aristotelian teleology. To this he credited all of his success as a comparative anatomist. Expressed as the notion of the "conditions of existence," it was his key to the understanding of organisms: "As nothing may exist which does not include the conditions which made its existence possible, the different parts of each creature must be coordinated in such a way as to make possible the whole organism, not only in itself but in its relationship to those which surround it, and the analysis of these conditions often leads to general laws as well founded as those of calculation or experiment" (Cuvier 1817, 1, 6; quoted in Coleman 1964, 42).

The key word here is *coordination.* The parts of organisms, argued Cuvier, are correlated in an overwhelmingly tight manner: "Hence none of these separate parts can change their forms without a corresponding change on the other parts of the same animal, and consequently each of these parts taken separately, indicates all the other parts to which it has belonged" (Cuvier 1813, 90–91). It was his skillful use of this principle ("correlation of parts") that gave Cuvier (especially to the public) his aura of scientific invincibility. Supposedly, he could take a bone or a tooth and construct the whole animal—and again and again on incomplete fossil remains he did just that, later being triumphantly vindicated when full specimens were found. Given the stomach of a carnivore, the jaws must be made "to fit them for devouring prey," the claws "for seizing and tearing it to pieces," and "the teeth for cutting and dividing its flesh," and so on (Cuvier 1813, 91).

Translating his philosophy into a practical tool for the taxonomist, from the conditions of existence Cuvier drew the "subordination of characters": "The parts of any animals possessing a mutual fitness, there are some traits of them which exclude others and there are some which require others; when we know such and such traits of an animal we may calculate those which are coexistent with them and those which are incompatible" (Cuvier 1817, 1, 10; quoted in Coleman 1964, 77). Once you have got certain basic characters in place, as it were, then other characters at the same hierarchical level are excluded, and the options for what can be fitted in further are circumscribed. Once given a backbone, for instance, you can still go the way of a fish or a bird. What you cannot now do is go the way of an invertebrate. Nevertheless, given the range of possibilities, it is obviously still an empirical matter as to what you find in the real world. Cuvier, in fact, thought there are four basic types or *embranchements:* Vertebrates, Molluscs, Articulata (insects, spiders, and so forth), and Radiata. (The last group, which contains such "radiating" organisms as starfish, was something of a grab-bag category.) With these groups and the nested sub-sets within them, Cuvier thought he had both theoretical and empirical authority to catalogue the whole of the animal kingdom.

What about evolution? Interestingly, Cuvier's very first writings hint at, if not a sympathy for, then at least an openness to evolutionary ideas. There is the claim that, between terrestrial and aquatic sowbugs and the marine mantis shrimp, "Here, as elsewhere, nature makes no leaps"; there is an (admittedly unpublished) sketch of the Chain of Being; and, in a joint publication with Etienne Geoffroy Saint-Hilaire in 1798 on the classification of monkeys and orangutans, there are some very Buffonian notions about the temporal degradation of species (Appel 1987, 50).

But, very quickly—certainly by the beginning of the new century—Cuvier solidified his position in other directions, and from then on his opposition to any kind of transformism was implacable. All excursions into evolutionism like Lamarck's were ridiculed or ignored, and massive empirical evidence was brought against any such hypotheses (Laurent 1987). In this second respect, Cuvier was particularly well qualified, for he was interested not only in the living world but also in the world of the past. He sought answers as revealed by the fossil record, especially the vertebrate fossil record. Opening up this unknown domain, Cuvier emphasized that one fact stands out, beyond all others: fossil types are extremely well defined, with sharp beginnings and ends and no intermediates. If evolution be true, then we should expect to find transitional fossils. And this we do not find: "no such discovery has ever been made" (Cuvier 1813, 114–115).

Backing this claim, Cuvier noted that domestic species (like the dog) are molded in all sorts of different ways but never into two different species. Additionally, Cuvier drew attention to the pertinent case of the Egyptian mummies. When unraveled, these relics—appreciated as being thousands of years old—were found to be indistinguishable from today's organisms. In the opinion of Cuvier, this could mean only one thing. There is no permanent change. The Egyptians "have not only left us representations of animals, but even their identical bodies embalmed and preserved in the catacombs" (Cuvier 1813, 123).

Empirically, therefore, the facts stood against evolutionism. But these empirical arguments lay just at the surface to his opposition. There were much deeper objections. At the most fundamental level, Cuvier took evolutionism to be in stark opposition to his teleology.[1] If organisms are tightly correlated, then transitional forms are not simply empirically impossible. They are conceptually impossible. A linking organism would be out of functional focus and simply could not exist. Think of an analogy: the internal angles of an n-sided polygon total $2n - 4$ right angles. This gives you lots of options, depending on n. What you cannot have is a polygon with an intermediate, odd number of right angles. Evolution from one form to another, or from a joint ancestor, is quite impossible.

What then did Cuvier propose to substitute in evolution's stead? What was *his* theory of origins? At one level, he abjured any such speculation. But in his one semi-popular work *(The Preliminary Discourse),* Cuvier did lay out his general position. He argued that the main agent responsible for the present state of the globe is water, which has at periodic intervals come flooding across those areas now dry. That these floods are rapid and "catastrophic" (to use his English translator's term) is shown by the fact that some animals, notably the mastodon, are found caught in a state of perfect preservation—flash frozen, as it were. There was some ambiguity about the causes of these floods, the most recent of which was but a few thousand years ago and identified by Cuvier as Noah's flood. To an English correspondent he wrote: "I would say that the most reasonable conjecture was that it was due to several ruptures in the crust of the globe which changed the level and position of the seas as they had already been changed at other periods and by other catastrophes" (letter to H. de la Fite, 17 April, 1824; quoted in Coleman 1964, 135). But he added that he did not really have too much of an idea about causes.

Cuvier and Progressionism

Let us turn now to some of the factors underlying Cuvier's scientific views. Although this is our first fully negative anti-evolutionary case, it can help us to confirm and define our main theses. As with last chapter's Germans, let us play things straight, using for the non-evolutionist the same measures as for the evolutionist: beliefs about life's history, social

Georges Cuvier (in the Académie uniform he designed).

Etienne Geoffroy Saint-Hilaire

beliefs and their relation to science, and extent to which claims outstrip empirical data.

First, then, there is the question of Cuvier's beliefs about the history of life. In fact, somewhat paradoxically, although he was no evolutionist, Cuvier was himself increasing the evidence for a progressionist reading of life's history. Thanks to his anatomical skills, he was able to map on the catastrophic record the history of life, at least with respect to the quadrupeds. He was able to show that, around Paris at least, a succession of forms is observed, from fairly primitive strange specimens to the most modern and recognizable of species. Moreover, humankind, as Cuvier explained at length, is not to be found in the fossil record, and hence is obviously very recent.

But, for Cuvier himself, there was no progress to be seen here. The clearing out of old forms is to be explained in terms of the catastrophic floods. The appearance of new forms is to be explained in terms of migrations from elsewhere on the globe: "I do not pretend that a new creation was required for calling our present races of animals into existence. I only urge that they did not anciently occupy the same places, and that they must have come from some other part of the globe" (Cuvier 1813, 125–126). Although there are hints in Cuvier's writings that the succession may have been a function of changing geology, and who knows what this might have meant ultimately for orders of appearance, the point is that the connections are contingent and not causal.

We move next to the possible significance of cultural values, particularly those centering on Progressionism. Where did Cuvier stand on Progress and did this affect his thinking about evolutionism? At one level—as one might expect from a successful European scientist—Cuvier was quite receptive to the idea of Progress. He believed that, although the last catastrophe set the clock back to zero, whites have outstripped Negroes ("the most degraded race among men") and that the human race has "resumed a progressive state of improvement since that epoch, by forming established societies, raising monuments, collecting natural facts, and constructing systems of science and of learning" (Cuvier 1813, 171–172). At best, however, statements like these give us a very restrained notion of Progress—certainly there is no hint that improvement will continue, or indeed that such as has occurred is truly a function of human effort. Let us therefore open up the discussion, and to see where Cuvier stood generally with respect to Progressionism, specifically (given his nationality and education) the French and German varieties, let us begin with the man himself.

Everything that we know of Cuvier's life suggests that it—his behavior, his thought, his work—was defined against the turbulent times in which he lived, and that his success came in conscious response to them. Above all else, he was scared by the appalling disruptions and excesses of the Revolution—the time when "by a reversal of ideas that will long be memorable in history, the most ignorant portion of the people had to pronounce on the fate of the most instructed and the most generous" (Cuvier 1827; quoted in Appel 1987, 255, n. 60). He felt that a repetition had to be avoided at all costs. Since he lived first under a dictatorship (Napoleon) and then under a conservative monarchy (the Restoration), it is sometimes suggested that he too was an arch-conservative. But this is only partially true. A better model for Cuvier is the Vicar of Bray. He was the consummate politician, responding always to stay on the side of those in power (Outram 1984). Or, rather, he was the consummate bureaucrat; it was this for which his training at the Karlsschule prepared him, and it was as this that his various masters employed him: on commissions and committees and the like. His sense of deference—by him to superiors, to him by inferiors—was natural.

In short, everything that Cuvier believed and did put him against the *idéologues* and all of the prophets of Progress, for with every good reason they were seen as a significant factor behind the Revolution. Moreover, although Cuvier certainly saw revolutions in life's past, his own world was one of rational functioning and stability. He faulted the evolutionism of Lamarck and others precisely because it proposed change and instability. This was almost calculated to turn science away from its true function, that of stabilizing minds and societies: "It consoles the unhappy, it calms hatreds. Once elevated to the contemplation of that harmony of Nature inexorably ruled by Providence, how feeble and petty are found to be those jurisdictions left to the free will of man" (Cuvier 1817, 1, xix–xx).

Cuvier's thoughts about French Progressionism incline us to think that he would dislike evolutionism and its progressionism intensely. He had little facility in English, and there is no evidence that Cuvier was touched by British social thought. But, as one educated in Germany (taught by Kielmeyer, no less), what of Cuvier's thoughts on Germanic notions of Progress, particularly those of the *Naturphilosophen* with their ideas of a world spirit working its way up the hierarchy? Coming, as he did, from a border province long under Germanic rule, Cuvier was a Protestant. There are questions about the depth of his faith—his personal life was crushed by the deaths of all four of his own children, and this may have marked him negatively. He was certainly a Believer, however, and although as a cultured French scientist he would never have invoked God as a causal factor in a scientific argument, Cuvier, the Believer, was one and the same person as Cuvier, the scientist. Moreover, Cuvier's science was framed by his religion—most notably, obviously, in his whole teleological approach to the organic world, even though the immediate influences were philosophical. His view of life's history in the *Discourse* was likewise influenced by religion, for all of the silence about direct links: "The oldest of books in Genesis, which for whatever reason always tallies with the geological record" (lecture notes taken in 1805; quoted by Corsi 1988, 184).

What bearing did any of this have on *Naturphilosophie?* At a direct level, Cuvier's functionalism led him away from one of the major supports of the *Naturphilosophen,* namely embryology. For Cuvier, what counted was the finished, adult organism—how does it work? Development was minimized, and so also was any theory which made it central.

At a less direct level, Cuvier's Christianity led him to abhor any science which promulgated one of the greatest heresies of the faith: pantheism. Again and again we find Cuvier raging against the identification of God with His Creation (Coleman 1964, 180). And, with every justification, Cuvier saw Germanic Progressionism as falling into this error—what else do you make of a philosophy that identifies matter with the World Spirit?! Hence, Cuvier was led to fight against "those who in recent years have sought to offer a new form of the metaphysical system of pantheism, which they call *Philosophy of nature*" (Cuvier 1825, 267). Here again, therefore, we have reason to conclude that Cuvier would find anathema any kind of biological developmental progressionism.

So we come to the question of evidence. It must surely be clear that there was nothing empirically unreasonable about Cuvier's anti-evolutionism. He did have a lot of facts—facts which he himself had discovered—on his side. Moreover, his teleological approach to the world apparently paid major dividends. Cuvier made sense of and brought order to the living world. In his hands, the doctrine of the conditions of existence was a very powerful tool, as he inferred what one might expect before one has found it. And yet! Cuvier was undoubtedly leaving some important questions dangling, most notably about the ultimate origins of organisms—their mode of creation, their sequence (if there be one), and so forth. Moreover, Cuvier was having to ignore or downplay the significance of aspects of the empirical world. Even though the starfish may not be homologous with the vertebrates, the facts of homology (within *embranchements*) were becoming increasingly pressing by the 1820s. There were costs to giving exclusive support for his version of teleology. The unwillingness to allow any significance to homology grew increasingly aberrant. Certainly, the flat refusal to take the compromise option that people like Kielmeyer were proposing, namely some kind of limited change within *embranchements*, is made the more striking by the fact that Cuvier himself might have been inclined somewhat this way in his early career.

The case is complete. Cuvier's opposition to evolutionism—which to him (as to everyone else) spelled progressionism—and his views on Progress are different sides of the same coin.

Philosophical Anatomy

Cuvier's great rival in morphology in early-nineteenth-century France was Etienne Geoffroy Saint-Hilaire (1772–1844). It is an ill wind that blows no-one any good, and ill though the French Revolution may have been for many, it blew very well for Geoffroy. Through being the right man with the right patrons at the right time, at the age of twenty-one he became professor of quadrupeds, cetaceans, birds, reptiles, and fish in the newly formed Muséum d'Histoire Naturelle. He was to remain lecturing for forty-seven years at this Muséum—for thirty-two years combining the work with courses at the Sorbonne.

Geoffroy was responsible for settling and encouraging Cuvier in Paris, and at first they not only collaborated but lived together. Soon, however, their ways parted and rivalry developed, ending finally in outright hostility. The romantic Geoffroy made the mistake of following Bonaparte in his Egyptian adventures (1798–1801), and while he risked life and limb up the Nile, the ambitious stay-at-home Cuvier built himself an impregnable reputation and consolidated power in the all-important Institut, something Geoffroy did not enter until 1807. The friendship was over, although the worst bitterness came only many years later (in 1831), when Geoffroy, by now an evolutionist, engaged the arch-anti-evolutionist Cuvier in open debate before the Académie des Sciences.

Yet, although their dispute over evolutionism was crucially important in this clash—it is certainly important to us—just as Cuvier's stand was part of an overall picture, so also Geoffroy's stand was part of an overall picture (Laurent 1987). Simply put, where Cuvier stressed function, Geoffroy stressed form. For Cuvier, the key to the understanding of organic parts lay in the ends or functions they serve. For Geoffroy, the key to such understanding lay in the relationships of the various parts to each other and, more significantly, in the relationships of the parts of one animal to the parts of another animal. In particular, Geoffroy thought that animals—initially vertebrates but he started to extend out to other animals—follow a common pattern. Thus, isomorphisms or analogues should and can be sought, often between animals of quite the most diverse kind. (Late in life, Geoffroy made it clear that his ideas were meant to apply to plants also.) There are strong similarities here with the views of the *Naturphilosophen,* a point to which I shall return in a moment. But although by the 1820s both sides were aware of and pushed the similarities—as did critics—with respect to straight anatomy, Geoffroy's thinking was primarily homegrown, with the greatest debt to suggestions in Buffon.

Perhaps the need to define himself and get out of the shadow of Cuvier was important to Geoffroy also. Certainly, the circumstances of Geoffroy's first essays into "philosophical anatomy" made for originality, for they came when he was in Egypt. Writing on the ostrich, he noted that although it had no need of a furcula (wishbone), being flightless, there are nevertheless vestiges of just such a bodily part. "These rudiments of the furcula have not been suppressed, because nature never advances by rapid leaps, and she always leaves the vestiges of an organ even when it [the organ] is entirely superfluous, if that organ has played an important role in other species of the same family" (from a memoir read to the Institut d'Egypt, 7 September, 1798; quoted by Appel 1987, 74). At this point, Geoffroy was happy to note explicitly that the bone has no analogue in the world of the quadrupeds.

Ten years later, however, Geoffroy was ready to go further, looking for similarities between different classes of animals. And he found them, in bones which have corresponding positions and connections, albeit possibly with quite different functions. Still, however, there was no full theory. Particularly worrisome were the bones of the gill cover, or operculum, in fishes, which seemed to have no direct analogue in other vertebrates. At most, Geoffroy had (what had now deepened into) a full conviction that there is some underlying plan linking organisms. Finally, in 1817, the bell rang—and Geoffroy heard it, to use a happy metaphor. The bones of the operculum are transformed into the tiny bones of the mammalian middle ear: the malleus, the stapes, and the incus. Everything falls into place, as correspondences can be shown to run throughout the vertebrate world: "Strictly, it will suffice for you to consider man, a ruminant, a bird, and a bony fish. Dare to compare them directly, and you will reach in one stroke all that anatomy can furnish you of the most general and philosophical [nature]" (Geoffroy 1818, xxxviii). In the *Philosophie anatomique,* Geoffroy did dare to make the comparison, following the similarities through the ear, the sternum, the hyoid, and bones of the lungs and related organs, the shoulders and arms. Always: "An organ is sooner altered, atrophied, or annihilated than transposed" (Geoffroy 1818, xxx).

Not that any of this was evolutionary *per se.* Indeed, Geoffroy presented his thinking in idealistic terms, with formal connections running through the vertebrate world: it is worth noting that Geoffroy, like the Germans and unlike Cuvier, was turning to embryology for help. Nor at this point was Geoffroy stepping out of a Cuvierian *embranchement.* But he was soon to do this, arguing for analogies between insects and molluscs, on the one side, and vertebrates, on the other (Geoffroy 1819)—a line which stimulated the predictable Cuvierian response: "Your memoir on the skeleton of the insects lacks logic from beginning to end" (Geoffroy 1820, 34). Cautious and blooded, but persistent, Geoffroy returned to the theme in the 1820s, endorsing a study favoring cuttlefish/vertebrate similarities. It was this which really proved too much for Cuvier, who responded with a detailed savage defense of his functionalism. To the delight of the intelligentsia of Paris—of Europe—the two great biologists aired their opposing positions across a number of public meetings and then in print. Goethe loved every moment of it! By now, however, there was more than anatomy at stake, for Geoffroy had become explicitly evolutionist: in his view, his own anatomical stand both supported and was supported by transformism.

This evolutionism may represent a much earlier commitment—Geoffroy's friendship with Lamarck went back to the previous century—although it was surely a development connected with the path taken by his anatomical theorizing. Even more immediately and significantly, probably, was the interest in embryology, backed by an ever more detailed fossil record (thanks to Cuvier!). But, whatever the proximate cause, by 1825 Geoffroy felt sufficiently emboldened to endorse Lamarck's views explicitly: "These changes [in the environment] are of a nature to have acted on the organs of which I have just spoken, and to have done so precisely according to the two laws posed by M. de Lamarck in his *Philosophie zoologique*" (Geoffroy 1825; quoted in Appel 1987, 132).

In fact, however, Geoffroy was far from being a clone or devoted disciple of Lamarck. With respect to the path of evolution, he never went in for the grand-scale theorizing to be found in Lamarck. There were no overall trees of life—at most, just a few fragmentary reconstructions. Then, for all of his favorable talk of Lamarck's laws, Geoffroy had little interest in the effects of the environment on the adult organism, nor was he much concerned to drive organisms up the Chain of Being through some vague force dependent on needs or whatever. Rather, he saw changes coming through environmental effects on embryos. Over the years the richness of oxygen in the atmosphere decreases, and this sets up changes which give rise (in one step) to "monsters." Some of these prove better fitted to the changing circumstances of the globe than do the older, parent forms. Clearly, therefore, what Geoffroy had in mind was less the gradual change envisioned by Lamarck and more a form of "saltationism" (evolution by jumps).

A Romantic Frenchman's Philosophy

What of progress and of Geoffroy's cultural values? As we might expect, we see a shift in the answer to this question as Geoffroy moved toward evolution. In the *Philosophie anatomique* (1818) he referred to the classic ordering of vertebrates—fish, reptiles, birds, mammals—as "an invention of our schools," adding that "already one no longer speaks of this progressive succession of beings, and at the same time without doubt one divests oneself of the old talk of 'the most perfect beings.'" However, a decade or so later he was happily talking of "a progressive series" (Geoffroy 1833a, 84) and of the molluscs being low on "the chain of being" because of "the inferiority of their nervous system" (Geoffroy 1830, 225). Moreover, Geoffroy certainly thought that change would be in the direction of increased complexity: "the organisation . . . only awaits favourable conditions to rise, by addition of parts, from the simplicity of the first formations to the complication of the creatures at the head of the scale" ("Sur la vertèbre," quoted by Russell 1916, 68). There is little doubt that we humans have a special place: "the appearance of man upon the earth coordinates and achieves the sublime arrangement of the things which concern our planet" (Geoffroy 1838, 350).

Next, we must ask of Geoffroy's views on Progress. Although nominally a Catholic—before the Revolution he had been intended for the Church—we know that Geoffroy was attracted strongly to some form of deism. This clears the way for the fact that emotionally, socially, and intellectually Geoffroy was drawn toward Progressionism, as Cuvier was repelled by it. Ever the romantic—the "eternal adolescent," as he has aptly been described (Bourdier 1969, 54)—Geoffroy was attracted to adventure, excitement, and liberalism. Everything about the two men is summed up by the fact that the daring Geoffroy followed Napoleon to Egypt while the cautious Cuvier stayed home. While the bureaucratic Cuvier flourished under the Restoration, with honors showered on him, the impetuous Geoffroy stood in danger of dismissal. While the reactionary Cuvier stood for species fixity, the radical Geoffroy opted for evolutionism and was hailed by Progressionists bent on social reform. The novelist Balzac even went so far as to dedicate to Geoffroy the second edition of his masterpiece, *Père Goriot,* a tale of the crucial significance of social factors on the development of character.

Although it is doubtful that *Naturphilosophie* made a direct contribution to Geoffrey's thinking about anatomy, it may have been more significant as embryology turned him toward evolution. This Germanic "philosophy of Nature" got the credit (in 1830) for the order which he now saw in the living world: "The philosophy [perhaps 'science' in a more modern reading] of Germany has shown well that organic beings increase in number and complexity down through the passage of time, or in the progressions of the zoological chain, according to the order and in direct consideration of the diverse degrees of organization" (Geoffroy 1830, 143). Yet there was really something very un-Germanic about Geoffroy's evolutionism. Even less than Lamarck was he friendly to in-built teleological forces or momenta.

We need to look closer to home, something easy to do, for an excellent exemplar of Geoffroy's Progressionism—his distinctively French Progressionism, with its emphasis on the development of intellectual ideas up through the ages—is to be found in the *Discours préliminaire* to his *Philosophie anatomique.* Having due regard for the fact that Geoffroy was about to offer his own thoughts as the culmination of the efforts of brilliant minds through the ages, we learn that his brand of transcendental morphology occurs as the apotheosis of a series of ever greater epochs. We begin with Newton (on the grounds of comments in the *Optics*) and end with Geoffroy. At each step, we build on the one previous, getting even better. Unambiguously this is change for the good. There is a time factor here, of course, for at the writing of the *Philosophie anatomique* Geoffroy still pulled back from evolution and progress. But he was on the brink, and as he slipped over the Progressionism intensified—as it had to, as he staked out his ground against Cuvier.

Thus, in an evolutionary piece ("On the degree of influence of the world about us on the modification of forms") read in 1831 we find Geoffroy first laying out the development of biology as something exhibiting increasing complexity and sophistication, and then linking this directly to the scale of being: "Hence, the philosophy of natural relationships [the science of biology] had provided the facts and had prepared the way to the analogy among beings; which, in its turn, based upon an understanding of the facts of the zoological scale, . . . becomes the starting point of the philosophical [scientific] system of differences" (Geoffroy 1833b, 67).

Enough has been said. In Geoffroy we do find Progressionism, and the signs are that the links are with progressionism—even putting Lamarck aside, we can discern an intellectual climb and at least hints that change when it occurs is rapid, occurring in one generation (in the intellectual world though the actions of genius). Finally, with respect to the question of evidence, Geoffroy did discover hitherto unremarked connections between organisms of kinds quite dissimilar. But, especially with respect to his evolutionism, Geoffroy went beyond the bounds of the given. Engaged though he was in the study of monsters, he had no definitive grounds for his speculations about the causes of change. And some of his more daring analogies owed more to fancy than to fact. Especially, one must allow that, much though Geoffroy may have known of the vertebrates, his knowledge of invertebrate anatomy was vanishingly small and was shown to be so when he tried drawing grand connections.

But my purpose here is not to criticize—or to praise even. My purpose is to establish a link between the scientific and the cultural, which I believe is now done. Cuvier and Geoffroy fell out. The question of evolution was bound up in their dispute, and the dispute included arguments over the problem of progress. My claim is that the stands taken by both men, with due respect to their background commitments, support this book's proposed links between Progress, progress, and evolution.

Robert Grant

Establishment Britain around 1830 was as hostile to evolution as it was to Progress, the two often being linked. Cuvier was a much-admired authority. Both sides in a big debate about geology, then gripping the scientific community, came together on this (Ruse 1979a). Charles Lyell, the "uniformitarian"—arguing that the causes of the past were identical in nature and intensity to those of the present, and that the geological world holds in a timeless "steady state"—turned from a progressionist reading (1826) of the fossil record when he sensed its implications for evolution; inspired by a work on humankind by the future Archbishop, J. B. Sumner (1816), Lyell ever regarded human nature in a Providentialist fashion. Our friend Adam Sedgwick, embracing "catastrophism," loathed evolutionism, interpreted the upward rise of life's history in an explicitly Providentialist fashion, and we know where he stood on Progress (Sedgwick 1831, 1845, 1850). It is a moot point whether he was more against the British industrial variety, the French intellectual version (in a sermon he referred to the *philosophes* as "moral fanatics"), or *Naturphilosophie* (Sedgwick 1833). Oken and Hegel were ridiculed and their science spurned: "I say we have successive forms of animal life adapted to successive conditions (so far, proving design), and not derived

The frontispiece to the first volume of the first edition of Lyell's *Principles of Geology* (1830). The clearly defined erosion on the lower parts of the columns suggests that, after their construction, the land first dropped (and the columns were immersed in water) and then rose again to its present level. The image thus supports Lyell's claim that there is no overall directional change to earth's history.

in natural succession in the ordinary way of generation" (Clark and Hughes 1890, 2, 86). For Sedgwick, evolution, Progress, and progress were one horrible, indigestible gallimaufry. "The opinions of Geoffroy St. Hilaire and his dark school seem to be gaining some ground in England. I detest them, because I think them untrue. They shut out all argument from *design* and all notion of a Creative Providence" (Clark and Hughes 1890, 2, 86).

Robert Grant

But Progress was in the air. Significantly, John Stuart Mill wrote a fiery critique of Sedgwick's sermon, arguing for Progress and criticizing him in explicitly Comtian terms (Mill 1859, 1, 110, fn.). Did no-one become an evolutionist, perhaps driven precisely by thoughts of Progress? To answer this question, we must move down the social scale a notch or two, to a less self-satisfied level of British society—to people, many of whom were part of the unfashionable segments of the medical world, who did not have the ear of the men of power. In London these people were often linked with the newly founded (by the utilitarian Jeremy Bentham and like-minded reformers) University College. These were the people who were associated with the "Radical" group in parliament, who wanted to break the power and privileges of the rulers and spread the spoils more evenly through the classes (Desmond 1989).

The Scots were a driving force in the group we consider now. Through their own interests in continental thought, especially that of the French savants, they were thus indirectly introducing such foreign ideas into England. One prominent representative of this set was the surgeon Robert Grant (1793–1874), born and educated in Edinburgh. When the Napoleonic wars were ended, he began traveling extensively in Europe, especially in France, where he came to know both Cuvier and Geoffroy.

Although impressed by the former's work—Who could not be?—it was the latter's transcendental anatomy, promoting homology, which captured his imagination, then and for always. It was at this time that Grant fell under the spell of Lamarck and became a life-long evolutionist—although he may already have been far down this path, since he referred to *Zoonomia* in his medical dissertation (1814) and late in life praised Erasmus Darwin for having triggered his thinking about those "laws of organic life" which can be "applied to explain the abnormal phenomena of the human body" (Desmond 1984, 196; quoting Grant 1861, v).

A decade of travel, practice as a physician, and then some teaching at a private medical school in Edinburgh did not preclude a growing research program of Grant's own, especially in comparative studies of invertebrates. This work was acknowledged by appointment to the chair of comparative anatomy at University College in London, in 1827. Grant held this post for the rest of his life, and ostensibly his career was marked by success—election to the Royal Society in 1836, Fullerian Professor to the Royal Institution in 1837, and more. However, the reality was otherwise, for his stipend at University College, despite Herculean feats in the lecture hall, was derisory—as a consequence, simultaneously Grant was forced toward penury and an overworked drudgery. Typically of the restrictive society against which he fulminated, as a graduate of a Scottish university Grant had no automatic right to practice medicine in England. Equally typically Grant refused to bend to authority and obtain such a right.

Sponges are readily available in the Firth of Forth, and it was on these animals that Grant built his early reputation. It was in discussion of them, also, that Grant began revealing his Lamarckian allegiances. "The *Spongilla friabilis* . . . has a more imperfect structure than any of the marine species." Hence: "From this greater simplicity of structure, we are forced to consider it as more ancient than the marine sponges, and most probably their original parent" (1825, 283–284). Moving to London rather put such ideas as these on the side burner, for everything Grant now did was connected to the work that he was appointed to do, namely the teaching of comparative anatomy to medical students. The promotion of ideas for or against evolution was not the main task in hand. It was for this reason, no doubt, that, of the many French biologists who got favorable reference in Grant's lectures, it was now Geoffroy (rather than Lamarck) whose ideas occupied center stage. It was the younger Frenchman whose work was of more immediate relevance. Nevertheless, familiar themes started to emerge: "Some eagerly look into complex structures, with the hope of discovering anomalous parts, or peculiar formations, to mystify the study, while the philosophy of anatomy is chiefly occupied in reducing apparent exceptions to the general laws of development, and in demonstrating a unity of plan throughout all the grades of animal organization" (Grant 1833–34, 2, 1).

And demonstrate Grant set about doing, including—especially including—showing that organizational unity crosses the barriers that Cuvier thought absolute. Most dramatically, it bridges the gap between the invertebrates and the vertebrates:

> There are no sudden transitions in the development of important parts and systems of organization, when we discover the various stages of their development, and the order followed by nature in their formation. As we find, in the class of fishes, remains of the external shells, in the form of calcareous scales, or plates, or solid spines, so we find in the cephalopods the first soft cartilaginous rudiments of the vertebral column, which lead us gradually to the still-imperfect condition of that central part of the osseous system which is met with in the *myxene*, the *lampreys*, and other of the lowest cartilaginous fishes. (Grant 1833–34, 1, 505)

Aiding us in our investigations we have the fossil record, not to mention some version of an embryological recapitulation law: "[The student] traces, in comparative anatomy, the human organs coming successively into being, and rising in complexness, from the monad through all the grades of animal existence; and discovers, by the close resemblance which exists between the transient forms presented by man's organs during their development, and their permanent or adult forms in inferior orders of animals, that the plan of organization is everywhere the same, and man the climax of its development" (Grant 1833–34, 1, 44).

Where did evolution fit into all of this? At one level, one can say truly that it was the underlying theme tying together the various ideas: "A slight inspection of the organic relicts deposited in the crust of the globe, shows that the forms of species, and the whole zoology of our planet, have been constantly changing, and that the organic kingdoms, like the surface they inhabit, have been gradually developed from a simpler state to their present condition" (Grant 1833–34, 2, 1001). However, naturally enough given the context, the evolution was peripheral—the lectures were systematic, but not the evolutionism particularly. From a teacher with a task in hand, there was no intrusive distraction from the main subject matter. As a professor of anatomy he had no brief to write a full-scale book exclusively on evolution: a sort of British equivalent to the *Philosophie zoologique*. Nor was he ever to do this.

A Radical's Philosophy

What now of questions of progress and Progress? Toward the end of his life, Grant embraced a pessimistic reading of the nebular hypothesis, predicting that all of life is fated to frozen sterility. As so often is the case, the God of the Scotsman, even when He may not exist, seems bent on a peculiarly vindictive genocidal destruction of His own creation. Obviously if all is destined for refrigerated extinction, there can be no ultimate progress or Progress. But, if you pull back from the cosmic picture and focus on the more limited but more tangible existence we enjoy today, you will have sensed already that Grant offers a beautiful piece of supporting evidence for the kinds of conceptual links I am trying to forge. As far as biological progress is concerned, we have already before us massive evidence that his evolutionism was progressionist through and through. Although Grant saw that there would be branching, one never senses that the branching is particularly significant. Organisms go from the most simple to the most complex and (in the animal chain) right up to our own species.

Turning next to Progress, one can say simply and categorically that Grant's whole life and labors were testaments to the need for Progress, the possibility of Progress, and—despite setbacks and disappointments—the occasional successes of Progress. Outside the established system, he and his fellows strove to change and improve aspects of British education, specifically as they centered on the stranglehold of Oxford and Cambridge and of the powerful and well-connected on medical education. "Reform" must have been the word that pathologists found engraved on Grant's heart. He was explicit about Progress and the barriers which stood in its way.

> Human knowledge everywhere advances with colossal strides; the wisdom of the philosopher is becoming the common sense of the vulgar. The unfettered Institutions of civil society make rapid progress in improvement; the light of science now illumines the path of the humblest avocations, and the universal race on man is up and stirring. The antiquated qualities which commanded admiration in the Institutions of

> ancient times, are no longer suited to the condition of mankind, and fail to procure even respect for their antiquity. The sanctity of their Charters and Statutes, and the exclusive character of their by-laws and policy, are barriers to their progress, and as they cannot advance, they must retrograde in the great tide of civil improvement. (Grant 1841, 7–8)

That Grant's progress and Progress were a package comes through clearly, especially in the way that (for instance in his opening address to incoming medical students) he mixed the two together. And, as a Lamarckian—not to mention an educator—he stressed the intellectual inheritance of acquired characters. "The education of youth . . . is the basis of individual happiness, and of the rank and usefulness of man in society; it develops and cultivates those powers which distinguish him from the brute; it stamps the character of an age, and constitutes the chief distinction among men, and among the nations that are, or have been, on the earth" (Grant 1833–34, 1, 42). Nor is there any danger here that we are confusing Progress with Providence. At most in Grant's case we seem to have a weak deism, probably close to non-belief. Expectedly, Grant saw organized religion as one of his greatest obstacles and—the true mark of the Progressionist—in word and deed saw the undying need of individual, human effort.

Finally, there is surely little need to dwell on the outstripping of the evidence by Grant's theorizing about biological progression. As his friend and fellow Scot Robert Jameson wrote, just at the time when Grant first embraced evolutionism: "Although it should not be forgotten, that this meritorious philosopher [Lamarck], more in conformity with his own hypothesis than is permitted in the province of physical science, has resigned himself to the influence of imagination, and attempted explanations, which, from the present state of knowledge, we are incapable of giving, we nevertheless feel ourselves drawn towards . . . these notions of the progressive formation of the organic world" (Jameson 1826, 297; see also Secord 1991). It was the idea that counted, not the evidence.

Grant the biological progressionist and Grant the reforming Progressionist were one and the same man. Alas, what Grant did not expect, and what Grant did not get, was gratitude: "I have long possessed the privileges of Apothecary (by means of the Edinburgh College of Surgeon's Diploma), Surgeon, Physician, Licentiate, and Fellow of a Royal College, though now prevented by the laws of my country, from even writing a prescription, to save a brother's life, or to support my own; so that I am doubly neutral; first, in belonging alike to every 'grade', and next in being alike excluded from all" (Grant 1841, 10).

Robert Chambers and "Vestiges"

Grant's most extended thoughts on the subject of evolution came in comparative anatomy lectures to lower-middle-class medical students—hardly general reading. Yet, if my book has any truth, since Progress was an ideology now starting to flood Britain, evolution had to be an idea that could not be suppressed. What it needed was an extended treatment of its own, preferably presented in an attractive way with striking and easy-to-grasp examples and implications, with at least a nod to British Progressionism (unlike Grant's Francophile enthusiasms), and (given the tenor of the age) for maximum effect it had to be surrounded by an upbeat endorsement of a recognizable theology.

In 1844, a book was published (anonymously) satisfying all of these criteria. With the *Vestiges of the Natural History of Creation,* Adam Sedgwick's worst fears were realized. The book was an instant success, popularizing and informing on transmutatory ideas as never before. From the anatomist's dissecting slab to the genteel drawing room in one easy—all-too-easy, groaned the critics—leap. Despised by authorities, avidly consumed by the general public, *Vestiges* achieved Victorian immortality when it was transformed into the concluding stanzas of the poet Alfred Tennyson's much beloved *In Memoriam,* suggesting that his remembered friend (Arthur Hallam) was a forerunner of a future, more highly evolved type (Millhauser 1959; Ruse 1979a; Secord 1989).

We now know that the author—the "Vestiginarian"—was one Robert Chambers (1802–1871). Another of the many Scotsmen filling our pages, his was a story of which nineteenth-century legends were made, for he (with his brother William) triumphed over family misfortunes and poverty to achieve financial success and middle-class respectability dreamed of only in the tales of Methodist Sunday Schools (W. Chambers 1872). When evil times hit the Chambers household, the two lads set themselves up as booksellers, initially with little more than the broken remnants of the family library. Hard work, self-denial, a modicum of luck, and the guts and imaginations to seize chances, saw them rise through the ranks of shopkeepers, bookbinders, publishers, and authors of edifying tracts; by the 1840s they were the producers of *the* popular weekly paper of the day—*Chambers's Journal*—printed by steam press, distributed *en masse* cheaply across Britain, and filled with worthy and elevating information and advice, of a length and seriousness that only a pre-television age could truly appreciate.

Precisely why a successful Scottish businessman should plunge into the turbulent waters of transformism will occupy us later. For the moment, let us start with Chambers's simple—critics would say "simplistic"—thesis. The basic claim was that the world is governed by law, by unbroken law, and that because it is so it is subject to ongoing development. Preparing the way for organic evolution, therefore, *Vestiges* started with a discussion of the nebular hypothesis, arguing that the universe as we know it, with its individual solar systems, came about by the natural process of change. This was the background for an assault on the fossil record, arguing (as the far-seeing Lyell had dreaded) that we see an upward transformation from fossils of the most primitive life forms to those of the most complex. We start with the humblest forms of invertebrates, work up to the fish, which within themselves go from simple to complex, move on then to the reptiles, and end up with the mammals, which go from the most primitive forms right up to *Homo sapiens*.

> In pursuing the progress of the development of both plants and animals upon the globe, we have seen an advance in both cases, along the line leading to the higher forms of organization. Amongst plants, we have first sea-weeds, afterwards land plants; and amongst these the simple (cellular and cryptogamic) before the more complex. In the department of zoology, we see zoophytes, radiata, mollusca, articulata, existing for ages before there were any higher forms. The first step forward gives fishes, the humblest class of the vertebrata; and, moreover, the earliest fishes partake of the character of the next lowest sub-kingdom, the articulata. Afterwards come land animals, of which the first are reptiles, universally allowed to be the type next in advance from fishes, and to be connected with these by the links of an insensible gradation. From reptiles we advance to birds, and thence to mammalia, which are commenced by marsupialia, acknowledgedly low forms in their class. That there is thus a progress of some kind, the most superficial glance at the geological history is sufficient to convince us. (Chambers 1844, 148–149)

What mechanism did Chambers offer for this tale of upward development? He drew on embryology and proposed a saltationary theory of jumps. He argued (undoubtedly following German sources at first or second hand, and giving his own version of the three-fold parallelism law) that organisms develop sequentially through the various kinds—fish, reptile, bird, mammal. If birth occurs early in development, one is born as an organism low down on the scale. But if for some reason birth is prevented or delayed, a higher form of organism naturally ensues. "To protract the *straight forward part of the gestation over a small space*—and from species to species the space would be small indeed—is all that is necessary" (Chambers 1844, 212, his italics).

The consequence is upward movement—however, it is upward movement that goes in steps, rather than continuously. A fox's birth is delayed, and so a dog, a member of a different species, is produced and born. Unlike Lamarck, Chambers inclined to think that the original life forms could be produced only when conditions were appropriate. Hence, he inclined away from a continuous process of spontaneous generation. Most or all life began at one point in Earth's history, and Chambers gave an assortment of helpful hints and analogies about how and why such life might have started. Notorious was a suggestion that there are significant analogies between living plants and "frost ferns," those patterns left on windows after a frost. Like Lamarck, however, Chambers inclined to the belief that, once started, life would begin an inevitable upward process. Hence, he saw the Australian marsupials as beings which had progressed rather less than European mammals, not as primitive forms which nevertheless have ancestors in common with the Europeans.

All in all, concluded Chambers, with a confidence bordering on the reckless, the case for evolutionism is overwhelming.

> The inorganic has one final comprehensive law, GRAVITATION. The organic, the other great department of mundane things, rests in like manner on one law, and that is,—DEVELOPMENT. Nor may even these be after all twain, but only branches of one still more comprehensive law, the expression of that unity which man's wit can scarcely separate from Deity itself. (Chambers 1844, 360)

From Progress to Evolution

Other than this desire to be seen as the Newton of biology—a fairly common phenomenon among evolutionists—what motives lay behind Chambers's advocacy of evolutionism? How far can these motives be linked to a metaphysical commitment to Progress, of some kind? And if

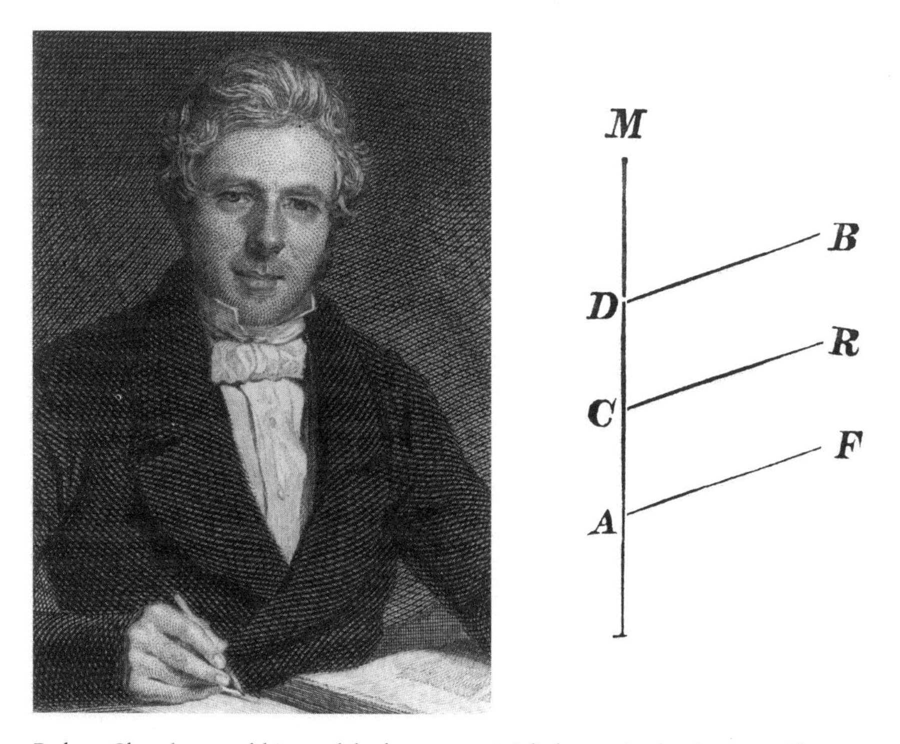

Robert Chambers and his model of ontogenetic/phylogenetic development (from the *Vestiges*).

Progress, then did progress follow? Let us run quickly through our three criteria, beginning with the question of biological progress.

The answer is easy. One simply cannot expound Chambers's theory without bringing in progress—generally in the overall cosmological course of history, and specifically in the fossil record, in embryology, in the link between the two. In all biological dimensions, there is a movement from less to more, from simple to complex, from the inferior to the superior. Most dramatically, in paleontology we have a climb up the stairway, from zoophytes to humankind. Admittedly, plants are put off on one side, but in the animal world even birds are included on the ladder of life. In later editions, after critics objected to his unidirectionalism, there is some little talk of branching (what Chambers called "stirpes"); but, even more than for Lamarck and Grant, Chambers was an "onward and upward" thinker. The point is that he did not care about detours.

It will come as no surprise to learn that, when we get to humans, we find an upward progression to the European type of human. Explicitly, Chambers denied that there has been a falling away from an initial perfect type. "Our brain goes through the various stages of a fish's, a reptile's, and a mammifer's brain, and finally becomes human. There is more than this, for, after completing the animal transformations, it passes through the characters in which it appears, in the Negro, Malay, American, and Mongolian nations, and finally is Caucasian" (Chambers 1844, 306). Chambers was happy to put humans only at some penultimate point on the stairway, awaiting the "crowning type." Just what one would expect from Progressionism, but hardly compatible with Providentialism.

Turn next to the direct evidence. A very strong bond can be forged between Chambers, Progress, and his science. First, there is the belief in Progress itself. Significant here is the testimony of his own life, one of successful triumph over life's adversities. It was one of Progress, marked moreover by his own personal transition from a romantic Toryism of his twenties (with adulation of the work of Sir Walter Scott) to a science-admiring Whiggism of his thirties. Chambers moved from looking back with nostalgia to looking forward with optimism. And he saw in his own story, the story of society: "[My early] works were an effluence of mental youth, analogous to a green phase of the studious mind of England at the present day, which shows itself in a love of patristic reading and of Gothic architecture. The mind, in progressive men, passes out of such affections at thirty, and the national mind will pass out of them when the time comes for its exercising its higher faculties" (Chambers 1847, 1, iii–iv).

Backing this imputation of Progressionism, there is the phrenological connection. This is the doctrine that claims the mind is related to distinct parts (faculties) in the brain and that one can read character from the shape of the skull. Never that respectable, it was nevertheless very popular in early-nineteenth-century Edinburgh. At least, it was popular among the middle classes, who used it as a weapon to promote forward-looking Progressivism against the establishment in Church and University (Shapin 1979). Proper environmental training (i.e., education) of the faculties will lead to a better person and better society. "My view of human nature is that men require, 1st knowledge, and 2nd. training of their moral and intellectual faculties, before they can be trusted with power or be made the arbiters of their own destinies with advantage to themselves; but I believe that men *collectively,* when enlightened and trained, will go right and promote their own happiness" (George Combe; quoted in Gibbon 1878, 1, 302).

Chambers was much bound up with the phrenological movement. He pushed and published its ideas. Indeed, it seems that the writing of *Vestiges* rose out of an uncompleted project to produce a book on phrenology (letter to John Ireland, late 1840s?; Chambers Papers, 341/110.9). And in the *Vestiges* itself, phrenology was the accepted philosophy of mind and was promoted. All of this fits in with Chambers being a key figure in the movement for Progress; for a British kind of Progress that is, one which sees science and industry and personal effort leading to a better—to a happier—society.

Moving from philosophy to science, one notes Chambers's use of the nebular hypothesis. The empirical backing for the hypothesis (especially by the 1840s) was ambiguous and doubtful at best, but it was being pushed by men in Chambers's circle, and again the underlying theme was Progress. What we get in the cosmological world we can expect in the social world. Thus, John Pringle Nichol, professor at Glasgow and chief enthusiast for the hypothesis: "In the vast Heavens, as well as among phenomena around us, all things are in a state of change and PROGRESS" (Nichol 1837, 206; quoted by Schaffer 1989, 131). In the human realm, Nichol—friend of George Combe (the leading phrenologist) and co-worker with J. S. Mill—wrote that the "elevation of man" is "the world's most determinate *First Cause;* in seeking to advance it by education, we therefore act in harmony with manifold resistless agencies, nor, if the task be understood aright, is it possible but that we must prevail" (Nichol 1847, xix; quoted by Schaffer 1989, 152). Chambers, who learned of the hypothesis from Nichol, knew full well the symbolic import of starting his book with the hypothesis.

Then, and this is surely definitive, we find that—in the face of the initial disapproval of *Vestiges*—Chambers *explicitly* tied his evolutionism in with Progress, arguing that organic evolution includes the human story and is simultaneously the human story writ large.

> The question whether the human race will ever advance far beyond its present position in intellect and morals, is one which has engaged much attention. Judging from the past, we cannot reasonably doubt that great advances are yet to be made; but if the principle of development be admitted, these are certain, whatever may be the space of time required for their realization. A progression resembling development may be traced in human nature, both in the individual and in large groups of men . . . Now all this is in conformity with what we have seen of the progress of organic creation. It seems but the minute hand of a watch, of

> which the hour hand is the transition from species to species. Knowing what we do of that latter transition, the possibility of a decided and general retrogression of the highest species towards a meaner type is scarce admissible, but a forward movement seems anything but unlikely. (Chambers 1846, 5th ed., 400–402)

This is simply the best passage of any that I know linking Progress, progress, and evolution. Privately also, Chambers linked biological and human change, even writing to Ireland of *Vestiges*' "leading doctrine, that of natural laws, as being so important to human welfare" (July 31, 1845; Chambers Papers, 341/110.142).

Finally, to wrap everything up, there is the question of religion. Chambers moved from the Presbyterianism of his youth to Anglicanism, as he trundled securely into the prosperous middle class. But formal religion had little hold on him, and he distrusted deeply the reactionary power of the clergy, especially after the attacks on *Vestiges* by Sedgwick and others. Chambers was certainly no non-believer. In print—in the *Vestiges* with a persistent ease that even the harshest critics acknowledged—and privately, Chambers spoke to a form of deism. But it was the deism of unbroken law: "The theory is called a dispensing with God and a putting of the act of creation on a senseless unmeaning law. I insist that it is only an attempt to establish law as the *manner* in which the deity has proceeded; being the manner in which he continues to rule the world" (letter to unknown correspondent, April 1847; Chambers Papers, 341/110, 237). There was no place for Providence.

The evidence points to Chambers as an enthusiast for Progress, and to the evolutionism of *Vestiges* as a manifestation of that enthusiasm (Secord 1989). Going on to ask our third question, therefore, is almost unnecessary. What was the relational standing between Chambers's science and the empirical evidence? The answer simply is that it was standing on very uncertain ground, which is precisely what one would expect were a cultural notion like Progress a main formative factor in the production of the science. The tenuous relationship between the nebular hypothesis and the evidence has been mentioned already. Even more tenuous was the base for Chambers's speculations on spontaneous generation. Critics like Sedgwick had a field day with the frost ferns. Chambers's strongest suit was the fossil record, for he got most of his facts from his critics! But here also he was running ahead of the evidence then known. There was no trace of early primitive forms. There were gaps galore in the record. And, as many pointed out, even if there were a general progression upward, within

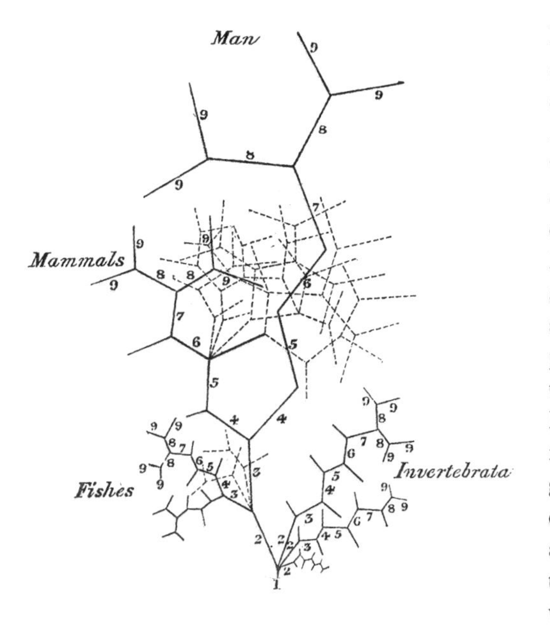

Developmental tree as envisioned by the Scottish physician Martin Barry (1837, 346), an attempt to illustrate von Baer's theory of embryology. Although von Baer was critical of the excesses of *Naturphilosophie,* Barry manages to give a firmly progressionist message, with Man prominently ensconced at the top. We do not know whether Darwin saw this picture before he drew his own tree diagrams (see Chapter 4), but clearly such ideas were in the air. There is no suggestion that Barry, unlike Darwin, was promoting evolution.

groups the more advanced often seemed to precede the less advanced. Much were made of various early fishes which informed opinion decreed to be of advanced form (Bowler 1976).

The case is complete. Erasmus Darwin and Robert Grant had put Progress right into their evolutionary theorizings. So also did Robert Chambers. And, as behooves a great communicator, it was a Progress of the times.

Louis Agassiz and the Three-fold Parallelism

German biology began to gather steam. Important for us was the embryological work of Karl Ernst von Baer (1792–1876), which laid out the basic laws of development and the growth path of all organisms, or at least of all animals. In a four-part analysis, von Baer claimed, first, that the more general characters in a large group of animals make their appearance in the developing embryo, before the more special characters appear; second, that from the most general forms the less general and finally the most special develop; third, that there is divergence, with embryonic development leading to branching, with different forms going different ways; and then fourth, that there can be no straight recapitula-

Louis Agassiz

Stephen Jay Gould describes the illustration at right as "Agassiz's favorite example of recapitulation (and an enduring classic for all later supporters). *(A)* The ontogeny of a 'higher' teleost (the flatfish *Pleuronectes*) showing transition from diphycercal to heterocercal (upper lobe larger than lower) to homocercal (equal-lobed) tail. *(B)* Comparative anatomy of adult tails in sequence of primitive to advanced fish (also paralleled by their order of appearance in the geological record); from top to bottom: *Protopterus* with its diphycercal tail, a sturgeon with a heterocercal tail, and a salmon with a homocercal tail." (From Gould, *Ontogeny and Phylogeny,* p. 67; originally from Schmidt 1909.)

tion, with the embryo of one form directly resembling the adult of another form (von Baer [1828–37] 1853; see also Russell 1916, 125–126).

Von Baer was a student of a leading *Naturphilosoph,* Ignatius Döllinger. Von Baer did agree that there could be the appearance of recapitulation, for some forms are more primitive (in the sense of being less specialized) than others—hence, the adults of these forms would be similar to their embryos, which in turn would be similar (if not identical) to the embryos of other (more specialized, that is, advanced) forms. He even went so far as to think there was a kind of progressionism from the general to the special, from the simple to the complex, from the *homogeneous* to the *heterogeneous:* "The more homogeneous the whole mass of the body is, so much the lower is the grade of its development. The grade is higher when nerves and muscles, blood and cell-substance, are sharply distinguished" (von Baer [1828–37] 1853; quoted by Ospovat 1981, 119).

But, in essence, his face was set against *Naturphilosophie.* In this, he differed from the Swiss-American comparative zoologist, Louis Agassiz (1807–1873), whose history was the recapitulation of Romanticism. The

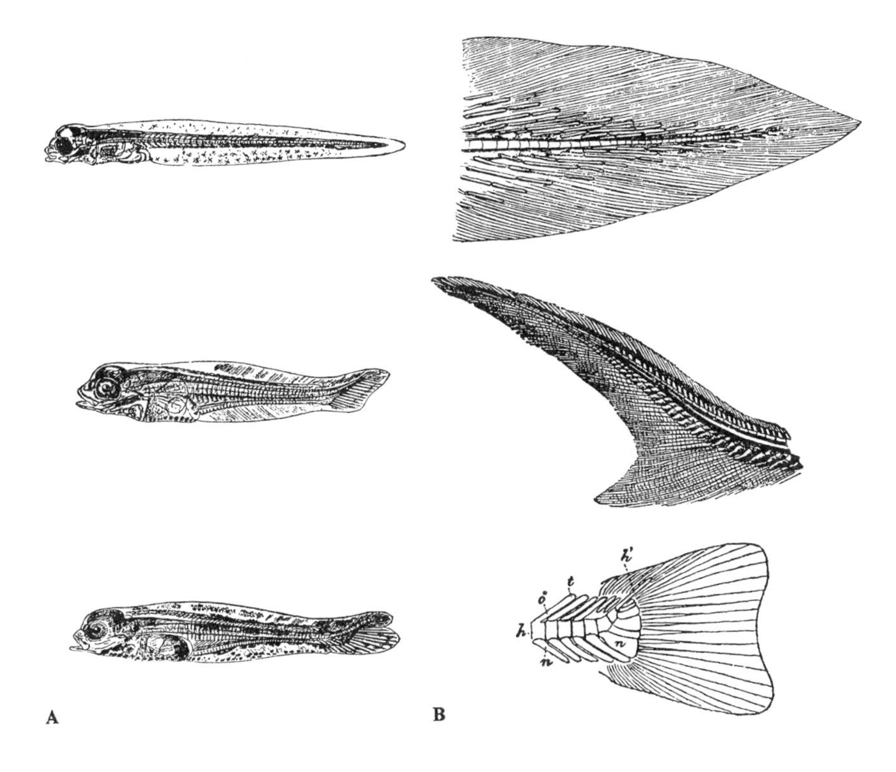

son of a Protestant minister, he spent the first half of his life in the Old World and the second half in the New. His first wife died shortly after his transition, and Agassiz then married a Bostonian. His religious enthusiasm—a major factor in Agassiz's psyche—was marked by a move from the pietism of his youth to a sympathy with the Unitarianism of his new helpmate. A great fundraiser, speaker, and organizer, Agassiz demanded emotional loyalty, which he returned with real love and support. This was a man of passion, and he pursued the vile doctrine of evolution passionately, first as professor in the land of his birth doing seminal analyses of fish as well as popularizing the notion of Ice Ages, and second after he had crossed the Atlantic, as professor at Harvard, founder of the Museum of Comparative Zoology, and bosom friend of the likes of Emerson and Longfellow. Yet at another level, we see in Agassiz's history progress, Progress, and connections between the two. Agassiz has a place in our story. Indeed, it will be my claim later that Agassiz has a much larger place than most of us know (Agassiz 1885; Lurie 1960).

A key influence on Agassiz, following university years (briefly) in his homeland and then in Germany, was time spent in Paris with Cuvier. That he was literally the last student of the great French anatomist, who passed on to Agassiz his own notes on fish, clearly had as strong a psychological as strictly a scientific effect on the younger man. It is no surprise, therefore, to find that the four Cuvierian *embranchements* became a central feature to Agassiz's thought. But there were other influences at work. In particular, he was an archetypal thinker in a way quite alien to Cuvierian anatomy. The "attentive observer" discovers that, within the branches and then at (at least) the level of class, "one single idea has presided over the development of the whole class, and that all the deviations lead back to a primary plan, so that even if the thread seem broken in the present creation, one can reunite it on reaching the domain of fossil ichthyology" (Agassiz 1885, 1, 241).

Agassiz was Germanic in more than just this. Again and again he reiterated the traditional three-fold parallelism: between the developing individual (embryology), the extant members of a group of organisms (comparative anatomy), and the history of life (paleontology): "One may consider it as henceforth proved that the embryo of the fish during its development, the class of fishes as it at present exists in its numerous families, and the type of fish in its planetary history, exhibit analogous phases through which one may follow the same creative thought like a guiding thread in the study of the connection between organized beings" (Agassiz 1885, 1, 369–370).

These "analogous phases" were not in an evolutionary mode, however. Writing to Sedgwick, at the time of the Vestiginarian scandal, Agassiz made his anti-transmutationism crystal clear: "Have fishes descended from a primitive type? So far am I from thinking this possible, that I do not believe there is a single specimen of fossil or living fish, whether marine or fresh-water, that has not been created with reference to a special intention and a definite aim" (Agassiz 1885, 1, 392–393). The empirical evidence—gaps in the fossil record, inadequate mechanisms, and the like—were enough alone to make evolution an impossibility. And this was a position to which Agassiz stuck to the end of his life.

What can we learn from him? Was Agassiz a progressionist, if the term is understood in a non-evolutionary sense? The Cuvierian four-part division made impossible any simple Chain of Being. Moreover, within *embranchements* Agassiz saw obstacles to simple progressionism. The complex often comes before the simple, although Agassiz stated his position in such a way as to give joy to an evolutionist, allowing that certain forms ("prophetic types") actually herald the arrival of new forms: "it seems to me even that the fishes which preceded the appearance of reptiles in the plan of creation were higher in certain characters than those which succeeded them; and it is a strange fact that these ancient fishes have something analogous with reptiles, which had not then made their appearance" (Agassiz, 1885, 1, 393). Also, at least for a long while, Agassiz refused to allow that any one of the four *embranchements* appeared before or after any other. Reluctantly, by 1857, he was starting to realize that special pleading was needed to make the vertebrates as old as the others.

When all is said, however, in Agassiz's eyes these objections would be but qualifications. Where it counts, in the vertebrates, there is progress, all the way up to humankind. Moreover, it is a human-centered progression, of an ilk quite alien (say) to one like Sedgwick, who would have seen the arrival of humankind as miraculous and in itself quite unpredictable. For Agassiz, the end justifies, and in a way informs, the means: "The history of the earth proclaims its Creator. It tells us that the object and the term of creation is man. He is announced in nature from the first appearance of organized beings; and each important modification in the whole series of these beings is a step towards the definitive term of the development of organic life" (Agassiz 1842, 399; also Agassiz 1859, 103–104).

Turning, as is our wont, to possible influences, these are not hard to find. It is true that Agassiz tried to put space between himself and the worst excesses of the *Naturphilosophen* (Agassiz 1885, 1, 388); but, in his archetypal thinking, in his three-fold parallelism, and above all in his progressionism, he was at one with them. Moreover, from his educational background, this is precisely what one would expect. Like von Baer, Agassiz studied with Döllinger (who by this time, 1827, had moved to Munich)—boarded with him, even. He had studied already with Friedrich Tiedemann, who argued that vertebrate skulls are modeled on certain basic ground plans—less complex forms being cases of arrested development.

Above all, Agassiz was a student at Munich in the lectures of the two most beguiling *Naturphilosophen:* the biologist Oken and the philosopher Schelling. "At four o'clock we go usually once a week to hear Oken on 'Natur-philosophie' (a course we attended last term also), but by that means we secure a good seat for Schelling's lecture immediately after. A man can hardly hear twice in his life a course of lectures so powerful as those Schelling is now giving on the philosophy of revelation" (Agassiz 1885, 1, 91; this letter was written by a fellow student). Oken was a good friend of the students, and Agassiz and others would spend a weekly evening at his house, talking, arguing, and drinking beer.

What, finally, of the evidence? This is perhaps a good moment to stress—or stress again—what might perhaps get lost or belittled in such a discussion as this. A man like Agassiz really was a first-class scientist. He truly was in touch (literally and metaphorically!) with the empirical world. That his position on the evolutionary question might today seem incorrect and dated in no wise detracts from his very solid contributions to biological science. He uncovered and coordinated a vast amount of material on the fishes: their embryology, anatomy, and paleontology. His laurels were earned. Yet this said—and it was important that it be said—there were clearly major gaps between evidence and theory, gaps that a transcendental Progressionism filled nicely. Already, thanks to von Baer's researches, Agassiz's embryology was (properly) judged old-fashioned, even as it appeared. And although Agassiz may have been right in his details against Chambers, it was he who was having to spend a lifetime in retreat against new paleontological discoveries, like the above-mentioned question of the age of the vertebrates.

Again, therefore, we can congratulate ourselves on having found links between progress and Progress—a Germanic form of Progress, that is. So, finally, why was Agassiz no evolutionist? A familiar pattern emerges, although with expected personal differences. On the one hand, Agassiz's religious roots were important, and the later Unitarian graft could not change his early influences. His father was a Protestant pastor, in a French, Catholic part of Switzerland. Agassiz and Cuvier shared a sense of the significance of one's religion as part of one's cultural identity. Not entirely separate from this was the direct connection with Cuvier and his virulent anti-evolutionism. It is clear that Agassiz felt it was his role to carry the banner of Cuvier. The ice-age theory was presented as an extension of catastrophism, and obviously anti-evolutionism was part of the package deal (Rudwick 1974).

Then, on the other hand, we have the simple fact that German idealism did not demand material progress. The notion alone is sufficient: "The leading thought which runs through the succession of all organized beings in past ages is manifested again in new combinations in the phases of development of the living representatives of these different types. It exhibits everywhere the working of the same creative Mind, through all times, and upon the surface of the whole globe" (Agassiz 1859, 175).

The student of Oken and Schelling had spoken.

Richard Owen

Von Baer and Agassiz were part of German biology, even though strictly speaking the former was an Estonian nobleman and the latter Swiss born and bred. I turn, to complete my pre-Darwinian survey, to an Englishman. He belongs in this company, for although in his lifetime he was known as the "British Cuvier," I would rather class him as the "British *Naturphilosoph*." Not, I hasten to add, that I do this with evil intent, even though traditional histories of evolutionary theorizing have tended to characterize Richard Owen (1804–1892) as the anti-Christ. Owen is portrayed as an ill-motivated Creationist, bringing every force at his power to block the way of evolutionism in general and Darwinism in particular. To say that Owen comes already saddled with a reputation is the understatement of this book. However, although not even the most revisionist of historians could turn Owen into a warm, cuddly human being, considered without prejudice, Owen emerges as a much more complex and interesting and significant thinker. Hence, for his sake and for ours, we must do him justice (Ruse 1979a; Desmond 1985, 1989; Richards 1987; Rupke 1994).

Richard Owen, born in Lancaster, was at school with William Whewell, a life-long friend. Apprenticed as a surgeon, he studied briefly at Edinburgh and then (in 1825) went down to London. There, Owen became a member of the Royal College of Surgeons, for which he worked for many years, cataloguing its large collection bequested by John Hunter, the eighteenth-century anatomist. In 1836 he was appointed Hunterian professor at the College, a post that called for an annual series of lectures, which Owen used as a vehicle to expound and develop his own work and ideas. In later life, Owen turned increasingly to science administration. He became (in 1856) superintendent of the natural history branches of the British Museum. He died, covered in glory by the scientific establishment and a grateful nation, in a grace and favor house in Richmond Park, granted to him by Queen Victoria—in every sense, a long way from his rather humble origins in the north of England.

As a trained surgeon, Owen's initial focus was upon comparative anatomy, although later his interests turned more and more toward

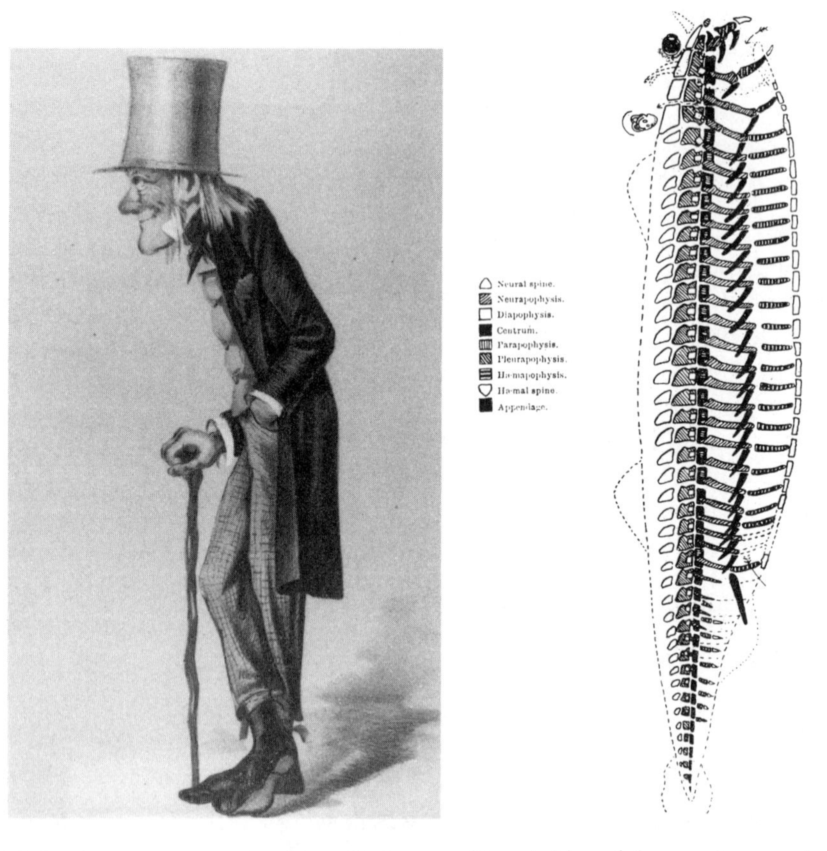

Richard Owen, as portrayed in a *Vanity Fair* cartoon.

Owen's idea of the vertebrate archetype. (From *On the Nature of Limbs.*)

geology. As one whose professional life was spent in England, he was naturally sensitive to Paleyite design. The appropriateness of this philosophy as a tool for understanding the organic world was reinforced by a warm if short connection with Cuvier, at the end of the latter's life. Yet Owen was never a simple and straightforward teleologist. There was ever a strong fondness for isomorphisms or homologies within his thinking, coupled with a keen interest in embryology, with Owen taking a von Baer–type perspective in this regard. "Thus every animal in the course of its development typifies or represents some of the permanent forms of animals inferior to itself; but it does not represent all the inferior forms, nor acquire the organization of any of the forms which it transitorily represents" (Owen, 1843, 367–371).

In the area of phylogeny, Owen initially (in the early 1830s) relied on the expertise of others, but increasingly he made it his own. Here, three kinds of fossils in the record are especially noteworthy: those which (appearing early) link forms which later diverge (phylogenetically); bridging fossils between different earlier and later forms; and those earlier (adult) forms which show features seen in later forms only at the embryological stage. Like von Baer, Owen argued for a move (phylogenetically) from general to special: "The horse is the swifter by reason of the reduction of its toes to the condition of the single-hoofed foot; and the antelope, in like manner, gains in speed by the coalescence of two of its originally distinct bones into one firm cannon-bone" (Owen 1851, 448–450). But, again like von Baer, he denied the possibility of simplistic recapitulation. Indeed, one must beware of such "baseless speculations" as "that the Human Embryo *repeats in its development* the structure of any part of another animal; or that it *passes through the forms* of the lower classes;— . . . that a Fish is an overgrown Tadpole" (Hunterian lecture 4, May 9, 1837; Owen 1992, 192).

Summing up, emphasizing both form and function, Owen saw two forces at work in organic development. To start, there is a "polarizing force," which (as it were) lays down the ground plan or *archetype* through "vegetation or irrelative repetition" (Owen 1848, 171). One sees this repetition, very obviously, in the vertebrate archetype. It is no surprise that Owen was a firm supporter of the vertebrate theory of the skull. Then, laid across the archetype, there is an adaptive force, which leads to all of the specific utilitarian contrivances of the natural theologian. The greater the effect of the adaptive force, the farther an organism is from the archetype. There is thus a creative tension within the developing organism between its underlying form and its surface (although hardly superficial) function.

What of evolution? For the first thirty years of his career—that is, up to the time that Darwin published the *Origin* at the end of the 1850s—there is much to suggest that Owen was against transmutation of any kind, simply and utterly. Much of his early research career was devoted to an attack on key pieces of evidence forwarded by evolutionists like Geoffroy (Desmond 1989). Against the claim that the platypus constitutes a bridge between reptiles and mammals (a claim by Geoffroy, an advance on a claim by Lamarck that it lay between mammals and birds), Owen showed that the animal has mammary glands and that therefore it is a mammal and no true link. Against the claim (by just about every evolutionist) that apes and humans have significant anatomical similarities, Owen argued that such claims can be made only on the basis of immature forms; they fall to the ground when one considers adults. And against the claim (by Grant) that the fossil record supports evolution, Owen created the concept of the dinosaur, a more advanced reptile than any existing today. "They were as superior in organization and in bulk to the Crocodiles that preceded them as to those which came after them" (Owen 1841, 200; quoted by Desmond 1979, 226). Taking full license with the paucity of fossil evidence, deliberately Owen made them as mammal-like as possible. There seems no happy upward climb here.

These arguments were backed by explicit denials, as in his first series of Hunterian lectures (1837): "The doctrine of Transmutation of forms during the Embryonal phases is closely allied to that still more objectionable one, the transmutation of Species. Both propositions are crushed in an instant when disrobed of the figurative expressions in which they are often enveloped; and examined by the light of a severe logic" (Lecture 4, May 9, 1837; in Owen 1992, 192). Privately, Owen seemed prepared to back up that position. In 1845, to Whewell, who was scavenging for anti-Vestiginarian titbits, Owen wrote that the idea of any kind of parallel law is just ridiculous: "The brain of the Human Embryo does not resemble, at any period, however early, the brain of any Mollusk or any Articulate" (February 3, 1845; Whewell Papers, Add Ms 210 70(1)).

The case seems to be complete. But this would be to reckon without the complex personality of Richard Owen, and—to be quite fair—the very difficult position in which he was enmeshed. Owen was not a rich man. He was making his way by his talents. He lined himself up with the conservative Royal College of Surgeons as well as with the Oxbridge network of professors, like Whewell and Sedgwick. This strategy paid off. Through their organizations, his patrons funded him generously. They even arranged that he get a civil pension from the government in 1842, of £200 per year. (Although compare this with the young Darwin, who started married life in 1839 with a family-endowed income of £1300 per year.) The obverse side was that Owen had to tread carefully. He had to decline a proffered knighthood, for fear of appearing to outstrip his station. Scientifically speaking, Owen had to bend to the will of his masters. They loathed evolution, particularly given the way that radicals like Grant were using it to promote change, and they expected Owen to provide the counter-blast. The point therefore is that he was simply not a free agent to express his own true feelings. Whewell, for instance, did not just want anti-*Vestiges* facts; he wanted Owen himself to hold forth in one of the leading quarterlies.

What were Owen's true feelings? A tantalizing comment occurs in one of Darwin's (private) notebooks, written at some point in 1837: "Mr. Owen suggested to me, that the <cas> production «of monsters» (which, Hunter says owe their origin to very early stage) which, follow certain laws according to species present an analogy to production of species—" (B 161).[2] And as interesting is the fact that, by this time, Owen was (at least) hinting that he did not consider Cuvier's embranchments quite as rigidly distinct as did the Frenchman. He was "surprized" to find that Cuvier nowhere acknowledged that Geoffroy's "comparison of a Cuttlefish with a Vertebrate animal bent double" was first made "by Aristotle himself, in his Philosophical review of the different plans of organization which the sanguineous molluscous and crustaceous animals presents" (Owen 1992, 106).

But none of this is anything of importance compared with what happened, admittedly privately, a decade later, when the evidence really does start to mount that Owen was edging toward some kind of evolutionism. Not only did he decline to take up public arms against *Vestiges,* in 1845 Owen wrote a letter of appreciation to its unknown author: "the discovery of the general secondary causes concerned in the production of organised beings upon this planet would not only be received with pleasure, but is probably the chief end which the best anatomists and physiologists have in view" (Owen 1894, 249–250). This is hardly full-blooded acceptance of evolutionism, but in the context of the time it is not so very far off. That the appeal was the Germanic element in Chambers's thinking is supported by Owen's reaction when the *Origin* was published, for he then praised *Vestiges* publicly, stressing its transcendentalism. And like Chambers (and Geoffrey before him), Owen showed sympathy for saltationism ("anomalous, monstrous births").

By the Light of the Archetype

If Owen were a Germanic-type thinker about development, even (whatever the politics of his situation) at the time of *Vestiges,* he may simply not have been looking for actual, physical evolution. But where did he stand on these matters? Pick up first on the question of P/progress. Was Owen a progressionist? From the 1830s on (by the latest), very few professional biologists were of the simple monad-to-man progressionist type, temporally or conceptually. The influence of the work of Cuvier and von Baer precluded this, and Owen was no exception. In the already quoted letter concerning *Vestiges* to his friend William Whewell, Owen wrote: "Animals in general cannot be arranged in a series proceeding from less to more perfect in any way, so many, in different natural series, being on a par; much less can they be so arranged as that the more perfect in their foetal condition pass through the successive stages of the less perfect, the characters being taken from the brain to the heart" (letter of February 3, 1845; partly in Owen 1894, 1, 252–253).

This said, however, Owen was certainly not against general notions of progress. "It is very true that, by tracing the progressive additions to an organ through the animal series from its simplest to its most complex structure we learn what part is essential, what auxiliary to its office" (Hunterian Lectures, 1843; quoted in Owen 1894, 1, 214). Moreover, his conclusion to a celebrated public lecture which he gave on limbs (in 1849) shows unambiguously that Owen put humans at the top: "we learn from the past history of our globe that [nature] has advanced with slow and stately steps, guided by the archetypal light, amidst the wreck of worlds, from the first embodiment of the Vertebrate idea under its old Ichthyic vestment, until it became arrayed in the glorious garb of the Human form" (Owen 1849a, 85–86). In Owen's own terms, progress seems to involve a move from the simple to the complex, and it just so happens that humans are the most complex of all organisms. But at a deeper level what we have is the simplicity of the archetype, which consists of basic units (vertebrae, in the case of vertebrates) repeated—what Owen called "serial homology." Then the homologous form is made more and more complex, as adaptive needs lead further and further from the archetype. Ultimately, in the case of vertebrates, these changes lead to *Homo sapiens*.

What about Progress? Here, as with evolution, we must tread carefully. Owen had hitched his star to Establishment Anglicans. Open sympathy with strong programs of reform was no more an option than was sympathy with evolutionism. Yet Owen did believe in social change—after all, he as much as anyone was making science an important part of Victorian culture. Even more, he believed in Progress, almost paradoxically the most dangerous variety, namely Germanic Progress. It would have been professional suicide for Owen to have endorsed such a philosophy openly. Even so obviously and evangelically (at that time) a Christian as Agassiz was regarded as suspect by Sedgwick because of the pantheistic taint of *Naturphilosophie.* But Owen's transcendentalism was there—his fondness for homology was not just a surface phenomenon—and it was the metaphysical tail which wagged the body of the scientific dog.[3]

Had Owen's fondness for homologies and archetypes been no more than a surface phenomenon, then our quest for influences might take us no further than Grant and Geoffroy. The young Owen knew Grant well—he may even have attended his lectures in Scotland. Most significant is the fact that, when Owen visited France in 1831 at Cuvier's invitation, his constant companion seems to have been Grant, and their interaction was cordial and intellectually exciting. During that visit Owen met Geoffroy and became fully acquainted with the French debates on anatomy. However, one suspects that, overall, Owen's French debt was more to the conservative Providentialism of Cuvier, with his emphasis on adaptation, than to the radical faction, French or Francophile British. The key influence in Owen's transcendentalism lies earlier, in the person of Joseph Henry Green (1791–1863), with whom Owen studied and for whom Owen acted as a teaching assistant in his early years in London. Green was an ardent Germanophile, a keen student of *Naturphilosophie,* and a close friend of Coleridge. In his lectures, which were from all accounts incredibly stimulating, we get all of the components of transcendentalism—especially the conflation of progress with Progress, which were simply considered different aspects of the same reality.

Thus, via Owen's own notes and reports on the lectures, we learn of the unity given through the archetype: "For the first time in England the comparative Anatomy of the whole Animal Kingdom was described, and illustrated by such a series of enlarged and coloured diagrams as had never before been seen. The vast array of facts was linked by reference to the underlying Unity, as it had been advocated and illustrated by Oken and Carus" (Owen 1865, 1, xiv; quoted by Desmond 1985, 31). We get the limited evolutionism of the Germans: "Considers nature as a series of evolutions not under the idea that the lower can assume the characters of the higher—" (Lecture 1, Tuesday, March 27, 1827; Owen Papers, 275.b.21, p. 131). We get the law of parallelism: "In my last Lect. I described the structure of the Brain in the Mammif. animals and I availed myself of what had been done by Tieddeman on the development of the brain in the foetus—as there is this advantage of considering the forma-

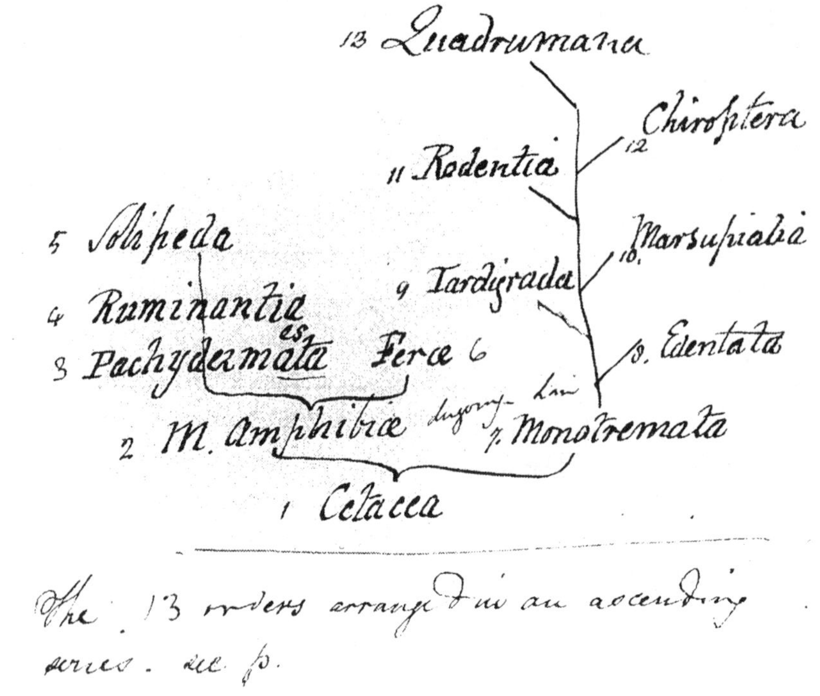

Joseph Henry Green's progressionist diagram of the thirteen orders of mammals. (From notes taken in 1828 by Richard Owen.)

tion of the brain in the Embryo, that what, in it, is imperfect and transitory, is an abiding form in the simpler animals" (Lecture 12, Saturday, April 19, 1828; Owen Papers). And we get progress up to humans: "In whole of the vertebrated class of animals, we find a nervous system appearing with its proper characters, and if I were [to abstract] one character in the ascending series it would be a gradual approach to the perfect system in man" (Lecture 11, April 19, 1827; Owen Papers).

All that we know suggests that Green's philosophy was Owen's philosophy, right down to warm comments made by the latter just after the former's death (in 1863): "that noble and great intellect" (Owen 1865, xiv; quoted by Desmond 1989, 275). Thus, though we must keep in mind a due regard for his social position, everything we have seen in Owen's science makes good sense: his enthusiasm for homologies and archetypes, his endorsement of the significance of embryology (filtered through von Baer's reforms), his progressionism, and above all his developmentalism, extending only cautiously toward evolutionism, at least in the pre-Vestiginarian years—not to mention the fact that his liking for *Vestiges* centered precisely on its most Germanic elements.

Moreover, we find that the Owen of the 1840s—who now had the added support of a growing friendship with Thomas Carlyle—became increasingly explicit and public in his liking for *Naturphilosophie.* He arranged to have Oken's major work translated into English and published, and then he himself came out in open praise of the German thinker: "Oken's famous 'Programm, Uber die Bedeutung der Schädelknochen' was published in the same year (1807) as Geoffroy's Memoir on the Bird's skull; but it is devoted less to the determination of 'special' than of 'general homologies': it has, in fact, a much higher aim than the contemporary publication of the French anatomist, in which we seek in vain for any glimpse of those higher relations of the bones of the skull, the discovery of which has conferred immortality on the name of Oken" (Owen 1848, 176, fn.; see also Oken 1847).

One should note that these pro-transcendentalist moves did not go unnoticed. Indeed, Owen felt the flick of the whip. The Oken translation, done under the auspices of the Ray Society, caused criticisms and resignations and Owen had to lie low. Sedgwick started to snort suspiciously. And at least one newspaper columnist attacked Owen savagely for fomenting radicalism at a time of social unrest (E. Richards 1987; Rupke 1994). But Owen persisted in his enthusiasm. In fact, some years later he was even to write a highly laudatory encyclopedia essay on Oken (Owen 1858). The transcendentalism had come early and struck deeply. Directly or indirectly, therefore, Owen's biological progressionism was connected to cultural Progressionism—in a German context.

Finally, I will take without argument the fact that Owen's goal-directed progressionism to our own species was not read from nature, directly. There was much in Owen's work that was empirical, but not this. If anything, he seemed better at supplying counter-examples! I will therefore conclude that he fits a familiar pattern. Owen viewed the organic world through the lens of his cultural presuppositions. He was a progressionist because, directly or indirectly, he was influenced by Progressionism—essentially, indeed, they were one. Moreover, Owen's somewhat ethereal view of evolution—that is surely neither too harsh nor inexact a term—was a function of the distinctively Germanic element in his thought. He stood where he stood because of his influences—as did all others discussed in this chapter. Because of his nature and because of his position, Owen was a lot more cagey than most in his allegiance to Progressionist ideals, but we can see the connections.

The Status of Evolutionary Thought

At the beginning of the nineteenth century, evolutionary thinking had low status. It fell on the side of popular science rather than professional science; there was no hint of a discipline forming around evolutionary studies; and most serious thinkers (with reason) regarded any form of transformism as pseudo- or quasi-speculation. The connection with Progressionism was reason both for some people to become evolutionists and for others to relegate evolutionism to so low a position. It was a long way from maturity.

Under Napoleon, France had a professional scientific framework basically in place. In the years following, although there was certainly extension and formalization, there was not a hugely significant change in the structure of this framework. With good reason, people now feel that French science peaked and slowly declined, ossified by the conservative hand of an established and overly satisfied scientific elite. Germany, to the contrary, stood on the threshold of its rise to power and influence (Crosland 1992; Turner 1971; Farrar 1975). The founding of the university in Berlin was a mark of the rejuvenation and growth of the whole higher education system. In science, Liebig showed the way forward, taking a post at Giessen and surrounding himself with a group of students who worked and learned in his laboratory, absorbed the mysteries and methods of their craft, and (touching a chord with our own day) earned the increasingly important doctoral degree.

In Britain also there was change, although science as a profession had to wait until the second half of the century before it really achieved any importance. In the early decades, despite the continued opposition of the Royal Society, new specialized societies were founded. Most lively and most influential was the Geological Society of London, which began in 1809 (Rudwick 1963). Reformers within the Royal Society itself were not immediately successful; but there was a growing sense that the body's days as a gentleman's club were numbered. And at a slightly different level, perhaps most important of all for British science, came the birth (1831) of the British Association for the Advancement of Science (Morrell and Thackray 1981). Meeting once a year in a different provincial city, the Association was a combination of professional and public science—information for the elite and spectacles for the hoi polloi. Very importantly in a country without state support for science, its success meant profits, and its profits meant research grants for the men of power and influence—and their protégés.

There was reform at Oxford and Cambridge, also. Not much, but some. The mathematics was upgraded; the science professors started to take an interest in their subjects; and although it was mid-century before the government forced science degrees on the institutions, the interested student could attend series of real, in-depth lectures on specific branches of natural philosophy (Winstanley 1935, 1940, 1947). There had always been the Scottish alternatives for those interested in science. Now there were English alternatives—University College and an Anglican-run rival, King's College, likewise in the nation's capital—though it took several decades for the University of London, formed from a combination of colleges and professional schools, genuinely to offer a scientific research/education alternative to the established places of study (Davie 1961).

The organizational developments went hand in hand, especially in Britain, with epistemological ferment. I have spoken already of the impact of the wave theory of light. People may have gone on talking of "Baconian" virtues—no-one approved of sweeping hypotheses in themselves—but there was serious rethinking about the epistemic values exhibited by good (what I call "mature") science. As people were trying to define and establish their status as scientists—it is no chance that the very word *scientist* was invented in 1833 by Whewell, precisely to articulate the new professional way of doing things (as opposed to the broader, flabbier *natural philosopher*)—so they were trying to articulate what it meant to do science (Fisch and Schaffer 1991; Yeo 1993).

Consistency and the like were taken for granted, but what else? Prediction was on everybody's list, usually toward the top, and often (thanks to Fresnel) linked with the ability to make surprising forecasts ("fertility"). Thus, the influential Comte: "all science has prediction for its goal" (Comte 1830–42, 1, 63; quoted by Laudan 1981, 141). Also important was some sort of unificatory power: "All *science* consists in the coordination of facts, if the diverse observations were entirely isolated, there would be no science" (Comte 1830–42, 1, 131; quoted by Laudan 1981, 142). These ideas were usually set in the context of the desirability of some sort of deductive system, and since the theory of Newton's *Principia* was taken to be the paradigm, unsurprisingly the British in particular had much to say. Among others, both the astronomer John F. W. Herschel and Whewell wrote major works on scientific methodology, struggling both to capture the best of the past (distant past, Newton; recent past, Fresnel) and to provide recipe books for future workers.

Naturally, there was much overlap (between the two good friends) on the desirability of consistency, simplicity, and predictability. Where they did differ somewhat was over causation, specifically over the interpretation of what Newton had called "true causes," *verae causae* (Kavalowski 1974). The more empiricist Herschel sympathized with the eighteenth-century thinkers and wanted (causal) hypotheses only if one could tie them firmly into analogies with the already known or sensed (Herschel 1827, 1831, 1841). Phenomena such as light-wave interference patterns could be explained and justified by analogies with like phenomena with water and sound waves. The more rationalist Whewell was more disposed to sweeping hypotheses, if they effected major unifications, explaining the already known or sensed (1837, 1840). This was what consilience was all about. Interference patterns were deductive consequences of an all-embracing causal wave theory. Expectedly, part as cause, part as effect, Herschel was sympathetic to Lyellian uniformity, stressing as it does that one must explain the past in terms of causes of a kind and intensity we experience today. Whewell favored catastrophism, since (although not experienced) catastrophes supposedly explain all the facts of geology (1831, 1832).

Returning now to the men at hand, our French scientists, Cuvier and Geoffroy, were fully professional. Cuvier is to this day an exemplar of what one means by "professional scientist." He held a post at the Muséum d'Histoire Naturelle from 1796 (age 27) and a full chair from 1802, he held a chair at the Collège de France from 1800, and he got into the Institut right at the beginning, becoming a permanent secretary (one of two senior administrative positions) in 1803. Geoffroy was in the Muséum at the start (1793), held a chair in the university (Faculté des Sciences) from 1808, and was hurt that he was not admitted to the Institut until 1807 (age 35).

In Cuvier's case particularly, we see in addition to his institutional achievements a deliberate and self-conscious effort to build around himself a group of people all working on his problems or those closely related. Notoriously, he had bevies of assistants, ready to take up projects under his direction. Being the founder of modern comparative zoology meant discipline building, and Cuvier was nothing if not good at that.

Agassiz may have been eternally grateful for the attention shown by the great French savant, but Cuvier's real virtue was in spotting the potential of the brilliant young Swiss scientist and in binding the newcomer to him forever.

Cuvier was alert to the epistemic demands of good-quality professional science, and he was very much aware that, as a *biologist* in an age when the running was made by the physical sciences, his work was cut out for him. There was no ready-made pattern waiting to be assumed or imitated. In the life sciences the established model was the charming, diffuse, non-rigorous but popular descriptive writings of Buffon. As Cuvier himself admitted, what passed for biology was intellectually speaking, pretty soft: "Placed between the mathematical sciences and the moral sciences, they [the life sciences] begin where the phenomena are no longer susceptible of being measured with precision, nor the results able to be calculated accurately; they finish, when one can no longer believe that what one is playing at is any more than a function of one's own mind and its influence on one's will" (Cuvier 1810, 5; quoted by Outram 1984, 128–129). It was this state of affairs that the doctrine of the conditions of existence was intended to address—to introduce some method, some rigor, some informed attention to the facts, some way of bringing *predictive* power to biology. Cuvier's flamboyant inferences had a meaning below the surface: "Experience alone, precise experience, made with weights, measures, calculations and comparisons with all the substances used and all the substances one can get, that today is the only legitimate way to reason, to demonstrate" (Cuvier 1810, 390; quoted by Outram 1984, 137).

Along with all of this in Cuvier's mind went an abhorrence of sweeping, uncontrolled, causal hypothesis. This was matched at the social level by a dread of the descent of science into the public forum. In that way science lost its esoteric purity, it became a vulgar entertainment for savages, and (unspoken implication!) it put the permanent secretary of the Institut at the heart of unwanted cultural controversy. We have seen how frustrating it is to the modern reader that Cuvier refuses to speculate on origins. This was not cowardice. It was prudence, certainly, in one who was system building—but also a conscious and heartfelt implication of his methodology.

I should add that Cuvier himself was only partially successful at and consistent in this policy of staying free of controversy. In the official publications of the Institut he kept a grave facade of disinterested objectivity, but he could be tempted into the popular gutter as need be, as in his lectures at the Collège de France. Especially, Cuvier was prepared to enter the public arena to combat what he saw as threatening pseudo-sciences. These included phrenology, which Cuvier abominated; *Naturphilosophie,* which (apart from the theological objections) he properly saw as an attempt to deduce *a priori* that which can be discovered only through careful observation of nature; and, obviously, evolutionism.

Which topic brings us to the other side of Cuvier's epistemological labors. It was not enough to push positively for epistemic values. Negatively, he had to argue against cultural values in biological science. His concerns here should not be misconstrued as unsullied epistemic fervor. As was firmly pointed out to Lyell when he visited Paris in 1823, Cuvier was a modestly born Protestant in an increasingly conservative, Catholic France (Lyell 1881, 1, 127). To make his way, he had to be in an area without taint of ideology—one which the state would heed and respect, and from which it would feel no threat. Science, a science of (cultural) value-free neutrality but attesting clearly to the highest reaches of the human intellect, was the perfect vehicle.[4]

The *Preliminary Discourse* was aimed (in major part) at evolutionism. Cuvier attacked (what he took to be) a pseudo-science, epistemically flawed, illicitly culturally laden. It was not predictive, it flagrantly contradicted known facts, and it stank of *idéologue* philosophy. I will not say that Cuvier was enough a cynic as to welcome evolution for the opportunity it gave to burnish his own worthwhile labors. But whatever his full motives, to Cuvier, evolutionary speculations epitomized unrestrained causal guesswork, at its dangerous pseudo-scientific worst. At the end of his life, with France again in social turmoil, Cuvier became so far incensed he was prepared to violate the sanctity of the Institut (again the Académie des Sciences, by then), for it was the site of the debate with Geoffroy on the topic—although, properly, Cuvier saw his whole life's work under attack. The fear was of Progressivism as a causal wedge opening the way to political unrest (Outram 1984). This being so, Cuvier was prepared to use his status and powers as a professional to fight for all that he stood. But it is important to note that, despite the site of the debate, it was now a public spectacle rather than a technical disagreement among full-time professional biologists.

Geoffroy was no Cuvier—he would never have been able to carry through a steady life-long campaign to define the status of science—but he too was sensitive about the need for biological science of the proper standing. It was just that he did not think that the key lay in the unadorned doctrine of the conditions of existence! The true empiricist pays attention to isomorphisms. Indeed, so far did Geoffroy take seriously the question of proper (professional) science that, when he became engaged in the study of monsters ("teratology"), we find him experimenting with batteries of eggs, trying to produce abnormalities through varied environmental pressures. This is just the sort of thing one would expect of the physical sciences. Biology should demand no less. But, for all this, Geoffroy, like Lamarck, had another side—a wild, speculative, non-professional side. The trip to Egypt at the beginning of the century had sparked a very odd flight of fancy about "imponderable fluids" (light, caloric, electricity, and the nervous fluid) that its author thought was going to explain all of the mysteries of the universe. Fortunately, when it produced the predictable response, Geoffroy had the good political sense to hide it away.

Yet, it is clear that by the 1820s, perhaps out of frustration from years of being regarded as the "other" French anatomist, perhaps because the country was getting steadily more conservative (until a revolution in 1830), Geoffroy's romantic, non-professional side was rising again. This was marked by his turn to explicit evolutionism, and by the 1830s to a diffuse, rambling, ever hypothesizing, never-ending verbosity that drove his friends, his enemies, and particularly his long-suffering son (also a biologist) to despair. A new principle appeared, and great things were promised of it: "Universal Law (Attraction of *Soi pour Soi* [like for like]), or key Applicable to the interpretation of All Phenomena of Natural Philosophy" (Geoffroy 1835, 125). If evolutionism was professional science, you would not have guessed it from Geoffroy's activities. "The anatomist championed the right of the genius to see beyond facts, to illuminate gathered data with his intuitions, and to make up for missing data with bold reasoning" (Corsi 1988, 265). This was not the road to mature science. Significantly, it was at this late period of his life that Geoffroy really became the darling of the radicals, as he and they hymned the praises of Progress. Cuvier was not mistaken in his target.

Britain presents a somewhat different picture. Simply put, the two overt evolutionists, Grant and the Vestiginarian, were just not up to the task of providing professional science, let alone discipline building. You hardly gather a group of people around you if nobody knows for sure who you are! Grant was certainly a professional scientist, or rather a professional teacher of medical students. But he was not after a body of mature science. He was in the business of changing people's hearts, not their minds. He really used his evolutionism more as background for his (professional) anatomy—it was an ideological tool to beat his opponents. We have seen that he had no interest in laying out evolution in a coherent theory. Chambers certainly wanted to be taken seriously—he was (to compound terms) a very professional popular writer. *Vestiges* went through several editions in the face of criticisms, and the science therein was much improved. But Chambers's audience was ever the general public. Even though (quite naturally) he was happy to get the endorsement of any professional that he could, his mission in life was to spread the glad tidings of Progress, to which conventional thought had better bow or be crushed. He was not writing for the Royal Society.

To the professionals, the greater the popularity of evolutionism, the more they insisted hysterically that it belonged at the pseudo end of the science spectrum. In fairness, they could (and did) point to the fact that it satisfied virtually none of the epistemic criteria of good science. Where was the predictive fertility of Chambers's theory? What efforts were there to make the ideas consistent with the rest of biology? Nice distinctions about *verae causae* were quite lost on the evolutionists. But there was more to their criticisms than this, and as in the case of Cuvier one senses that, pure theory aside, judgments were based in large part on personal needs and fears. Lyell, writing at the beginning of the 1830s, was under some pressure, since he was (for a short time) professor at King's College in London (L. Wilson 1971). But he did have independence and it showed. In 1831, in the second volume of his *Principles of Geology,* he treated Lamarck's ideas with respect, even though he did ultimately fault them (in part) because, the evidence being that domestic breeding leads to no lasting full-time change, they fail the empiricist test for a *vera causa.* Were we able to change species, then we would have direct knowledge of such change. Lamarck had not produced top-quality science, but the impression nevertheless was certainly not that he was propounding something akin to Mesmerism.[5]

As the decade stretched, the critics' tone got more shrill. In part this was because for people like Sedgwick and Whewell, evolution posed a deep threat, not only to Christianity as doctrine but also to Christianity as foundation of the social fabric that they as ordained Oxbridge professors personified. But there was more than simply self-interest in this direct way. The very act of trying to open a space in British (especially English) society for professional science, especially by men who generally had to be ordained priests in an Establishment church in Establishment universities, was fraught by difficulties and critics from all sides. They faced resistance from below, by extreme evangelicals, who saw that the likes of Whewell and Sedgwick were playing fast and loose with the Bible taken absolutely literally, and from above, in the 1830s, by proponents of the Oxford Movement, who saw science as the thin end of a secularizing wedge rending education and the very fabric of society. The latter knew that Whewell's linguistic suggestions were not theory neutral: "Physical science, the science, in other words, of matter and material things, now arrogates in effect the name 'science' exclusively to itself" (Bowden 1839, 8).

Hence, all of the other arguments apart—and the British Association made much of the utilitarian virtues of science—a Cuvier-like strategy was much favored in England. It was crucial to present good-quality science, the mature science of professionals, as something beyond culture in an important way. The religious were happy to argue that science testifies to the glory of God ("she says nothing, but she points upwards"; Whewell 1837, 3, 588). Whewell (1833) went so far as to say that doing science is our intellectual task as doing good is our moral task. But these critics argued that science in itself is neutral. This is why one absolutely must dismiss evolutionism as pseudo-science, since with all of its Progressionism it so gravely disturbs the delicate balance. It fails to fit the standards. It is epistemologically barren and culturally corrupt.

Next, consider briefly those under the spell of Germany. Agassiz was a fully professional scientist and would have thought (with reason) that his rejection of evolutionism stemmed from his expertise in that direction. I have argued that his philosophy was as, if not more, important. But that does not contradict points made already in this section. Here, apart from noting that we have not yet finished with Agassiz and shall be rejoining him after his trip to the New World, I will point out that there was an interaction of cautious respect between him and the British. They were deeply suspicious of his progressionism and yet they needed his expertise and authority. Probably for the ongoing happiness of all, it was as well that he went West. Dangerous tensions could be kept at arm's length.

What, finally, of Richard Owen, if one accepts (as I have argued) his deep attraction to *Naturphilosophie*? All of Owen's actions in the crucial 1830s and into the 1840s show what a dangerous attraction this was. He was a man who simply had to watch his step every inch of his way, else his career was destroyed. As it happens, Owen did watch his step, and his career as a professional scientist did flourish. He continued on the path opened by Cuvier and von Baer and others, and as one who did not (at the time) declare openly for evolution, he did not offend his patrons. He knew what he simply could not do or say. He sailed close to the wind, but never so close that he capsized.

Yet what makes Owen interesting and more than a pathetic marionette, jerking in London to strings pulled in Cambridge, is the fact that there are strong hints that he had thoughts of upgrading *Naturphilosophie,* including the evolutionary aspects. Although his most detailed exposition of his pre-*Origin* quasi-evolutionism came in his *On the Nature of Limbs* (1849a)—a Friday-evening lecture at the public-welcoming Royal Institution—he may have had visions of evolution as professional science. For instance, by mid-century we find that Owen was very interested in questions about parthenogenetic reproduction and alternative life-cycle forms—good models for saltationary change (Owen 1849b). This is just the kind of enthusiasm one would expect of someone who was trying to launch an experimentally based working science, a science with the epistemic marks of maturity.

Not that anything amounted to much. No major research program was opened. Nor, remembering a key factor about professional science and its disciplines, do we see a group forming around Owen, to develop and to promote new ideas. Whether this was entirely his fault or mainly a function of circumstance it is hard to say. It is surely a major reason why Richard Owen never became more than he was. Yet it is easy for us to be critical. Looking at Owen's work, one wonders what might have been the case had he had more (and less encumbered) money, and more nerve, and more . . . !

The notion of science as a profession is solidifying at this point in our story, and with this the articulation and cherishing of some familiar-looking epistemic values. At the same time, we are starting to learn how one might have reasons—paradoxically, non-epistemic reasons like a desire to succeed as a professional—for pushing cultural values like Progress out of one's science. This will surely be a point to look out for in the future. For now, returning to our story, let me point out that there is no more revealing fact about the status of evolutionary thought at the end of its first century than that Robert Chambers set out first to write a book on phrenology and changed his topic to evolution as he got under way (Secord 1989). One bogus science led naturally into another. It was

Progress which made the transition natural, as it was Progress which was a major factor in keeping evolutionism at the low end of the respectability spectrum. Yet Progress had to be also a major factor in evolution's continued, ever strengthening existence. *Vestiges* sold 20,000 copies by 1860 (Williams 1971, 192). Evolution may have been the fast food of the scientific world; but, like the best examples of the genre, what it gave up in status it made up in sales. Our question now is whether a master chef could change the picture significantly.

4

Charles Darwin and Progress

We come now to Charles Darwin, author in 1859 of *On the Origin of Species by Means of Natural Selection, or the Preservation of Favoured Races in the Struggle for Life.* I shall discuss his ideas in some detail, for he is the key figure in the history of evolutionary thought.

The Route to Discovery

Charles Robert Darwin was born, in 1809, to a distinguished and rich family (Darwin 1887; de Beer 1963; Bowler 1990a; Desmond and Moore 1992). His paternal grandfather was Erasmus Darwin, already familiar to us. His father, Robert Darwin, was likewise skilled at medicine, with a well-deserved reputation throughout the (English) Midlands. His maternal grandfather was Josiah Wedgwood, fellow member (with Erasmus Darwin) of the Lunar Society, and he who was primarily responsible for bringing the Industrial Revolution to the pottery trade. Young Charles Darwin was to make even more sure of his share of the Wedgwood wealth when (for love, I hasten to add) he married another of old Josiah's grandchildren, his cousin Emma Wedgwood.

Tradition has it—a tradition founded primarily on his own self-deprecating remarks in his *Autobiography*—that Darwin's childhood and

Charles Robert Darwin

youth were intellectually undistinguished. He was second-rate at school, he was (after two years) a drop-out from the medical program in Edinburgh, and he was then an idle undergraduate at the University of Cambridge. However, although no-one would suggest that Darwin was an infant prodigy, it is clear that he showed an early interest in and aptitude for natural science. Even before he went to Cambridge he had considerable training, theoretical and practical. Through the tutelage of his older brother, Erasmus, and through first-rate lectures at Edinburgh, Darwin received a thorough grounding in the chemistry of the day. Additionally, in Scotland, thanks to ardent pursuit of extra-curricular activities, he developed considerable knowledge of and skill at marine invertebrate zoology, as well as favorable exposure to evolutionary ideas from that articulate spokesperson for the cause, Robert Grant.

At Cambridge, Darwin fell into the company of the science clique, the members of which (including such professors as Adam Sedgwick, of geology, John Henslow, of botany, and William Whewell, then of mineralogy) saw real talent in the young man. Continuing his informal science education, for the full three years of undergraduate life Darwin attended

Henslow's very comprehensive course in the principles of botany. Although by the time of his graduation in 1831 Darwin was hardly a fully trained scientist, there is therefore little surprise that, through his university connections, he was offered and accepted the *de facto* post of naturalist on HMS *Beagle,* which was to spend five years surveying in the Southern Hemisphere, primarily around South America. It was on the basis of his diaries written during this trip that Darwin was to write a still-charming travel book which established his reputation firmly with the Victorian public (Darwin 1839).

On the *Beagle* voyage, Darwin developed rapidly into a more-than-competent natural philosopher. Trained by Sedgwick, he was however much impressed by Lyellian uniformitarianism, from which stance he made his own contribution to geology when he gave the (still-accepted) reasons behind the formation of coral reefs, showing how they are a function of the coral growing upward as the seabed beneath sinks gradually (Darwin 1842). Work like this brought the newly returned Darwin right into the heart of the Oxford- and Cambridge-linked, middle-class scientific community. Nevertheless, it is clear that, from the very start, Darwin's horizons were fixed above geology. He wanted to solve what Herschel referred to as "the mystery of mysteries": the origins of organisms, in all of their variety (Cannon 1961). It was this that led to evolution (Oldroyd 1984).

Hindsight often makes it difficult to discern the exact course and importance of events. However, with respect to his conversion, there is little reason to doubt Darwin's own recollection of the stunning impact of a Pacific island group visited by the *Beagle* in 1835 (Sulloway 1982b, 1985). He could not get over the fact of "the South American character of most of the productions of the Galápagos Archipelago, and more especially by the manner in which they differ slightly on each island of the group" (Darwin 1958, 118–119). This is not to say that there was an immediate, on-board conversion to evolutionism; rather, full insight came later (spring of 1837), back in Britain, when Darwin learned that the birds of different islands belong to different species (Sulloway 1982a).

Why should this (somewhat trivial) piece of information make so much difference? By this stage of his life Darwin, who had earlier been a practicing Christian, had moved toward his family's deism. It simply did not make sense for God to have put similar but different animals on different islands, within sight of each other, especially since the habitats were essentially identical. Even less did it make sense for God to have ensured carefully that the denizens of the Galápagos were South American–like rather than European-like. Who would care? The explanation had to lie elsewhere and Darwin found it in evolutionism. Ancestors had come to the islands and then evolved as they moved from island to island (Barrett et al. 1987).

This takes us to the *fact* of evolution. What of causes? What of evolution as *theory?* Initially, Darwin seems to have supposed that there might be limited terms for species, ending with sudden switches to new forms. Then, more confidently, he favored some kind of Lamarckian causal process, in which the inheritance of acquired characters played a major role. He always kept this as a secondary mechanism, but soon queried its adequacy as a general cause. Continuing his search, Darwin next realized that animal and plant breeders effect great changes by picking or selecting those organisms with desired characteristics and breeding from them alone (Ruse 1975b). But, how could such a process occur in nature? The insight came late in September of 1838, when Darwin read the *Essay on the Principle of Population* by Robert Malthus. He realized at once that the central doctrine of that work on human societies applies throughout the living world. There is an ongoing, potentially explosive population pressure, against an invariable background of limited supplies of space and food. Hence, there will necessarily be a struggle for existence. Yet, this struggle need not be entirely negative, for success is not chance. It is rather a function of the possession of various peculiar, useful features—and the non-possession of such features by the losers. And so there will be a process akin to the breeders' picking, a process Darwin labeled "natural selection."

Given enough time and an ongoing supply of new variation, there will be full-blown evolution. But more than this followed from Darwin's line of thought. As one who had been reared at Cambridge on an almost unrelieved diet of the works of Archdeacon Paley, Darwin felt strongly—with every established middle-class scientist in Britain at that time—that any adequate force must be able to explain that all-pervasive facet of the organic world: its teleology, its adaptedness. Every organism, in just about every way, seems as if it were designed to be what it is. And to this, Darwin thought his new cause could speak. Natural selection is as much a force directed toward organization and functioning as is Paley's God (Ruse 1979a).

A mechanism is not a full-blown theory. Darwin realized this, and so he worked furiously to embed his central notion in a complete framework. In 1842, that is four years after discovering natural selection, Darwin wrote up a short (private) sketch of his ideas, and two years later he expanded this into a full-length essay (Darwin and Wallace 1958). He did not publish. We shall probably never truly know the full reason for this reticence—assuming that there is a full reason. We know that Darwin had fallen sick by this time, from a mysterious illness that was to plague him for the rest of his life. We know his wife, whom he married in 1839, would most probably have been upset by a public airing of his ideas, and Darwin was not about to cause trouble in that direction. We know that *Vestiges,* published in 1844, caused tremendous controversy, and Darwin had no wish to alienate his fellow scientists—especially given that Chambers's greatest critics, like Sedgwick, were not only Darwin's friends but often his sometime teachers.

I myself doubt that Darwin realized that the delay would become as long as it did—but long it became, as Darwin buried himself, physically in the countryside and intellectually in a massive study of barnacle taxonomy (Darwin 1851a,b, 1854a,b). Finally, in 1858, when Darwin had started to write an immense tome on evolution—overwhelming with fact and footnote—his hand was tipped. A young naturalist, Alfred Russel Wallace, sent to Darwin (of all people!) a short essay containing Darwin's own ideas in perfect cameo. Thus spurred, Darwin wrote up his theory in a full yet manageable volume, and *On the Origin of Species* appeared toward the end of 1859.

The Origin *and Its Aftermath*

The *Origin* was controversial. But let me begin by stressing the positive side. After the work's publication, in a very few years for the educated person a belief in evolution became the norm, not the exception (Ellegard 1958). And this was in major reason a consequence of Darwin's great skill as a promoter of ideas. He was highly sensitive to the fact that he had—to put it mildly—a job of selling to do. Not only had he to brush past all of the religious sensitivities of the mid-Victorians, he had also to persuade people of a thesis which, by its very nature, is essentially unobservable. Even if evolution be true, we shall never see it—except in a limited and trivial sense. That people did accept that evolution is true is to Darwin's everlasting credit (Oldroyd 1986).

His strategy in the *Origin* was to make the mechanism of natural selection the central focus. Rather than presenting the case for evolution as fact, and then going on sequentially to evolution as cause, Darwin went straight to the heart of his case. Thus, on opening the *Origin* we find that Darwin did not use artificial selection simply as a guide to discovery (heuristically). He used it also to persuade the reader through analogy—the analogy of such "obvious" things as cattle and sheep breeding—of the plausibility of natural selection (as justification). Then, having discussed the struggle for existence—a well-known notion in the post-Malthusian era of mid-Victorian Britain—he began an intentionally seductive presentation of the central mechanism:

> Can the principle of selection, which we have seen is so potent in the hands of man, apply in nature? I think we shall see that it can act most effectually . . . Can it . . . be thought improbable, seeing that variations useful to man have undoubtedly occurred, that other variations useful in some way to each being in the great and complex battle of life, should sometimes occur in the course of thousands of generations? If such do occur, can we doubt (remembering that many more individuals are born than can possibly survive) that individuals having any advantage, however slight, over others, would have the best chance of surviving and of procreating their kind? On the other hand, we may feel sure that any variation in the least degree injurious would be rigidly destroyed. This preservation of favourable variations and the rejection of injurious variations, I call Natural Selection. (Darwin 1859, 80–81)

Now, with the mechanism made explicit, Darwin could turn to evolution as fact, and so the rest of the *Origin* (a full two-thirds) consists of a journey through the various branches of biology, as evolution through selection is applied to major claims and theory problems. Hence, on the one hand the reader is persuaded of the value of such an approach, and on the other hand facts that we know about the living world point us in one shared direction, namely toward evolution through natural selection.

Opening the survey, Darwin went first to instinct and the problems of behavior. How can it be that hymenopterans (the ants, the bees, and the wasps) go through such delicate and sophisticated motions as building their intricate nests and hives? Darwin showed how one can break down the behavior into components, and that thus understood the behavior as a whole is clearly a function of selective forces. Likewise paleontology and the fossil record: "We can thus understand how it is that new species come in slowly and successively; how species of different classes do not necessarily change together, or at the same rate, or in the same degree; yet in the long run that all undergo modification to some extent" (p. 343).

Biogeography, unsurprisingly (given its key role in Darwin's becoming an evolutionist), also got major treatment. Why are the distributions of organisms around the globe as they are? Can they be the products of natural forces of motion and of the power of evolution? Darwin argued that they can be, and gave considerable evidence to that effect.

Then, moving along, Darwin rounded off his discussion by covering topics like systematics and embryology. The latter, particularly, shows how Darwin kept subtly reminding readers of the familiar, everyday side to his thesis. Why are the embryos of organisms of different species frequently so similar when the adults are so different? The answer lies simply in the fact that selection only tears apart the adults. Conditions in the human womb and the canine womb are not so very different. If proof be needed, Darwin was happy to provide it, fleshing out his case by referring to the work of animal breeders. As in nature, it is the adults that most excite breeders, and Darwin was able to show that in the world of the domestic dog the differences between adults of different breeds are far greater than the corresponding differences between puppies. This may have surprised some readers, but there is nothing which could not be checked. And so, on such comfortable notes as these, the *Origin* drew to its close. Homologies, behaviors, fossils, distributions, embryos, and more—these are the footprints, the blood stains, the fingerprints. And they point uniquely to one culprit: evolution through natural selection!

Notwithstanding this confident conclusion, I shall later be stressing what I take to be the ambiguous nature of the *Origin;* here, we must note that, without at all undercutting this section's opening comments about the work's importance and success, the immediate response shows there was also a negative side to the work's reception. Acceptance of the main idea of change is not to say that people were as enthused as was Darwin about the mechanism of selection. No one denied it, or that it could have some effect—perhaps even that it could take organisms across species barriers. But, as we shall learn in later chapters, selection was usually supplemented with (at times virtually replaced by) other mechanisms. Given the significance that Darwin gave to his mechanism in the *Origin,* this might suggest the failure of his strategy of presentation and counter my claim about the work's success. But this is too quick a judgment. Because selection was so prominent, it was an obvious target. Hence, the reader could concentrate fire on the mechanism—on Darwin's suggestions for evolution as *theory*—thus demonstrating orthodoxy, as the message of evolution as *fact* was accepted quietly and completely. Darwin wanted more; but that, we shall learn, is part of our story.

Not that this is to deny that some of the criticisms made of the *Origin* had real scientific worth. Two things in particular (rightly) gave Darwin great trouble. First, there was the age-of-the-earth question (Burchfield 1974, 1975). The physicists, led by William Thomson (later, Lord Kelvin), were able to show, from theoretical and empirical studies on the earth's cooling and the like, that apparently the earth is just too young for the demands of evolutionary biology—a point Darwin appreciated and fretted over. Ultimately, however, there was not much he could do—which was just as well, since the physicists had omitted one crucial factor: the heat generated by radioactive decay and the consequent lengthening of earth's cooling period. Once the effects of radioactive heat were understood, there was quite enough time for Darwinian evolution.

The second major worry for Darwin was the problem of variation and heredity (Vorzimmer 1970). Where do new variations, the "raw stuff" of evolution, come from and how do they get transmitted from one generation to the next? Darwin was convinced that the crucial variations had to be small, minute even, else selection could not go to work. Large variations, saltations, tend to be maladaptive. But, as to the real reasons behind heredity, Darwin had to confess ignorance. And, despite various desperate suggestions and theorizings (including "pangenesis"), that essentially remained that: a point which was seized on by critics who (rightfully) sensed a major lacuna. Now, today, we think we can go far toward solving Darwin's problems (Ruse 1982). This is a topic that will occupy us in later chapters.[1]

What of the status of that troublesome organism *Homo sapiens?* Given the agonizing that so many of his fellows were doing over our status, Darwin's position is quite noteworthy. He was always convinced, absolutely and utterly, that we humans are part of the natural world, and as such we are the products of evolution (Ruse 1979a; Gruber 1981; Herbert 1974, 1977). A major reason he was so sure was that he had had experience of pre-literate folk in their native habitat. The *Beagle* had spent time in Tierra del Fuego, at the bottom tip of South America, and Darwin had been much impressed with the primitive life of the inhabitants. This impression was greatly reinforced by the fact that the *Beagle* was returning three natives to the area after a short sojourn in Europe. Almost at once, they reverted from civilized beings to near animals—at least, as then judged. The ship's naturalist learned a lesson that was never to be forgotten.

Indeed, it is interesting to find that the very first explicit use of the idea of natural selection, by Darwin in a private notebook in the fall of 1838, centers on humans—not on arms and legs either, but on our brains and our power of thought. Darwin was convinced that differential reproduction in the struggle for existence is the key to human intelligence (M. 154, in Barrett et al. 1987). You cannot make your position plainer than that, although, of course, you can conceal it—which is precisely what Darwin did. In the *Origin,* we get but the briefest mention, right at the end. It was not until some twelve years after the *Origin* that Darwin did finally speak at length on humankind, in his *Descent of Man.*

On the surface, this later book is rather curious, practically ill-balanced. Most of its concern is not with humans and natural selection at all! It is with a secondary selective mechanism, sexual selection, which occurs always within a species and involves competition for breeding partners. This was not a new idea, but for years Darwin never really made anything of it. Then, in the *Descent of Man,* the topic fills more than half of the book. Yet, there was a reason for so lopsided a treatment. By the end of the 1860s, the co-discoverer of natural selection, Wallace, had fallen into apostasy over human origins, arguing that many aspects of human nature—our hairlessness and our thinking capacity, for instance—just could not have come through natural selection. The savage has the potential for deep thought, but because in the general course of life he or she never uses it, it could not have been produced by natural selection (Wallace 1870, 332–371; Kottler 1974; Smith 1972). Agreeing in part with the premises of Wallace's argument, Darwin turned in search of an alternative conclusion. In sexual selection he found his answer. Rivalry for mates centers on human notions of beauty, which bring on the evolution of those human attributes for which Wallace supposed we must have a non-natural cause. Hairlessness, intelligence, and the rest come from intra-specific struggle for mates. Hence, like all other organisms, humans are an entirely natural product of the evolutionary process.

I shall have more to say shortly about the *Descent of Man.* For now, though, we can conclude our direct general treatment of Darwin and his work. He was to live to 1882. When he died he was buried just a few feet from the great Newton, in Westminster Abbey. He had become part of the national fabric (Moore 1982).

Darwin's Biological Progressionism: Before the Origin

I come now to the question of Darwin and progress, and the related question of Darwin and Progress. Does he, in any sense, show progressionist tendencies? Was he influenced by ideas of Progress? Did he have an evidential base?

Begin with the question of biological progress, and turn to the first jottings on evolutionism. Right at the beginning of one of the earliest notebooks, when Darwin was edging toward transmutationism but had not yet grasped selection, we see that he was thinking in terms of high and low with mammals at the top.

> The living atoms having definite existence, those that have undergone the greatest number of changes towards perfection (namely mammalia) must have a shorter duration, than the more constant: This view supposes the simplest infusoria same since commencement of world. (RN, inside front cover)

Note that he was already speculating on causes, suggesting that the more change you need to account for (in order to explain the appearance of) higher organisms, the more turnover you presuppose.

Then as Darwin really got under way on his speculating (still pre-selection), the progress comments came thick and fast:

> Each species changes. does it progress.
> Man gains ideas.
> the simplest cannot help . . .—becoming more complicated,; & if we look to first origin there must be progress (B, 18)

> There must be progressive development; for instance—?, of the <vebtetrata> <<vertebrates>> could exist without plants & insects had been created; but on other hand creation of small animals must have gone on since from parasitical nature of insects & worms.—In abstract we may say that vegetables & mass of insects could live without animals but not vice versâ (B, 108–109)

And, specifically, with respect to humans:

> We see gradation to mans mind in Vertebrate Kingdom in more instincts in rodents than in other animals & again in Mans mind, in different races, being unequally developed.—is not Elephant intellectually developed amongst Packydermata. like Man amongst Monkeys—or dogs in Carnivora.— (C, 196)

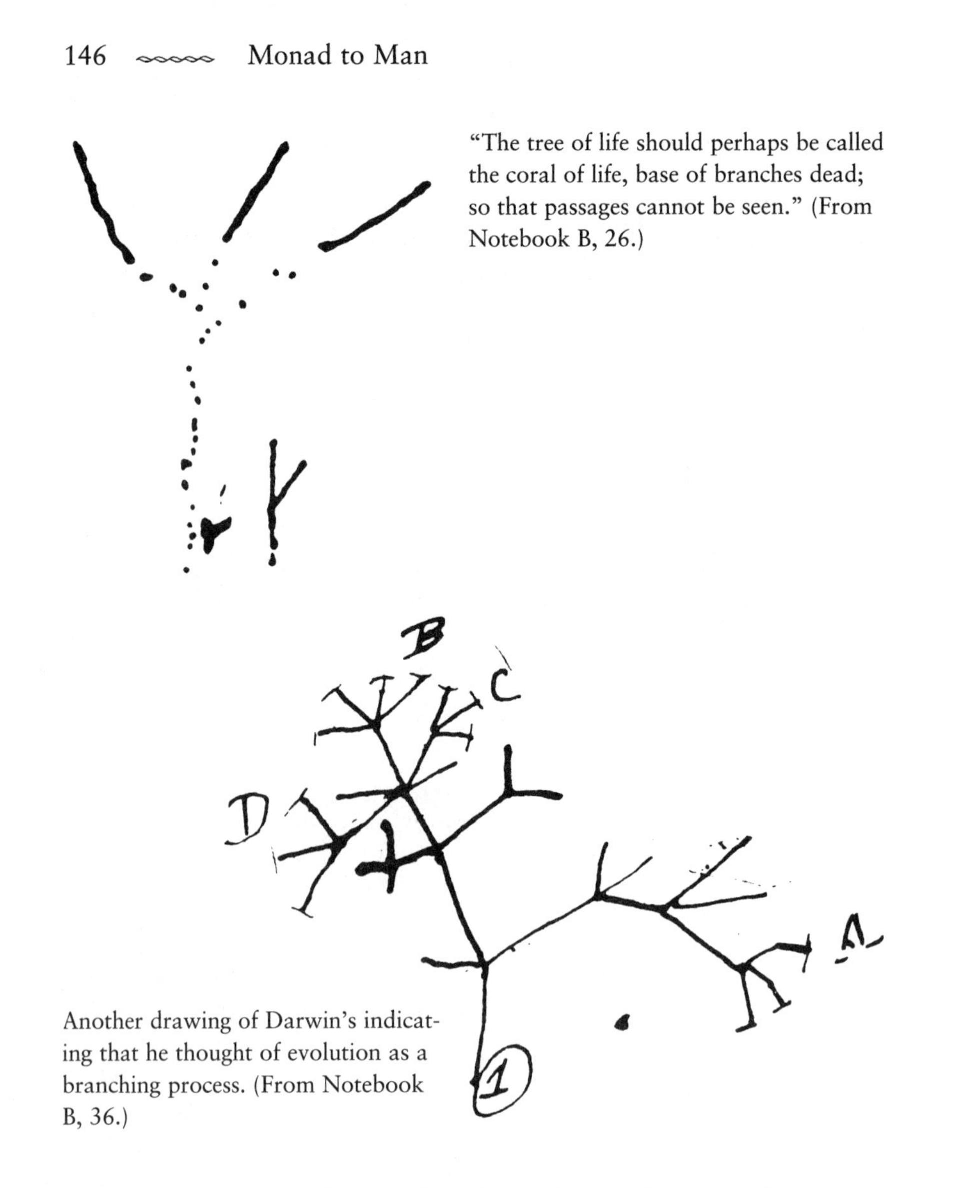

"The tree of life should perhaps be called the coral of life, base of branches dead; so that passages cannot be seen." (From Notebook B, 26.)

Another drawing of Darwin's indicating that he thought of evolution as a branching process. (From Notebook B, 36.)

Moreover, even in these earliest days—just at the time when he was courting his wife-to-be!—we see the seeds of plants that will fruit fully only in the *Descent*.

> Sept. 9th. It is worthy of observation that in insects where one of the sexes is little developed, it is always female which approaches in character to the larva, or less developed state. (D, 76)

Yet even from the beginning, Darwin was struggling with progress. He believed in it, but he was not quite sure what he believed in. Most definitely, Darwin knew what he did not want to believe in, namely any kind of simple, unilinear, monad-to-man progressionism. The Galápagos factor was crucial here, for it led Darwin to think of change as sparked by geographical isolation (as on islands), with consequent evolution to differing forms. In other words, transformation always has splitting at its heart. For Darwin, the tree of life (or "coral" of life, as he speculated on calling it) was fundamental.

In addition, however progress may be defined, Darwin always appreciated that an organic descent from higher to lower might be a smart life-strategy move. That first (pre-selection) mechanism, a kind of Lamarckian adaptation of structure, sparked by new habits, certainly opened the path for successful degeneration. Moreover, according to the outside influences, such degeneration was indeed a possibility. Others more knowledgeable than he were telling Darwin that simple progressionism was paleontologically false. Owen, at this crucial point in time working on Darwin's *Beagle* fossils, was uncompromising: "the different organized forms which have succeeded each other do not display regularly progressive stages of complication, or perfection of Structure. Plants and animals exhibiting different degrees of Complication of Structure have co-existed at different periods" (Owen 1992, 222).

Ladder-like progressionism is impossible, and Darwin was thus led into comments about why we should expect to find insect instinct happily thriving alongside human intelligence. Nevertheless, for all the qualifications, ultimately one branch of the tree of life leaves all others behind. Why? Already, we have some of Darwin's speculations. In a way, progress is an artifact of the very fact of evolution itself. You have to start from the bottom up—necessarily you go from simple to complex, and you cannot have complex organisms (vertebrates, for instance) without simple organisms for them to subsist on. Yet, obviously, these are more necessary than sufficient conditions for progress. Once given plants, why should vertebrates evolve? Why is there more on our planet than thick, green jungle? It is at this point, when he is trying to generate a positive force for progress, that we start to see Darwin running into real difficulties: "there is no <<NECESSARY>> tendency in the simple animals to become complicated" (E, 95). I assume that this comment referred not only to Lamarck but also to the spirit forces of the *Naturphilosophen*.

Unfortunately, Darwin's own general alternative was inadequate. In the early days, he saw change (specifically adaptation) as coming in response to environmental changes. A domino effect does operate in the organic world, but can it account for progress? At times Darwin thought it could: "It is another question, whether whole scale of Zoology may not be perfecting by change of Mammalia for Reptiles, which can only be adaptation to changing world:—I cannot for a moment doubt, but what cetaceae & Phocae now replace Saurians of Secondary epoch: it is impossible to suppose such an accumulation at present day & not include Mammalian remains" (B, 205–206). Essentially, however, Darwin was caught on the horns of a Lyellian dilemma. If geological change be directional, hotter to cooler (say), then organic progress might follow. But Lyell (and Darwin following him) interpreted uniformitarianism as implying a non-directional, steady-state earth history. Hence, overall directional organic change seems impossible.

The early Darwin rested uneasily on the question of progress. He needed both definition and cause. With respect to the first, the need for an adequate understanding of what one might mean by "progress," not to mention "higher" and "lower," intensified when, with the publication of *Vestiges,* critics lambasted Chambers's suggestions. It is no wonder that we find Darwin at once starting to disavow any such problematic terms. Yet, at the same time, he started to inquire of his friends as to their understanding of the notions! Thus to Hooker: "on what sort of grounds do Botanists make one family of plants higher than another" (Darwin 1985– , 3, 301). Response: "Generally speaking in Botany highness and lowness are synonymous with complexity and simplicity of structure." Unhelpfully, Hooker then added: "I can hardly conceive either simplicity or complexity of one particular organ indicating the rank of a being in the scale of creation" (3, 306, letters, March 1846).

Gradually, however—thanks particularly to the work of Germanic scientists—Darwin began to see his way to an adequate (and by the 1850s, fairly conventional) notion of highness and lowness. A rise up the scale required increasing differentiation, specialization, division of labor. Remember von Baer: "The more homogeneous the whole mass of the body is, so much the lower is the grade of its development. The grade is higher when nerves and muscles, blood and cell-substance, are sharply distinguished" (von Baer 1828–37, 1, 207). Darwin began to see how ideas like these could be incorporated sympathetically into his own thinking.

Causally speaking with respect to progress, Darwin needed a more directed, a less passive and more active, force at the center of his evolutionism. He had to move away from seeing change as just a reaction to outside forces and more as a process with its own internal dynamic.

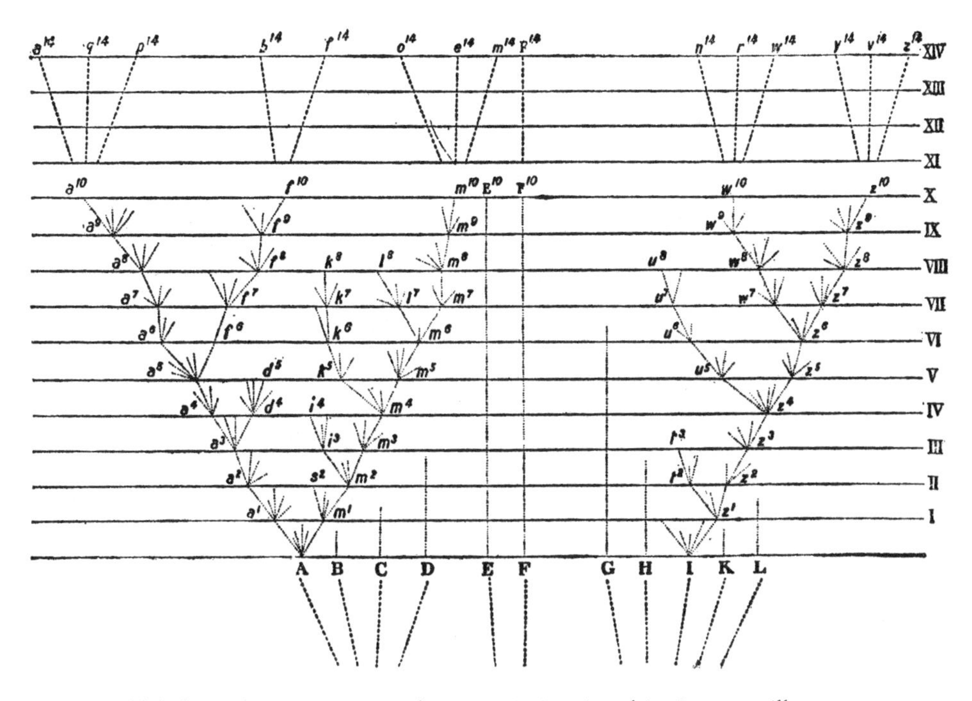

The tree of life from the *Origin*. Note that Darwin is using this picture to illustrate his principle of divergence, and one should not be misled by its branching nature into thinking that he is against or indifferent to progress. The point of a good metaphor is that it can be used to emphasize different things at different times; on the other hand, it may be that, on this occasion, Darwin was trying deliberately to put a distance between himself and progressionists like Chambers and the *Naturphilosophen.*

Grasping the significance of selection did not do this *per se*—the adaptation it produces might occur merely in response to external forces—but selection does open the way to inter-organic tensions. In other words, it might be an activity which in itself alone could fuel change—even progressive change.

Indeed, soon after he had discovered selection, Darwin saw the possibilities it raised:

> The enormous *number* of animals in the world depends on their varied structure & complexity.—hence as the forms became complicated, they opened *fresh* means of adding to their complexity.—but yet there is no *necessary* tendency in the simple animals to become complicated al-

> though all perhaps will have done so from the new relations caused by the advancing complexity of others. (E, 95–97)

It was not really until the 1850s, however, that Darwin truly started to develop and exploit this kind of thinking. He had always believed in divergence and splitting, but his growing sympathy for the continental characterizations of advance and improvement led Darwin to a broadening out of his thinking from his first speculations. Increasingly he became dissatisfied with isolation as the sole cause of change—no doubt in major part because of the rather special circumstances that this seems to demand before evolutionary change can take place. He therefore started the move away from a picture of selection working only to keep organisms on top, responding to geological/geographical change, toward a picture where selection plays a more activist role, making splitting (what today we would call) an internal ecological phenomenon. Organisms themselves, in relation to each other, cause change and splitting (Ospovat 1981).

Thus Darwin was led to develop—or, rather, given the early speculations, to elaborate—what he called his "principle of divergence": "The same spot will support more life if occupied by very diverse forms . . . This I believe to be the origin of the classification and affinities of organic beings at all times; for organic beings always *seem* to branch and sub-branch like the limbs of a tree from a common trunk, the flourishing and diverging twigs destroying the less vigorous—the dead and lost branches rudely representing extinct genera and families" (Darwin and Wallace, 1958, 266–267, from a letter written to Asa Gray, September 5, 1857).

The point is that you get more success by diversifying. It must be emphasized that Darwin always thought of selection as working for the benefit of the individual rather than the group (Ruse 1980). So he was not claiming that selection was working to produce divergence and variety, merely that more organisms (taken as a whole) can be supported on the same resources. And this was the kind of total life history within which Darwin's progressionism was framed.

Darwin's Biological Progressionism: The Origin

These various thoughts bring us to the Darwin of the *Origin.* After twenty years of hard work, he was beginning to think that he was getting on top of the problem of progress. He wrote on the topic, in some detail, to Hooker, at the end of 1858. First, on December 24, he promoted the claim that "species inhabiting a very large area, and therefore existing in large numbers, and which have been subjected to the severest competition with many other forms, will have arrived, through natural selection, at a higher stage of perfection than the inhabitants of a small area" (Darwin 1985– , 7, 221). And then, on December 31, he expanded and qualified his thinking:

> Your letter has interested me greatly; but how inextricable are the subjects which we are discussing! I do not think I said that I thought the productions of Asia were *higher* than those of Australia. I intend carefully to avoid this expression, for I do not think that any one has a definite idea what is meant by higher, except in classes which can loosely be compared with man. On our theory of Natural Selection, if the organisms of any area belonging to the Eocene or Secondary periods were put into competition with those now existing in the same area (or probably in any part of the world) they (*i.e.* the old ones) would be beaten hollow and be exterminated; if the theory be true, this must be so . . . I do not see how this "competitive highness" can be tested in any way by us. And this is a comfort to me when mentally comparing the Silurian and Recent organisms.—
>
> Not that I doubt a long course of "competitive highness" will ultimately make the organisation higher in every sense of the word; but it seems most difficult to test it. (Darwin 1985– , 7, 228–229)

And yet, for one who had worked for so long on the topic, the tone of these letters was not as confident as one might expect. All was not entirely well. The problem of progress has been moved sideways in a not entirely satisfactory manner. As Darwin wrestled to preserve—or, rather, strengthen—the notion of advance, while at the same time stressing even more the tree-like nature of his evolutionary thought, he now offered us what we might call *relative* or *comparative* progress. This was what Darwin called "competitive highness": a process whereby organisms pursue a particular route, trying to better their rivals as they do so. But, in itself, this was not an analysis of an *absolute* notion of progress, which (culminating as it does in humans) was Darwin's original goal—and which appeared still to be his hope.

There is a gap here: Darwin skipped over from relative progress to absolute progress. His implicit argument seems to be that, if you compared specializations between branches, perhaps even thinking of who would beat out whom, you could get some idea of which organisms come out on top on an overall scale. Implicit or not, this was now the official line; notwithstanding the tensions between the two notions of progress, inasmuch as any tack was taken in the *Origin,* it is this. The tree of life is the fundamental metaphor, and—for all of the branching—through it the very picture of evolution is of one stretching ever upward. To support this idea Darwin made only one rather sweeping statement, which presciently, effectively blurred relative and absolute progress: "The inhabitants of each successive period in the world's history have beaten their predecessors in the race for life, and are, in so far, higher in the scale of nature; and this may account for that vague yet ill-defined sentiment, felt by many paleontologists, that organization on the whole has progressed" (Darwin 1859, 267). Apart from a repetition of this sentiment in the conclusion, there was no detailed discussion.

By the third edition of 1861 (Darwin 1959), when it was clear that he was going to be treated as a responsible scientist—perhaps spurred also by a correspondence he had just had with Lyell on the topic—Darwin felt somewhat more confident, at least on the question of relative progress, arguing that selection "will, I think, inevitably lead to the gradual advancement of the organisation of the greater number of living beings throughout the world." He made appeal to the authority of von Baer and referred specifically to the significance of "the completeness of the division of physiological labour." Then, he drew a connection between efficiency, specialization, and highness:

> If we look at the differentiation and specialisation of the several organs of each being when adult (and this will include the advancement of the brain for intellectual purposes) as the best standard of highness of organisation, natural selection clearly leads towards highness; for all physiologists admit that the specialisation of organs, inasmuch as they perform in this state their functions better, is an advantage to each being; and hence the accumulation of variations tending towards specialisation is within the scope of natural selection. (Darwin 1959, 222)

But, although there was a final promise about linking relative to absolute progress, as it turns out this was virtually an end to the matter. The geological section simply reprinted the original passage about the "vague yet ill-defined sentiment" with a comment about specialization thrown in. Somehow, relative progress leads to absolute progress—and if you are not now convinced you will have to take it on trust.

Darwin's Biological Progressionism: Post-Origin

As with evolution in general, there was one species about which Darwin showed no ambiguity: "Amongst the vertebrata the degree of intellect and an approach in structure to man clearly come into play" (Darwin 1959, 221; addition to the 1861 edition of the *Origin*). This sets us up for the preparation and reception of the *Descent of Man,* the work that occupied the last decades of Darwin's life. The theme of progress, particularly at the absolute level, was endorsed and driven home further, as Darwin tried to show how humans represent the high point of the evolutionary picture—and how certain white Anglo-Saxon males represent the high point of the human picture.

In basic theory, there was not a great deal of difference between progress in the *Descent* and Darwin's general treatment of progress; although there was a sense that, toward the end of human evolution, the action started to switch almost completely from the physical to the psycho-social. Enormously impressed by an article written by Wallace in 1864, Darwin wrote: "The great leading idea is quite new to me, viz. that during late ages the mind will have been modified more than the body; yet I had got as far as to see with you that the struggle between the races of man depended entirely on intellectual and *moral* qualities" (letter to Wallace, May 28, 1864; in Marchant 1916, 2, 127). This emphasis on the mind got reflected right into the *Descent of Man,* as did all sorts of Victorian racial and sexual sentiments. "Man is more courageous, pugnacious, and energetic than woman, and has a more inventive genius" (Darwin 1871, 2, 316). Woman, on the other hand, has "greater tenderness and less selfishness" (p. 326). There has simply been more intense sexual selection between men than there has been between women. It is men who have to battle with the elements—and with each other. Naturally, here as elsewhere in life, the effort pays off. "The female . . . ultimately assumes certain distinctive characters, and in the formation of her skull, is said to be intermediate between the child and the man" (p. 317). Remembering Darwin's comments in the early notebooks, we can truly say that the child was the father of the man.

Yet, there were fresh causes for concern and insecurity. Biology may have carried the day, but pervading and unsettling the whole discussion were nagging doubts (in Darwin's mind) about the distorting effect of our technology and the influence of our moral sense, as we turn it back on ourselves. The consequences of modern medicine, combined with acts of sympathy, highlight the problem in stark relief: "With savages, the weak in body or mind are soon eliminated; and those that survive commonly exhibit a vigorous state of health. We civilised men, on the other hand, do our utmost to check the process of elimination; we build asylums for the imbecile, the maimed, and the sick; we institute poor-laws; and our medical men exert their utmost skill to save the life of every one to the last moment" (Darwin 1871, 1, 168).

Fortunately, Darwin was able to comfort himself that not all of the effects of modern society are necessarily bad. The grandson of Josiah Wedgwood felt that he could comfortably endorse the chief features of capitalism. "In all civilised countries man accumulates property and bequeaths it to his children. So that the children in the same country do not by any means start fair in the race for success. But this is far from an unmixed evil; for without the accumulation of capital the arts could not progress; and it is chiefly through their power that the civilised races have extended, and are now everywhere extending, their range, so as to take the place of the lower races" (p. 169). Hence, although in the long term there are challenges to be met and no guarantees of success, in the short term the course of modern society seems fair. All those years of evolution led us right up to the best aspects of middle-class, commercial, Victorian Britain.

Early Influences

Let us move to our second question. Is there reason to think that Darwin was a Progressionist and that this inclined him to biological progressionism? The first part is answered readily. Charles Darwin was virtually predestined to be a Progressionist. He came from the family which worshipped at the shrine of Progressionism. Both of his grandfathers were Lunatiks, and we know what an obsession Progress was with Erasmus. Robert apparently followed very much in the tradition—interested in science, free-thinking, committed to change. And Uncle Jos (Josiah Wedgwood the second), later Darwin's father-in-law, a man who was much respected by Darwin for his integrity, was an ardent Unitarian and so far committed to the philosophy of Progress that he became one of the reforming members of parliament after the Bill of 1832.

Then, perhaps more influential than any, there was Darwin's older brother Erasmus. It was he who inducted Charles into the joys of chemistry—hardly a theory-neutral enterprise, given the importance of that particular discipline in the Industrial Revolution. It was he who went before Charles in shucking the trimmings of conventional Christianity (and may well have had a strong influence on his younger brother, judging from worried comments by Emma Darwin). And it was he who, after the *Beagle* voyage, introduced Charles to intellectual London, including such luminaries as Harriet Martineau and Thomas Carlyle.

None of this is to deny the overlay of the Cambridge years—more Providential than Progressivist, although note the point to be made shortly about Whewell—but Darwin's intellectual and social background was one of Progressionism. Moreover, this was a major factor feeding into his becoming an evolutionist. Evolutionism for Charles Darwin *meant* progress, linked to Progressionism. In the 1830s, that was the very heart of the doctrine. And we see Darwin, whose introduction to evolutionism came precisely through the ideas of Lamarck and Erasmus Darwin and others, accepting that background without question. Indeed, at times Charles Darwin went so far as to echo his grandfather. Thus Erasmus, in a passage scored by Charles:

> Perhaps all the productions of nature are in their progress to greater perfection! an idea countenanced by modern discoveries and . . . consonant to the dignity of the Creator of all things. (Darwin 1803, 54)

And Charles, on progress to "perfect" adaptations:

> How far grander than idea from cramped imagination that God created. (warring against those very laws he established in all <nature> organic nature) the Rhinoceros of Java & Sumatra, that since the time of the Silurian, he has made a long succession of vile Molluscous animals—How beneath the dignity of him, who <<is supposed to have>> said let there be light & there was light.— (D, 36–37)

Finally, note that although Darwin saw man as the top of the chain, he thought it quite possible that some other being, intelligent but non-human, might well have reached the peak.

> What a chance it, has been, (with what attendant organization, Hand & throat) that has made a man.—[any monkey probably might, with such changes be made intellectual, but almost certainly not made into man. (E, 68–69).

This possibility is simply ruled out by the Providentialist, who sees us as made, uniquely, in God's image. It is quite what is expected by the

Progressionist. There can be no unique, predestined candidate for occupancy of the top rung.

Let us move on now to the kinds of influences felt by Darwin after he became an evolutionist and started working toward natural selection. Again, Progress rates high. The awful memories of the French Revolution and the Napoleonic threat had faded; it was clear that the Reform Bill was not paving the way toward general disaster; the railway boom was under way; and the talk was optimistic. "Find Mankind where thou wilt, thou findest it in living movement, in progress faster or slower: the Phoenix soars aloft, hovers with outstretched wings, filling Earth with her music; or, as now, she sinks, and with spheral swan-song immolates herself in flame, that she may soar the higher and sing the clearer" (Carlyle [1834] 1937, 248). Later in life, in his *Autobiography,* Darwin wrote somewhat negatively of Carlyle. But his response of the time was much more flattering: "Erasmus's dinner yesterday was a very pleasant one: Carlyle was in high force, & talked away most steadily; to my mind Carlyle is the most worth listening to, of any man I know" (letter to Emma, 2–3 January 1839; in Darwin 1985– , 2, 155).

Other influences were also at work. In particular, Comte was at full steam and ready to mold British thinkers. Just before he hit upon natural selection as *the* key to evolutionary change, Darwin read a long (and almost generally) favorable review of Comte's views, by David Brewster (1838). This moved Darwin deeply, making him so excited that he got a headache, and led him to write to Lyell: "By the way, have you read the article, in the 'Edinburgh Review' on M. Comte 'Cours de la Philosophie' (or some such title)? It is capital; there are some fine sentences about the very essence of science being prediction, which reminded me of 'its law being progress'" (September 13, 1838; in Darwin 1985– , 2, 104). (See Schweber 1977 for discussion of the origin of the phrase "its law being progress," which comes from Macaulay *via* Lyell. See also Manier 1978.)

But did Darwin learn of Comte's views on Progress? He certainly did, for Brewster first quoted and then endorsed Comte's theory about the three stages of change: theological, metaphysical, and positive: "Although M. Comte has reserved his demonstration of this fundamental law, and his discussion of the results to which it leads, for that part of his work which treats of social physics, yet we have no hesitation in admitting its general accuracy" (Brewster 1838, 281). Underlining the importance of Darwin's reading, we find that in Brewster's discussion of Comte's views on earth history, with its origin in the nebular hypothesis and its eventual end in an engulfment by the sun's flames, there is a remarkable anticipation of a famous picture with which Darwin ends the *Origin.* Brewster wrote:

> In considering our own globe as having its origin in a gaseous zone, thrown off by the rapidity of the solar rotation, and as consolidated by cooling from the chaos of its elements, we confirm rather than oppose the Mosaic cosmogony, whether allegorically or literally interpreted . . .
>
> In the grandeur and universality of these views, we forget the insignificant beings which occupy and disturb the planetary domains. Life in all its forms, in all its restlessness, and in all its pageantry, disappears in the magnitude and remoteness of the perspective. The excited mind sees only the gorgeous fabric of the universe, recognises only its Divine architect, and ponders but on its cycles of glory and desolation. (Brewster 1838, 301)

The concluding words of the *Sketch* of 1842 (repeated with little change in the *Origin*) echo the words of Brewster's conclusion:

> There is a simple *grandeur* in the *view* of *life* with its powers of growth, assimilation and reproduction, being *originally* breathed into matter under one or a few *forms,* and that whilst this our *planet* has gone circling on according to fixed laws, and land and water, in a *cycle* of *change,* have gone on replacing each other, that from so simple an *origin,* through the process of gradual selection of infinitesimal *changes,* endless *forms* most beautiful and most wonderful have been evolved. (Darwin and Wallace, 1958, 87; I have italicized echoing words)

Consider also the above-quoted passage (D, 36), written August 16, four days after reading Brewster. The ideas of Erasmus Darwin were being reinforced with the ideas of Comte.

Paradoxically, Brewster himself was more of a Providentialist than a Progressivist. He was praising Comte to get under the skin of an arch-rival, Whewell. Not that Darwin felt any great need to choose between Brewster and Whewell, for the latter was pushing his strongly Progressionist program in his just-published *History of the Inductive Sciences*—a picture of the development of science from dark, pre-analytic eras to the present and future times of success, understanding, and development. Darwin had read Whewell's work twice by the end of 1838, praised it to the author, and by all accounts found it highly stimulating. Hence, even from his Cambridge connections there was little need to doubt that Darwin accepted the view that humankind is moving toward a state of improvement from any known previously. And this is an impression that would have been strengthened, in that crucial period after the *Beagle* trip, by the growing friendship with Owen. Although the latter was no simplistic P/progressionist, he would surely have stressed the man-directed nature of life's history to his new chum.

Progress was the philosophy of the day; Darwin was submerged in it; and the indications are that it seeped over—more precisely, flooded—into his science. Moreover, at that time people were thinking hard about causes of Progress, and this too is Darwin's concern, especially as reflected by the distinctively Darwinian, relativistic notion of progress. Here we have competition between organisms, striving to stay one step ahead, with consequent divergence and development into all kinds of specializations. Although it took years to develop fully, the idea is there from the beginning in Darwin's thought, and it surely has Malthusian connections since complexification emerges "from the new relations caused by the advancing complexity of others" (E, 95–96). Significantly, countering Malthus's pessimism would have been the optimists who argued that competition leads to specialization and specialization leads to Progress. I refer, of course, to the members of the highly influential Adam Smith school of political economy. And, just as Darwin was influenced in his thinking about upward progress—that is, of a more traditional absolutist kind—by views about an absolute form of Progress, so we find he was influenced in his thinking about relativistic progress by thoughts on a relativistic notion of Progress.

In particular, again just before he hit on natural selection (August 1838), Darwin read and was much impressed by Dugald Stewart's *Account of the Life and Writing of Adam Smith*. What was Darwin to find there? An explicit statement of the Smithian view of Progress!

> I shall content myself, therefore, with remarking in general terms, that the great and leading object of his speculations is, to illustrate the provision made by nature in the principles of the human mind, and in the circumstances of man's external situation, for a gradual and progressive augmentation in the means of natural wealth; and to demonstrate, that the most effectual plan for advancing a people to greatness, is to maintain that order of things which nature has pointed out; by allowing every man, as long as he observes the rules of justice, to pursue his own interest in his own way, and to bring both his industry and his capital into the freest competition with those of his fellow citizens. (Quoted in Schweber 1977, 280)

This was meat and drink for those who derived their riches from trade, and never do forget that Darwin was the grandson of one Josiah Wedgwood and about to become the son-in-law of another. Relativistic progress was nigh tautological.

Later Influences

Let us turn the calendar to the period between the discovery of natural selection and the appearance of the *Origin.* Here, tied in with the "principle of divergence," the notion of splitting or branching was crucial. Organisms compete against each other and success in the competition frequently comes to that which is somewhat more specialized—that is, success comes less often to the all-purpose being, and more often to the being which has specially suited adaptations. Darwin connected this view with the then-current view of progress, which saw specialization as a mark of advance. And the end result is competitive highness or relative progress, or, more precisely, a rethinking and refinement of this notion. Note incidentally that there is an ambiguity at some point here between specialization within a *single organism* and specialization of individual organisms within a *group.* The principle of divergence requires the latter, but Darwin seemed to think that the former obtains also.

Now, what of influences, actual or potential? Darwin himself (in his *Autobiography*) spoke strongly and enthusiastically of the importance of Henri Milne-Edwards, a Belgian/French biologist who had argued that organisms function more efficiently when the various parts do different things. "When, on the contrary, life begins to manifest more complicated phenomena, and the final result produced by the interplay of the different parts of the body becomes more perfect, . . . the life of the individual, instead of being the sum of a larger or smaller number of identical elements, results from essentially different acts produced by distinct organs" (Milne-Edwards 1827; quoted in Schweber 1980, 250). Darwin picked up on this and generalized to a view of the relations *between* organisms.

However, given this influence, even as Darwin was rethinking divergence and relative progress, the cultural input to his thinking remains high. The work of Milne-Edwards, itself, was in language and intent quite explicitly modeled on a theory of (British) Progress. "The principle which seems to have guided nature in the perfectibility of beings, is as one sees, precisely one of those which have had the greatest influence on the progress of human industry and technology: *the division of labour*" (Milne-Edwards 1834, 1, 8). Interestingly, when Darwin did read one opponent of the British view of Progress through competition and the division of labor (de Sismondi 1847), he labeled it "poor"! Through and through, Darwin was as keen on Progress as the next man.

We come to the time after the *Origin,* when Darwin was leading up to the work on humans at the beginning of the 1870s. Relativistic progress took a back seat to absolute progress, as Darwin tried to show that we humans—especially Anglo-Saxon humans—are the apotheosis of the evolutionary process. Once again, extra-biological views on Progress were crucial (Greene 1977). For instance, on his copy of Wallace's important essay, Darwin carefully highlighted some typical Victorian sentiments on P/progress: "the better and higher specimens of our race would therefore increase and spread, the lower and more brutal would give way and successively die out" (Wallace 1864, clxv). Consequently: "The red Indian in North America, and in Brasil; the Tasmanian, Australian and New Zealander in the southern hemisphere, die out, not from any one special cause, but from the inevitable effects of an unequal mental and physical struggle" (Wallace 1864, clxv). This got double line marking in the margin by Darwin and a scribbled comment confirming the point, quite apart from a letter to Wallace saying that this part of the paper was "grand and most eloquently done" (Marchant 1916, 2, 127).

Like sentiments were read and noted on (Darwin's copies of) other writings. Consider the views of one W. R. Greg, essayist and sometime classmate of Darwin: "Here the abler, the stronger, the more advanced, the finer in short, are still the favoured ones, succeed in the competition; exterminate, govern, supersede, fight, eat, or work the inferior tribes out of existence" (Greg 1868, 356). So far, so good. However, in the *Descent* Darwin felt obliged to quote Greg on the dreadful story of the Irish and the Scots.

> The careless, squalid, unaspiring Irishman multiplies like rabbits: the frugal, foreseeing, self-respecting, ambitious Scot, stern in his morality, spiritual in his faith, sagacious and disciplined in his intelligence, passes his best years in struggle and in celibacy, marries late, and leaves few behind him. Given a land originally peopled by a thousand Saxons and a thousand Celts—and in a dozen generations five-sixths of the population would be Celts, but five-sixths of the property, of the power, of the intellect, would belong to the one-sixth of Saxons that remained. In the eternal 'struggle for existence,' it would be the inferior and *less* favoured

> race that had prevailed—and prevailed by virtue not of its good qualities but of its faults. (Darwin 1871, 1, 174)

Fortunately, things are, perhaps, not always quite as bad as they may seem at first. In this particular case, Darwin consoled himself that high breeding rates do not always produce the most lasting children. The Scots may win after all.

Then, there is Walter Bagehot telling us that civilization is an advantage in life's struggles: "what makes one tribe—one incipient tribe, one bit of a tribe—to differ from another is *their relative faculty* of coherence" (Bagehot 1868, 456, Darwin's emphasis). This was likewise noted by Darwin and he wrote in the *Descent:* "Obedience, as Mr. Bagehot has well shewn, is of the highest value, for any form of government is better than none. Selfish and contentious people will not cohere, and without coherence nothing can be effected . . . Thus the social and moral qualities would tend slowly to advance and be diffused throughout the world" (Darwin 1871, 1, 162–163).

Finally, let me refer to one David Page, author of *Man: Where, Whence, and Whither; Being a Glance at Man in His Natural-History Relations.*

> Bound by the obligations of enlightened humanity, the white man may and must endeavour to civilise and ameliorate the condition of his less enlightened and coloured brethren; but no humanising scheme, however anxious or earnest, can ever arrest that law which has destined the progression of the human race—the extinction of the inferior, and the rise and spread of the higher varieties. Humanly speaking, it is only in this way that the progressive advancement of mankind can ever be attained; rationally, it is the only method the human mind can comprehend and appreciate. (Page 1867, 92)

This passage was marked by Darwin and annotated "Refer to." Page's tension between our obligations to the "lesser races" and the inevitability (and biological desirability) of the struggle just about sums up Darwin's work. By this point, biological progress and social Progress had become, if not virtually identical, so intertwined as to be impossible to separate.

The Evidence

We come to the question about the evidence on which Darwin based his theorizing. First, we ask on what foundation rested the belief that biological progress of some kind actually occurred: in Darwin's case that there was a kind of branching process with some branches proving more significant than others, and that ultimately one branch—that leading to humankind—triumphed above all others. Here, obviously, paleontology is the most significant potential source of evidence. Although Darwin tended to use fossils for geological ends (proving the Lyellian steady state), he always kept well abreast of modern research. Yet, it seems not unfair to say that the Darwin of the 1830s was making houses without stones, or rather progressive speculations only with stones (Bowler 1976). We know that there was some evidence of a sequential succession of fossils—Darwin himself was impressed by the fossils of South America—but there was not so much connection that Lyell could not argue against any kind of progress or evolution. Certainly, it was not primarily the fossils which made Darwin an evolutionist, or a progressionist.

Empirically more pertinent was comparative morphology, particularly as represented by Darwin's massive study on barnacles. One must move with caution here since, given that this study occupied the years at the end of the 1840s and the beginning of the 1850s, publicly Darwin was working in a non-evolutionary mode. There are no overt phylogenetic reconstructions, and even when later scholars have provided such reconstructions based on Darwin's work, they are bush-like and not at all obviously indicative of progress (Ghiselin and Jaffe 1973; Newman 1993). Yet it is clear that Darwin was using the barnacle work to articulate and confirm his progressionist beliefs, though not so much at the general level. On reading Milne-Edwards (at the end of 1846) he commented in a private reading notebook: "Barnacles in some sense, eyes and locomotion, are lower, but then so much more complicated, that they may be considered as higher" (quoted in Richmond 1988, 392). At this point, Darwin was sufficiently puzzled as to conclude that perhaps one had better "leave out the term higher and lower." Indeed, this ambiguity went right through to the published work, dedicated incidentally to Milne-Edwards. Note, however, that Darwin's worries were more with the barnacles as a subject than with the propriety of the notion of progress as such: "On the whole, I look at a Cirripede [barnacle] as a being of a low type, which has undergone much morphological differentiation, and which has, in some few lines of structure, arrived at considerable perfection,—meaning, by the terms perfection and lowness, some vague resemblance to animals universally considered of a higher rank" (Darwin 1854a, 20).

At the particular level, however, Darwin thought the barnacles to be brilliant confirmation of his progressionism. He had always taken sexuality as a progressionist triumph of what he came to see as a division of labor, with a move (analogous as he thought to fetal development) from hermaphroditism to full sexuality (C, 167). Working on the barnacles, Darwin discovered what he called "complemental males" (minute, parasite-like males on hermaphrodites) that he took to be the evidence of the evolution of full (male/female) sexuality, something which occurs (albeit rarely) in some species of barnacle and which likewise involves very small and primitive males. In barely guarded language, Darwin wrote of "how gradually nature changes from condition to the other,—in this case from bisexuality to unisexuality" (p. 29). Many years later, when his evolutionism was open and public, Darwin tied in his discovery explicitly with the fact that "a division of physiological labour is an advantage to all organisms" (Darwin [1873] 1977, 2, 180).

The barnacle fossil evidence was sparse, and in any case would hardly prove a general point. At this level, through the 1840s, the evidence for some sort of overall sequence—call it "progressive," if you will—went on growing. But, non-evolutionists could continue to point to the supposed inconsistencies with any developmental hypothesis, especially the gaps and the high status of the earliest representatives of new forms. Then, however, through the 1850s there was a growing realization (from and through people like Owen) that perhaps the fossil record is not so very antithetical to evolutionism—which, remember, everyone equated with progressionism in one form or another. It is true that the gaps in the fossil record remained. Yet, some holes were plugged. The greatest find came after the *Origin,* when the bird/reptile *Archaeopteryx* was discovered and identified (Owen 1863; Huxley 1867–68, 1868). Of course, more significant than any of this was (by the 1850s) the unambiguously branching nature of the fossil record, even if it was branching with—if not a purpose—at least with a rhythm. One starts with a fairly general form and then goes off in various directions, adding adaptive specializations. Furthermore, the early forms tend to be more embryo-like, a point that Darwin seized on: "If it should hereafter be proved that ancient animals resemble to a certain extent the embryos of more recent animals of the same class, the fact will be intelligible" (Darwin 1859, 267).

What of our own species? There was some evidence, recognized as such by the end of the 1850s, that the human species is very old—far older than allowed by the traditional 6,000 years of Genesis (Oakley 1964;

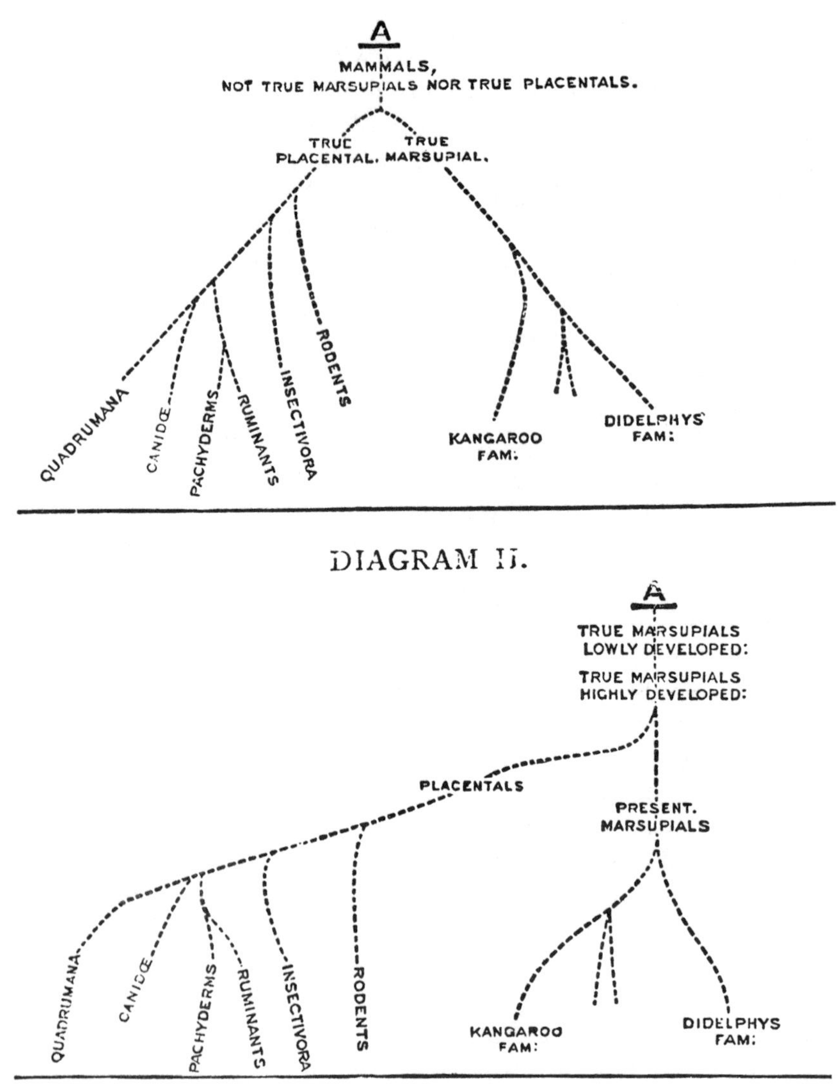

Two hypotheses by Darwin on the ancestry of humans. (Redrawn from a letter to Charles Lyell, September 23, 1860; in Darwin 1887.)

Gruber 1965). Human remains had been discovered along with the remains of extinct organisms, and people appreciated that this points to an early human beginning. The fossil evidence of human evolution was less helpful, and really did not make the case at all. Neanderthal man was discovered (in Germany) in the 1850s, but the opinion even of Darwin's supporters was that it is part of our own species rather than a genuine "missing link" (Huxley 1863).

For Darwin this absence of human fossil evidence was hardly devastating to his general case for human evolution; it merely pointed (as he and everyone thought) to our comparatively recent arrival on this globe. Darwin's main argument for human evolution, and for our close relationship with the apes, was grounded in comparative anatomy and in embryology and so forth. He did indeed speculate (in discussion with Lyell) about the actual course of human evolutionary history; but the details of this were never that significant for his overall position. Yet, as we swing next to look at Darwin's thinking on the causes of progress, we should not let him escape too easily here. *Qua* progressionist, he held to the claim that humans come top. If we think purely in terms of the evidence available to Darwin about paths and histories, it is clear that his human-centered progressionism transcended the available data. Even if it is agreed that we humans are among the final products of the evolutionary process, this in itself does not prove progress. Degenerate barnacles are also among the last in the process. At most, therefore, Darwin could make a rough claim for consistency between the evidence and his theory. There was certainly no overwhelming proof.

What about the question of causes? Darwin thought that evolution results in relative progress, marked in terms of the division of labor (and flagged by such things as complexity), and that this somehow shifts over into absolute progress. Humans—particularly white, European, male, capitalist humans—come top. The issue here is whether natural selection, with a little help from secondary causes, particularly sexual selection, can do all that is asked of it. Or, rather, whether the evidence behind selection can do all that is asked of it. And the candid answer is that, judged by today's standards and even more by the standards of Darwin's day, to arrive at the conclusions that were reached there has to be a huge amount of (shall we say) "stretching" of the evidence to meet the theory. Beginning with the idea of a struggle, going on to selection, speaking to adaptation, and triumphing through a division of labor—all this meant going beyond the given, in dramatic ways. I am not criticizing Darwin from today's perspective and even less from that of his own day. The mark of the genius is to see a pattern among the scattered pieces. I am just stating a fact.

Start with the struggle, which Darwin got from Malthus. From the beginning of the nineteenth century, a steady stream of critics of Malthus had argued that there was neither the explosive potential in population growth, nor the pessimistic limitations on food or space, to justify his conclusions about human society (Inglis 1971). Even Malthus himself had pulled back somewhat, agreeing that it is possible that humans control their growth and avoid the struggle (Malthus [1826] 1914). Truly, Darwin could point to specific instances of potential growth and constraint. But, as he himself was forced to admit, this does not support a literal claim: "I should premise that I use the term Struggle for Existence in a large and metaphorical sense, including dependence of one being on another, and including (which is more important) not only the life of the individual, but success in leaving progeny" (Darwin 1859, 62). And if the concept was difficult to grasp, the evidence for its universality was even more elusive. Who is to say that organisms might not regulate their own population numbers?

If cautionary notes are needed in treating of the evidential basis for the struggle, they are required even more when we turn to the variation Darwin postulated to make the struggle biologically meaningful and (especially) to provide an ongoing supply of fuel or "raw stuff" for evolution. By the time of the writing of the *Origin,* thanks especially to his work on barnacles, Darwin was starting to feel quite confident that large-scale variation of the needed type does indeed exist; but, again, this was hardly a simple function of reading off the "facts" from the book of nature. Nor was natural selection itself such a simple function. Indeed, Darwin had virtually no direct evidence of the process. And, as numerous critics pointed out, the analogy from artificial selection was shaky (to say the least). No matter how much you select, you are not about to turn a cow into a pig.

Darwin was pushing out into the unknown, even with his metaphor of selection: "It has been said that I speak of natural selection as an active power or Deity; but who objects to an author speaking of the attraction of gravity as ruling the movements of the planets? Every one knows what is meant and is implied by such metaphorical expressions; and they are almost necessary for brevity" (Darwin 1959, 164–165; from the third edition of the *Origin,* 1861). Of course, what was at stake here was Darwin's commitment to the design-like nature of the world, together with his equally strong commitment to a God who works only through unbroken law. But, whether this is legitimate or not, it was a vision that Darwin was imposing on the organic world, not a deduction made by reasoning from the evidence.

Finally, brushing over such matters as Darwin's division of selective forces into natural and sexual—surely primarily a result of his seeing human selection as being directed toward both utilitarian and recreational ends—we see how, given the view that organisms develop in accordance with a design or function, Milne-Edwards's division of labor recommended itself so strongly to Darwin. He needed a solution to divergence. But the particular metaphor he embraced is part and parcel of the picture of the world as a machine or a factory or a humanly created society. It so happens that Darwin ran a number of experiments proving to his own satisfaction that the same plot of land can support far more life when it carries a diversity of organisms than when it carries organisms of one or a few types. Essentially, however, Darwin took from this experiment a vision imposed rather than a vision gleaned. Whatever the pertinence of the discoveries about sexuality, the barnacle work confirms this point: "One naturally wishes to ascertain how far Cirripedia are highly or lowly organised and developed; but in all cases this, as it seems to me, is a very obscure enquiry" (Darwin 1854a, 19).

So much for Darwin's mechanisms. What of the uses to which he put them? Relativistic progress may be an important aspect of reality. We shall learn that many today believe that it is. But, whatever its current status, Darwin gives us little reason to believe it a common empirical phenomenon. My hunch is that his faith in its existence was primarily a function of his knowledge of what occurs in the breeders' world, where competition can lead to ever-exaggerated features. His experience with breeding was spliced with an understanding of the course of manufacturing and technology, where again one finds the "evolution" of ever-developed contrivances—combined of course with an ever-greater reliance on a division of labor.

The evidential case for the shift from relativistic progress to absolute progress is even less strong. Judged from his references, Darwin (1859) clearly had in mind the supposed overall superiority of organisms forged from a greater struggle over those organisms forged from a lesser struggle. He specifically mentioned the triumph of Asian organisms over Australasian organisms (p. 106). But there is not much in the way of definitive proof here, and certainly not of why we humans come out on top of all. What we can say specifically in the case of humans is that Darwin did not do any original research of his own, preferring to rely on his—admittedly extensive—literature search. Moreover, while it is true that Darwin's view on humankind and its nature was the common position in mid-Victorian Britain, it was certainly not the exclusive position. John Stuart Mill, for one, gave a major role to environmental and social factors in determining racial and sexual differences (in his *Subjection of Women*). Darwin explicitly considered Mill's thinking and then rejected it (Russett 1989, 2).

Of course, his rejection does not in itself imply that Darwin exclusively was guided by cultural factors. Even less does it imply that Mill was right and Darwin was wrong. But in the absence of references to detailed personal empirical investigations into human nature, one can be forgiven for suspecting that non-epistemic values were busily at work in Darwin's thinking on this matter.

Darwin and National Strands of Progress

As in all things, with respect to Progress Darwin wove his rope from many strands. First, from France: although Darwin kept Lamarck at arm's length, one should not underestimate the influence of that particular author. Darwin took in Lamarck's ideas indirectly from Grant and from the *Principles of Geology* (despite Lyell's intentions), and more directly from personal reading of Lamarck and others, like Geoffroy. Apart from anything else, Darwin always accepted the key Lamarckian mechanism of the inheritance of acquired characters. And this is not to mention additional French influences, like Comte, Milne-Edwards, and others. Likewise, although Darwin was no great Germanophile—he always had great trouble with the language—socially and scientifically there were influences. People in brother Erasmus's set, like Thomas Carlyle, were great enthusiasts for German culture. Also there was Whewell and his almost Hegelian view of the history of science, not to mention Owen and, either directly or through him, the important work of such scientists as von Baer. The unity of type, so important in Darwin's thought, clearly owes much to German idealism—as do many other elements that found their way into the *Origin*. In fact, at times Darwin wrote as though he were a full-blown recapitulationist: "As the embryo often shows us more or less plainly the structure of the less modified and ancient progenitor of the group, we can see why ancient and extinct forms so often resemble in their adult state the embryos of existing species of the same class" (Darwin 1959, 704, 6th edition).

Yet, one must beware of over-stressing the foreign influences. Darwin was no crypto-*Naturphilosoph*. His causal explanation of the relation between ontogeny and phylogeny had to yield an Agassiz-like conclusion—with development through adult rather than embryo forms—for he believed that evolution often involved the adding of another stage onto the adult and squashing all of the earlier forms back somehow in the growth period (Richards 1992). But this conclusion was not something imposed by Darwin's metaphysics. Owen was a transcendentalist at core. Darwin never was. Again and again he denied the teleological upward momentum of the German idealist. So strong were the denials, in fact, that people have often (mistakenly) thought Darwin to be no true progressionist at all.

When all is said, other contributions to Darwin's ideas pale beside the British element. Whatever the foreign sources, the notion of relative Progress was deeply embedded in his own culture. Darwin's picture was one of hard, slow, physical grind—of a real groping gradually through much sweat and toil, often unsuccessfully, toward a better state. And this state, ultimately as absolute as one could wish, was mid-Victorian middle-class society. In the fullest sense of the word, Charles Darwin was the heir of the eighteenth-century British Enlightenment—David Hume, Adam Smith, and above all others, Erasmus Darwin and his circle. As Marx famously remarked in a letter to Engels, Darwin's genius was to transfer the British industrialist's philosophy—which had at its heart a faith in Progress—right into the biological world (June 18, 1862; in Marx and Engels 1965, 128).

Darwin's (Limited) Achievements

Let us put Darwin and his achievements into context, beginning with the divide between professional and popular science. Already, you will have sensed some ambiguity and tension here. Darwin was no mere amateur but, undoubtedly, he was in respects an old-fashioned gentleman naturalist, a "position" made possible by his great family wealth. He could afford to spend five or six years on the *Beagle,* taking time for land trips as he felt the urge. He could devote time and effort to thinking about questions on such topics as geographical distributions simply because they interested him. On his return to England, there was no pressing need to find employment. Although he had to work privately, Darwin had the time and independence to push ahead on his evolutionary speculations. He was not torn as Owen was, having to feed to his masters grist for their mill, however he may have disagreed with it. Darwin had the independence and psychological disposition of the privileged to plough his own furrow. Additionally, most significantly, he had the space to hone his literary skills writing up his diary as a travel book—and because he was so very clearly trying to please people like his father and his uncle, who were sponsoring him, he fell naturally to trying to address a popular audience. This was where his patronage lay.

But Darwin also had a place in the world of professional science, and this also went into the success of the *Origin*. His training at Edinburgh and Cambridge showed through again and again—quite apart from the general Scottish connection which may have fueled his initial enthusiasm for evolutionism itself. Chambers was a powerful writer, but he was just not scientific in the way that Darwin was. Nor was he sensitive to the nuances of (such topics as) embryology, as was Darwin. It is true that Darwin was doing things with the ideas of von Baer and Agassiz that they could never accept; nonetheless, they and others knew that Darwin's use demanded the respect one gives to a professional.[2]

Darwin's professional standing was enhanced by the barnacle study, as well. One might question the worth of eight years of detailed taxonomy, but it did rivet Darwin's position as a major biologist, complementing his position as a geologist. He became a Royal Society prize winner and council member. Not that, by the 1850s, this was a matter of special note. From the very beginning Darwin had been socially centrally integrated within the professional realm. He had trained with the right people, and he had entered the right network when he returned from the *Beagle* voyage. He was accepted straight into the Geological Society, onto the council, no less; he was elected to the Royal Society and published a major geological paper in its *Transactions* (thanks to a very favorable review by Sedgwick); and his career was promoted by powerful figures like Whewell and Henslow (Rudwick 1974, 1986; Ruse 1979a). Even though these people almost all failed to become evolutionists, they launched Darwin and gave him status as a serious scientist. This in itself gave the *Origin* an authority that *Vestiges* never had. Significantly, when he published, Darwin sent off complimentary copies (with nice notes) to Sedgwick, Whewell, Herschel, and Henslow. He did not waste his books on popular writers or bishops.

Darwin's ambiguous status translates immediately into epistemological factors, especially in the *Origin*. On the one hand, it is a popular book, the product of one skilled at playing on human interests and emotions. I have noted the brilliant use of friendly metaphors and analogies. Moreover, like his grandfather's work, the *Origin* is deeply anthropocentric: "One reason is that Darwin constantly humanizes animals, in what used to be called the 'pathetic fallacy': male alligators have courtship rites 'like Indians in a war dance'; frightened ants 'took heart' and . . . the surviving animals are not only vigorous and healthy, but 'happy'" (Hyman 1962, 31). Just the sort of thing that the regular reader likes, and marred by none of the things that the regular reader does not like—such as mathematics.

On the other hand, the book was as skillfully constructed in a professional fashion. Content apart, the norms of good (mature) science were ever foremost. The *Origin* may not have presented an axiom-law system, but there is a nod in that direction, particularly in the way that Malthus's ideas are integrated within the text (Ruse 1975b). More particularly, there was a clever playing off of the empiricist methodology of Herschel against the rationalist methodology of Whewell. Thanks to the influences at Cambridge, Darwin knew and had read the chief theorists on the subject of scientific methodology (Ruse 1975c). He knew of the debate between Herschel, who said we must argue *from* sense experience, and Whewell, who said we must argue *to* sense experience (Herschel 1831, 1841; Whewell 1831; 1840). And Darwin knew of the divide over *verae causae,* Herschel arguing that we know a *vera causa* only when we have experienced it or something closely like it (that is, through analogy), and Whewell arguing that we know a *vera causa* through its effects (that is, through the circumstantial evidence or "consilience").

Darwin so took the methodological dicta to heart that he incorporated into the *Origin* arguments suggesting that natural selection is a *vera causa* both *qua* empiricist criteria and *qua* rationalist criteria! The argument from artificial selection was an argument from the immediately experienced, and the argument from selection across the total biological range was an argument to the experienced: a paradigmatic example of a Whewellian consilience. Moreover, Darwin was fully aware that the methodologists had been forced back from their ban on hypotheses by the success of the wave theory of light, and as a result he was ever ready to justify the slip of his own theorizing into the gap which had been thus opened: "In scientific investigations it is permitted to invent any hypothesis, and if it explains various large and independent classes of facts it rises to the rank of well-grounded theory. The undulations of the ether and even its existence are hypothetical, yet every one now admits the undulatory theory of light." Likewise for natural selection: "If the principle of natural selection does explain . . . large bodies of facts, it ought to be received" (Darwin 1868, 7–8).

Conceptually, the *Origin* is a work of genius, and yet, with respect to status, one must allow also that the *Origin* is deeply ambiguous. Its ambiguity is most evident in the treatment of progress. Because of the anthropocentrism, not to mention metaphors like the tree of life, progress comes with the territory: "Man is always on Darwin's mind as he talks of the lower orders. The criteria for an 'advance in organization' among the vertebrata are 'the degree of intellect and an approach to structure of man'" (Hyman 1962, 31). Nevertheless, Darwin the professional knew that generally cultural values have no place in science, and that up to this point P/progress had been professional poison—in Britain particularly. In marginal notes he made on his copy of *Vestiges,* he warned himself specifically against using the idea publicly. This explains the attempt to conceal the idea, and the fact that Darwin spent more time than anyone else we shall encounter in this book in attempting to provide a culture-neutral analysis of the idea. Why else strive so hard to reduce the value notion of absolute progress to the evaluative notion of comparative progress?

These comments, of course, apply most particularly to the first edition of the *Origin.* I do not want to anticipate the discussion of future chapters, but it is a highly suggestive fact that, by the third edition of the *Origin,* Darwin himself became more relaxed about overt progressionism. Perhaps his change of heart may be explained by the ambiguity of the reaction to this ambiguous book. Even though Darwin slid together the case for the *fact* of evolution and the case for the *theory* of evolution, we have seen that the common response was to separate these two, with much more enthusiasm being shown for the former than the latter. But why was there any positive response, given the negative factors? In part, we have the full reason. People, including professional scientists, became evolutionists because single-handedly Darwin had moved evolution up from the status of pseudo-science, or from its epistemological equivalent. No longer was evolution just a theory that ignored, flagrantly, the norms of good science, an idea that existed only because of a cultural value. But the value—the value of Progress—was not absent from the *Origin,* and this suggests that the enthusiasm for evolution lay also in the truth that *qua* fact the book was written at the level of popular science—a level which could include culture.

Perhaps Darwin himself sensed this and, always willing to grab what he could when he could, exploited the situation to push a value which (after all) he shared as much as anyone else. Not that this is to deny that, in the long run, he himself wanted more. That was the point of natural selection. Darwin was after a fruitful and ongoing mature science, a world where natural selection is the tool of the working biologist. Which fact raises the question of *discipline* building, for Darwin was sufficiently professional to know that you need a group to take your ideas and work with them as a functioning theory or paradigm. And here we see that, for all the years he spent as a semi-recluse, often on his sickbed, the social side to Charles Darwin was no less significant than the intellectual side.

In the years before the *Origin,* he knew that his ideas would need powerful friends and supporters—and if they were to be taken seriously as professional science, they would need powerful professional scientific friends and supporters. The botanist Joseph Hooker was a firm ally and already close to being a recruit. Lyell was always doubtful, but better sympathetic than hostile, and so was treated appropriately. Asa Gray, fast becoming the dean of American botanists, was approached and encouraged. And then, above all, there was Thomas Henry Huxley, bursting onto the scientific scene in the 1850s. In respects, Owen ought to have been Darwin's spokesman, but Darwin shrewdly saw that intellectually, emotionally, and socially this could never be. The young Huxley was therefore petted and flattered—even though he had just written an absolutely scathing review of *Vestiges,* and through it a cruel attack on Owen, and despite the Darwin family's opinion that there was something a little crude and ill-bred about Huxley. And, significantly, the insecure but ambitious Huxley responded, coming into Darwin's camp. His enlistment paid an immediate dividend on Boxing Day, 1859, when Huxley wrote a highly laudatory review of the *Origin* in that bastion of the establishment, *The Times* (Huxley [1859] 1894). It was things like this which helped to lift evolutionism out of the slough of quasi-science.

But Darwin knew that he could not stop here. Reviews in the *Times,* even when penned by professional biologists, do not make for professional/mature science. Darwin had to keep working. First, he had to develop the theory at the conceptual level. After the *Origin,* Darwin's best-known production was the *Descent of Man.* That Darwin should have written a book on our species was hardly surprising, given the general interest in the topic. Nor is some of Darwin's other work all that surprising. For instance, he produced a two-volume tome on *Variation in*

Animals and Plants Under Domestication. This was part of an effort to flesh out the message of the *Origin.* But even setting aside the already-noted huge discussion of sexual selection in the *Descent*—huge to the point of oddity, even if you accept the reasons why Darwin provided it—some of Darwin's other work seems downright strange. Straight after the *Origin,* he wrote a little book on orchids (1862). Then there were books on climbing plants (1865) and insectivorous plants (1875) and earthworms (1881), among others. The choice of topics seems almost perverse.

Yet, there was order in the apparent randomness. These books, together with related papers and perhaps even the bloated discussion of sexual selection, were part of Darwin's plan to show his fellow professionals what an active, ongoing program of selection-based evolutionary biology would look like. He wanted to give exemplars drawn from a potential discipline of evolutionary science—akin, say, to comparative anatomy or (in our own day) biochemistry—which would use natural selection, much as Newtonians use gravity, as a foundation or point of departure. Consider a letter Darwin wrote to his publisher, John Murray, justifying his orchids book. Revealingly, Darwin had earlier characterized the book as a kind of secular *Bridgewater Treatise,* in that he was showing how selection could effect what had previously been ascribed to Design (letter of September 21, 1861; in Darwin 1985– , 9, 273).[3] Now he said: "I think this little volume will do good to the Origin, as it will show that I have worked hard at details, and it will, perhaps, serve [to] illustrate how natural History may be worked under the belief of the modification of species" (September 24, 1861; in Darwin 1985– , 9, 279).

In a way, therefore, although the orchids book was backward-looking in its teleology, it was also forward-looking in a manner that something like the *Variation* was not. And this is a point backed by the other side to the post-*Origin* Darwin, the political Darwin. Most particularly, we see this in the way Darwin cherished and supported others who actually used selection as a tool of inquiry. Henry Walter Bates, Wallace's traveling companion in the Amazon in the 1840s, wrote a paper showing how selection explains why certain non-poisonous butterflies mimic others which are poisonous (Bates 1862). Darwin's enthusiasm for the piece knew no bounds. He himself went so far as to write an anonymous advertisement for Bates's paper in the *Natural History Review,* a short-lived magazine then being edited by Huxley (Darwin 1863). And Darwin could not have been more explicit in his belief that Bates's virtue was in showing just how a naturalist using selection could solve nature's puzzles:

> Mr. Bates concludes that in every case the *Leptalis* originally varied; and that when a variety arose which happened to resemble any common butterfly inhabiting the same district (whether or no that butterfly be a variety or a so-called distinct species) then that this one variety of the *Leptalis* had from its resemblance to a flourishing and little persecuted kind a better chance of escaping destruction from predacious birds and insects, and was consequently oftener preserved;—"the less perfect degrees of resemblance being generation after generation eliminated, and only the others left to propagate their kind." This is Natural Selection. (Darwin 1977, 2, 91)

Darwin's enthusiasm was paralleled by other factors, particularly where he was able to use his great financial power to further his own ends. He was ever happy to dollop out sums of cash to support the kind of science of which he approved. There was a subsidy to work on the selection of hardy potatoes (Darwin and Seward 1903, 1, 374). There was money for the fare to India of a valued but impoverished experimenter (2, 327). There was more than one subvention to print favorable reviews or defenses of his work (Darwin 1887, 3, 86 and 145). There was a grant for equipment at a marine station. There was a very nice sum to help Huxley through a breakdown (Huxley 1900, 1, 366). The ample pocket of Charles Darwin lay ready and open.

A functioning, professional, selection-based discipline of evolutionary studies was Darwin's dream. Unfortunately, it was a dream which was not realized. Darwin, on his own, could not make a new professional discipline. Certainly he did not make such a discipline. In the early post-*Origin* years, we do not find the creation of societies, journals, academic units, professorships devoted to evolutionism—the mark of the birth of a professional discipline. In part, perhaps, this failure was a function of the state of the art, or an indication that people did not find selection compelling as a mechanism. But not entirely. After all, which is cause and which is effect? If there had been such a discipline, selection might have been more compelling. Rather, the failure in execution stems from the same ambiguity as before: Darwin was part of the professional community, and yet, both a strength and a weakness, he was rich enough to stay aloof from it.

On the one hand, conceptually, the very success of the *Origin* as popular science took the edge off its merits as professional science. It did not compel professional attention exclusively. The lack of mathematics, for instance, meant that it was not opaque. It was not that awesome either. Huxley's comment about natural selection was: "How extremely stupid not to have thought of that!" (Huxley 1900, 1, 170). The work was greeted with envy, not wonder filled almost with fear. Owen's review combined jealousy at Darwin's favored station—"Of independent means, he has full command of his time for the prosecution of original research"—with a tainting nod at Darwin's (earlier) non-scientific writings—"Mr. Charles Darwin has long been favourably known, not merely to the Zoological but to the Literary World, by the charming style in which his original observations on a variety of natural phenomena are recorded in the volume assigned to him in the narrative of the circumnavigatory voyage of H.M.S. Beagle"— and a sneer at the non-professionalism of the new book—"The same pleasing style which marked Mr. Darwin's earliest work, and a certain artistic disposition and sequence of his principal arguments, have more closely recalled . . ."—etc., etc., etc. (Owen 1860, 487).

On the other hand, socially, there was likewise something missing from Darwin's handling of his own achievement. One does not have to deny the authenticity of Darwin's illness, although one is much aware that he was rich enough to indulge it, to say that he made the illness work for him. It gave him an excuse to avoid meetings and time-consuming commitments and the like and granted him guilt-free opportunities to follow his own intellectual interests, as and how they pleased him. But, with the freedom from certain worries went a lack of that bloody-minded strength and organizing ability that one finds in the best academic empire builders. It tells much that, when the idea was raised that the already mentioned (later India-bound) experimenter, one John Scott of the Glasgow Botanical Gardens, might move to Down to work directly under the Master's supervision, Darwin veered away in terror at the possible strain (Darwin and Seward 1903, 2, 39).

Darwin would join in the work of discipline-building, but only if he could keep his distance. And this was a pattern, the down side to Darwin's financial independence. He never had to go out and fight for his own piece of the turf—his own supported place in the world of science. Darwin did not have to build his own unit with backing for himself and his staff and his students—a unit where he was just simply going to impose his own will and set up selection experiments and like studies which would then show the scientific world how selection-based evolutionary science can be done. Unfortunately, because he did not have to, he did not do so. Perhaps, ultimately, for all his professionalism, Darwin had that crippling English disdain for winning at any price.

Still, it must be recognized that after Darwin, evolution was pseudo-science no longer. But it was not yet mature science; it still had Progressionism right at its heart; and there was no professional discipline. What happened next? What of professionalism? What of evolutionism? What of the role of values, epistemic or otherwise? Let us move at once into the second half of the nineteenth century.

5

Evolution as World View

Before we go further, I must tackle a major methodological problem. To date, within bounds it has been possible to discuss, if not every major evolutionist or critic, a good cross-section. Now as we move toward the closing decades of the nineteenth century, this will be impossible. Evolution at some level or another became, by that point, the background belief of almost all working biologists. Selection is essential, and to this end, I am going to be guided by the fact that my goals are philosophical. My story is intended to test theses about the status and persistence of Progress as a value in science. The endpoint to this study must be with those areas of evolutionary biology, active today, which we judge to be the best and strongest. Any other destination renders the whole exercise worthless. But, what is the best of today's evolutionary work? Here, at least, the answer is easy, for no one could dispute that Britain and America, by far, lead the way. It is true that other countries make significant contributions—Japan, for instance, in the theoretical world—but Anglo-American evolutionism is the biggest, the best, the most mature. It is here that our study must end.

This being so, on the grounds that the work in these two countries since Darwin is the most direct route to the present, I am now going to exclude discussion of activity elsewhere in the world, except inasmuch as it impinges on British and American evolutionism. Note that this is essentially a pragmatic decision and does not judge of worth at all points and at all times; although, fortunately, after Darwin published, some national differences rather support our new criterion of selection. In the world of evolution, there was a significant shift sideways—literally as well as metaphorically! France almost entirely took itself out of the picture (Conry 1974; Stebbins 1974; Mayr and Provine 1980; Corsi and Weindling 1985). Balancing this loss, America started to matter. The exception is Germany, for in the second half of the nineteenth century it was indisputably the world leader in the biological sciences, and this excellence extended to evolution. "Darwinismus" became virtually a religion within Bismarck's newly unified state (Richards 1987; Bowler 1984; Nyhart 1986, 1995).

Not that I want to suggest that German evolution became merely a flag for social philosophy—although it certainly was that. Ernst Haeckel, who has fair claim to being the world's most important (certainly, most influential) evolutionist in the post-Darwinian period, was not just a spokesman for evolution as general world view and philosophy but, particularly in the early days (the 1860s), was a serious scientist in his own right, doing detailed and praiseworthy studies of marine invertebrates (Haeckel 1862). As Haeckel's ideas make their way to Britain, I shall be saying more about them; already, in fair extent we can grasp their essence, the main point of which is that, despite the propaganda, his work was not desperately Darwinian. Having had as teachers Johannes Müller, who preached an updated Kantianism, and the botanist Alexander Braun, ardent transcendentalist—"The individual essences of Nature are links in the development of that Kingdom of nature to which they belong, and in the widest sense, links in the development of the totality of natural life" (Braun 1853, 413)—brother-in-law of Louis Agassiz, and likewise student of Schelling and Oken, Haeckel himself stood very much in the footsteps of the *Naturphilosophen* (Richards 1987; Russell 1916; Boelsche 1909). It was he who, more than anyone, pushed the identity in the laws of growth of the individual and of the group, expressed in his so-called biogenetic law: "*Ontogeny is a short and quick repetition, or recapitulation, of Phylogeny, determined by the laws of Inheritance and Adaptation*" (Haeckel 1866).

Let me stress that I do not thus emphasize Haeckel's roots in *Naturphilosophie* to insinuate that he and his followers were not genuine evolutionists. Obsessed with phylogeny tracing, they were nothing if not that. In the land of its origins, the stage of ethereal German idealism had passed. Nor do I decline to treat of post-*Origin* Germanic evolutionism in

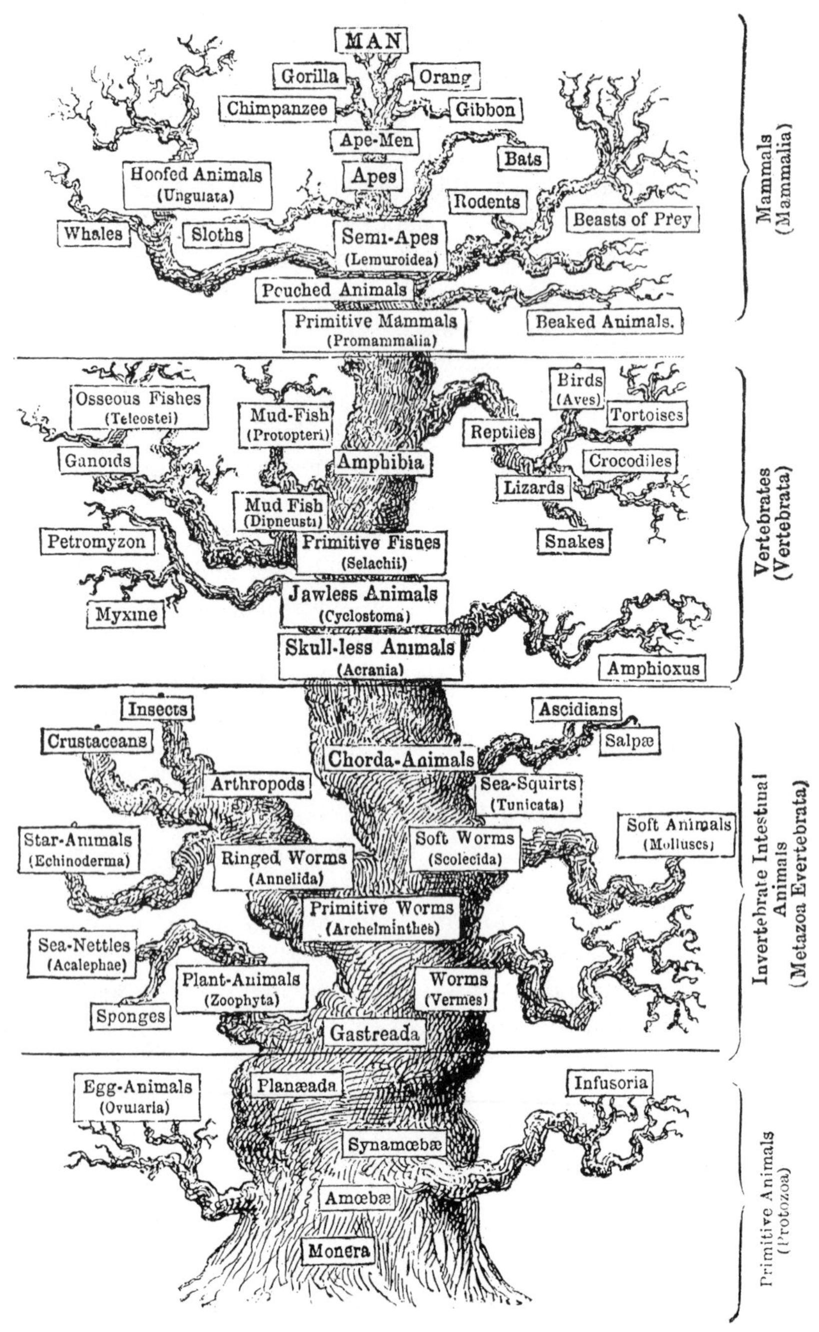

The tree of life as drawn by Ernst Haeckel in *The Evolution of Man* (1896, 2, 188).

its own right because Haeckel and friends were not true Darwinians. That, we shall learn, would take out most of Anglo-American evolutionism for the next fifty years, and perhaps later (Bowler 1988). Especially, I am not side-stepping them because they had nothing to say on progress. In this respect, I am pouring water on my own altar. As neo-*Naturphilosophen,* post-*Origin* German evolutionists were fanatics on the subject of upward development. Haeckel's first public words on evolution were to identify himself with "development and progress" as against "creation and species" (Nyhart 1986), and he was ever an ardent painter of upwardly reaching, evolutionary trees. And finally, unlike the Germans themselves, I am not supposing that there is some kind of Hegelian inevitability to history, and that thus what occurred in England and America was bound to be what occurred in Germany. As it happens, for reasons I shall touch on, in this respect I do feel comfortable with my exclusion of Germany, because the story in the various countries does have a similar resolution. But the details were very different.

Understanding, then, that my criterion of selection is not a judgment on the science of the past, but a practical necessity guided by my philosophical inquiry, I turn now to consider post-*Origin* evolutionism in Britain and America. In fact, let me say straight off that, as soon as we go beyond the fact of evolution, much of nineteenth-century British evolutionism and most of nineteenth-century American evolutionism owes far more to German transcendentalism than it does to anything in Darwin, and I shall be the first to acknowledge this fact. But I get ahead of myself. Beginning with Britain, my subjects in this chapter are two who came to evolution independently of Darwin. In the next chapter, I shall turn to those who were supposedly carrying his torch.

Herbert Spencer

In 1863, the Scottish philosopher and psychologist Alexander Bain wrote to Herbert Spencer (1820–1903): "You have certainly constituted yourself *the* philosopher of the doctrine of Development, notwithstanding that Darwin has supplied a most important link in the chain" (November 13, 1863; Spencer Papers, Ms 791/67). This assessment was one shared by many, so let us start with him, although to be candid it is not easy today to write about Herbert Spencer. Immensely popular and influential in the early years of his writing, he lived a very long life into this century—long enough to see that his reputation in Britain as a creative

Herbert Spencer

thinker had sunk to hitherto unfathomed depths. Yet, whatever else may be true, there can be no doubt that his contemporaries took Spencer seriously. And in this group—much more so than with Chambers—we must include professional scientists. For instance, one of Spencer's closest friends—a man with whom he discussed evolutionary topics endlessly—was T. H. Huxley. "It seems as if all the thoughts in what you have written were my own and yet I am conscious of the enormous difference your presenting them makes in my intellectual state" (letter of September 3, 1860; in Huxley 1900, 1, 212). And this was written *after* the *Origin*.

Spencer's intellectual upbringing was, to say the least, unconventional, for his father more or less allowed young Herbert to study where and what he pleased (Spencer 1904; Duncan 1908). In his mid-teens, with some mathematics drummed into him by a patient and gifted uncle, he took up surveying, a skill much in demand given the quickening pace of railway development in Britain in the late 1830s and 1840s. On and off, his surveying engagements lasted for over a decade and led to his becoming an evolutionist in 1840. Fascinated by the fossils thrown up by the earthworks for the laying of railway lines, Spencer bought Lyell's *Principles of Geology*. At once, he was a convert: "my reading of Lyell, one of whose chapters was devoted to a refutation of Lamarck's views concerning the origin of species, had the effect of giving me a decided leaning to them." Spencer explained that, for him, unlike so many others, religion was never a barrier. The family tradition was very strongly one of nonconformism, principally Wesleyan Methodism, laced with some Quaker elements. He himself went to the deist limit: "Supernaturalism, in whatever form had never commended itself" (Spencer 1904, 1, 201).

Entering the 1850s, Spencer was writing regularly for the *Westminster Review*, the organ which represented the best of radical opinion. Boldly, he began promoting an unequivocal organic evolutionary line. Never very generous with acknowledging sources, he made few references to the work of others and we cannot easily pinpoint all of the influences. Lamarck (as expounded by Lyell) was certainly there, as also was von Baer, as explained by W. B. Carpenter (1839). Chambers had been read, although apparently more as collateral confirmation, and the same seems to go for the smatterings of knowledge that Spencer picked up from more conventional scientists like Owen. These and others, rarely read in the original and never read in whole, were soaked up and processed by Spencer's unconscious. Then they came tumbling out, yielding a grand world picture—the writing of which (the so-called Synthetic Philosophy) was to occupy the next forty years of Spencer's life.

"The development hypothesis" (1852b) blew a fanfare for the cause: "Those who cavalierly reject the Theory of Evolution, as not adequately supported by facts, seem quite to forget that their own theory is supported by no facts at all" (p. 377). One private admirer of this piece was Richard Owen, who seems to have been given to the writing of warm, albeit secret letters to evolutionists: "There is no subject in the perusal of which I shall relax with more pleasure than that of your work which concurs so nearly with some of the trains of thought that have been suggested to me by the comparisons of dry bones" (letter to Spencer, April 28, 1851; Spencer Papers, Ms 791/27).

In the same year, Spencer backed up his announcement by turning to Malthus. Like Darwin before and Wallace to come, he rose to the challenge (Spencer 1852a). What consequence or sense can one draw from the pressures of potential population growth? Happily, a ready answer lay at hand, in the form of a universal law of organic life, namely that there is an inverse relationship between an organism's ability to maintain itself and its fertility. Apparently, this law is an *a priori* truth—you only have so much vital energy to expend, and if you use it one way (as in making brains, a key component in successful maintenance), you cannot use it in other ways (as in making sperm cells, a key component in successful reproduction). As always, however, Spencer was happy to back up his deductive intuitions with carefully chosen empirical examples. In our own species, he chose as an example the well-known fact that savages are far more prolific than are civilized folk. And brain sizes are directly (although inversely) correlated, with the Australian having 75 cubic inches of brain space and the Englishman having 96 cubic inches. This point was underlined by the personal testaments of the parents of any adolescent boy: "undue production of sperm-cells involves cerebral inactivity. The first result of a morbid excess in this direction is headache, which may be taken to indicate that the brain is out of repair; this is followed by stupidity; should the disorder continue imbecility supervenes, ending occasionally in insanity" (p. 263).

Now, how does this all tie in with Malthusianism? The answer is simple. Population pressures bring on a struggle. This demands that one exercise one's faculties, in order to overcome the hardships one faces in competition for resources. The faculties most needed are precisely those contributing to personal maintenance (like brains). These enlarged faculties are passed on through heredity, in a Lamarckian fashion. There is thus a development of superior organisms, taking us from primitive forms, up eventually to man. But, with growth and development comes the inevitable loss of fertility. Eventually, evolution will have run its course, with a happy balance between superior beings and just-sufficient reproduction to replace lost members. Malthusianism will have brought about its own demise: "in the end, pressure of population and its accompanying evils will entirely disappear; and will leave a state of things which will require from each individual no more than a normal and pleasurable activity" (Spencer 1852a, 501).

The common starting point notwithstanding, the perspective is fundamentally non-Darwinian. It is true that right at the end of his essay, Spencer did skate tantalizingly close to natural selection. Drawing attention to the dreadful example of Ireland (remember, the potato famine was just a few years earlier), Spencer hypothesized that there would be other

In line with general beliefs about the progressionist nature of evolution, this cartoon from 1882 depicts an Irishman as ape-like. The point is reinforced by linking the Irish with black Africans through the caption: "The King of A-Shantee." (From *Puck,* 1882.)

forces reinforcing the effects of Lamarckian-type changes: "For as those prematurely carried off must, in the average of cases, be those in whom the power of self-preservation is the least, it unavoidably follows, that those left behind to continue the race are those in whom the power of self-preservation is the greatest—are the select of their generation" (p. 267). Yet, this was but an aside, and even after the *Origin* was published, Spencer always thought of selection as a somewhat secondary mechanism.

The year 1855 saw a major work on psychology (Spencer 1855). Here, in a project which he modestly confessed to his father he felt sure history would place next to the work of Newton, Spencer tried to offer an evolutionary solution to a dispute between the philosophers Whewell and Mill over the nature of necessary truth. To this end, he espoused a form of evolutionary Kantianism; one should add that, typically, Spencer's knowledge of Kant was patchy, mainly secondhand, and more than tinged with scorn. More pertinent to us is the essay of 1857 which laid out the overall world picture: "Progress: Its law and cause." Everything, argued Spencer, is part of the one all-embracing evolutionary process. All that exists is in a state of becoming or development. Picking up on ideas expressed somewhat fleetingly earlier, Spencer argued that von Baer's view of embryology holds universally: "It is settled beyond dispute that organic progress consists in a change from the homogeneous to the heterogeneous." This is "the law of all progress" (p. 3). And in usual fashion, Spencer set off at breakneck speed through a range of examples, proving his thesis. Thus, for instance, the nebular hypothesis clearly implies a transition from a uniform diffuse state to the highly complex system we see around us. Likewise, at a more specific level, in the fossil record we see transition from homogeneity to heterogeneity: "The earliest known vertebrate remains are those of Fishes; and Fishes are the most homogeneous of the vertebrata. Later and more heterogeneous are Reptiles. Later still, and more heterogeneous still, are Mammals and Birds" (Spencer [1857] 1868, 8–9).

Humans also show the workings of the law: "in the relative development of the limbs, the civilized man departs more widely from the general type of the placental mammalia than do the lower human races." Whereas the Papuan had limbs all of the same size, "in the European, the greater length and massiveness of the legs has become very marked—the fore and hind limbs are relatively more heterogeneous" (pp. 10–11). Analogously, when it comes to intellectual abilities, we have the interesting phenomenon of language: "And it may be remarked, in passing, that it is more especially in virtue of having carried this subdivision of function to a greater extent and completeness, that the English language is superior to all others" (p. 17).

But what is the cause of this process of change? The answer is almost trivially simple—though it took a Spencer to spot it. "*Every active force produces more than one change—every cause produces more than one effect*" (p. 32). The law of evolutionary development turns out to be a direct consequence of the most fundamental nature of causation—the multiplication of effects is a one-way process. You always get phenomena exploding or rippling outward, and thus homogeneity necessarily evolves into heterogeneity: "No case can be named in which an active force does not evolve forces of several kinds, and each of these, other groups of forces. Universally the effect is more complex than the cause" (p. 33).

The picture is almost complete. All that remained, at the end of the decade, was the forging of a connection between Spencer's evolutionism and modern physics, particularly what came to be known as the Second Law of Thermodynamics, that rule which argues that physical systems are moving always to a state of equilibrium (maximizing "entropy"). It may seem strange that an enthusiast for evolution should want also to connect his thinking to notions of equilibrium, since they—implying a kind of balance between forces—seem the antithesis of onward change. For Spencer, however, equilibrium was important in two ways. First, in a *dynamic* fashion it appears that the tendency to move toward stable equilibrium (and, more particularly, to fly *from* unstable equilibrium) is the move from homogeneity to heterogeneity by another name. The "truth is, that *the condition of homogeneity is a condition of unstable equilibrium*" (Spencer [1857] 1896, 1, 81; his italics). Then secondly, *static* equilibrium was important in that this was the final state that the universe, including the organic/human universe, tended. Delicate footwork was needed to show why the physicists' predictions of a uniform, maximum-entropy, final equilibrium (a state referred to as "heat death") was not quite the gloomy endpoint that it seems.

With this argument, the props were in place. Essentially, all else was a matter of filling in details. Spencer was now about to devote the remainder of the century to laying out his world picture, in mind-benumbing Victorian detail. Unmarried, touchy, fussy about his food as only the English know how to be, hypochondriacal beyond the point of neuroticism, fond of little girls in that peculiarly repressed Victorian manner, determined to refuse rank and honor in a way far more troublesome than if he had simply demanded the throne of England, worshipped by devoted disciples who tried to satisfy his every whim—and they were many—Spencer lived out his life in a series of dreary lodgings chosen specifically for their dreariness, that his surroundings and companions not excite him and detract from his great task. Yet, for all its Wagnerian dimensions, the Synthetic Philosophy was a set of variations on already established themes. (See reminiscences by W. Troughton, Spencer Papers, Ms 791/355/3.)

Spencer: The Paradigmatic Progressionist

Herbert Spencer has no rivals when it comes to open, flagrant connections of social Progress with evolutionary progress. As my exposition of the last section shows, it is impossible to talk of Spencer's evolutionism without bringing in the word *progress* somewhere. The very attempt to do so is, in respects, deeply distorting. But, if only for the sake of comparison, let us continue to follow our usual pattern of analysis. And this means, first, asking about the idea of progress in Spencer's biological theorizing.

For Spencer, evolution was progress and progress was evolution. Biological progress involved going from the homogeneous to the heterogeneous, and the greater the heterogeneity the greater the value and degree of highness. "It is in the greater complexity of the co-ordination—that is, in the greater number and variety of the co-ordinated actions—that every advance in the scale of being essentially consists" (Spencer 1852a, 252). Evolution is therefore an upward climb—"from the monad up to man" (p. 267)—via the invertebrates, the fish, the reptiles, the mammals, and the apes. Moreover, when we get to humans, there too we find progress, from savages to Englishmen. Atypically, Spencer had sufficient modesty to forbear remarking that his own evolution had progressed thus far that his fertility had apparently failed entirely.

Is this a value-claim? Does heterogeneity imply that there is greater worth? "Higher organisms are distinguished from lower ones partly by bulk, and partly by complexity" (p. 259). Who cares about bulk and complexity? In themselves, perhaps no-one, but they are indicative of greater efficiency of a desirable kind. Preceding Darwin in date of publication, Spencer tied in his thoughts with Milne-Edwards's "division of labour": "The complexity essentially consists in the mutual dependence of numerous different organs, each subserving the lives of the rest, and each living by the help of the rest . . . And this 'physiological division of labour,' as it has been termed, has the same effect as the division of labour amongst men" (pp. 259–260).

Spencer's picture of the evolutionary process was progressive, and the mechanisms he endorsed were designed to keep it that way. At the broader metaphysical level, the asymmetry between causes and effects (with the latter outnumbering the former) meant that complexity and heterogeneity were necessary consequences of any developing system—and any system had to develop because homogeneity was unstable. At the narrower biological level, the effects of the Malthusian pressures meant that organisms had to strive to survive and succeed, and thanks to the medium of Lamarckian heredity, the good results would be collected, magnified, and passed down to future generations. For Spencer, biological progress was not an accidental side effect but a necessary outcome of the very possibility of change.

Going straight to the second of our questions, Spencer was so far a Progressionist that he has already been used as our exemplar of British Victorian views on Progress. Moreover, his Progressionism fed straight into his progressionism. Indeed, the latter was a sub-branch of the former: the "law of organic progress is the law of all progress. Whether it be in the development of the Earth, in the development of Life upon its surface, in the development of Society, of Government, of Manufactures, of Commerce, of Language, Literature, Science, Art, this same evolution of the simple into the complex, through successive differentiations, holds throughout" (Spencer [1857] 1868, 3). Despite the almost Germanic inevitability about the move upward that resulted from the fragmentation of the homogeneous into the heterogeneous, the vision was truly one of Progress, with struggle and effort having their own reward.

Let us dig back into Spencer's past and see if and how thoughts of Progress led into thoughts of progress. The key factor is that Spencer came from a family of dissent, of provincial thinkers and workers on the outside looking in—at landowners, Tory squires, the established church, and all of the other privileged people and institutions of early-nineteenth-century Britain. Even a Spencer uncle who became an Anglican clergyman shared the family creed of opposition to authority. But these people were not crushed or downtrodden. Far from it. They had their own inspirational philosophy: self-help, extreme dislike of government institutions (which, with very good reason, they saw as simply serving the interests of the stronger), and an unquenchable belief in the better state to come—a mixture of John Wesley's salvationism ("God helps those who help themselves") and Adam Smith's economics.

Herbert Spencer was brought up on this radically liberal philosophy and, for all that he threw out the theology, quickly became the most articulate family spokesman. In early letters (appropriately written to the *Nonconformist*) he decried all attempts at government regulation—since the letters appeared in 1842, they came at the height of the campaign against the notorious Corn Laws—and he showed just how intimately were his beliefs tied to hopes of Progress. "If it be granted that man was created a progressive being, it must be granted also that the constitution given to him by his Creator was the best adapted to secure his progression" (Spencer 1842, 256). It was this position that was repeated in Spencer's first full-length work, *Social Statics* (1851).

By mid-century, Spencer was already a Lamarckian evolutionist. To this he added a somewhat idiosyncratic understanding of phrenology—another support of Progressionism. Mixing this all in with his extreme dislike of any kind of regulation or government support, Spencer argued that life's struggles will lead to improvement of human faculties or abilities, and ultimately we shall all Progress to a state of happy, non-militaristic harmony. Which is, of course, precisely the view of upward development that Spencer was to read into, or out of, his biology. It is often thought that Spencer endorsed a crude ("Social Darwinian") *laissez-faire* social philosophy, where Progress came through the failure and elimination of society's unfortunates. But although there are certainly hints of this in *Social Statics,* and even more in later post-Darwinian writings—he definitely thought that change would be painful—the selective process is never crucial. The chief cause of change is upward striving, preserved for posterity by Lamarckism: exactly the Spencerian position in biology.

In Spencer's social thought, likewise we see early hints of the importance of the division of labor and of the move to heterogeneity. "The development of society, as well as the development of man and the development of life generally, may be described as a tendency to individuate—to become a thing. And rightly interpreted, the manifold forms of progress going on around us are uniformly significant of this tendency" (Spencer 1851, 408). Significantly, among his possessions Spencer kept some letters, to that uncle who had taught him mathematics, from the heroes of the struggle over the Corn Laws. Illustrating how the division of labor was a commonplace in Spencerian circles, we find Richard Cobden joking that: "I think a division of labour is necessary for success in political as in industrial life" (letter to Thomas Spencer, April 23, 1849; Spencer Papers, Ms 791/9).

We need not bother the point to death. For Spencer, the guiding factor behind his evolutionism was the cultural value of Progress—British, although certainly backed by the progressionism of the German biology on which he drew, and reinforced by his Lamarckism. Let us therefore move to our final question, namely about the relationship between Spencer's ideas and the empirical world. Care must be taken here. At one level, it is very tempting simply to dismiss contemptuously virtually everything Spencer said. One is reminded of Huxley's quip that Spencer's idea of tragedy was a beautiful deduction destroyed by an ugly little fact—and in respects that joke seems an appropriate epithet (Duncan 1908, 502).

Yet at another level there is more to be said, including, perhaps, more to be said in Spencer's favor. Spencer was not so far out of the mainstream, including the empirical mainstream, as he is often taken to be—quite apart from the fact that in respects he defined what was meant by "mainstream." At a level—admittedly, usually an indirect level—Spencer was responding to the empirical findings of the biologists of the mid-century.

Eclectic, opinionated, and full of (invariably half-digested) ideas—although in a brilliant combination—he was weaving his web from facts that others had discovered about the biological world: von Baer, Milne-Edwards, even the physicists—not that their views on entropic death were much comfort, or that Spencer took advice readily, as Clerk Maxwell discovered when he tried to correct Spencer on a couple of points (Duncan 1908, 428–431).

For all this, appropriate dues having been paid, given what we have seen from the work of others, this all only goes to confirm our main conclusion that the backbone of Spencer's progressionism was put into the world rather than taken from it. It was the theme of Progress which ran through all Spencer's work and gave his evolutionary biology the distinctive shape it possessed. His biological progressionism was rooted in social ideals rather than deduction from the phenomena.

Alfred Russel Wallace

I turn next to one of the most remarkable but engaging characters in my whole story, the co-discoverer with Charles Darwin of the mechanism of evolution through natural selection: Alfred Russel Wallace (1823–1913). His story begins with a kind of mishap that would become typical for him: the idiosyncratic spelling of his second name was the result of a mistranscription on his birth certificate. (For general information about his life and work, see Wallace 1905; Marchant 1916; McKinney 1972.)

Wallace will always be linked with and contrasted to Darwin, and perhaps that is no bad way to introduce the man and his ideas—for it is truly striking that, whereas Darwin was so completely a man within and benefiting from successful Victorian science and society, Wallace was so completely not such a man. There was no silver spoon (or, should one say china cup?) in his mouth at birth, there was no expensive education leading to friendship with all of Britain's men of science, and there was certainly no comfortable support for a life of upper-middle-class ease. Wallace had to make his own way, and this meant that—right through his very long life—he was always in respects on the outside looking in. Without the neurotic egoism of a Herbert Spencer, he yet had a childlike innocence and moral firmness that altogether disarmed even the most hardened cynics.

The eighth child of a father who so mismanaged the family finances that he enjoyed "comparative freedom from worry about money matters,

Alfred Russel Wallace

because these had reached such a pitch that nothing worse was to be expected" (Wallace 1905, 1, 13), Wallace was pushed out from school at the age of fourteen. A few months in London with one older brother were followed by apprenticeship to another older brother, as a land surveyor. Unfortunately, no sooner had he finished his training than, owing to lack of work, Wallace had to switch to a brief spell of school teaching. By this time a firm love of natural history had developed and so, with a friend, Henry Walter Bates—that very Bates who was so to excite Darwin with his work on mimicry—Wallace set out on a collecting trip to South America.

Wallace's fortunes seemed to have changed, but they had not. On the journey home, the ship sank beneath him and his valuable specimens were lost. Undaunted, Wallace set out on a second trip, to the Malay Peninsula, which he crossed and recrossed for eight long years. Returning to England in the early 1860s, he was not only jilted by his girlfriend but he so mismanaged the proceeds from his new collections that for the next fifty years he was forced to live as a freelance writer. And so, from Wallace's desk flowed a stream of articles, pamphlets, and books on subjects as diverse as land nationalization (for it), vaccination (against it), spiritualism (for it), life on Mars (against it), and vegetarianism (for it, but compelled for health reasons to live on an uninterrupted diet of lean beef).

Wallace became an evolutionist in 1845, converted by the newly published *Vestiges of the Natural History of Creation:* "I have a rather more favourable opinion of the 'Vestiges' than you appear to have. I do not consider it as a hasty generalization, but rather as an ingenious hypothesis strongly supported by some striking facts and analogies but which remains to be proved by more facts and the additional light which future researches may throw upon the subject" (letter to Bates, 28 December 1845; in McKinney 1972, 11). Indeed, it was the unfinished program that *Vestiges* started that spurred the two young men to set off for the Amazon. Quite simply, they wanted "to gather facts . . . towards solving the problem of the origin of species, a subject on which we had conversed and corresponded much together" (Bates 1892, vii).

What made Wallace so special is the dogged persistence he showed, through long years in the Amazon, and then on to even longer years in the Far East. The effort started to show through, however. By 1855, stimulated by his reading of Lyell's *Principles,* Wallace had his own contribution to make, when he formulated a "law" governing the introduction of new species on this globe: "Every species has come into existence coincident both in space and time with a pre-existing closely allied species" (Wallace [1855] 1870, 25). Cagily, Wallace left his language just sufficiently ambiguous that he could be read in a non-evolutionary fashion, but no-one was meant to be deceived—nor indeed were they. The causal explanation came three years later when, during a fever fit, Wallace remembered the long-since-read *Essay on a Principle of Population* by Malthus. Generalizing, as had Darwin some twenty years previously, Wallace went from the struggle for existence to natural selection. At last he had a mechanism and, like Darwin, Wallace was sensitive to the way in which he had also an explanation of adaptation (Wallace 1858).

There are, nevertheless, some interesting differences of emphasis between Darwin and Wallace. Perhaps most significantly, Wallace showed clear signs of what was to be a persistently maintained holistic attitude to natural selection and its results—as opposed to the more individualistic, perhaps "reductionistic," stance usually taken by Darwin, especially in the *Origin.* Wallace clearly saw that evolution through selection must entail different survival and reproduction rates between fellow species members, as well as between species. However, whereas Darwin openly expected and stressed conflict between species members, Wallace's emphasis was away from competition within groups (Ruse 1980). The threat to members of small sub-specific groups comes not from within but from without. Thus, to use modern terminology, right from the first we see Wallace inclining toward a "group selectionist" perspective on nature and its mechanisms of change, as opposed to the "individual selectionist" perspective of Darwin.

These different ways of looking at the central force of evolution carried over into the 1860s. There was a divide over sexual selection. Although prepared initially to accept Darwin's (necessarily individualistic) secondary mechanism, before long Wallace began to have doubts about sexual selection through female choice. Dubious as to the propriety of ascribing human-like feelings for beauty to the dumb brutes, Wallace argued that much sexual dimorphism is due to dictates of camouflage. There was a clash over the origins of inter-specific sterility. Darwin claimed that such sterility (hardly aiding individuals) is just a by-product of natural selection—barriers come down simply because organisms have evolved away from one another—and Wallace argued that selection (working for the group) positively initiates and reinforces the sterility. And, most divisive of all, there was the matter of our own species.

We have seen how Wallace initially gained Darwin's praise and admiration for his suggestions about the significance of human mental and social evolution. Yet, congratulated though he was for these ideas, Wallace soon grew dissatisfied. Primarily because of a growing belief in spiritualism—but also because of a distaste for any kind of struggle *within* the human species—as the decade (the 1860s) drew on, Wallace turned increasingly to some unseen world force as the primary guide for human evolution (Kottler 1974; Smith 1972). He denied that certain key aspects of human nature—specifically our large brains, but also such subsidiary features as our hairlessness—could be naturally caused, and so he looked elsewhere: "The inference I would draw from this class of phenomena is, that a superior intelligence has guided the development of man in a definite direction, and for a special purpose, just as man guides the development of many animal and vegetable forms" (Wallace 1870, 204). There is little wonder that Darwin was driven to despair at Wallace's apostasy.

Interaction with Darwin apart, in the years following the *Origin* Wallace looked in detail at butterflies as a source of information on the mechanism of selection. In 1866, he published a major (Bates-like) paper on the variability and geographical distribution of Papilionidae, a family of butterflies found in Malaysia. This work led him to more general concerns with issues of geographical distribution, culminating in an attempt to show how division of the world's organisms into (six) basic geographical districts can serve as the basis for a full evolutionary explanation of the development and flow of life across the globe's surface (Wallace 1876). But he could never stay away from grand speculation, and before long Wallace was back to wild social dreams. Although staying true to his spiritualist beliefs, he began to suspect that humans—women in particular—might have some significant effect on the course of our own evolution. If not now, then there exists the possibility of such effect in the future.

While standing strong against sexual selection through female choice in the animal world, Wallace now hypothesized that possibly for our species such selection might prove crucial. If we can remodel society to the point where women might develop their full potential, who could tell what the end result might be?

> In such a reformed society the vicious man, the man of degraded taste or of feeble intellect, will have little chance of finding a wife, and his bad qualities will die out with himself. The most perfect and beautiful in body and mind will, on the other hand, be most sought and therefore be most likely to marry early, the less highly endowed later, and the least gifted in any way the latest of all, and this will be the case with both sexes. From this varying age of marriage, . . . there will result a more rapid increase of the former than of the latter, and this cause continuing at work for successive generations will at length bring the average man to be the equal of those who are now among the more advanced of the race. (Wallace 1900, 2, 507)

Wallace was a holist to the end. Uniquely in the human case, sexual selection through female choice promotes cohesiveness. The perfected man and his choosy mate will live a life of harmonious bliss with their fellows, with nary an aged degenerate bachelor or spinster to mar the social scene.

Wallace and the Promise of Progress

Let us begin in usual fashion by asking about possible thoughts of progress in Wallace's work. This is an easy question to ask and an easy question to answer: for Wallace, evolution was synonymous with progress. Or rather, one should qualify this slightly, for as with Darwin we do see some doubts toward the end about the onward progress of evolution—doubts which will grow and magnify as we move to later evolutionists.

The overall perspective from which Wallace always worked was stated clearly in his first major essay of 1855: "The admitted facts seem to show that there has been a general, but not a detailed progression. Mollusca and Radiata existed before Vertebrata, and the progression from Fishes to Reptiles and Mammalia, and also from the lower mammals to the higher, is indisputable" (Wallace [1855] 1870, 11–12). Indisputable, yes, but even so Wallace realized that the good old days of a unilinear progression up a Chain of Being are gone. There are far too many anomalies for that: "it is said that the Mollusca and Radiata of the very earliest periods were more highly organised than the great mass of those now existing, and that the very first fishes that have been discovered are by no means the lowest organised of the class" (Wallace [1855] 1870, 12).

In the famous essay on natural selection, Wallace ([1858] 1870) really said very little about his criteria of highness and lowness or about how selection might be expected to lead to the former rather than the latter. With everyone else, Wallace seemed to presuppose some sort of vague notion of complexity. However, when Wallace turned (later) to problems of human evolution, it became very clear that intelligence is important and bound up with notions of progress. Specifically, humans are at the top of the list because of their intelligence, which is in some way flagged by complexity. Interestingly, Wallace, who unlike most of our group had firsthand acquaintance with "savages," sometimes suggested that primitive folk are truly no less primitive than we. To Spencer, he recommended a work showing "the great change induced in savages by living with civilized and educated men, who treated them on a *footing of perfect equality*" (letter of November 13, 1875; Spencer Papers, Ms 791/112; the book was the *Life of Bishop Patterson*). But usually, Victorian standards prevailed. Remember the ratings of the Red Indians and others: "The intellectual and moral, as well as the physical, qualities of the European are superior" (Wallace [1864] 1870, 177–178).

Intelligence aids in the struggle, because it makes us tougher and stronger than our rivals. But in this claim we see a source of tension for Wallace which was to remain with him for many years. In Wallace's holistic eyes, the superior being is not necessarily the brightest, and certainly not necessarily the most aggressive. It is rather the one with the highest social sense, the one most ready to work and interact with fellows. Yet, beings of this nature seem not to be the necessary end result of a bloody struggle for existence. "I *doubt* if evolution alone . . . can account for the development of the advanced and enthusiastic *altruism* that not only exists now, but apparently has always existed among men" (letter to Spencer, July 2, 1879; Spencer Papers, Ms 791/138). Here was a major reason why spiritualism was such a relief to Wallace. The unseen forces bring on change to the highest social point without the need for conflict, especially without the need for intra-specific conflict. Where the natural fails, the supernatural succeeds.

Wallace was still not truly satisfied. His confidence perhaps shaken by the economic depressions of the 1870s, he sought further reasons for progress—reasons to expect future progress. By 1880, Wallace had convinced himself finally that a Malthusian struggle does not really hold among humans, and so he felt free to seek a more positive force of creative change. He thought that the success of some individuals would directly encourage the improvement of the species, but this answer hardly guaranteed progress. Then it was that he found the true power of sexual selection through female choice. Effort is needed to get the process moving—for all his dabbling with spiritualism, Wallace was no real Providentialist—but at some point nature does take over. With appropriate social reform, we shall have young people "ready to respond at once to that higher ideal of life and of the responsibilities of marriage which will, indirectly yet surely, become the greatest factor in human progress" (Wallace 1900, 1, 509).

Next, what of Progress? Did Wallace believe in the social ideal? Is there reason to think that he used it as a lens through which to view the world of biology? Start at the beginning, with the fourteen-year-old Wallace. He learned more than surveying during his London sojourn, for his brother took him to the meetings of the followers of Robert Owen, the factory owner who had built a socialist Utopia around his mill at New Lanarck. The fellowship and optimism of these people moved the impressionable lad greatly. Undoubtedly, they were the chief inspiration of the holistic philosophy that colored his life's work, and the source of Wallace's deep conviction that human nature is in essence good and selfless—an essence to be realized given the right social conditions, and to be intensified through benevolent forces of selection. Right at the end of his life, Wallace devoted much space in his autobiography to the Owenites, writing: "I have always looked upon [Robert] Owen as my first teacher in the philosophy of human nature and my first guide through the labyrinth of social science. He influenced my character more than I then knew" (Wallace 1905, 1, 104). He was immensely thrilled that once he had heard Owen himself speak: "I . . . consider Owen one of the *best* as well as one of the greatest men of the 19th Century, an almost ideally perfect character" (letter to S. S. Cockerell, August 23, 1904; Wallace Papers, Add mss 46442).

Owenism is not only holistic and socialistic but also optimistic and Progressivist. It looks to a better future, as also does phrenology. *Vestiges* likewise was Progressionist as well as progressionist—for Chambers, the two were really indistinguishable, as was truly the case for many of his readers, including Wallace. There was of course a problem when Wallace took on board the ideas of Malthus, especially when he made them so central to his biology. The pessimism somehow had to be countered; however, Wallace thought he could do this. Probably a factor which played a key role here was his reading in 1853 of Spencer's *Social Statics* (Wallace 1905, 2, 235). Wallace was wildly enthusiastic about the book—later he was even to name his son "Herbert Spencer Wallace"—and we have seen good reason why this should be so. Spencer was Progressionist and optimistic, and he wrote from the perspective of the outsider. He was against land-ownership and speculation, and Wallace had particular reason to abhor these prominent aspects of Victorian society. His training as a surveyor often involved work for those who were enclosing and taking hitherto common pastures.

This philosophy lasted with Wallace through the first writings on humankind, although by now there was clearly a nagging doubt as to whether Spencerian selfishness was truly compatible with universal brotherhood. Fortunately, spiritualism, with its central role for the future of humankind, was available to take up the overload. "I admire and appreciate the philosophical writings of Mr. Lewes, of Herbert Spencer and of John Stuart Mill, but I find in the philosophy of Spiritualism something that surpasses them all,— . . . something that throws a clearer light on human history and on human nature than they can give me" (unpublished letter to editor of *Pall Mall Gazette,* May 1868; Wallace Papers). To this, Wallace was prepared to add the belief that not only are we the uniquely intelligent denizens of this universe, but we are located in an almost pre-Copernican manner at the center of the universe. (See Wallace 1903, 1907.)

By the end of the next decade, the 1870s, Wallace was growing increasingly uncomfortable with Spencerian libertarianism. And, increasingly, he felt the need of some theory to counter the gloomy specter of decay and decline that so worried his fellow biologists like Darwin. Very significant at this point was his reading of *Progress and Poverty* by the American writer Henry George, who (like Spencer and Wallace) saw the roots of deterioration and inequality to lie in land-ownership. George thought it possible to turn decline around, and the Progress he envisioned was certainly one to appeal to Wallace: "Men tend to progress just as they come closer together, and by co-operation with each other increase the mental power that may be devoted to improvement, but just as conflict is provoked, or association develops inequality of condition and power, this tendency to progression is lessened, checked, and finally reversed" (George 1879, 508).

George's vision of human nature was not one to appeal to Wallace, however, for it put all of the burden and hope of change on nurture and none on nature. "The advances in which civilization consists are not secured in the constitution of man, but in the constitution of society" (p. 562). What Wallace did get from George was his argument against the efficacy of Malthusian forces in human society, for George argued in detail that the struggle plays little or no role. This freed Wallace to look for other, yet still biological, reasons for advance. It is obvious why neither eugenics nor Lamarckism could appeal. The people against whom the eugenicists would turn their forces were people of Wallace's class. He wanted no part of this. Likewise with Lamarckism and its similar implication that the people of Wallace's class and below are biologically inferior—their lowly status being due to personal inadequacy and not to unjust social conditions. Hardly surprisingly, Wallace wanted even less part of this. He argued adamantly that people of low social status would perform as well as others when given the chance. A favorite example was the success of dissenters in the Cambridge mathematical exams, when once the religious restrictions were lifted (Wallace 1900, 1, 512).

The causes of successful, Progressivist societal change had still to be found. Wallace was always willing to try out utopian social schemes, including assumption of the Vice Presidency of the "British Freeland Association," dedicated to setting up a colony of Europeans at the base of Mt. Kenya, to implement cooperative ideas proposed by the German author Theodor Hertzka (in his *Freeland: A Social Anticipation*). For Wallace, however, the real breakthrough came at the end of the 1880s, and he produced "the most important contribution I have made to the science of sociology and the cause of human progress" (Wallace 1905, 2,

209). This time the creative spark came from a novel, *Looking Backward,* a utopian fantasy by the American author Edward Bellamy: "Every sneer, every objection, every argument I had ever read against socialism was here met and shown to be absolutely trivial or altogether baseless" (pp. 266–267).

When we turn to Bellamy's work, an idealistic vision of the world of 2000 from the year of 1887, we find so exact a model for Wallace's own position, the wonder is that he thought he was making any advance on it. In Bellamy's new society, thanks to the emancipation of women, "for the first time in human history the principle of sexual selection, with its tendency to preserve and transmit the better types of the race, and let the inferior types drop out, has unhindered operation." Hence: "The gifts of person, mind, and disposition; beauty, wit, eloquence, kindness, generosity, geniality, courage, are sure of transmission to posterity" (Bellamy 1887, 218). Wallace transposed this scenario into his own biology. Forward advance was possible, and would promote universal harmony, without the need for Spencerian strife. Progress and progress had become one, for the forces of change would come from within rather than without: "I believe that the majority of thinking socialists today would, as I do, entirely repudiate the idea of compulsion, either in the first establishment or the ultimate working of the Cooperative Commonwealth" (letter to John Morley, October 20, 1890; Wallace Papers).

And so finally we come to the third of our questions. What of Wallace and the empirical evidence? At this point, question and answer seem so easy that perhaps one should tread lightly, remembering first Wallace's brilliant contributions to the history of evolutionary thought. That his speculations inevitably went beyond the given surely now needs no argument. Nor do we need further to document the fact that his overextensions in the empirical realm (to use a friendly euphemism) were such that one might think a commitment to Progress at work. To take only his final suggestions—turning away from such tempting areas as Wallace's touching beliefs in the integrity of spiritualists—it is surely remarkably naive to think that young women will invariably choose the wisest and best in society as their mates. Or that young men will stand around idly and let this happen. Wallace's children must have been as strange as their father if he truly thought this a real possibility.

We must conclude that, despite worries about the present and future state of his society, Wallace was a Progressionist, albeit in a distinctive fashion, tinged especially with holism and socialism. This Progressionism was reflected into a biological progressionism, albeit distinctive in exactly the same ways. Culture rode high.

Triumph of a Popular Science

It seems fair to say that both Spencer and Wallace contributed to the cementing of evolution's basis as a respectable popular science—a popular science which for their readers, as for them, by virtue of its P/progressionism went some good way of providing a secular alternative to conventional Christian belief. (Perhaps in Wallace's case, only semi-secular!) It seems equally fair to say that neither Spencer nor Wallace much advanced Darwin's dream of building a professional discipline of selection-based studies. Indeed, they advanced it not one whit. And this, I must add, extends (in a connected causal chain) to the matter of values, epistemic or cultural.

Prima facie, Spencer had little or no scientific standing. He was never a member of any society, especially not the Royal Society. But, this alone is a little misleading. His formal absence from professional science was, as much as anything, caused by his steadfast refusal to stand for membership of any professional organization. He was, however, a member of the X-Club, a dining club of nine powerful scientists, including Hooker and Huxley, which held much of the power in British science through the end of the century (Jensen 1991). Thus, although on the one hand Spencer was ostentatiously very much a man of the people, outside the halls of official science, on the other hand he knew just what went on in those halls (and in the corridors also) and was respected personally by the scientific élite. This is a fact.

But it is a fact also that Spencer was just not prepared to settle beneath the constraints of a profession, and the status of the man reflects through to the status of his work. Spencer wrote for the general public. His work appeared in the quarterlies, and then through a subscription which he sold to willing buyers. His evolutionism was not and was never intended to be professional science, at a level that someone like Cuvier would have appreciated. He certainly never sullied his hands with detailed experimentation or that meticulous firsthand observation that had so occupied Darwin through the years of barnacle work. His work displayed a massive gathering of facts, but always these facts were presented such that, as in Chambers's work, the grand idea was the thing that counted. Spencer was spinning a world view—he had convinced himself of that, and for a long time he convinced many others. He was after bigger game than the paltry prey of the professional. His system encompassed all of the traditional areas of religion, from alpha to omega, and through its message of Progress could speak to all human needs, from politics through literature to moral philosophy.

Within the context of all this, quite apart from anything Darwin was doing to raise the status of evolutionary thought, Spencer was fixing evolution as a topic that one could and must take seriously. He drew in his idiosyncratic way on respectable science, and he did hit a general chord heard by the public—a public which included respectable scientists and like thinkers. Thus although today Spencer's evolutionary ethics is generally viewed with amused contempt, this was not so for his contemporaries. The logician W. S. Jevons wrote: "I have been thinking much of late concerning the theory of morals and feel perfectly convinced that your views afford a complete solution of the ancient problem" (letter to Spencer, June 27, 1873; Spencer Papers, Ms 791/85). For many, conventional Christianity as a religion seemed no longer adequate, especially as it was represented by the conservative Church of England. Progress did seem to be a fact of life, and Spencer was its prophet. But a prophet of the people.

Wallace was a different man, and yet essentially his effect on the professional/popular divide was similar. It is true that, unlike Spencer, Wallace did contribute to (Darwin's ideal of) professional science, both in his seminal paper and in his later work on Lepidoptera—not to mention the interaction with Darwin himself. He was respected for his work in biogeography, and later in life he did become a member of the Royal Society. Yet, typically, he entered some forty years after the younger Huxley, and then only with diffidence and much urging: "all the work I have done is more or less amateurish and founded almost wholly on other men's observations, and I always feel myself dreadfully inferior to men like Sir J. Hooker, Huxley, Flower, and scores of younger men who have extensive knowledge of whole departments of biology of which I am totally ignorant" (Marchant 1916, 2, 221).

The truth is that Wallace's self-assessment was quite accurate. His claim to being a genius is as least as secure as that of Spencer—and for reasons which professional scientists today would value more highly. Wallace's discovery of selection was not a gift of the gods, but a goal he sought when all around him were looking the wrong way. His debate with Darwin about the level of selection could and should be read with profit even to this day, however one judges the proper answer. And his work on biogeography influenced generations. But, for all this, Wallace always was something of an amateur. Less kindly, one could say he was a bit of a crank who would follow any daft idea if it were sufficiently appealing—a man whose interests just so happened to have fallen on evolution. Nor was there any surprise in this, for as a rather ignorant young man of the people, he was precisely the audience at which *Vestiges* was directed. In those days, evolution was another of the half-baked fantasies. And as time went by, Wallace's continued fascination with the unorthodox—phrenology, socialism, feminism—continued to keep him on the edge of respectability. Revealingly, in the 1860s, we find Wallace and Chambers engaged in a cosy, supportive correspondence about spiritualism (Wallace 1905, 2, 285–286).

Significantly, although the scientific community was prepared to pressure Gladstone into giving Wallace a pension, it denied him every professional post for which he applied. Even Wallace's attempts to support himself backfired, for in the 1870s he became embroiled with a madman over a challenge about the sphericity of the earth. Wallace won the wager, but only he would have forgotten that wagers in England are not legally enforceable. Before long, he was embroiled in lawsuits and lost money heavily. This got Wallace sympathy but no respect. His predicament was just not the kind of very public mess in which a professional scientist found himself.

But ultimately, perhaps, Wallace was really like Spencer, in that he was not aiming to produce a professional science. One has to agree that he would take epistemic values far more seriously than was ever dreamed of in the Spencerian philosophy. But his attitude in the 1860s toward our own species shows only too well how readily they would be brushed aside, if the urge came upon him. And as with Spencer, it was the very lack of professionalism in his training which made Wallace willing to devote his energies to the unorthodox, and it was the lack of professional constraints in his maturity which made Wallace willing to live with the unorthodox, even if all around him judged him to have moved beyond the domain of genuine science.

These traits are apparent in Wallace's involvement with broader, more concrete matters. Living by his pen, forever enthused by some quasi-religious or social movement, fearless in his championship of despised and unfashionable philosophies, self-deprecating about his abilities and true worth, Wallace was as far from discipline building as he was from finding financial security. Like Spencer, he was promoting visions, and like Spencer his unifying vision was one of progressive evolution toward Progressive goals. His was a vision more tinged with the supernatural than was Spencer's. But, in the end, like Spencer, Wallace was using evolution to promulgate a Progressionist world view. Popular science was what he wanted and popular science was what he produced.

6

The Professional Biologist

On December 31, 1856, T. H. Huxley (1825–1895) sat in his study, as his wife lay in labor with their first child. He wrote in his journal:

> 1856–7–8 must still be "Lehrjahre" to complete training in principles of Histology, Morphology, Physiology, Zoology, and Geology by *Monographic Work* in each Department. 1860 will then see me well grounded and ready for any special pursuits in either of these branches . . .
>
> In 1860 I may fairly look forward to fifteen or twenty years "Meister-jahre", and with the comprehensive views my training will have given me, I think it will be possible in that time to give a new and healthier direction to all Biological Science.

Then he added: "Waiting for my child. I seem to fancy it the pledge that all these things shall be" (Huxley 1900, 1, 162–163). A son was born just as the old year ended. His parents called him "Noel" because it was the Christmas season.

This is a very different man from those of the last chapter. This is not one to remain content with spinning dreamy metaphysical webs for the general public. Here is a professional scientist laying out his program (Caron 1988). His successes and those of his students and followers are the topic of this chapter. What did the move to professionalism do to evolution and thoughts of both progress and Progress?

Thomas Henry Huxley

It is important to understand the man (Huxley 1900; DiGregorio 1984; Desmond 1994). Huxley was born in Ealing, the seventh and youngest surviving child of a schoolmaster. His formal training was sparse, and right from the beginning there was internal conflict that was to persist to his final day: "My life from eight years up to manhood—was made up of two sets of feelings—joys and anticipations derived from the inward world—that much as a student only knows—(and however ill directed my energies or misspent my time still I *was* a student from my childhood);—and sorrows and misfortunes coming from the outward world—upon all those whom I had reason to love and value most" (letter to Henrietta Heathorn, December 31, 1848; T. H. Huxley Papers, File 40). His father was hopelessly unstable and his adored mother rejected him emotionally: "There was absolutely nothing to bring me into contact with the world—and I hated and avoided it."

Apprenticed to a brother-in-law physician, Huxley began his journey up the scale through scholarships at the Charing Cross Hospital. This was followed by commission into the Navy and a lengthy trip to the southern seas, as an assistant surgeon aboard HMS *Rattlesnake:* the social gap between Darwin and Huxley is well illustrated by the fact that the former messed with the captain and the latter with the midshipmen. Bursting with intelligence and ambition—he was a compulsive worker whose health was forever breaking down into depression—Huxley laid the foundations of his scientific career as he dissected and studied the fragile marine invertebrates that he was in the unique position to dredge up (Huxley 1898, especially volume 1). The work was to precede him back to London, earning him entry into the scientific community, fellowship in the Royal Society, and a medal to boot.

For Darwin and Wallace, global travel had been a significant factor in their development as *naturalists,* with consequent implications for their embracing of evolutionism. Huxley, however, was basically indifferent to his journeying. If he could have had his invertebrates delivered to a lab at home, he would have been just as happy. He was at heart a *comparative anatomist*—a morphologist and, increasingly after reading von Baer, which his self-taught knowledge of German enabled him to do, an embryologist: "what I cared for was the architectural and engineering part of the business, the working out the wonderful unity of plan in the thousands and thousands of diverse living constructions, and the

Thomas Henry Huxley, as caricatured in *Vanity Fair*

modifications of similar apparatuses to serve diverse ends" (Huxley 1900, 1, 7–8).

With such interests, Huxley was at once brought up against Britain's leading anatomist, Richard Owen (Desmond 1982). At first, Owen was friendly and helpful; but, soon the two men fell out and for the next two decades Huxley's work must be understood, in major part, as defining itself against stands taken by Owen. This is not to say that all their differences were simply a matter of personalities. Intellectually, Owen was drawn to speculative transcendentalism. By the early 1850s, his fascination with the ideas of Oken was no secret. Intellectually, the Huxley of this period looked to the developmentalism of von Baer. He translated parts of the great embryologist's major writings into English, and at this point in his career he turned against the parallelism law. Truly, "the progress of a higher animal in development is not through the forms of the lower, but through forms which are common to both lower and higher: a fish, for instance, deviating as widely from the common Vertebrate plan as a mammal" (Huxley [1855] 1898, 1, 301). To this, Huxley added an attack on Agassiz's claims about the embryonic nature of the heterocercal fish tail (p. 302).

Huxley had not long returned to London before he struck up a very close and long-lasting friendship with Spencer—who was nothing if not candid in *his* evolutionism. But it was Huxley's growing antagonism with Owen that fashioned his first public foray into the evolutionary arena, in the form of a vitriolic review of a late (and, in the opinion of many, much improved) edition of Chambers's *Vestiges.* Denouncing the work as "the product of coarse feeling operating in a crude intellect" (Huxley [1854] 1903, 5), the flotsam which might have been picked up from reading *Chambers's Journal*—a nice touch this, given the volume's supposed anonymity—Huxley left no doubt that the real target was not Chambers but Owen. In the edition under review, Chambers (1853) had cited Owen's (1851) progressionist critique of Lyell's (1851) latest stand against paleontological developmentalism. Chiding Chambers for having presumed that Owen was the author of the unsigned critique (a fact that Huxley, like everyone, knew to be perfectly true), Huxley (also anonymous but also known!) proceeded to quote the public Owen against the "ludicrous classification" of the Lyellian critic. The fossil record is flatly against progressionism and hence flatly against any form of evolution.

There was more Owen-baiting down through the 1850s, including a full-scale attack on the vertebrate theory of the skull—a point at which Huxley (1858) clearly demonstrated the great effect that German embryological thought had upon him. But, as the decade drew to a close, another influence was making itself felt upon Huxley. He was at last finding a niche within which he could belong—within which, given his insecurities, he desperately needed to belong. This he found in Darwin and his intellectual circle—Hooker, Lyell, and others. They gave him friendship and a sense of self-worth, and in return, Huxley was more than happy to act as their spokesman and to take on opponents. Membership in the "Darwin club" meant acceptance of evolutionism, but reading the *Origin* did that trick, and so Huxley swung half-circle and became as ardently for the idea as he had shortly before been against it.

But, note that this was ever evolution as *fact.* We find that the Germanic-leaning Huxley was never so very enthused by Darwin's key mechanism of natural selection (Huxley 1893a). He allowed its existence and importance but hankered always after species-changing saltations or jumps. This skepticism was partially a function of Huxley's empiricism—until selection actually brought about species isolation, he would not be truly convinced. It was partially a function of selection's irrelevance. Huxley as a comparative anatomist could go on without it, unlike such people as the mimicry-studying Bates, for whom a mechanism like selection was crucial (Ruse 1979a). In other respects, however, Huxley gave value for money, most lastingly through a brilliant essay on the status of our own species. On comparative and embryological grounds, we are firmly located within the animal world, as primates. Does this mean evolution? Most certainly it does, even for those with worries about mechanisms: "I can see no excuse for doubting that all [living things] are co-ordinated terms of Nature's great progression, from the formless to the formed—from the inorganic to the organic—from blind force to conscious intellect and will" (Huxley 1863, 128).

At the time of the *Vestiges* review, Huxley knew little of the fossil record. But then, thanks to an appointment at the School of Mines, and no doubt spurred by his competition with Owen, Huxley's interest and knowledge exploded. At first, there was tension. Although Huxley was edging over to evolutionism, he yet denied that the fossil record gives much comfort. Indeed, just after the *Origin* we find Huxley conceding, in a presidential address to the Geological Society in 1862, that in the world of plants: "The whole lapse of geological time has as yet yielded not a single new ordinal type of vegetable structure." Moreover: "The positive change in passing from the recent to the ancient animal world is greater, but still singularly small" (Huxley [1862] 1894, 290). By 1870, however, in another presidential address to the same society, Huxley had shifted ground in a major way. With respect to the fossil evidence for evolution, there is now enough information for a far more positive assessment: "when we turn to the higher *Vertebrata,* the results of recent investigations, however we may sift and criticise them, seem to me to leave a clear balance in favour of the doctrine of the evolution of living forms one from another" (Huxley [1870] 1894, 348). The full causes of evolution may still be shrouded in mystery, but the fact is beyond doubt and the paths are becoming much clearer.

Huxley and the Question of Progress

Turn now to Huxley's views about progress. Like everyone else, Huxley always linked evolution and progress. Indeed, his early critique of evolutionism came through an attack on progressionism. Against *Vestiges*, concerning "the lowest discovered remains [the oldest fossils] of animals" Huxley wrote that they "are anything but the lowest in the scale of organization of their class; they have a well-developed intestine and well developed hearts, a nervous system, and long, peculiarly organized arms" (Huxley [1854] 1903, 7–8). It is true that, in some respects, the conversion to evolution made no difference to this line of argument. Huxley went right on (until the late 1860s) attacking progressionism. Note, however, that this is progressionism *as revealed in the known fossil record*. It is clear that this attack had nothing to do with evolutionism *per se*. On other occasions, where he could be positive about evolution, Huxley's arguments—especially those resting on comparative anatomy—showed that he considered this to be a package deal along with progressionism. For instance, we have seen that for our own species, there is "Nature's great progression, from the formless to the formed—from the inorganic to the organic—from blind force to conscious intellect and will" (Huxley 1863, 128).

Huxley tied this all in with some fairly conventional views on race and sex: "It may be quite true that some negroes are better than some white men; but no rational man, cognisant of the facts, believes that the average negro is the equal, still less the superior, of the average white man" (Huxley [1865] 1893, 66–67). Likewise with women. White males stand at the top of life's order, and Huxley was even so far Spencerian as to think that certain among these are higher than others: "I utterly repudiate the Frenchman as a higher and a scientific man—they are logicians and not thinkers and their defects as men of science arise exactly out of their want of moral insight" (letter to F. Dyster, January 30, 1859; T. H. Huxley Papers, 15.106).

All of this goes to show that progressionism was evident in Huxley's writing, fossil evidence or not. By the end of the 1860s, with the fossil evidence coming on side, there was also a new enthusiasm for the writings and ideas of Ernst Haeckel and other German evolutionary morphologists (especially Haeckel's colleague, Carl Gegenbaur). This approach struck a responsive chord in Huxley. All of the worries of the past were forgotten, and we find the evolutionist Huxley, of the 1870s, trumpeting Germanic progressive parallelism. And, having gone this far, he even had the nerve to invoke the Chain of Being: "For though no one will pretend to defend Bonnet's 'échelle' at the present day, the existence of a 'scala animantium' is a necessary consequence of the doctrine of evolution; and its establishment constitutes, I believe, the foundation of scientific taxonomy" (Huxley [1880] 1898, 4, 461). In twenty years, Huxley's position on progress was spun right around, from progressionism being the greatest threat to scientific biology ("ludicrous classification") to its being the essential condition.

Turn next to the question of influences. Huxley read broadly and had great facility for languages. Hence, continental ideas readily made their way into his thinking. For all that he ran down the French, at a theoretical level he certainly believed in the idea of Progress in civilization, especially in the world of thought and most particularly in the world of science—with advances in science feeding back to and supporting the advance of civilization. In respects, this belief (as in biology) translated into saltationism: "The advance of mankind has everywhere depended on the production of men of genius; and that production is a case of 'spontaneous variation' becoming hereditary, not by physical propagation, but by the help of language, letters and the printing press" (letter to C. Kingsley, 5 May 1863; in Huxley 1900, 1, 259). For once, however, a Darwinian perspective also applied: "The struggle for existence holds as much in the intellectual as in the physical world" (Huxley 1893a, 229).

German thought was even more influential. Notwithstanding his sneers at Owen on this score, Huxley was ever tempted toward idealism. When in Australia, he had made friends with William Macleay, author of an idealistic taxonomy based on groupings of five (Macleay 1819–21). For a while, Huxley (like Chambers before him!) became a quinarian, and even though this enthusiasm vanished, in the heat of the anti-Owen battle, Huxley still openly admitted to non-material causes—"the phaenomena of life are dependent neither on physical nor on chemical, but on vital forces" (Huxley [1854] 1893, 64). This was combined with enthusiasm, especially via the influence of Carlyle's writings, for Progressive, idealistic German thinkers. Goethe was always a Huxley hero. Later, with Owen vanquished, with evolution overcoming and sanitizing transcendentalism—"After all it is as respectable to be modified monkey as to be modified dirt" (letter to Dyster; T. H. Huxley Papers, 15.106)—and with the fossil record beginning to add support for his side, Huxley could relax more openly into Germanic P/progressionism, where

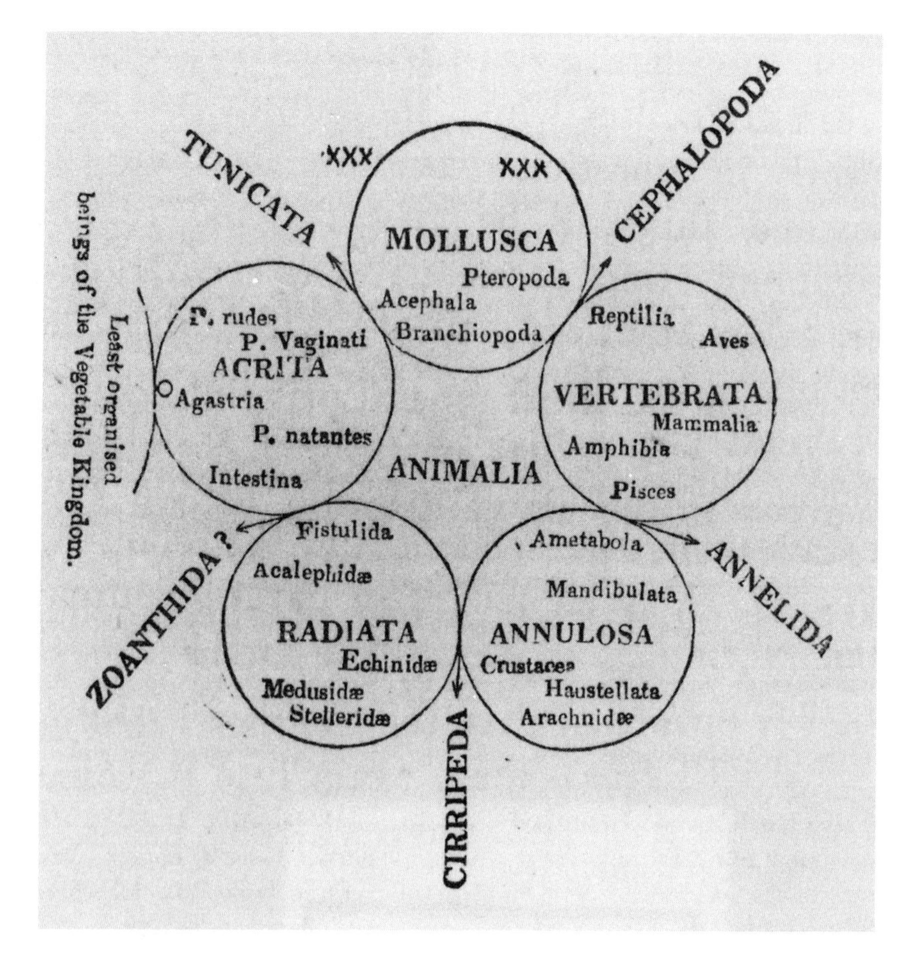

The basic pattern of Macleay's quinary system. Organisms are grouped in nested sets, each with five members. Certain groups, like barnacles (Cirripeda), are "osculating," for they are squashed between adjoining circles. (From Macleay 1821.)

the social and the biological are facets of one higher force. But, let no-one deny that his new debt ultimately was to precisely those *Naturphilosophen* who had so excited his arch-rival. Haeckel's biogenetic law did not spring from thin air.

Concerning the British component to Huxley's progressionist thinking, we must not be misled by the Darwin connection. Given their close intimacy, not to mention the conception of evolution with which Huxley started, one looks first to some influence by Herbert Spencer. This is complemented by a reading of Carlyle's *Sartor Resartus,* with the transcendental Calvinism of this work being transformed into what Huxley (who invented for himself the label "agnostic") called "scientific Calvinism," meaning that the laws of nature are inexorable and we must submit to them (Huxley 1900, 1, 328). Fortunately, nature is so ordered that "the wicked does *not* flourish nor is the righteous punished." However: "we must bear in mind what almost all forget, that the rewards of life are contingent upon obedience to the *whole* law—physical as well as moral—and that moral obedience will not atone for physical sin, or *vice versa*" (letter to Kingsley, 23 September 1860; in Huxley 1900, 1, 236).

Every Calvinist has trouble reconciling the necessity of law with the freedom of action. Huxley himself did not achieve what he achieved by passive acceptance of nature's laws or of its products. He became a force in science education; he served on Royal Commissions; and he was made a Privy Counsellor. He arrived where he was by bringing intelligence and guts and caring sympathy against the baser forces of nature, within and without his own skin. Moreover, given such a program for one's life, *laissez faire* proved inadequate. Huxley's success was due to his being the compleat and enthusiastic servant of the state: "The higher the state of civilisation, the more completely do the actions of one member of the social body influence all the rest" (Huxley [1871] 1893, 261). Thus, at the end of his life, he reformulated and restated his personal philosophy in his greatest essay, on the relationship between evolution and ethics (Helfand 1977; Paradis 1989). Progress is possible, but not in an easy, slick Spencerian way. For the late Huxley, the civilized man is he who strives to conquer the old Adam within himself: "Social progress means a checking of the cosmic process; the end of which is not the survival of those who may happen to be the fittest, in respect of the whole of the conditions which obtain, but of those who are ethically the best" (Huxley 1893c, 81).

Does this mean that, at the final point, humans have escaped their biology and hence there can be no true feedback from Progress to progress? Spencer saw that Huxley veered this way and feared that he was reverting to "the old theological notions, which put Man and Nature in antithesis" (Duncan 1908, 336). One suspects that there is truth in this, but Huxley did try to escape the dilemma. The ethical sense itself, the feeling of sympathy we have for other humans—Huxley acknowledged the influence of Adam Smith here, and he had earlier written a book positively inclined toward Hume—is as much a product of evolution as is the selfish brute within each and every one of us. In addition, Huxley linked the human sense of sympathy to an analogous sense one can find among lower social animals. "Wolves could not hunt in packs except for the real, though unexpressed, understanding" (Huxley [1893c] 1989, 56–57). This was clearly a biological instinct for Huxley, for he was even prepared to take his case down to the ants.

For Huxley, therefore, at the highest levels biological progress does mirror cultural Progress and indeed in humans eventually the two processes become one. This is a complex web of belief spun from many different strands—spun with such sincerity that it seems almost cruel to go on to ask about the evidence. By demanding proof one seems to be turning the knife, given how Huxley made honesty to nature a part of his religion: "My business is to teach my aspirations to conform themselves to fact, not to try and make facts harmonise with my aspirations" (letter to Kingsley, 23 September 1860; in Huxley 1900, 1, 235). However, like most religious claims, secular or otherwise, Huxley's self-avowed attitude should be taken with a pinch—a very large handful—of salt. It is true that new evidence was coming in from the fossil record, but it is far from obvious that the evidence was itself sufficient and decisive to cause a sea change in its significance. Similar points pertain to the transcendentalist metaphysics which Huxley took on board from Haeckel. It is a virtual truism to note that the past one hundred years have not been a happy time for those inclined this way. Indeed, there are few more reviled claims in the history of biology than the biogenetic law. But my point here is simply that the evidence did not change so very much between the time when Huxley was critiquing the position and the time when he embraced it. Hence, one doubts that naked reality was the crucial factor.

Analogous queries about the use of evidence to support a theory arise in other areas of Huxley's work, perhaps most particularly in the area of social evolution. Huxley may well have been right. In fact, many of today's evolutionists would find much to praise in his stand (Williams 1988). But Huxley certainly does not give a firm evidential base for the evolution of an ethical sense. I do not write to criticize. Indeed, many philosophers (myself included) would applaud Huxley's recognition that that which has evolved is not necessarily that which is good. I simply conclude that Huxley, like his fellow Englishmen—although in his own distinctive way—thought progressively in major part because he was a Progressionist.

Discipline Building

In 1860, little Noel Huxley died from scarlet fever. His distraught father yet had the strength to add to the earlier happy passage, in his journal: "My boy is gone, but in a higher and a better sense than was in my mind when I wrote four years ago what stands above—I feel that my fancy has been fulfilled" (20 September 1860; in Huxley 1900, 1, 163). And truly it had, for Huxley had positioned himself to make a profession of biology as never before in Britain—a task which he and his fellows accomplished in the next quarter-century, making careers for themselves and a discipline for the science.

When he returned to Britain at the beginning of the decade, Huxley had to struggle hard to keep afloat. Science, he learned, gave much praise but "no pudding" (letter to sister Lizzie, May 3, 1852; in Huxley 1900, 1, 100). It was not long, though, before he was eating full-course meals, so much so that he could remain in London, at what he perceived as the center of power, spurning jobs in Edinburgh and Oxford. For thirty years Huxley worked in various branches of what was becoming London University, ending as dean of the science college in South Kensington. He was able to make a career in a secular setting, as a scientist—researcher, teacher, administrator. At the same time, virtually throughout his career, he made money from writing and lecturing as well as through his government work.

And always Huxley was working to support those who labored alongside him, placing eager followers and students in key, influential posts. George Rolleston became Linacre Professor of anatomy and physiology at Oxford in 1860, thanks to Huxley's influence. "I suppose you will have heard the result to which you yourself have largely contributed. I shall set to work so as never to give you cause to regret the share you have had in my promotion" (letter to Huxley, n.d. [1860]; T. H. Huxley Papers, 25.148). Then, in 1870, at Cambridge, it was the turn of the physiologist Michael Foster, student, assistant, and close friend: "I read your letter to the Seniority yesterday. Your suggestion as to the Praelectorship of Physiology and the man to fill it [Foster], was most favourably received" (letter from W. G. Clark to Thomas Henry Huxley, April 2, 1870; T. H. Huxley Papers, 4.172).

In the scientific societies, also, Huxley gained and used power. He and his friends in the X-Club ran the Royal Society as their own personal fiefdom, for years taking the chief elective offices, from president down.

The specialist societies, like the Geological and the Linnaean, were similarly invaded and conquered. Nor was Huxley loathe to use his power: he gave his views directly in official addresses and indirectly through the refereeing process, as he strained out only the acceptable for publication in prestigious organs: "I have no hesitation in very strongly recommending the publication of the chief part of this memoir"; "The memoir is . . . well worthy of publication in the transactions of the Royal Society"; "The paper is a very valuable one" (Royal Society Papers, Referees' Reports: RR.5.173, RR.4.97, RR.6.200). This is apart from Huxley's personal involvement in the production of various professional publications, culminating (in 1869) with the successful launching of *Nature.*

How exactly did Huxley implement his plans for biology? Professionalizing science meant discipline building, and I remind you of a crucial point about discipline building. As in any commercial activity, you have got to find a market. You must find people willing to buy your product, or your schemes will wither on the vine. In professional science, you have to find employment for your teachers/researchers, and that means their activities must be valued by the public—a public that will produce students who want training or will need your specialized knowledge. In terms of employment potential, Huxley and his friends were working at the right time, for this was just the point at which the ancient universities were having secularization forced upon them; London University was consolidating and expanding, and provincial universities were being formed; a civil service based on merit was being formalized and developed; universal primary education was being enforced; traditional and newer professions were insisting on formal criteria for training and admission; and more. In short, a modern-looking state was being formed (Webb 1980).

It was Huxley's brilliance to create and seize the opportunities and see that professional biology could include two disciplines—speaking now essentially about the animal side, which was his real interest. The one was the fairly obvious subject of *physiology,* that part of biology dealing with the mechanisms and workings of the body. Huxley himself was not really that much of a physiologist—it is a science that relies heavily on experimentation, which was not his strength. But through his students, H. N. Martin in London and then in America, and especially his favorite, Michael Foster in Cambridge, Huxley could do his work. And there was a ready market for the product, for Huxley and his associates persuaded a (willing) medical profession that a thorough training in physiology is just what is needed by the top-flight physician of the future (Geison 1978). So before long, we find researchers in physiology, honors degrees in physiology, and eager and talented students queuing to take the examinations.

Huxley himself was a comparative anatomist. How could he also justify a discipline of *morphology?* Quite simply, by making it a crucial component of school education, he ensured a consequent need for qualified teachers. It was for this reason that he ran and served on the London School Board—by all accounts, in (for him) an incredibly non-confrontational manner—and it was for this reason that South Kensington was famous for its summer courses for school teachers. But, why should a schoolchild study morphology? What was the virtue in cutting up a crayfish or a frog? Explicitly, Huxley saw this course of study as the equivalent in modern terms to the training hitherto supposedly conferred by the study of the classics. Getting directly in touch with nature was a moral experience. Only half-joking, he wrote of "a course of instruction in Biology which I am giving to Schoolmasters—with the view of converting them into scientific missionaries to convert the Christian Heathen of these islands to the true faith" (Huxley 1900, 2, 59–60).

Hands-on comparative anatomy was the secular equivalent of a sacrament: "Science seems to me to teach in the highest and strongest manner the great truth which is embodied in the Christian conception of entire surrender to the will of God. Sit down before fact as a little child, be prepared to give up every preconceived notion, follow humbly wherever and to whatever abysses nature leads, or you shall learn nothing" (Huxley 1900, 1, 219).

Huxley was not kidding when he ran down Frenchmen:

> What I earnestly maintain is, that thoroughly good work in science cannot be done by any man who is deficient in high moral qualities—
>
> It is the moral which is essential to the right working of the intellect—and the value of science is that it compels men to know that such is the case. (Letter to F. Dyster, January 30, 1859; T. H. Huxley Papers)

What makes Huxley important is that he sold this vision to his fellow Victorians.

The professional-science discipline building was matched, and connected causally, with an emphasis on epistemological concerns. Physicians needed to stop killing people and to begin curing them. Hence, in physiology values like predictive accuracy started to take on supreme importance. Experimentation became the norm. It is no chance that mathematics began to assume significance. Morphology, particularly as Huxley saw it, was more descriptive. The need for consistency and the like was presupposed. Prediction was somewhat less forward and, to be candid, his empiricism made Huxley suspicious of prediction-alternative epistemic values like consilience (Huxley 1879). What was crucial was the elimination of cultural values. Huxley was playing the same game as Cuvier. For professional science to succeed, it had to be (ostensibly) free of cultural values. A radical science would be as ineffective as if its promoter were an acknowledged adulterer.

Which brings us to the key question: Where did evolution stand in all of this discipline building? Here we come to the most incredible and significant fact in my whole history. In major part, Huxley did not want evolution to have any part in his professional science! "Darwin's bulldog" excluded it, keeping it firmly down at the popular level—at least inasmuch as professional science was a matter of the day-to-day work within the discipline. There was essentially no place for evolution, either in physiology or morphology. As Huxley grew in power, and as he developed biology, the profession of biology and the subject of evolution became badly estranged.

A point noted by students: "One day when I was talking to him, our conversation turned upon evolution. There is one thing about you I cannot understand," I said, "and I should like a word in explanation. For several months now I have been attending your course, and I have never heard you mention evolution, while in your public lectures everywhere you openly proclaim yourself an evolutionist" (Father Hahn, S.J.; quoted in Huxley 1900, 2, 428).

This recollection is backed by notes taken down in Huxley's very thorough anatomy courses. For instance, the lectures for 1869–71 (160 in all) have but one brief mention of selection/evolution, concluding: "There is no positive evidence that selective breeding alone will produce forms infertile with one another" (November 16, 1869, notes taken by P. H. Carpenter; Zoology Library, Oxford University). Darwin and the *Origin* are not mentioned by name. The lectures for 1879–1880 are little different. There is now (in literally just one or two lectures) an explicit acknowledgment that evolution has occurred, but still no discussion whatsoever of natural selection. There are, of course, lots of things (everything!) which could be interpreted in an evolutionary fashion—Huxley dealt at length with the comparative physical anatomy of humans ("The most brutal race and that nearest the lower animals is the Australoid")—but basically nothing to give offense to Agassiz (lecture of November 15, notes taken by H. F. Osborn; Osborn Papers, AMNH).

The textbook Huxley wrote (with Martin) bears out the exclusion (Huxley and Martin 1875), as do the examinations given in the 1870s at Oxford, Cambridge, and London.[1] London was never evolution-friendly, right from the first honors biology exam in 1862. (With Huxley as examiner, a typical question was: "Describe the structure of the Eye in (1) a mammal, (2) a fish, (3) a cephalopod, (4) an insect.") At Cambridge, after an initial enthusiasm for evolution in the 1860s, coverage subsided. Exams generally included one question (among a possible forty) a year—perhaps a concession to the fact that Francis Darwin, a son of Charles, was one of the examiners. (1877: "The development of the individual is an epitome of the history of the species. Discuss this proposition." Of course, Agassiz would also have liked this question.)

Oxford is the most revealing of them all. Evolution and selection do not make an exam appearance until the mid 1870s. Then there is about one question a year. Typical is a question (the only such question) which appeared on an 1880 morphology paper: "How do you account for the existence of 'flowers'? Give a hypothetical sketch of the gradual evolution of a phanerogamous inflorescence." This question caused a tremendous row, with the Linacre professor (Rolleston) contemptuously scribbling "H. Spencer" in the margin of his copy and protesting that it went beyond the accepted area for examination as defined by the set text: "Not in Huxley and Martin"! His ire had been raised already by earlier questions of similar ilk. (Information from a manuscript in Oxford University's Zoology Library; no location given, except previous page says "Minutes of Board of Nat. Sci. Studies Nov. 16, 1880.")

Evolution was excluded from Huxleyite professional biology. Why? First, there is the fact that, at some deep level, evolution was incompatible with Huxley's ontology and his pedagogy. "It may indeed be a matter of very grave consideration whether true anamorphosis [progress] ever occurs in the whole animal kingdom. If it do, then the doctrine that every natural group is organized after a definite archetype, a doctrine which seems to me as important for zoology as the theory of definite proportions for chemistry, must be given up" (Huxley [1853] 1898). This was written in the pre-*Origin* days; but, whatever he may have said in the enthusiasm of the later moment, Huxley always thought in typological terms, and his teaching—focusing on exemplars: earthworm, crayfish, frog, etc.—was based on such thinking, explicitly. Notwithstanding his popular philosophy, his professional philosophy was static.

Second, there was Huxley's ardent empiricism and his religious attitude to unadorned nature. For him, particularly for him as a teacher, what counts first and foremost is the anatomist and the subject; nothing (including theories) is to intervene. To his puzzled student, Huxley responded: "Here in my teaching lectures (he said to me) I have time to put the facts fully before a trained audience. In my public lectures I am obliged to pass rapidly over the facts, and I put forward my personal convictions. And it is for this that people come to hear me" (Huxley 1900, 2, 428). This was the point. Start with facts and leave opinions out of it. Remember that for Huxley the morphologist, as for his physiological supporters and students, natural selection as a mechanism held no great attractions. None of them were working on issues of adaptation, as were Bates and Wallace. We have a feedback here, from the epistemological to the sociological. Huxley's vision of mature science did not need natural selection, and would probably bar it (as epistemically inadequate). Hence, there was no urge to study it at the professional level.

Third, and most important of all—certainly the point which impinges most directly on our inquiry into the history of Progress—think about where Huxley learned his evolutionism. Darwin made an evolutionist of Huxley, but it was Herbert Spencer who set the framework, both before and after the *Origin.* This was the man with whom Huxley took long argumentative walks, every Sunday afternoon. Their conversations obviously had an effect on Huxley, for of the very un-Darwinian prolegomenon to the Synthetic Philosophy, *First Principles,* he wrote that: "I agree in the spirit of the whole perfectly" (Huxley 1900, 1, 230). And, in this context, how else could Huxley regard evolutionism, other than as popular science? As a philosophy of Progress? As a metaphysic? And note, incidentally, how this point ties in with the second point, echoing the resemblance to Cuvier. Huxley the teacher, Huxley the teachers' teacher, purveying his wares as essential material for the young mind, could not afford to have people critiquing him as one trying to push dangerous doctrines into the minds of the youth of Britain. By the 1870s, in Germany, Haeckel was running into just these problems, precisely because he would so mix up his science and his philosophy (Nyhart 1986). The Englishman knew better than to fall into such a trap. It was because

Huxley's morphology was theory-free, something he wanted anyway, that it was suitable for the classroom. Culture be gone!

For these various reasons, Huxley excluded evolution from professional biology. It was not appropriate for a discipline and not a candidate for mature science. And since he equated evolution with P/progress, that meant Huxley was excluding P/progress also, which is precisely what he meant to do. Professional, mature science had no place for Huxley's "personal convictions." Yet, as we prepare to move on now from Huxley, note that there are reasons—internal and external—why his position might not have been very stable and why it might not have persisted. Internally, in speaking of evolution as "metaphysical," I do not imply that Huxley regarded it as pure speculation or as doctrine essentially beyond reason and evidence. It was rather a background picture, a theory which made sense of the world.[2] But it could not just exist in splendid isolation. If it was to have authority at any level, including the popular level, it would have it because of the status of Huxley and other evolutionists, *as professionals.* And it had to be as professionals that they were qualified to speak on evolution, whatever its status.

In other words, even though one might not have wanted to work on evolution within one's professional disciplinary science, it was precisely that science which would give the theory the stamp of approval. It was for this reason that Huxley was prepared—willing, even—to talk on matters evolutionary in such fora as presidential addresses to learned societies or when he was speaking by virtue of his authority as a scientist to a general audience, as in his Rede lecture in the University of Cambridge Senate House in 1883 (Huxley [1883] 1903). You may think that at such moments the line between popular and professional is being crossed, or at least muddied, and perhaps you are right. What one must recognize is that, whatever lines of demarcation Huxley himself might have drawn, someone without his background and emotional commitments might not feel compelled to draw exactly the same line as he.

This point is backed by reasons external to his own activities for instability within Huxley's position. In Germany, Haeckel may have been something of a wild man, but after the *Origin* he and then (even more) his friend and co-worker Gegenbaur were making a serious effort to create a mature science of evolutionary morphology. This would be a professional evolutionary discipline: professorships, journals, students, and the like (Nyhart 1986, 1995). Their efforts made people no more Darwinian than they had ever been, as they concentrated on transcendentalist morphological tracing of phylogenies, relying heavily on embryological analogies; but the movement as a discipline did exist and thrive, for a time at least. If not Huxley, then some of the young men around and influenced by him might have been expected to respond favorably to this German evolutionism as professional science and attempt to replicate it—and its disciplinary status—in Britain.

Indeed, for all of his emotional commitments otherwise, there is some suggestion that Huxley himself, by the time he came to the end of his career as an active scientist (around 1880), was edging a little that way. In the interests of getting the best of German thought into the English professional market, not only was he prepared to let through references to evolution in papers he championed, but his own publications were starting to include—at least hint at—evolutionary ideas. Very much under the influence of American fossil discoveries, in 1880 Huxley did try his hand at classifying the mammals, using a variant of the biogenetic law (with a favorable reference to Haeckel), at the Zoological Society (Huxley 1880). This was not the most prestigious of societies—it ran the London Zoo and was very much a gentlemen's club (Scherren 1905), as Huxley's grandson was to find out in this century when he crossed the governing board—and the paper does seem to have been an end to this inclination. But it does support a suspicion that although Huxley himself may have had his face turned away from a professional discipline of evolutionary biology, others following him might have felt and acted otherwise.

To test these suspicions, and to see how this all relates to questions of evolution and P/progress, I turn now to two of the key biologists at the end of the nineteenth century.

E. Ray Lankester and the Problem of Degeneration

E(dwin) Ray Lankester (1847–1929) was born to the scientific purple (Lester 1995). The son of a scientist, he grew up knowing the leaders of British biology as family friends. He met Darwin when he was a mere child, and the intimacy with the Huxleys was such that he referred to T.H. as his "father-in-science," never hesitating to draw on him for a reference. A fellowship at Exeter College Oxford was followed by professorships at University College London—where he was Grant's successor (!)—and at Oxford—where he was Rolleston's. Forever embroiled in controversy, Lankester inspired both exasperation and affection in his friends. "There is a mixture in him of the most bare-faced conscience-less selfishness with a certain good nature—a vigorous power of work, and meek acceptance of rebuke which completely fascinates me, and leads me to do things for him that my judgment does not approve of" (letter from Michael Foster, August 2, 1885; T. H. Huxley Papers, 4.256). To cap his career, he went off to head the British Museum (Natural History). Lankester's early retirement from this post was an event much desired by others (Stearn 1981). Thereafter, he became a very popular science columnist in the *Daily Telegraph* and elsewhere. Having been pushed out of the BM(NH), his enemies at Oxford blocked his reappointment as Linacre Professor (Howarth 1987).

Lankester's good qualities were of the highest. A brilliant teacher, he made important contributions in biology from parasitology to embryology. A Huxley-type biologist in the fullest sense—he was Rolleston's student at Oxford and had himself assisted in Huxley's summer lectures for schoolteachers—Lankester did his best work, from a developmental perspective, on the comparative anatomy of invertebrates. Especially noteworthy was a study of the king crab, showing its closer affinities with the scorpion (an arachnid) than with its hitherto assigned class-mates, the crustacea (Lankester 1881). Lankester was also a Huxley-type biologist when it came to his picture of evolution. He was happy to talk of natural selection, but he thought, acted, wrote biology from an evolutionary transcendentalist perspective. This was a given, not even to be questioned.

Hence, to bring together our discussion of Lankester's evolutionism and the first of our usual questions (namely that about progressionism), we find that as with Huxley and as with the German mentors—it was Lankester who supervised the translation into English of Haeckel's popularization of his ideas (Haeckel 1876)—a backbone of progressivism ran through Lankester's writings on evolutionism. He certainly thought of life's history as a branching tree, even (of the animals) speaking of "six or seven great lines of descent—main branches of the pedigree" (Lankester 1880, 26). But a main line does extend upward: "the general doctrine of evolution justifies us in assuming, at one period or another, a progression from the simplest to the most complicated grades of structure; [and] that we are warranted in assuming at least one progressive series leading from the monoblast to man" (Lankester 1877, 400). Apparently there is a similar line for plants: "The organic forms present two series Animal and Vegetable which in their structure ascend from the simple to the complex" (Lecture 2, 1876, notes taken by D'Arcy Power; Royal College Papers, 42 C8, p. 6).

Needless to say, Lankester was prepared to accept a (tempered) version of the biogenetic law and to use it in the game of phylogeny tracing: "The ontogeny of the individual development has a tendency to recapitulate the phylogeny." Nor will you be surprised to learn that, when it comes to humans, the usual ordering prevails.

Anthropi (Sub order Catarrhini, order Simiae) Sole genus *Homo sapiens*
Ulotrichi (curly haired races)
Negroes (incl. Zulus)
Negritoes of Papua
Bushmen and Hottentots

Leiotrichi (straight haired races)
Australians
Mongolians

Mediterraneanes
Jews
Hindoos, Persians
Celts, Latins, Greeks,
Teutons, Slavs

(These are lecture notes, and hence complexity increases as we go *down* the page—corresponding to the order of presentation.) As one might expect, lower races and children show more clearly our simian origins: "gorilla [is] potbellied, in fact, children, esp. of lower races of men are somewhat in this condn at first." (Notes kept by H. J. Harris, 1889, in D. M. S. Watson Library, University College, London, Ms Add 95.)

The distinctive twist in Lankester's thought comes from the fact that he was a good friend of the German biologist Anton Dohrn, student of Haeckel and founder of the Naples Marine Biological Station (Oppenheimer 1978). Against a general background of evolutionary progressionism, Dohrn argued strongly that often we find organisms degenerating, ontogenetically and phylogenetically. Lankester picked this up immediately and made it a major part of his own theorizing. Defining "degeneration" as "a loss of organisation making the descendent far *simpler* or *lower* in structure than its ancestor," he gave detailed examples illustrating the widespread occurrence of the phenomenon (Lankester 1880). A prime instance is the ascidian or sea squirt, which begins its existence as a tadpole-like vertebrate but which ends life as a sponge-like invertebrate stuck to objects on the sea bottom. Whereas the tadpole goes forward and upward into froggy-hood, the ascidian goes backward and downward into a vegetable-like state.

Complementing the examples, Lankester argued that causal explanations lie close at hand. "Any new set of conditions occurring to an animal which render its food and safety very easily attained, seem to lead as a rule to Degeneration" (Lankester 1880, 33). By illustration, Lankester drew on *Homo sapiens*—"an active healthy man sometimes degenerates when he becomes suddenly possessed of a fortune"—which pushes us to ask the second of our questions, about the influences Progressionism had on Lankester. Speaking generally, Lankester both young and old was a Progressionist. Interestingly, his sympathies, at least at first, seem to have lain rather more with Herbert Spencer than with Huxley. Writing to Dohrn about religious tolerance, he said:

> I confess I have more sympathy with [Spencer's] views than those of Huxley on this matter. Huxley has some liking for Carlyle and you too—you would like to *make* people do what is good for them to do. I have a strong feeling against this—and even if there is some delay in progress—a true natural development of institutions by the recognition on the part of the people concerned of their necessity—is it seems to me far more healthy and enduring than a growth pushed on by the strong hands of government. (Letter of March 26, 1872; Dohrn Papers)

The theme here is Progress, and it is this same theme which Lankester continued to endorse, right down to one of his final popular essays (Lankester 1923).

What of the talk of "degeneration"? For Lankester the biologist, degeneration was a theme to be played against a background of progressionism; similarly for Lankester the social thinker. Despite Progress: "The traditional history of mankind furnishes us with notable examples of degeneration. High states of civilisation have decayed and given place to low and degenerate states" (Lankester 1880, 58). The writing, friends, is on the wall even for us, Europeans: "Possibly we are all drifting, tending to the condition of intellectual Barnacles or Ascidians" (p. 60).

You can see how Lankester shifted easily back and forth between the biological and the social, using the one to support the other and conversely. Behind such writing as this were layered concerns holding up the argument, rather like a supporting column of fossiliferous rock. At an upper, more public level, Lankester as a science professional was worried

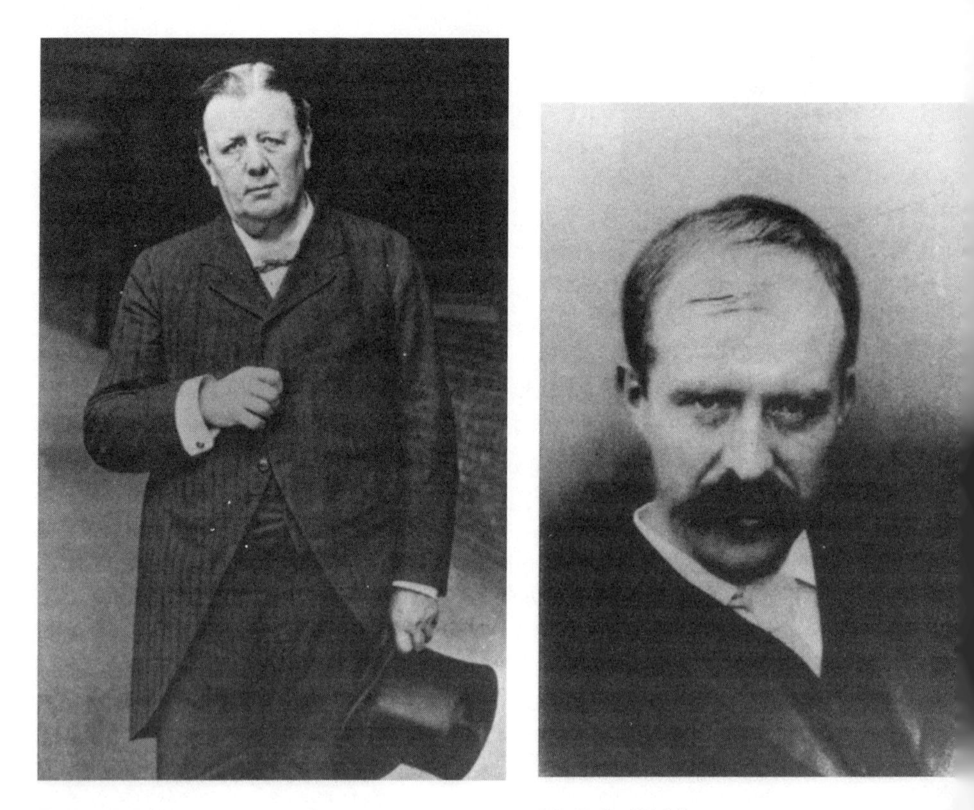

E. Ray Lankester

W. F. R. Weldon

desperately about the inadequate treatment of and funding for science in Britain: "no attempt is made in this country to raise scientific research, and especially biological research, from the condition of destitution and neglect under which it suffers" (Lankester 1880, 65). For a while during the First World War Lankester feared that the point of no return had been passed, but Allied victory brought a renewal of faith. At a lower, less public level, Lankester had personal concerns about degeneracy, because he felt that he himself, through the ill effects of English society, had become a victim of the trend. As a fellow of an Oxford college, ostensibly a privileged success, Lankester was both adding to England's troubles and suffering from them:

> No one knows who does not live in the place—the inextricable mess of mediaeval folly and corporation-jealousy and effete restrictions which surround all Oxford institutions . . .

> It is not my own opinion alone, but I hear it from many other fellows—that many men have their lives ruined by taking resident fellowships. They stay up making large incomes (for bachelors) and spending a great deal—till it is too late for them to take to anything else. Their fellowships are forfeited by marriage so they don't marry. They don't work—for why should they? Their time is sufficiently occupied with routine duties and routine amusements such as riding, dining and whist. To these may be added an occasional visit to London—for unknown purposes. That is the society of Oxford and one has the prospect of becoming one of these social parasites. (Letter to THH, December 18, 1872; T. H. Huxley Papers, 21.39)

Unfortunately, in speaking of London's "unknown" attractions, Lankester was not being quite truthful. If not to Huxley, then at least to his own friends he admitted candidly that because of his background he was unable to relate sexually to women of his own class. "The 'Eternal Question of Woman' is always driving me this way or that. If one could only love a whore the problem would be solved—but I hate them and yet I can not touch the others" (to Dohrn, July 17, 1872). This attitude led to two broken engagements, a public brush with the police in London, a reputation as a "womanising drunkard" (letter from E. R. Lankester to E. B. Poulton, February 17, 1896; Poulton Papers), and a lifetime of lonely Christmases spent with other bachelors at assorted golf clubs. Convinced that his own emotional capacities had been driven back to the pre-vertebrate level, Lankester's views on the process of evolution were steeped in bitter personal emotion. Degeneracy may be the way into which one is easily led. It is not the happy way or the right way.

There is hardly need now to linger over our third question, about the evidence. What was already said about Huxley and earlier transcendentalists holds here. Without in any sense denying the really solid productions which issued from Lankester's pen, the value-impregnated status of his thinking on evolution is there for everyone to see. Certainly, the discussion of degeneration is not backed by the empirical evidence that Lankester himself would have demanded in the case of straight morphology. The definition offered was hardly that formal or fully developed; there was no systematic survey, just a few conveniently chosen examples; and the supposed causal underpinnings were understated, to say the least. At first, Lankester inclined toward Lamarckian-type factors behind degeneration, although later—in line with general opinion as the century closed—he grew to suspect that these could not be significant. But, either way, there was no real attempt to find an adequate mechanism.

All the facts point to one conclusion, namely that Lankester's thinking on progress and (biological) degeneration was a direct function of his social thought. I will hold putting this in context until I have discussed my other example of a biologist in the post-Huxley years.

W. F. R. Weldon and the Mathematics of Selection

W. F. R(aphael) Weldon (1860–1906) was one of the leaders, together with his good friend and sometime colleague Karl Pearson, of the so-called biometric school (Pearson 1906; Provine 1971; Mackensie 1981; Gayon 1992). These were people bound by a belief in the importance of a mathematical approach to biology—Pearson was an architect of modern statistics—who also, notoriously, opposed strenuously the arrival at the start of this century of the Mendelian theory of inheritance. As it happens, although posterity judges the biometricians as losers, looking at the details of the debate I think they had good reasons to be suspicious of early Mendelism. But my interest lies elsewhere, namely in Weldon the selectionist—the man who actually put natural selection to empirical test—and in the implications that his work had for his philosophy of biological research.

Weldon's career fits a pattern one would expect in the making of a professional biologist in the late Victorian era. Schooldays, at a small private school for the sons of non-conformists, were followed by undergraduate years, first at University College, London, where Weldon studied under Lankester, and then at Cambridge, where he studied under the great embryologist Francis Balfour—meanwhile slipping back up to London for dissection classes with Lankester. There was a postgraduate lectureship in invertebrate anatomy, at Cambridge, and then Weldon succeeded in 1891 to Lankester's chair at University College, London. It was now that Weldon forged his close friendship with Pearson, who was the professor of applied mathematics and mechanics at University College. An intense man, Weldon lived in constant conflict, thinking that all who differed from him were deliberately obtuse. In 1899 he again succeeded Lankester, this time as Linacre Professor at Oxford. Controversy with the Mendelians increased at this time, a fact made that much more bitter because William Bateson, the leader of that group, was Weldon's sometime student (of morphology) and friend. Tragically, the strain proved too much. In less than a decade Weldon collapsed and died.

For all of his morphological and embryological training, Weldon's true interests lay in mechanisms of biological change, and his passion—one that would have gladdened the heart of Darwin—was natural selection. Moreover, unlike fellow enthusiast E. B. Poulton, Hope Professor of entomology at Oxford and expert on butterfly mimicry, Weldon was determined to bring work into the laboratory. Although he was no trained mathematician, Weldon was convinced that it is through and only through mathematics that one can hope to grasp nature's causal mechanisms:

> It cannot be too strongly urged that the problem of animal evolution is essentially a statistical problem: that before we can properly estimate the changes at present going on in a race or species we must know accurately *(a)* the percentage of animals which exhibit a given amount of abnormality with regard to a particular character; *(b)* the degree of abnormality of other organs which accompanies a given abnormality of one; *(c)* the difference between the death rate per cent in animals of different degrees of abnormality with respect to any organ; *(d)* the abnormality of offspring in terms of the abnormality of parents, and *vice versa.* These are all questions of arithmetic; and when we know the numerical answers to these questions for a number of species we shall know the direction and the rate of change in these species at the present day—a knowledge which is the only legitimate basis for speculations as to their past history and future fate. (Weldon 1893, 329)

Preparing the way for controlled studies, Weldon reasoned that if there is no selection acting on a population of organisms, then by the regular laws of heredity—whatever these may be—one ought to find specified variable characters satisfying that so-called normal curve of distribution, which data collectors in the nineteenth century had found to hold true of the widest range of phenomena, natural and artificial. However, if selection is at work the normal curve should be shifted (through successive generations), and the shift and its extent should reveal the nature and magnitude of the operating forces.

With this hypothesis as guide, Weldon turned to the measurement of shrimps and crabs and found that many of their features did show the expected normal curve (Weldon 1890, 1892). Not all did, however. The frontal breadth (the distance between two points on the upper front shell)

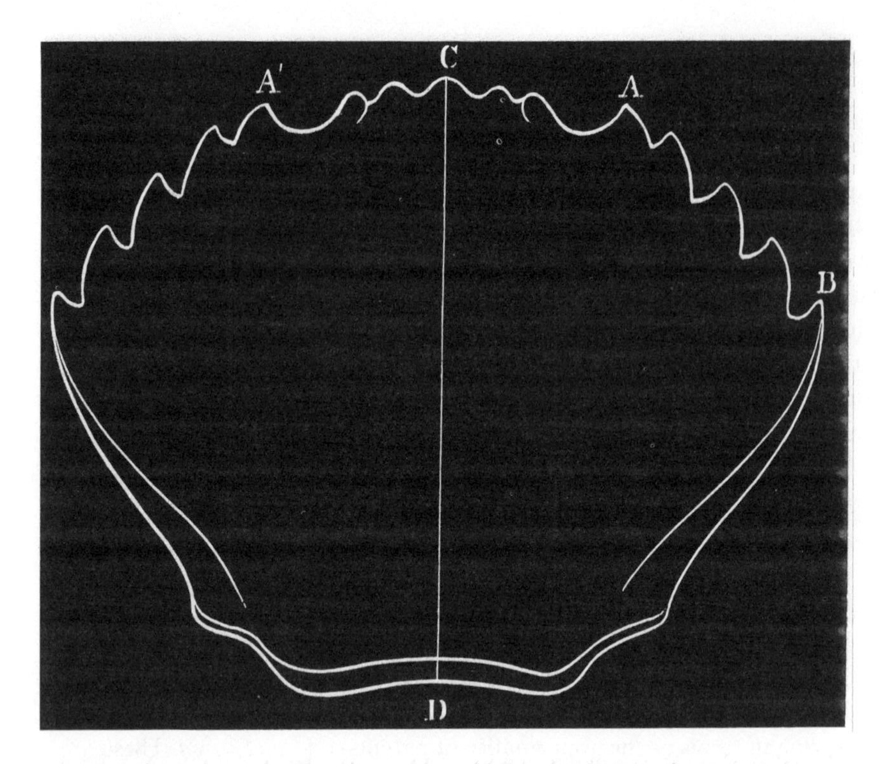

Stylized outline of a crab. The "frontal breadth" is the distance from *A* to *A*′. (From Weldon 1895.)

of crabs plucked from the Bay of Naples showed a skewed curve (Weldon 1893). Weldon now argued that perhaps two populations, pushed apart by selection, were represented in the sample. If this be so, then the skew in the measurements should be resolvable into two normal curves (one for each population), and by a method of trial and error Weldon did in fact find two such curves: "In the last few evenings I have wrestled with a double humped curve, and have overthrown it . . . If you scoff at this I shall never forgive you" (letter to Pearson, November 27, 1891; Pearson Papers).

Weldon did not have the ability to provide a general solution to the problem, and it was here that (at Weldon's urging) the incredibly talented Pearson stepped in (Pearson 1894). Finding just such a solution, Pearson demonstrated that his work could be applied profitably to precisely those empirical findings determined by Weldon. Considering the Naples crab measurements, Pearson was able to show readily (by theory rather than by trial and error) that although their distribution is asymmetrical, the curve breaks down into as neat a pair of normal curves as one could wish. Conversely, considering another series of measurements which yielded a normal curve, Pearson was happy to show that (in this particular case) there was no good reason to dissect it into two. Selection therefore seems to be acting to break up a population in the first example but not in the second.

As far as Pearson was concerned, this was but the start of a dazzling series of papers in which he pushed to the limit the task of exacting information from crude empirical data (Mackensie 1981; Pearson 1936, 1937–38). For Weldon, our subject, this was the plank by which he could move into experimentation, of a kind we shall not meet again for another half-century. Through careful long-term study, Weldon found that the shore crabs around the Biological Station at Plymouth were changing significantly in physical appearance (Weldon 1895). Specifically, their frontal breadth was getting narrower. He was able to link this change with the increased silt in the water caused by a huge (and relatively new) breakwater across Plymouth Sound (Weldon 1898). Putting his morphological knowledge to good use, Weldon hypothesized that the increased sediment clogs the crabs' filtering apparatus, and that hence decreased frontal breadth was a source of adaptive strength. Then, to complete his work, in a stunning program of experiments, Weldon showed first that crabs kept in conditions of artificially high sediment do show the expected differential death rate. Conversely, crabs kept in clean water have a definite tendency overall to an increase in frontal size (Weldon 1898; Pearson 1906).

With triumph Weldon concluded that:

> I hope I have convinced you that the law of chance enables one to express easily and simply the frequency of variations among animals; and I hope I have convinced you that the action of Natural Selection upon such fortuitous variations can be experimentally measured, at least in the only case in which anyone has attempted to measure it. I hope I have convinced you that the process of evolution is sometimes so rapid that it can be observed in the space of a very few years. (Weldon 1898, 902)

Enough of the science. What of progress? If one looks back into Weldon's earliest papers, those on morphology and embryology written while he was still at Cambridge, one finds clear evidence that—as one would expect—he was working within a progressionist framework. He talked

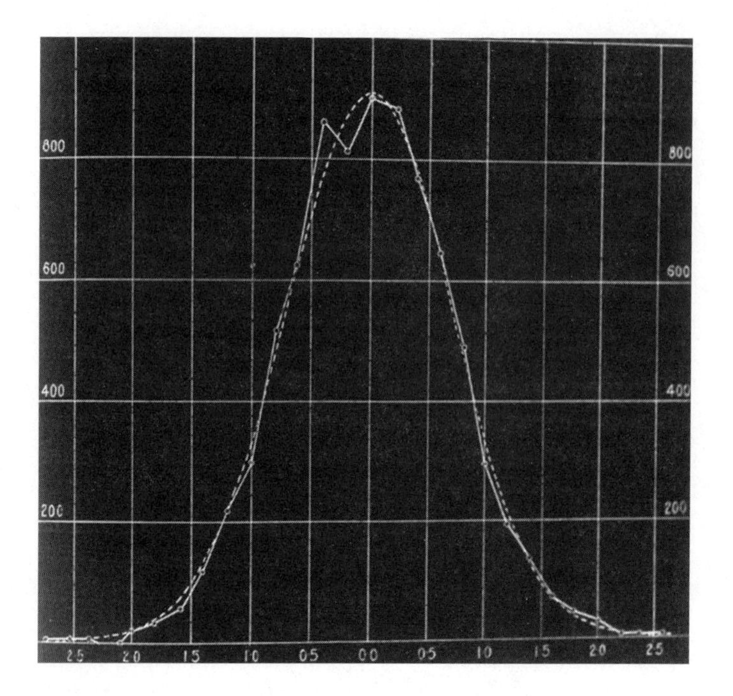

"Distribution of Frontal Breadths in 8069 Female Crabs from Plymouth Sound, old and young." The curve is normal and hence there is no reason to look for selection. (From Weldon 1893.)

confidently of "higher" and "lower" forms of animals and assumed all of the usual correlations—for instance, about the more specialized organs of birds and mammals (Weldon 1884, 177). Biological progress was a given. Yet, I have to confess that once he branched out on his own and turned his attention to natural selection, I find not one mention at any point, by Weldon, on the subject of progress. He was not against it. He was just silent. Read the crab work as you will, whether it be the technical papers or the general survey for the British Association, there was nothing.

Remarkably, in an unfinished, unpublished textbook (written when at Oxford, about 1901), Weldon defined evolution without a hint of progress:

> The theory [of natural selection] asserts that the characters of all existing species of animals or of plants are either being maintained in the same condition, generation after generation, by the operation of this process,

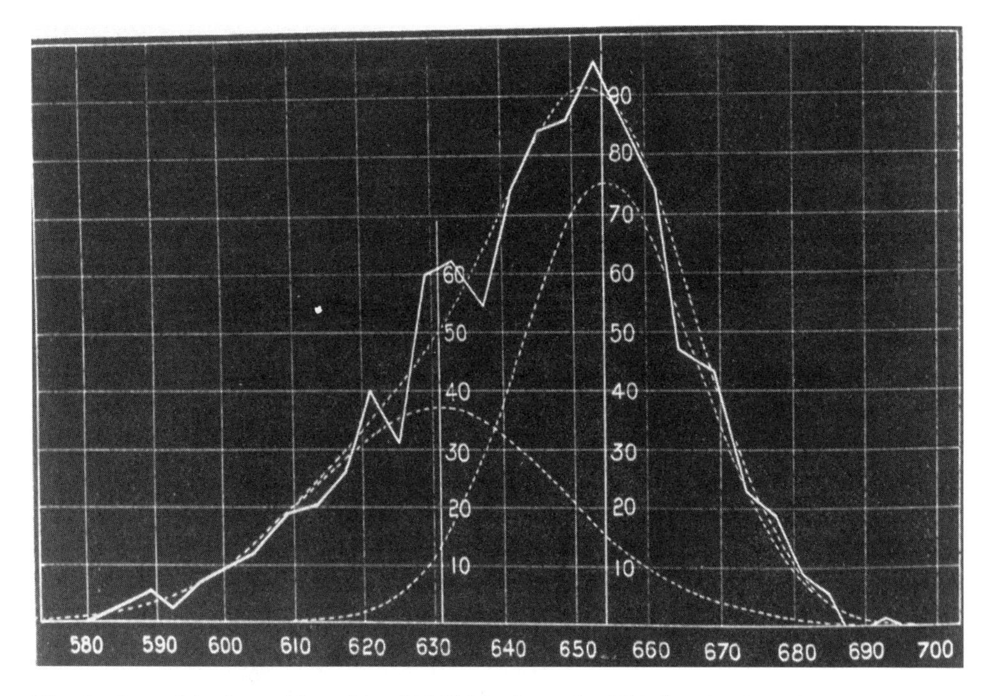

Distribution of the frontal breadth of 1000 Naples crabs. This breaks down into two normal curves (summing the ordinates), suggesting that the species is dimorphic and that natural selection may be at work. (From Weldon 1893.)

> or that they are being more or less rapidly changed by its means, and the same process, which maintains or modifies species now, is said to be sufficient to account for the very great modifications which most organic species have undergone since the first appearance of life upon the Earth. (Pearson Papers, 264/1, Clip VI)

It is always easier to prove a presence rather than an absence, but a strong case can be made—judged on a survey of all of Weldon's written work, the very extensive extant correspondence with Pearson and others, and biographical memoirs—that Weldon's lack of interest in biological progress mirrors a lack of concern with social Progress. I do not mean that he was indifferent to societal change and well-being—as a young man he campaigned for the liberals, and he worked hard to improve things at London University. If it is a mark of Progress to disapprove of King Edward the Seventh, then he certainly qualifies: "As for the English people, with the man who is perhaps more responsible for their slackness than any other single person proclaimed as their king, God help them" (letter to Pearson, January 25, 1901; Pearson Papers). I am talking now more at the level of theory.

If not Progress, was there something else at work? Perhaps. Weldon's father was a Swedenborgian, although he himself professed agnosticism. But he did have an emotional feeling for religion (Pearson 1906). What we can say with assurance is that Weldon was a fanatical pan-selectionist. He saw the marks of adaptation everywhere. For instance, in that same unfinished textbook, referring to the neck markings of species of birds, Weldon wrote:

> It is not ridiculous to suppose that there is some environmental factor, present everywhere from here to Japan, to which Marsh Tits are especially sensitive; it is not ridiculous to suppose that this factor produces such an effect upon Marsh Tits that those individuals with a black neck have a better chance of living and breeding than others. In the same way it is not ridiculous to suppose that over the whole area inhabited by Coal Tits there is some common environmental condition which so affects these birds that they have a better chance of living and breeding if their necks are grey or white. (Pearson Papers, 264/1, Clip I)

Right or wrong, such pan-selectionism is certainly what one would expect of a Believer, or one who had been trained as such. (Poulton, another pan-selectionist and a practicing Christian, went to the same school for the children of non-conformists as did Weldon.)

Yet, at most one can say that Providence was an ontogenetic relict in Weldon's thinking. He was not using his science to push a Christian line, any more than he was using it to push a Progressivist line. That was just not the way that Weldon did biology. It was not for him a religion-substitute—which point leads to our analysis of post-Huxley professional attitudes to evolution and P/progress.

Two Strategies

Huxley was inclined to exclude evolution from professional biology, but, as I have hinted, one might expect to see the next generation move evolution right over to the professional domain, with the creation of a solid (British) discipline. For this, one could not imagine a better man than Lankester, especially given the fact that he inherited from his father the editorship of the key publication, *The Quarterly Journal of Microscopical Science,* a post which he held from 1869 to 1920. And, in truth,

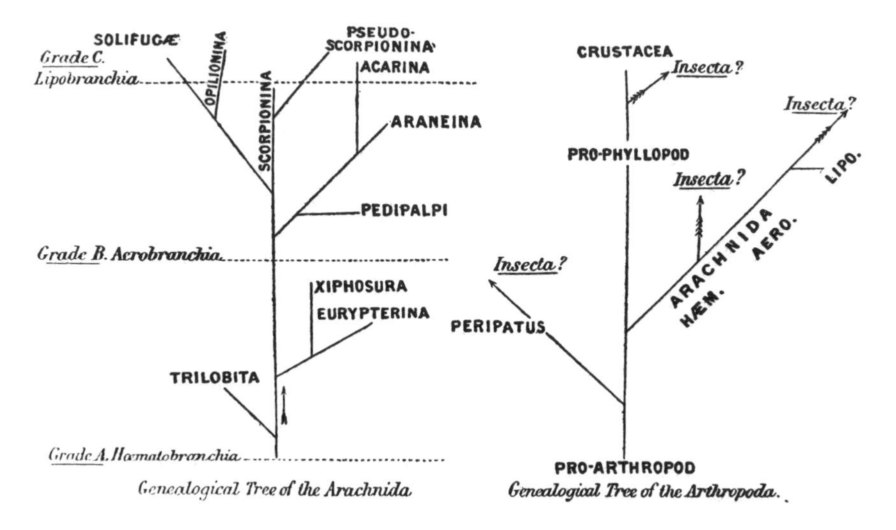

Lankester's genealogical trees. (From Lankester 1881.)

one must say that Lankester *qua* professional was a lot more relaxed about evolution than was Huxley. It was he, for instance, who so excised Rolleston at Oxford, for he was the one who slipped questions on evolution into the honors biology papers. Moreover, the *QJMS* was indeed used as a vehicle to promote a German-type professional evolutionism. Key papers by the Germans—Haeckel on Monera, for instance, in 1869—were translated and republished. Lankester himself was prepared to go in for some phylogeny tracing, including a couple of trees in his key paper on arachnids (Lankester 1881). And others were permitted to talk of evolution and to try to fit their work, usually detailed studies of comparative invertebrate morphology, into phylogenetic scenarios.

One starts to see Huxley's labors broadened into a discipline of evolutionary morphology. Yet, one must not exaggerate. The conservative Huxley influence loomed large. In Lankester's teaching, for example, evolution still got a very minimal treatment, and selection even less so. The recommended reading in Lankester's courses may have been Haeckel's *History of Creation,* but the required reading was Huxley and Martin (1875). For Lankester, as with Huxley, evolution was first and foremost an offering for the popular forum. His most detailed discussions of the topic come in encyclopedia entries, newspaper articles, and public addresses. And, as with Huxley, evolution played the same secular religious role in Lankester's life, for he too was committed to a form of agnosticism—"it seems to me that we are *limited* and can not arrive at a knowledge of 'the ultimate nature of things' and the 'beginnings' "—and he too felt that science can fill the gap: "The spirit of resignation to our limitations and the joy of gaining more and more what we *can* know—is what corresponds to the religious peoples' phrases as to the 'humility of faith' 'to walk humbly with thy God' 'to submit to God's will'—and to 'believe in' God" (letter to J. Reid Moir, January 28, 1914, Add MSS 44969, British Library, London).

Indeed, like Huxley, Lankester was happy that evolution had the status that it had, because then it could so readily and properly carry the message of Progress. Faced with the failure of "the dreams and aspirations of the youthful world [i.e., Christianity]," we must look for a substitute: "The faith in science can fill this place—the progress of science is an ideal good, sufficient to exert this great influence" (Lankester 1880, 110–111). Where Lankester went beyond Huxley was to pick up on those end-of-the-century worries about decline and to work them into his science—specifically, into the science he thought appropriate for presidential addresses to sections of the British Association. There is, after all, a time and a place for everything.

There is an ambiguity about Lankester's discipline building. In the context of his caution about evolution's place in his professional biology, his move to the British Museum in 1898, a post which he made strenuous efforts to obtain through mobilizing the scientific community to his corner, was very significant. He may have irritated his masters but all report him as having been an incredibly hardworking and inspirational director. From our perspective, the move to the museum had two key consequences. On the one hand, positively, Lankester was able to put his own stamp of science on the institution, the biggest permanent showplace for popular science in Britain. Thus he was able to reinforce his views about evolution and progress and decline. Indeed, this was precisely what was expected in a state-owned public place of instruction and education. The museum was not, nor was it intended to be, value-neutral. As his predecessor, Huxley's protégé Sir William Flower, had said of the effect of natural selection on the organic world: "it is a universally acting and beneficent force continually tending towards the perfection of the individual, of the race, and of all living nature" (Flower 1889, 23–24).

On the other hand, somewhat less positively, Lankester moved from being primarily a professional scientist to being primarily a professional museum administrator. His aims were changed, for he was no longer chiefly in the business of promoting a professional scientific discipline, evolutionary or otherwise. And that meant that the place of progressivist sentiments, or any other cultural values, in an ongoing professional science were really no longer his major concern. Where he was now, progress was allowed and important; but he was no longer where he had been. (Here, I speak briefly of this event and its significance. In the next chapter, I shall discuss in some detail analogous happenings in the United States.)

What then of those remaining in academic professional science, especially those who grew to scientific maturity as Lankester was moving out? If Weldon is anyone to judge by—or Poulton, or Bateson for that matter—something unexpected was happening. They all trained as evolutionary morphologists, happily progressionist, quoting Haeckel, Gegenbaur, and Lankester, and most of them got a start publishing in the *QJMS* (Weldon 1884; Poulton 1884; Bateson 1884). But then they went off in other directions: Weldon and Poulton into selection studies (Weldon 1898; Poulton 1890); Bateson into what was to become genetics (Bateson 1894). It was pattern that could be found also in Germany, for around 1880 people began to realize with dismay that the evolutionary morphology program just would not work—at least, it would not be the basis of a confident, fully functioning discipline that could occupy the front ranks of science.

Most urgently, there were money problems. Unlike physiology, or what was to become the new trendy science of experimental embryology—eagerly embraced and supported by the medical profession, for its analogies to human development—no-one had a direct use for evolutionary morphology. Even Huxley's purified, non-theoretical morphology, suitable for schoolteachers, never attracted the numbers that did physiology.[3] But, more important than money problems, no doubt in part bringing on the money problems, the program was conceptually and evidentially flawed. Epistemologically, evolutionary morphology simply did not cut the mustard.

It was certainly not that people turned their backs on epistemic norms. Listen, for instance, to an early German morphologist (in a piece whose translation was arranged by Darwin) pushing his science both on grounds of consistency and of predictive fertility: "If the absence of contradictions among the inferences deduced from them for a narrow and consequently easily surveyed department must prepossess us in favour of Darwin's views, it must be welcomed as a positive triumph of

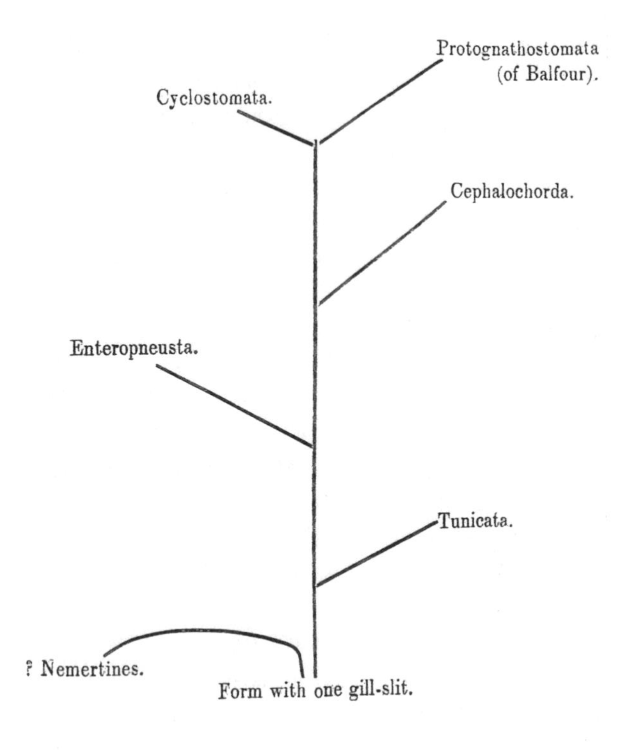

William Bateson's "phylogeny," which by his own admission may not have been an accurate reflection of reality. (From Bateson 1886.)

his theory if far-reaching conclusions founded upon it should *subsequently* be confirmed by facts, the existence of which science, in its previous state, by no means allowed us to suspect" (Müller 1869, 13). Unfortunately, these brave words found no true response in reality. It was hopeless trying to get ongoing internal coherence from the data. There were not enough fossils, not enough good analogies between ontogeny and phylogeny. And there was too much reading in of what you wanted to find (including progress) and too little reading out of what was really there.

Bateson, in old age, remembering a trip to Hampton (Virginia) and a summer in a marine laboratory, told all:

> Morphology was studied because it was the material believed to be most favourable for the elucidation of the problems of evolution, and we all thought that in embryology the quintessence of morphological truth was most palpably presented. Therefore every aspiring zoologist was an embryologist, and the one topic of professional conversation was evolution. It had been so in our Cambridge school, and it was so at Hampton . . .

> Discussion of evolution came to an end primarily because it was obvious that no progress was being made. Morphology having been explored in its minutest corners, we turned elsewhere. (Bateson 1922, 390–391)

Nor was this an old man's fiction. Back in 1886, the brilliant young biologist had published in the *QJMS* a paper on the ancestry of chordates, complete with phylogeny. Not only did he feel obliged to apologize profusely for what he was doing—"That the results of such criticism must be highly speculative, and often liable to grave error, is evident" (Bateson 1886, 536)—but when he had finished he suggested that perhaps his diagram should not be "meant so much as a genealogical tree as to serve as an exhibition of the logical relation of the various forms, showing their points of divergence" (p. 571)!

Evolutionary morphology stalled. It did not come to an end, but it acquired the taint of the second-rate. Hopes of the best kind of maturity began to fade, and with the shift into the museum, the opening of a place for culture was made precisely at the cost that top-flight professionals would not pay: for note that none of this repudiates the professional science/popular science dichotomy. The very opposite. Weldon more than anyone had taken to heart the division between professional science and popular science—so much, in fact, that he would probably not have considered the latter science at all. Huxley was one of Weldon's great heroes. The only time Weldon quarreled with Pearson was when the latter published, in the *Times,* a letter (justifiably) critical of Huxley. "I may as well say now what I must say sooner or later, and so get it over. I am very sincerely grieved by the publication of your letter" (letter to Pearson, December 3, 1892; Pearson Papers). Yet, interestingly, the enthusiasm was not for Huxley as a man—Weldon thought Huxley was too given to controversy, that he exaggerated the opposition (Owen excepted) that he had to face, and that there were questions about his treatment of his wife—but because he had fought and expelled extraneous metaphysics from biological science (letter to Pearson, December 2, 1900; Pearson Papers). This is as professional science should be, and Weldon was incensed at Leonard Huxley's *Life and Letters* of his father for not dealing simply with the straight science, thereby simultaneously detracting from the professional science and violating the professional/popular divide (letter to Pearson, November 7, 1900; Pearson Papers).

Hence, Weldon's greatest reticence with even the closest of scientific friends like Pearson about his own extra-scientific beliefs—perhaps Weldon thought that a scientist should not have such beliefs. So strongly did Weldon feel on these matters that he was even prepared to criticize Huxley himself (not to mention Lankester) when he stepped beyond the bounds.

> Are you going to find energy for a tilt with Lankester? Why must every one who gives a Romanes lecture talk such utter rubbish? Has the spirit of a founder some abiding influence after all? Huxley ["Evolution and ethics"] was weak enough: but Lankester's description of Man rebelling against the Order of Nature was palaeolithic in its anthropomorphic childishness. Of course, some of the theologians up here are rather pleased: but all the serious people are amused. (Letter to Pearson, June 21, 1905; Pearson Papers. Note how Weldon turns to the polished popular writer, Pearson, for help.)

When he and Pearson jointly set up their own journal, Weldon made it very clear how he regarded proper professional science—and the rest.

> I think a real change has come over people. The man of Galton's youth who tried to know about everything,—the man who tried to be like Goethe or Huxley—in the first place never did tolerate mathematical treatment, and in the second place has long ceased to exist.
>
> I think people now want very technical journals, *plus* Rudyard Kipling and the evening papers. (Letter to Pearson, November 25, 1901; Pearson Papers)

The Huxley move was essentially to exclude evolution from the inner domain. The Lankester move brought evolution to professional science, somewhat. But then the drive got diffused with the move to museum work, where the aims were different from those of the professional scientist, building or working within a discipline. Weldon's move was to bring evolution firmly over into professional science—but it had to be an evolution shorn of all elements that would serve to qualify it as popular science. "The curse of biology has been that every idea about evolution has had to be presented in a popular forum, with the inevitable shirking of difficulties, the inevitable dominance of the windbag [Lankester!]" (letter to Pearson, November 25, 1901; Pearson Papers). This meant that ruthlessly he dropped all of the value-introducing components, especially anything which might introduce P/progress. In writing on evolution, Weldon ignored his own background in embryology, and he was positively contemptuous of paleontology. "I began this afternoon . . . a course of lectures on the way to attack the problem of Evolution,—and I took the Horse series as an example of the way in which fossils do *not* help, when you have too many of them" (letter to Pearson, January 25, 1901; Pearson Papers). The student simply must not be exposed to anything which tends to sloppy metaphysics.

Conversely, Weldon's careful selection studies were intended as exemplars of how to produce a professional evolutionary biology, how to put oneself on the road to mature science. Measurement, experimentation, and above all the use of mathematics were deep and essential signs of this move toward acceptance of the epistemological norms which had proven so successful in the physical sciences. Here, Weldon was very much at one philosophically with Pearson, who in 1892 published his highly influential positivist work, *The Grammar of Science,* extolling the virtues of finding empirically based laws—a "*formula,* which in a few words resumes the whole range of facts" (p. 93, his italics)—and of avoiding foggy metaphysics. Given the timing, it is improbable that Weldon set out explicitly to satisfy Pearson's program; but, undoubtedly, Weldon was happy to continue work in the philosophical mode which his friend advocated. As was Pearson happy to use Weldon as an exemplar in the second edition of his work.

Darwin should have been as delighted at Weldon's work as was Pearson; whether he would have willingly paid the cost, though, is another matter. To make a professional science, a discipline-based activity, not only did Weldon expel progressionism, but he did it by dropping virtually the whole range of areas studied in the *Origin:* embryology, paleontology, biogeography, and more. Nothing must detract from the purity of one's central causal investigations. Hence, although Weldon's evolution crossed the professional/popular divide, most of the troops were left behind.

Toward the Future

We stand at the beginning of this century, and of major movements in our understanding of heredity. Before we continue further with our story, we must cross the Atlantic and see what was happening in the New World. We leave Lankester promoting a full (much Germanized) evolutionism, although he has moved from being a professor in Oxford to being director of the Natural History Museum at South Kensington. His was an evolutionism where a society-driven progress flourishes, except as it is marred by an equally society-driven decline. And we leave Weldon with a much truncated (thoroughly pan-selectionist) evolutionism, as firmly in the professional realm as (increasingly) Lankester's was not. His was an evolutionism where there was no place for progress.

Or, to point out a contingent fact which is significant, we leave a Weldon who died in 1906. I shall argue that his ideas and example did have effect, but what he did not leave was a school or a discipline to carry forward his vision. In the decade before, he had been outmaneuvered in London at the Royal Society, when Bateson and his supporters had taken over an "Evolution Committee" which Weldon and his supporters had started (Provine 1971). Then came failure in this decade. Time and again, Weldon confided to Pearson that he had hoped to found a school of mathematical selectionists at Oxford: "I wish you had come up here too! We would have made Animal Evolution hum" (letter to Pearson, April 12, 1899; Pearson Papers). (Pearson had applied, unsuccessfully, for a job at Oxford.) It was not to be. There was violent opposition to anything mathematical, from students: "I have not got one man to care for anything I say outside a textbook" (letter to Pearson, July 11, 1900; Pearson Papers); faculty: "Their tutors all tell them one is an amiable crank" (ibid.); and fellow evolutionists: "I told Lankester about these snails, and he wrote me an earnest letter, urging me to return to the pleasant way of describing beautiful beasts for the delight of the faithful" (letter to Pearson, October 11, 1900; Pearson Papers). He concluded sadly: "This place is the greatest danger to England I know. I hate it, and I hate myself because I have sold myself to it for money" (letter to Pearson, October 21, 1904; Pearson Papers).

Not that there is any sense that Weldon realized, as Huxley had certainly realized, that a successful discipline needs patrons and a market. Back in London, Weldon's friend Pearson appreciated the economic realities of science, but by now he was grinding a somewhat different axe, for he was off full-time into eugenics and the related statistics and thus seeking support for different ends. And up at Oxford, Poulton was little help. Inevitably, Weldon made him an enemy rather than an ally, squabbling over research funds: "Poulton, the richest man in Oxford,—biscuits *and* tin, as they say here,—wanted it all" (letter to Pearson, December 3, 1904; Pearson Papers). (Poulton had married an heiress of the Huntley & Palmer wholesale bakery firm. "Tin" was English schoolboy slang for money.)

Hence, at the beginning of this century, although one might truly say that people had articulated different strategies for locating evolution on the professional/popular divide, Darwin's dream was far from realized. There is masses of interest in evolution and people are firmly committed to the idea as fact. It is a major component of science in the popular domain, and as such it is firmly progressivist. This connection is cause and effect. In the professional realm, there is an insecure German-type evolutionary morphology, progressionist it is true, but as a discipline it is increasingly dismissed or deserted. And we have seen no evidence that those who thought differently about evolution, progress, and professionalism—and there were those—were more effective. Or rather, they may have been more effective about exemplifying mature science, but they were no more effective at the task of making a functioning, top-quality discipline. We shall return to this point.

7

Evolution Travels West

Racked and torn asunder by a bloody civil war of unprecedented ferocity, the American Union was transformed in the mid-nineteenth century. Slowly, painfully but definitively the Northern states gained ascendency over the rebellious states of the South, and as they did, America's gaze turned about-face. No longer did the country look back to the eighteenth century and the era of its founding; the people now looked ahead to the twentieth century and the era of its supremacy (Tindall and Shi 1989). Progress became the philosophy of the day. Nevertheless, though similar movements had already taken root in England and the Continent, it has always been a mistake to think of America merely as a pale and somewhat backward reflection of Europe. It is true that the U.S. Constitution separates Church and State in a fashion, deliberate and distinctive, still not attained in England and many other countries. But, for genuine, all-pervasive belief, America has had few equals. Progress may be the philosophy of the day, but Providence is the philosophy of the centuries. A sophisticated Frenchman like Cuvier would cringe with embarrassment at much which still passes for normal in today's political discourse. We must not expect that the Progress/Providence dichotomy will ever be as clear-cut in America as it can be in Europe. As they set about rebuilding and expanding their nation, self-respecting Americans did not doubt that they were carrying out God's Work.

Acknowledging America's distinctive nature does not deny its debt to the Old World, in ideas as much as in people. Embodying both, there is Louis Agassiz. He came to America just before mid-century, and he rapidly and comfortably settled into a Harvard professorship. Around himself he gathered a platoon of assistants and students, who would work with him, learn from him, and contribute to the greater glory of American science and its most famous import. This industry was located in and sprang from Agassiz's own quarters and collections, and these—aided by both private and public funding of a magisterial nature—developed into the Museum of Comparative Zoology, a wonderful edifice, intended to serve both public and professional alike (Winsor 1991). To glorify and promote the work that he and his fellows elsewhere were doing, Agassiz and his close chums persuaded Congress to give official seal to the formation of a National Academy of Sciences.

Yet, Agassiz himself was Janus-faced, looking backward as well as forward. Try as he might—and after his fashion he did try—he could never stomach evolutionism. Hence, to tell our story we must turn to the lives and works of others.

Asa Gray: Darwin's American Lieutenant

Asa Gray (1810–1888), long-time professor of botany at Harvard and author of the definitive texts in the nineteenth century, trained first as a physician. Introduced by his teachers to the joys of science, especially botany, Gray was soon using medicine as a (very) part-time support aid. Increasingly, he became involved with the then-leading botanist in the United States, John Torrey, on the production of a definitive *Flora of North America.* Although this work was never to be finished by Gray, his reputation was launched (Gray 1894; Dupree 1959).

At first no friend of evolutionism, Gray found the need for some sort of overall transmutationist background when, in the 1850s, he started comparative botanical studies and tried to explain significant similarities between the plants of Asia and those of the eastern United States (Gray [1859] 1889). But Gray was never much of a field botanist—he preferred to do his traveling from friend to friend in Europe—and one senses that after the *Origin,* somewhat paradoxically, he reverted away from biogeography toward orthodox taxonomy. It is clear that what really tipped Gray into the evolutionist camp, and made him a Darwinian (of a kind), was the personal factor. No-one in the history of science has exceeded Charles Darwin's networking abilities. Coming into contact with Gray

Asa Gray

through the mutual friendship of the Hookers, father William and son Joseph, Darwin rightly sensed that in Gray he might have a suitable North American lieutenant. Thus, the American botanist was cultivated (to use an appropriate metaphor!) and was one of those privileged, before the *Origin*, to be let in on the great selection secret.

This effort paid off, for Gray saw that the *Origin* would be reviewed favorably and distributed fairly in the United States. More than this, there was a fight to be waged on behalf of Darwinism against Agassiz, with Gray having to take on the self-confident ichthyologist before increasingly large audiences in the Cambridge orbit, as well as in print. Not that this was an entirely unwelcome task, for the longer-established Gray must have resented the lionized Swiss outsider. Even a saint would have been sore at the fact that comparative zoology was getting $100,000 from the state (Massachusetts) government alone, while botany had to grub around for $10,000 (Winsor 1991).

What adds an ironical edge to this story is that Gray was a Darwinian only in a limited respect. He was an evolutionist, certainly. He believed in natural selection also. But he was an ardent Christian, and it was part of his religion that nature shows design—or, rather, Design. To this end Gray maintained always that the raw stuff of evolution, the new variations, must in some sense be directed: "we should advise Mr. Darwin to assume, in the philosophy of his hypothesis, that variation has been led along certain beneficial lines" (Gray [1860b] 1876, 121–122). And although Gray and Darwin argued this matter at some length, in the end they had to agree to disagree. As far as the Englishman was concerned, putting direction into the raw stuff of evolution was essentially to gut the force and worth of natural selection. However, Gray was a distinguished scientist, fighting the good fight. The savvy Darwin knew that it is better to give a little and have two-thirds of the cake than to remain stiff and unbending and get no cake at all.

What of progress, generally, and specifically in botany? The early thoughts of Gray on the subject were not promising. In a review (in 1836) of John Lindley's *A Natural System of Botany,* the old idea of an uninterrupted Chain of Being was firmly dismissed—even though it was admitted that the artifact of writing about a classification tends to put groups into an apparent line: "nothing is more evident than that almost every order, or other group, is allied not merely to one or two, but often to several others, which are sometimes widely separate from each other; and, indeed, these several points of resemblance, or affinity, are occasionally of about equal importance." An upwardly directed natural chain is therefore impossible, even though this pattern is "the only one that can be followed in books" (Gray 1836, 7–8).

This philosophy seems to have guided Gray in his writing of his first texts. Nor was there any change when Gray commented critically on the supplemental *Explanations* (of the *Vestiges*). Although sneering that "we need not our author's hypothesis to tell us why it was so," he did agree that a general progressionist line "well comports with our general notions respecting the stages through which our earth passed ere it became habitable for man" (1846, 479). Yet, when it comes to particulars, especially botanical particulars, the author of *Explanations* is plunged into deep trouble: "He admits that the traces of a historic progress in the vegetable creation are less clear than he could wish, and well he may. For in 'the first great burst of land vegetation,' as he phrases it, not only are the lower forms absent, so far as we can tell, but all the higher classes of plants are actually represented" (pp. 480–481).

By the time of the Darwin debate, although the facts had not really altered that much in kind, Gray was prepared to think more positively about progress, in the plant world as well as the animal: "It seems clear that, though no one of the *grand types* of the animal kingdom can be traced back farther than the rest, yet the lower *classes* long preceded the higher; that there has been on the whole a steady progression within each class and order; and that the highest plants and animals have appeared only in relatively modern times" (Gray [1860b] 1876, 95). He was willing to allow some kind of Agassizian parallelism, in time and form "from simple and general to complex and specialized forms." There is even: "the parallelism between the order of succession of animals in geological times and the changes their living representatives undergo during their embryological growth, as if the world were one prolonged gestation. Modern science has much insisted on this parallelism, and to a certain extent is allowed to have made it out" (Gray [1860b] 1876, 97).

However, Gray's endorsement of progress in the plant world, at least, was lukewarm, and less than three years later he was dampening Darwin's hopes.

> In reply to your question:—
>
> If oak and beech had large, colored corolla, etc., I know of no reason why it would be reckoned a low form, but the contrary, quite. But we have no basis for high or low in any class, say, dicotyledons, except perfection of development or the contrary in the floral organs, and even the envelopes; and as we know these may be reduced to any degree in any order or group, we have really, that I know of, no philosophical basis for high and low. Moreover, the vegetable kingdom does not culminate, as the animal kingdom does. It is not a kingdom, but a commonwealth; a democracy, and therefore puzzling and unaccountable from the former point of view. (Letter to Charles Darwin, 27 January 1863; in Gray 1894, 496)

Here, the worry does start to be hardening against the very *idea* of progress in botany, and although he did always allow some kind of relative progress, this seems to have been Gray's final position. What lay behind Gray's thinking? He was at best a tepid progressionist, if that, so we should not expect to find him expressing an ardent enthusiasm for Progress. And in this, we are not disappointed. Cultural Progress really seems not to have been a burning issue for Gray. We do not encounter in his writings, public or private, espousal of one of the various philosophies of Progress. He had little liking for the writings of Herbert Spencer and even less for those of positivism (Dupree 1959, 302; Gray 1894, 592). Analogously, although Gray toyed with Agassiz-like thoughts on the direction of change in the plant as well as the animal world, there was never any empathy for the underlying philosophy.

Significant here was Gray's attitude toward "phyllotaxy," or the arrangement of leaves on the stem of a plant. The *Naturphilosophen,* starting with Goethe, had made great efforts to understand the phenomenon. Alexander Braun, Agassiz's brother-in-law and Haeckel's teacher, had devised mathematical formulae to quantify the distributions, supposedly showing underlying unifying principles. But although Gray (1879) discussed the topic (including the mathematics) in detail, and although as an earlier quotation surely shows ("*grand types*") Gray was prepared to accept a Cuvier/Agassiz division of animals into basic forms, there was never a whiff of Germanic archetypes or any such things. This was just not to Gray's taste, any more than was the progressionism that Braun found in the plant world. Indeed, given his intense rivalry with Agassiz, one cannot but believe that Gray felt real satisfaction in being able to challenge, from his own undisputed knowledge of the plant world, some of the more expansive claims of *Naturphilosophie.* The non-progress of plants was a nasty thrust into the Swiss underbelly of Germanic philosophy.

There remains, however, the major non-scientific reason why we ought expect no genuine progressionism in Gray's vision of the plant world. Given his comment to Darwin, it is tempting to think this would be a political reason: Gray the American republican, citizen of the *Commonwealth* of Massachusetts, deploring the monarchies of the Old World. This is probably fanciful thinking. Gray was an Anglophile who approved of the monarchy—for the British! The real reason rather lies in Gray's religion. Remember, we are dealing with one who grew to maturity in the first part of the nineteenth century rather than the second. Providence before Progress. Again and again in Gray's correspondence, it is this to which he appealed, especially during the dark days of the War: "To the North the war, with all its sad evils, has been a great good, morally and politically. The end is in the hands of Providence, and we humbly wait for it" (letter to R. W. Church, Cambridge, 25 December 1863; in Gray 1894, 518). These are not the words of a man controlling his Progressive destiny.

It is true that God, inasmuch as possible, works through the agency of law, but Gray still saw His Handiwork in everything. Consider the following passage from Gray's review of the *Origin,* and then reflect on the way he supplemented Darwinism, teleologically:

> But there is room only for the general declaration that we cannot think the Cosmos a series which began with chaos and ends with mind, or of which mind is a result: that, if, by the successive origination of species and organs through natural agencies, the author means a series of events which succeed each other irrespective of a continued directing intelligence—events which mind does not order and shape to destined ends—then he has not established that doctrine, nor advanced toward its establishment, but has accumulated improbabilities beyond all belief. (Gray [1860a] 1876, 48)

The fact of the matter is that, at heart, Gray was not looking for progress, and it is little wonder that he did not find it. As a Providentialist, he did of course think of humans as special and was happy to see them at the head of creation. But there was no urge to see an unfurling upward chain in every part of God's creation. It was adaptation which spoke to His glory, and He could do this in any order He so pleased.

Turning to the third of our usual questions, about the empirical base for Gray's position, with respect to the issue of progress, we must conclude that, perhaps more than for most of our scientists, the brute facts were significant. Or, if not the facts alone, then the undoubted truth that in the plant domain there is not a favored winner as are humans in the animal domain. But one doubts that the facts were the exclusively determining factor. In the plant world, the evidence for progress may not have been overwhelming, yet Gray himself admitted to some such evidence. In the common course of things, he never hesitated to speak of "highness" and "lowness" in plants. The point is that, ultimately, one does not see an empirical case absolutely dictating matters one way or another. And do note how Gray's espousal of directed variation was not thrust upon him by the evidence, for all of the genuine problems that Darwin had left on this score. It was primarily a function of his natural theology, as Gray strove to justify and explain the design-like nature of all organisms.

Gray's scientific thinking reflected cultural values that he held dear. It would, perhaps, forge the central link of this book a little too readily if one were to say that, since he was an indifferent P/progressionist, he was an indifferent evolutionist. But, if we qualify that we are talking of a *scientific* position on evolution, such, apparently, would be one person's judgment of Gray's position: "The view that each variation has been providentially arranged seems to me to make Natural Selection entirely superfluous, and indeed takes the whole case of the appearance of new species out of the range of science" (Darwin and Seward 1903, 1, 191).

Alpheus Hyatt and the Question of Degeneration

Agassiz never grasped that the American dialectic had evolved beyond the master/slave relationship, and his much put-upon students eventually founded the "Society for Protection from Foreign Professors" and deserted him. Yet, although the new generation broke from Agassiz physically, it never left him emotionally or intellectually. People became evolutionists, so in this sense Gray won the battle. But, in the long haul, Agassiz won the war. The evolutionism shown by Agassiz's students—and others touched by his orbit—owed far more to the transcendentalism of German *Naturphilosophie* than it ever did to the natural theology of pure Darwinism. In America, by the 1870s, we are in a world of *Baupläne* (groundplans) and homologies and guiding forces and—above all—parallelisms between embryonic and phylogenetic development. Natural selection is conspicuous only by its absence. The legacy is that of Munich, not Cambridge.

This point is made with great force by the best student of them all, custodian of the Boston Society of Natural History, Alpheus Hyatt (1838–1902). Born into a rich Maryland family, conventionally schooled at a military academy, Hyatt went to Harvard, where he soon came within Agassiz's orbit (Morse 1902; Brooks 1909; Mayer 1911; Jackson 1913; Dexter 1954). Although he was to rebel against Swiss autocracy, Agassiz's influence was deep and lasting. In the words of Hyatt's son-in-law: "Louis Agassiz's lucid exposition of von Baer's law and his own additions thereto, and his high praise of the philosophy of Oken, produced a profound effect upon young Hyatt's mind" (Mayer 1911, 131). So much so, in fact, that although apparently Hyatt was an early (1860) convert to the idea of evolution, he revealed his hand so gradually that his first publications really can be read as much as in the line of Agassiz-like orthodoxy as supportive of transmutation. More important than the physical details was the crucial analogy between embryological development and past history. As the developing individual goes through its phases, so it reflects the story of the group. Where Hyatt really came into his own—although even here he was following up suggestions by Agassiz about "the contortions and death struggles preceding the extinction" of a particular group of ammonites—was his suggestion that, just as the individual eventually declines into old age and senescence, so also the group declines into old age and senescence (Brooks 1909, 320). Before

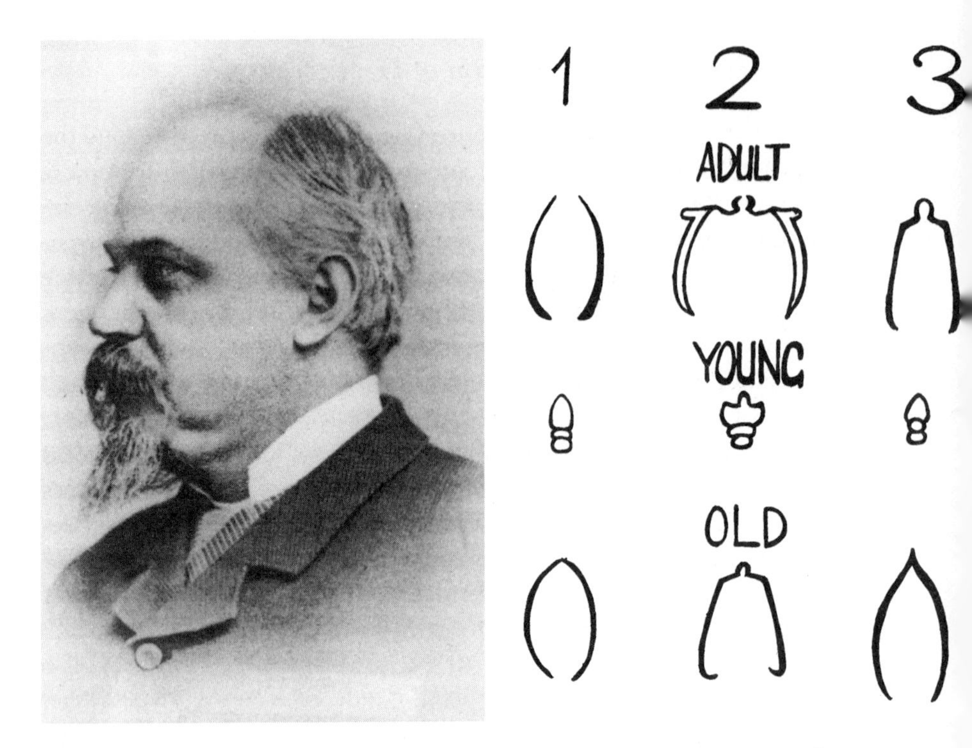

Alpheus Hyatt and his illustration (for Darwin) of his law of acceleration and retardation (redrawn from a letter, November 1872).

extinction, we get degeneration of its members (Hyatt 1889, 1893, 1894).

Hyatt was a man much loved; reading between the lines of his obituaries, however, one gathers that organization was not his forte. This is certainly true of his evolutionism, as when he wrote at length on the subject of degeneration to (the very puzzled) Darwin: "terminal forms are at the same time the highest of their series in their organization and development and yet like the most immature in many characteristics. Again these terminal forms have not only these resemblances but they also resemble the old age of earlier species of their own series" (letter to Darwin, November 1872; Darwin Papers, and partly published in Darwin and Seward, 1903). He tried to show what he was about with three examples of organisms, taken in order from the fossil record. Change is a function of the speeding ("acceleration") or slowing ("retardation") of development. Most significant, because the most recent specimen (3) has had a new stage added on to its old age, what was previously old age (in 2) is now the adult form (in 3), and there is even a suggestion that senescence is a return to childhood (as we see by comparing the old age of 3 with the youth of 3 and the old age of 1).

> The development of the form of the adult and the smoothness of the shell in nos 1 & 3 and the great difference shown by the middle form into which is keeled on the abdomen with deep channels and has tuberculated ribs on the sides. Compare this with the development of no 2 and we find that the adult of No. 3 is like the old of No. 2. Thus not only is adult no. 3 like adult of no. 1 but it is also like the old of no. 2.

What is the causal power behind all of this exemplification of the "law of acceleration and retardation"? Essentially, it is Lamarckism: "I have found, or thought I have, that this resemblance was due to the direct inheritance of degradational characteristics, not as an unusual but as a usual mode of development, among the higher or rather later occurring species of different series." Trying to justify himself to Darwin, Hyatt really did attempt to tie in his own views with selection. But he had little success, for ultimately he could not "get a clear idea of how Natural Selection can bring about such a series of phenomena or on the other hand how these, as I formerly thought, show some higher and more comprehensive law" (letter to Darwin, November 1872; partly published in Darwin and Seward, 1903).

Hyatt had great respect for Darwin as a man and as an evolutionist, but basically he had no sympathy for Darwinian mechanisms. In explaining the changes in the vertebrate limb, Hyatt returned to old beliefs: "The habit of using an organ is known to possess the power of producing modifications or variations" (Hyatt and Arms 1890, 40–41). Likewise, apparently, with the invertebrates. Degeneration can then be seen as a virtually inevitable outcome of the evolutionary process, because any organism tends to collapse into old age, and it is just a matter of time before this decay gets pushed back in development into adulthood.

The vast distance which separated Hyatt from pure Darwinism came through strongly in a little book on insects which he wrote (with a junior author), for schoolteachers, in 1890 (Hyatt and Arms 1890). This was prepared at exactly the time when the English Darwinian, Poulton, was engaged in studies of animal coloration, carrying on the work of Bates and Wallace, showing especially how selection is the key factor in explaining the adaptively crucial markings of insects (Poulton 1890). Not only was there no mention of natural selection in Hyatt's book, but the whole issue of mimicry and camouflage got but one passing paragraph in three hundred pages. We are just not in the world of Darwin's problems, let alone the world of his solutions.

What of progress, and of our other questions about Progress? Clearly, overall, Hyatt worked within a progressionist framework—an Americanized *Naturphilosophie,* which added Lamarckism to give it a directed evolutionary mechanism. Possibly reinforced by Spencer, his very characterization of change reeked of German progressionism: "Nature leads us along lines of modification which sometimes rise through continuous progressive specialization to more and more differentiated structure with correspondingly increased functional powers, or larger or different fields of work" (Hyatt and Arms 1890, 287–288). Then, as a kind of variation on a theme, Hyatt incorporated his thoughts on degeneration. Even humans are included here, for although *Homo sapiens* undoubtedly comes out at the top, we do seem to have peaked in respects: "The Caucasian type, in losing the prognathism of the Anthropoids, which is certainly a highly specialized characteristic of the adult forms among the apes, has in a morphological sense made a step backwards instead of forwards" (Hyatt 1889, 45–46). In a sense, that is, for "degeneration" is "not necessarily of 'low grade' in any scientific scheme of arrangement founded upon the principles of evolution" (Hyatt and Arms 1890, 288).

Was Hyatt a Progressionist—more pertinently, was he a Degenerationist—who was feeding his views back into his biology? As with all transcendentalists, one cannot deny that there is an air of inevitability about everything that Hyatt wrote. That odor is virtually built into the embryological analogy. Perhaps also there was a dash of Providentialism, for in his youth Hyatt certainly had strong religious convictions. Before Harvard, he spent a year in Europe with his Roman Catholic mother, and there was serious talk of his becoming a priest. But Hyatt was never a man to sit back and let others control his destiny. Both in word and in action, he showed that he thought human effort and action are demanded for forward change. In the Civil War, he put his childhood training at the service of the Union forces, even though Agassiz was opposed to his students getting engaged, and—more seriously—he thereby alienated his own family, who supported the South. This was a sacrifice he never regretted. Social Progress, in America, at least, is possible and actual: Man's past history "and the social laws of nature" "hold out hopes for the maintenance of progress through an indefinite time, if he is capable of controlling his own destiny through the right use of experience and of the wonderful control over nature that his capacities have enabled him to attain" (Hyatt 1897, 93).

Yet, what of decline? Why do we find Hyatt pushing biological degeneration? Was this, in some way, linked to thoughts of social Degeneration? The context was always Agassiz's and as expected, therefore, Hyatt was led to stress that what happens elsewhere in nature happens with full force in the human realm. For Hyatt as for Agassiz, the biological and the social were as one. Right in the middle of a major paper on ammonites, we find an elaborate analogy drawn between the history of invertebrates and the history of architecture.

> The buildings of primitive times would necessarily be substantial, plain, and suitable to the limited wants of the people; then, as wealth increased, the architects would respond with showy structures, having more ornamentation, and more complicated interiors . . . As time progressed these structures would assume vast proportions and would be built in ever increasing numbers, until at last the nation, having outgrown its strength, would begin to decline. The vast buildings would have to be abandoned, and smaller habitations would arise, in answer to the requirements of a poorer population. The architects, faithful to their inherited canons, but forced into simplicity, would gradually follow the decline, and record it in the structures of the decadence. (Hyatt 1889, 79)

Yet, why make a theory out of all of this? Why make decline virtually the trademark of one's evolutionism? One might wonder if the Civil War and its dreadful carnage could have been the trigger. But Hyatt regarded it as a morally uplifting struggle, and—judging from his letters (Hyatt Papers, Syracuse University)—seems personally rather to have enjoyed it. More promising as an influence, especially given that Hyatt was always fascinated with Gibbon's tale of the decline of the Roman Empire, was that teenage trip that he took to Italy with his mother. In the diary he kept for that trip, we find every social equivalent of the mature biological theory. Most significantly, degeneration was the prominent theme, for Hyatt's meeting with the local inhabitants shocked the rather sheltered and privileged young man beyond belief: "The lazzaroni live, beg, starve, make love and shit upon the church steps and along the quai, which last being the most public is the place generally preferred for the last picturesque action" (Alpheus Hyatt's Travel Book, p. 6; Hyatt Papers, Syracuse University Library, Syracuse, New York).

What made the whole experience so traumatic was the dreadful contrast with the glory that had been Rome: "I have never enjoyed anything more than this siesta on the old Roman wall. All nature seemed to be lazily inclined and as my body lay upon the ancient structure and my fingers and motions held converse with the *degenerate* descendants of those who built that solid but shattered structure, my mind gradually wandered into most delicious vagaries" (p. 8a, *his italics!*).

How could this have all come about? There was certainly a flavor of Lamarckism in Hyatt's causal speculations: "Rome and the Italians are subject similar and as if it were suited to each other, both worn out and rusted with the accumulated bad customs of ages" (p. 61). Not to mention premonitions of developmental retardation in the childlike nature of today's Italians: "Such is the strange mixture of their character, having great strength of affection and capable of powerful attachment yet steal and lie to an unheard of degree" (p. 109). Thank God for America: "the universal respect for departed worth had made one regret this past grandeur as if it were that of his own native land. And then there in the old Roman forum how quickly crowded the great glorious future of my own native land" (p. 51).

It is all there: a youthful mind, ready to take on Agassiz's transcendentalism, to throw in a dash of evolutionism, and to produce the theory of phylogenetic senescence. There is even hesitancy about whether simplicity truly means decline. Unfortunately, as the young traveler had not then realized, even America—"the free light to penetrate the rottenness of the institutions of this land as the sun does the clear water of this sea" (p. 119)—might itself someday bend beneath the inexorable forces of nature. It is precisely this fear that we find troubling the mature evolutionist in later years. Any good biologist knows that one of the first signs of degeneration is identity of the mature adult sexes. We can see for ourselves that old men grow to look like old women. Old women grow to look like old men. But: "Co-education of the sexes, occupations of certain kinds, and woman suffrage may have a tendency to approximate the ideals, the lives, and the habits of women to those of men in these same highly civilised races" (Hyatt 1897, 90–91). With readily predictable consequences, since "women would be tending to become virified and men to become effeminised, and both would have, therefore, entered upon the retrogressive period of their evolution."

The case for the value-impregnated nature of Hyatt's progressionism/degenerationism is nigh complete. I need hardly say—especially in an era when the Old World bids fair to surge ahead of the New—that Hyatt's thinking about progress and decline owed a minimal amount to the facts of the case. Even if some evolving lines show senescence or bizarre final forms, there is little reason to think that there is therefore evidence of non-adaptive degeneration. *Tyrannosaurus rex,* one of the last of the dinosaurs, was a superbly engineered carnivore—not at all a geriatric lizard. The gap between theory and evidence yawned deep and wide. Thus, we have now all the support we need to draw our conclusion: in expressing his views on progress (and degeneration), Hyatt followed many other evolutionists in revealing his cultural commitments rather than in offering reflections on disinterested reality.

E. D. Cope: From Quinarian Creationist to Lamarckian Evolutionist

The true justification of America's place in the nineteenth-century story of evolution lies in the breathtaking views it offered into the past, with the fabulous fossil finds revealed as the frontiers were pushed back. A leader of the paleontological pack was Edward Drinker Cope (1840–1897), son of a rich Quaker family. Setting his course westward, in a decade of frantic collecting Cope spent $70,000 of his father's money in pursuit of fossil remains. (To put matters in perspective, Gray felt himself fortunate to get about $1,500 a year.) Unfortunately, he then lost his share of the family inheritance, to the tune of $250,000, in a mining fraud. His financial wings clipped, Cope took up a professorship in mineralogy and geology at his home university of Pennsylvania (in 1889); he was able to escape from penury only by selling his mammalian collection to the American Museum of Natural History, albeit for a fraction of what it had cost him. Apparently Cope's personal failings were not confined to business. Alexander Agassiz wrote to Huxley of him that "whatever his ability which is great he has such qualities as unfit him to come into personal contact with anyone who lays the least claim to be guided by honorable feelings in his dealings" (letter of July 14, 1874; T. H. Huxley Papers, 6.144). There were also well-founded suspicions of sexual laxity. Syphilis brought on his relatively early death. Dashing good looks can be a two-edged sword.

Cope is generally linked with Hyatt, the vertebrate paleontologist complementing the invertebrate paleontologist (Osborn 1931). Indeed, reference is often made to the "Hyatt-Cope" position or school, and there is

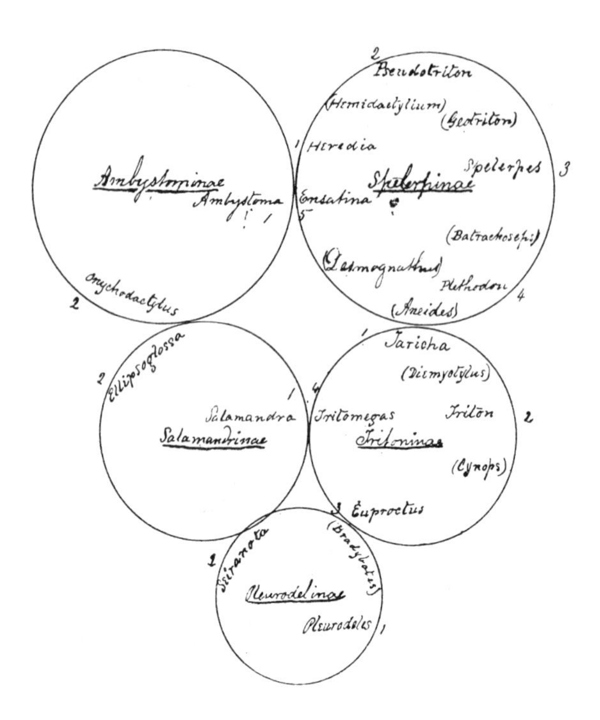

good reason for this (Cope 1886). Like Hyatt, Cope made much of parallels between embryology and phylogeny. Like Hyatt, Cope saw significant evolutionary change coming through the speeding and slowing of rates of development. The very name of "the law of acceleration and retardation" seems to have been his. And, like Hyatt, Cope had little time for natural selection, saw adaptation as but part of the picture, and plumped for Lamarckism.

For all the later similarities, however, Cope's early evolutionism was distinctively his own. This is especially true of his most famous paper, "The origin of genera," published in 1868. There are indeed Agassiz-like features, and even reference to Hyatt's first musings on organic change through time. But, apart from the absence of any Lamarckian speculations in this early piece, the main thesis Cope pushed—that groups of similar species seem to evolve upward, in unison or parallel, as it were—is quite distinctive. Among the members of any one of the groups (which we consider genera) there are differences which could well be adaptive, but although these differences are enough to make for species barriers, in the biological scheme of things they are relatively insignificant. Much more important are the differences which occur across time; these differences

Edward Drinker Cope

Left: The young Cope tried his hand at a little amphibian classification along quinarian principles. (From a student notebook, dated "Summer 1857.")

not only take all the members of one genus into another genus, but take the species within the first genus isomorphically across to corresponding species in the second genus. (Species *A* evolves into species *B*, and *C* into *D*; within the new genus, *B* has the same relationship to *D* as *A* had to *C* in the old genus.) There is nothing particularly adaptive about these macro-changes, which apparently can occur quite quickly.

It goes without saying that, although this is not really yet the world of Hyatt, even less is it the world of Darwin and Wallace. Mimicry says nothing of adaptation and everything of "the independence of generic and specific characters of each other, which may suggest the possibility of the former being modified without affecting the latter" (Cope [1868] 1886, 106). It all seems very peculiar, especially to the modern reader; but, it is not so peculiar as to be entirely original. Take the number of species in Cope's first major example of "homologous genera," two groups of lizards. There were *five* species in each homologous group, and if the number "five" does not trigger memories of systems past, then a friendly reference to William Swainson must. In the early part of the century, Swainson (1835) had been the ardent spokesman for the quinarian system of William Macleay—he of classification-by-fives fame. Detailed student note-

Cope described the "Esequibo Indian women" as "showing the following peculiarities: deficient bridge of nose, prognathism, no waist, and deficiency of stature through short femur." (From Cope [1881] 1886.)

books (filled in 1855, when Cope was fifteen) confirm that Macleay's system *via* Swainson was a real influence (these notebooks are in the Quaker Collection, Haverford College, Haverford, Pennsylvania). And, in fact, by 1857, even though he was now reading Agassiz as well, Cope was trying his own hand at quinarian classification (of amphibia).

It is a nice point as to just how persistent these idealistic views proved to be in Cope's work, taken as a whole. By the 1870s, quinarianism was something of a joke. But, by then, almost certainly influenced by Herbert Spencer's writings, Cope had started to augment his thinking with Lamarckism. Still no enthusiast for adaptationism of a Darwinian (selective) ilk, he nevertheless made significant the inherited response to environmental pressures. Moreover, Cope now suggested that (in some sense) organisms have a control over their destiny, due to a kind of "growth force" (bathmism), and he even went so far as to suggest that primitive forms of organisms have a kind of consciousness and choice (Cope 1896).

As the years went by, much more of the same was produced. Yet, perhaps in line with Agassiz's transcendentalism, perhaps in line with his own boyhood quinarian idealism, it seems always to have been the case for Cope (as surely was also the case for Hyatt) that there was more to evolution than mere Lamarckian response to random environmental factors. Ever central to Cope's vision of the biological world was the way in which organisms fall into trends. By this stage of the century, of course, the branching pattern of the fossil record was taken as fundamental. (See Cope 1896, 143.) Nevertheless, the emphasis was heavily on the way that a group of organisms, once started in a certain direction, tends to stay on path: "The method of evolution has apparently been one of successional increment or decrement of parts along definite lines" (Cope 1896, 24). This is now a version of the theory which became known as "orthogenesis," the view that evolution has a kind of momentum of its own that carries organisms along certain tracks.

What of Cope's views on our own species? Much that he had to say was familiar. "The Indo-European race is . . . the highest by virtue of the acceleration of growth in the development of the muscles by which the body is maintained in the erect position (extensors of the leg), and in those important elements of beauty, a well-developed nose and beard." Although, of course, one must beware of sweeping generalizations: "The strongly convex upper lip frequently seen among the lower classes of the Irish is a modified quadrumanous character" (Cope 1886, 291). Likewise with human intelligence—"In the lowest races there is a general deficiency of the emotional qualities, excepting fear, a condition which resembles one of the stages of childhood of the most perfect humanity." The comments on the female sex were equally predictable. The surprise perhaps is not that females are more embryonic than males in five features, but that males are more embryonic in as many as two features (narrow lips and short hair). Cope highlighted "the structure of the generative organs, which in all Mammalia more nearly resemble the embryo and the lower vertebrata, in the female than in the male" (Cope 1886, 290). All experience is an arch to build upon.

From Providentialist to Progressionist

Very obvious is the underlying progressionist nature of Cope's evolutionism. Overall complexity, behavior, intelligence—these (combined with facial hairiness) are the marks of the superior, the higher, and they are increasingly manifested in the course of organic history. Moreover, the valued qualities are not just contingent happenstances. They are features that emerge naturally from the evolutionary process. At the top we find humankind—white, English-speaking, male humankind. It is true that, in certain respects, we humans might properly be judged rather primitive. Our limbs are of "the primitive type," and the same holds of our teeth: "[Man] possesses, in fact, the original quadri-tuberculate molar with but little modification" (Cope 1886, 278). But impressions mislead. An idea to prove very influential in the history of evolutionary thought, the "law of the unspecialized," tells us that when an organism has gone too far down a particular phylogenetic path, especially a particular adaptively specialized path, it can never pull itself out and evolve into an altogether new and fruitful form (Cope 1896, 172–174). The horse's hoof is excellent for its immediate ends, but useless for further change. True progress stems only from the general, from the unspecialized. Our apparently primitive human form, therefore, is proof of our high status, rather than a denial of our success.

With the fact of progress established, we come naturally to questions about Cope's social/political/economic/religious beliefs and their possible connections to biology. Cope's Quaker background at first inclined him to an absolutely literal reading of the Bible. The exact point when these views began to fade is in some doubt, but the early years of biological inquiry were hardly rank atheism: "In Batrachia the . . . divisions are not as some think humanly invented, but are divine creations down to the lowest groups" (letter to Jacob Stauffer, March 22, 1863; Stauffer Papers, American Philosophical Society, Philadelphia). Even as Cope became an evolutionist, it was always part of his belief that life and the world have meaning, and obviously he tried to build this sense of meaning right into his evolutionizing. There was ever a vitalistic element to his thought. The course of evolution is not random in a Darwinian sense: "I believe that 'the very hairs of our head are all numbered' and that we are of much more value in the sight of God than the lilies and other works of his hand for which he provides bountifully" (letter to Jacob Stauffer, June 29, 1865; Stauffer Papers, American Philosophical Society).

This all sounds very much more Providential than Progressive. Cope may have gone beyond the literalism of his childhood training, but he certainly did not go so far as to cause any significant queries. In fact, not only was Cope's early theorizing firmly embedded in the numerological idealism of quinarianism, but those innovations taking organisms across generic boundaries were openly laid at God's door: "If . . . descendants have attained to successive stations on the same line of progress, in subordinate features of the nervous and circulatory systems, constituting the 'synthetic' predecessors of the orders in each class, the type finally reached seems to rest on no other basis than the pleasure of the Almighty" (Cope [1868] 1886, 106–107).

As his Lamarckism came into sight, however, we find that increasingly Cope was willing to rely on human (and animal) effort rather than simply God's intervention as an agent of change. It was not so much that Providentialism faded, but rather that it became a unifying background for the front-stage drama of Progressionism. In particular, Cope saw a most definite Progressive rise in the history of humankind, from early beginnings to present Western civilization: "Man as a species first appears in history as a sinful being. Then a race maintaining a contest with prevailing corruption and exhibiting a higher moral idea is presented to us in Jewish history. Finally, early Christian society exhibits a greatly superior condition of things. In it polygamy scarcely existed, and slavery and war were condemned. But progress did not end here" (Cope [1870] 1886, 166). And, so on and so forth; although: "The Women Suffragists say if the blacks can vote we should also. Perhaps—only I would reverse the proposition and say—since white women cannot vote colored people oughtn't to either" (letter to daughter, February 5, 1887; in Osborn 1931, 572).

Cope himself was quite open in drawing attention to the parallelism between the biological and the social: "We find a marked resemblance between the facts of structural progress in matter and the phenomena of intellectual and spiritual progress" (Cope [1870] 1886, 154). Moreover, like Lamarck who started it all, Cope explicitly identified the force for biological improvement with that for social advance: "From the first we see in history a slow advance as knowledge gained by the accumulation of tradition and by improvements in habit based on experience" (p. 155). And the same is true of decline and decay, for "the history of the animal man in nations is wonderfully like that of the type or families of the animal and vegetable kingdoms during geologic ages. They rise, they increase, and reach a period of multiplication and power. The force allotted to them becoming exhausted, they diminish and sink and die" (p. 170).

It is hardly necessary to move on to ask about the evidence. History judges Cope to have been an important and talented paleontologist—and in his day he was recognized as such. It is true that some of his crucial claims have now been revealed as fictional—this holds particularly of his key claim about the law of the unspecialized (Cope 1883, 23)—but at the time and for years afterward they were thought sound. Yet, even on his own terms, much in Cope's progressionism went beyond the empirically given. By the time Cope died, one generally sympathetic fellow scientist wrote (to Poulton) that Cope was virtually alone in maintaining that Lamarckism is *the* mechanism of evolutionary change (letter from H. F. Osborn, August 27, 1891; Osborn Papers, NYHS). This is an exaggeration. For all that, by the century's end, the critics were numerous and noisy, Lamarckism had a shelf-life well into this century (Bowler 1988). But Cope was against modern trends here, and even more so with his speculations about "growth forces." People recognized that the empirical evidence for these theories was simply not there. That Cope should have persisted in giving them so central a role suggests that he was driven by non-epistemic motives. It is obvious now that these centered on his belief in Providence-*cum*-Progress. Cope needed his mechanism to support his culture-fueled picture of life's past.

H. F. Osborn and Orthogenetic Evolution

The family of Henry Fairfield Osborn (1857–1935) was rich—very rich (Gregory 1937; Colbert 1980; Simpson 1978; Rainger 1991). There was "old money," and then Osborn's father was the successful president of the Illinois Central Railroad. Socially his was among the leading families of New York City, literally the friends of presidents. Theodore Roosevelt was a childhood pal of one of Osborn's younger brothers, and later in life he himself and the president grew close. Sent as an undergraduate to Princeton—a maternal ancestor was a founder—Osborn finished his training in Britain. First he studied embryology in Cambridge under Frank Balfour, and then he studied comparative anatomy in London under Huxley. One day Osborn was introduced to the visiting Darwin, the memory of which no doubt graced as many after-dinner speeches as it did books of reminiscences (for instance, Osborn 1924). Returning to Princeton, now as a teacher, Osborn gave more and more attention to his primary love, paleontology. This had become his full-time avocation by the early 1890s, when Osborn answered a call from Columbia University to found a department of biology. At the same time, Osborn was appointed a curator at the American Museum of Natural History.

This latter post grew in relative importance. By the end of the century, Osborn admitted that he was spending five of his six working days on Museum business (Rainger 1991, 108). Eventually, in 1908 Osborn became president of the Museum—a position he was to hold for nearly thirty years, guiding the Museum's growth and persuading his rich friends to give generously. Of course, by this time he had become a full-time administrator, but Osborn kept up a stream of publications—relying heavily on the labors of assistants and underlings. With one home on Madison Avenue—"His house is a grand one and full of fine things. Open fires everywhere and a colored man in buttons, etc. etc. etc."—and another up the Hudson, Osborn would have agreed with the sentiment that he was "a big frog in the scientific puddle and social waters" (letter from F. W. Putnam to father, January 31, 1897; Osborn Papers, NYHS).

Cope was the early scientific influence on Osborn. Indeed, Osborn's evolutionism was always in spirit an extension of Cope's orthogenetic underpinnings, even though Osborn thought that Cope was often mistaken about the particular paths of evolution. Where Cope was right was in his view of the *pattern* of change. First, apparently, there is an *adaptive radiation* of organisms, an explosion out from an ancestral form: "According to this law each isolated region, if large and sufficiently varied in its topography, soil, climate, and vegetation, will give rise to a diversified mammalian fauna" (Osborn 1910, 23). This idea goes back at least to Darwin, if not to Lamarck. But Osborn set it in a Copian framework, with the potential for radical change lying only in the general: "A finely specialized form representing a perfect mechanism in itself which closely interlocks with its physical and living environment reaches a *cul de-sac* of structure from which there is no possible emergence by adaptation to a different physical environment or habitat zone" (Osborn 1917, 159).

Then, following this adaptive radiation, with related but already different organisms happily set up in their new ecological niches, *trends* set in. In parallel, the distinct groups go along certain evolutionary paths, generally developing identical new features in bigger and better forms. The titanotheres (extinct rhinoceros-like mammals), for instance, broke into sub-groups but then evolved in parallel and developed their identical new horns independently: "evolution, even in any one geographic region, seldom moves along a single line of descent; more frequently it moves along many lines—it is polyphyletic; in other words, it radiates, following the principles of local adaptive radiation" (Osborn, 1929b, xix). This is from the late stage of Osborn's writings, but the sentiment was there forty years earlier. And of course it goes right back to Cope's earliest thought.

Where Osborn did break from Cope, and try to inject his own theoretical component, was over the issue of causes. Initially, Osborn was attracted to Lamarckism (Cope's mechanism), but his English experiences—and particularly the withering anti-Lamarckism of Poulton (a

Henry Fairfield Osborn

The titanothere, rhinoceros-like but evolved into a baroque form, was used by Osborn as an example of orthogenetic evolution beyond the adaptive optimum. (From Osborn 1929b.)

good friend)—made Osborn realize that it is just not a plausible causal influence. At the same time, although Darwinian with respect to such phenomena as adaptive coloration, for the production of trends Osborn felt obliged to reject the blind chanciness of pure natural selection. For a while, he toyed with a kind of Darwin/Lamarck hybrid, which he called "organic selection" (now better known as the "Baldwin effect") (Osborn 1896, 788; see also Rainger 1991, 127). This was never abandoned, but it started to appear less relevant to Osborn's needs as his paleontological studies began to push him toward an updated version of a theory—"one of the most important in the whole history of biology"—of the Austrian paleontologist Wilhelm Waagen (Osborn 1917, 138–139). Apparently, lying inherent in the heredity of an original pre-radiated group are potentialities—dormant qualities waiting for environmental factors to trigger their development. These are minute variations which show themselves only after organisms are split into new groups. These "rectigradations," as Osborn first called them, give rise to important new adaptations, like titanothere horns. And because they are shared across groups, they lead to parallel effects in different lines.

Later, Osborn was to label these new features "aristogenes," because they are the best one could possibly have or hope for. Note that they were more than the functional units of the geneticist. He thought, at a more holistic level, they are both cause and effect: "*Aristogenes* are new adaptive units originating directly in the geneplasm and slowly evolving into important functional service" (Osborn 1934, 210). Of course, some lines show certain features more strongly or distinctively than do others. Hence, Osborn supposed the existence of distorting factors that, even more than the aristogenes, affect the relative proportions of the new functional features. Aristogenes may have brought on the nasal horns of the titanotheres, but it was "allometrons" which made them grow bigger or broader or whatever.

And here we have a crucial finding, a link between Osborn and other orthogeneticists: "Certain characters of proportion, such as extreme broad-headedness or extreme long-headedness, seem to interfere with adaptation; they appear to be carried so far in one direction as to render the animals less adapted to survive than its less specialized ancestral forms. In other words, certain tendencies of evolution may carry a phylum beyond its requirements in adaptation" (Osborn 1929a, 28).

Osborn realized that he carried virtually no-one with him, but he had no hesitation in applying his theorizing directly to us. Humans are the result of an original radiation out from the other primates, and it was believed that the transformation began in Asia (Bowler 1986; Osborn 1916, 1927). Then, evolution proceeded in separate lines. The reason *Homo* advanced and others did not is that they continued an easy life in the trees, whereas we had to struggle on the plains. But do note that, proximate causes aside, there is no reason to think that there is an immediate ape/human ancestor. Our similarities with the apes are "homoplasies," that is to say similarities caused by shared rectigradations, and not "homologies," pointing to recent, shared forefathers.

Once on our own, humans split again—and again. Be not deceived about our interfertility. This is no true mark of species difference: "We now subdivide *Homo sapiens* into three or more absolutely distinct stocks, which in zoology would be given the rank of species, if not of genera; these stocks are popularly known as the Caucasian, the Mongolian, and the Negroid" (Osborn 1927, 169). Indeed, Osborn was prepared to subdivide even further: "The European variety of man . . . includes three very distinct subtypes, races, or stocks, namely, the Scandinavian or Nordic, the Alpine or Ostro-Slavic, and the Mediterranean, each distinguished by racial characters so profound and ancient that if we encountered them among birds or mammals we should certainly call them *species* rather than *races*" (Osborn 1927, 171).

The Worries of a Public Man

Move on now to the question of Osborn's views on P/progress. With respect to biological progress, meaning the development of some kind of specialization or complexity, Osborn was open and positive.

> The earth speaks not of a succession of distinct creations but of a continuous ascent, in which, as the millions of years roll by, increasing perfection of structure and beauty of form are found; out of the water-breathing fish arises the air-breathing amphibian; out of the land-living amphibian arises the land-living, air-breathing reptile, these two kinds of creeping things resembling each other closely. The earth speaks loudly and clearly of the ascent of the bird from one kind of reptile and of the mammal from another kind of reptile. (Osborn 1925, 5–6)

Osborn made it very clear that he considered us humans to be the crowning pinnacle of evolution.

Yet, we must hedge the straightforward claim that Osborn was committed to a relentless biological progress. First, Osborn did not have much by way of a mechanism for his progressionism. As far as we humans are concerned, our success is due to intelligence and the like (Osborn 1925, 7), and this is in important respects a reward for effort. "*The moral principle inherent in evolution is that nothing can be gained in this world without an effort*" (Osborn 1925, 52, his italics). Also, for humans, some kind of organic selection mixed with a dash of sexual selection seems important: "Those who attain the greatest skill and facility are naturally the most successful members of the tribe: they are the best climbers, the best fishermen, the best hunters, and they are rewarded by the first choice of wives and blessed with the first crop of offspring" (Osborn 1927, 173). But other than this, we fall back on the rectigradations or aristogenes. All of the important developments are built in from the beginning.

Second, note how at the center of Osborn's progressionism must be the claim that there is not just one line going upward but several. This comes straight out of his evolutionism. This is not to say that every branch of the bush of life reaches equal height or develops to the same extent. Some parts out-achieve others. Of the key human races (or genera!), Caucasian, Mongolian, and Negroid, for all their moral equality (Osborn was brought up on *Uncle Tom's Cabin*), Osborn would insist that Negroes are clearly inferior. Their skulls close over more rapidly because their brains cease developing earlier. There are all sorts of hints in Osborn's writings about how being close to the equator makes life a bit too easy for advanced development (Osborn 1927). Presumably, therefore, there are good reasons for thinking the Nordic peoples among the best of the best.

Third, there is the question of decline and degeneration. As part of his general orthogenetic vision of evolution, Osborn expected an eventual toppling and a decline away from adaptive perfection. How does this impinge on progress? Specifically, did Osborn expect decline for us humans? Have we indeed already started to decline? Coupled with gloomy speculations at the macro-level, Osborn felt obliged to conclude that, at the micro-level, for the last fifty years (since 1875), even the best of the human species have peaked. They are now on the biological decline, with respect to intelligence and other valuable characteristics. Only in certain isolated regions like Scandinavia are there pockets of excellence, for "when man begins to specialize and human races begin to intermingle, Nature loses control" (Osborn 1927, 184–185).

Why should the collapse of barriers be so deleterious?

> Every race has a different kind of soul—by soul is meant the spiritual, intellectual and moral reaction to environment and to daily experience—and the soul of the race is reflected in the soul of the individual that belongs to it. This racial soul is the product of thousands or hundreds of thousands of years of past experience and reaction it is the essence or distillation of the spiritual and moral life of the race. (Osborn 1927, 186)

Mix up these racial souls, and you are in big trouble. And it was Osborn's fear that this was precisely the fate of America, flooded as it was with southern and mid-European immigrants. These would intermarry with the already established stock. And the next generation would be—is already on the path to being—a degenerate mongrel race.

Where did Osborn get his views? Were thoughts of Progress behind Osborn's thoughts of progress? As with the other American evolutionists of this period, it really is very difficult (and probably ultimately pointless) to distinguish between his views on Providence and those on Progress. He had a very devout mother and he particularly was the focus of her spiritual attentions. Also highly influential were the teachings of James McCosh, president of Princeton both when Osborn was a student and when he was a professor. One of the first American divines openly to embrace evolutionism, McCosh's belief that science and religion can exist harmoniously together became Osborn's life credo (Moore 1979). Indeed, in the 1920s Osborn became one of the leaders in the fight against fundamentalism, because of its attempt to close down the teaching of evolutionism in schools. Such an idea was anathema to Osborn (1925).

It was not just religion, but religion of a particular kind, that can be counted as a strong influence on Osborn's thinking. Although Osborn was later to worship as an Episcopalian, the family and university background of Presbyterianism left its distinctive mark—as it did on virtually every other major figure in this period. The Calvinist notion of predestination was the theological equivalent of Osborn's biological theory of rectigradations (Moore 1979; Crunden 1982). The really valuable properties, whenever they may actually manifest themselves, are ultimately built in right from the start—and there is nothing anyone can do to alter this fact. Overlaid on all of this, obviously, was the particular social situation that Osborn and his fellows found themselves in. This point applies not just to the orthogenetic theory, as applied to humans—Osbornian sheep and Lower East Side goats (notions that led Osborn, incidentally, to assume presidency of the International Congress of

Eugenics)—but to the very language itself. The very talk of "aristogenes" bespoke the social roots of the theorizer. Osborn was much influenced in his thinking on race by the writings of his friend, the lawyer and AMNH trustee Madison Grant. Grant's *The Passing of the Great Race* contained most of Osborn's views on the racial underpinnings of society. Editions of the book carried enthusiastic forewords by Osborn.

Osborn himself would certainly not have distinguished Progress proper (in any way strictly) from Providence. For him the Bible spoke less of the Fall and more of "the spiritual and moral progress of man" (Osborn 1925, 24). But what of the secular factors behind his Progressionism? As far as biologists are concerned, Osborn was a generation beyond the direct influence of Agassiz. However, there was the undoubted influence of Cope. The idea of rise, with eventual decline, is precisely what one would expect from a disciple of the Hyatt-Cope school. Moreover, although it was certainly not a prominent part of Osborn's theorizing, we do find some use of the biogenetic law—which Osborn may have got first from Balfour. Of the titanotheres, he wrote: "In the earliest geologic stages in which horns have been observed they are found only in adult individuals, but through ontogenetic acceleration they are gradually pushed forward into younger and younger ontogenetic stages until finally they appear on the skull before birth" (Osborn 1929b, 814–815). This reasoning supported the notion that females are undeveloped males.

Osborn certainly believed in intellectual Progress. He was forever telling how evolution as a fact is now firmly established, whereas it was but a hypothesis at the time of Darwin and Huxley. With respect to social Progress in some broad sense (material goods and the like), Osborn was more ambivalent. There had been such Progress in the nineteenth century, but as the years went by, Osborn inclined to think that society had moved toward a period of decline. Recent pessimistic writers on the subject "are no doubt under the influence of the shock of the great World War, which they regard, and in a measure rightly so, as a calamity of the first magnitude in contrast to the optimism of the Victorian period" (Osborn 1927, 185). Osborn's biological progress and his Social Progress mirrored each other. Indeed, the basic ideas of progress and Progress were so obvious to Osborn, he never really felt the need to establish their truth and connection in any particular sense. He did state clearly that "The Social Progress of the last 75 years has arrested and killed racial progress" (Notes for Columbus Address, p. 2; Osborn Papers, AMNH).

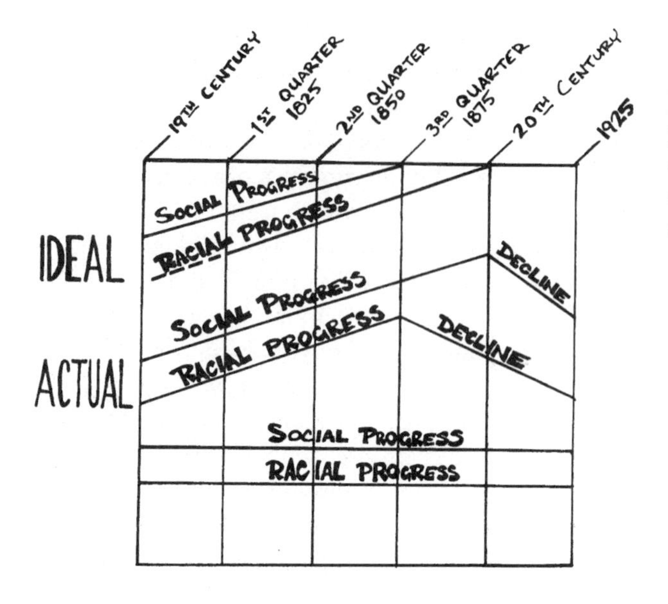

Osborn's diagram (about 1925) of the connections between racial and social progress suggests that decline in the former leads to decline in the latter. (Redrawn.)

As the accompanying figure shows, there seems to be a feedback mechanism here. Social Progress stops racial progress—presumably through allowing unrestricted racial mixing. Then, in turn, racial decline stops social Progress. As marks of such decline, Osborn noted "irreligion, denial of a divine order of things; individualism; feminism; anarchy." It is small surprise that by the 1930s, Osborn was praising Hitler and Mussolini.

We hardly need say more. The cultural underpinnings to Osborn's evolutionism are there for all to see. His progressionism was part and parcel of his Progressionism. And this is amply confirmed when we turn to the question of empirical evidence. One should not underestimate Osborn's formidable skills as a paleontologist, whether it be in recreating a single organism or in looking at the broad panorama of life's evolution. Nor should one neglect his prodigious capacity for hard work. But Osborn was making facts fit theory, rather than conversely. And this was a point which became increasingly obvious to his biological contemporaries, within and without paleontology. He had no evidence for his theory of rectigradations. He had even less for his insistence on parallel lines of change. And his division of humans into separate genera was simply not twentieth-century science. Osbornian culture was so very obviously the aristogene driving the titanothere of Osbornian science.

Evolution Marginalized

We turn now to the social and political side of science. It was Agassiz rather than Gray who was the real intellectual influence at the beginning of our period. The same can be said, even more so, in the social realm. It was Agassiz who set the stage for the evolutionary debate and who thereby defined the kind of status that evolutionism—even for the evolutionists—would have. And what we have to remember is that the Agassiz of America was no longer the young Agassiz of Switzerland. Even if he had been inclined to accept evolution, which he was not, by about 1860 he was no longer a front-line formal scientist. He was in the business of fund raising and institution building. One senses that he was basically tired of professional science—he wanted to go out and preach to large audiences of non-professionals, especially audiences larded with rich potential donors. Students and assistants could stay at the bench and graft (Winsor 1991).

The evolution controversy was a godsend to Agassiz. Not only could he show his own learning, and to his credit fight the good fight for ideals he thought important, but he could demonstrate to *his* public why they must support the science he favored. Agassiz had no interest in moving the evolution question out of the public domain. Not that Gray wanted things otherwise. Since he disagreed with Agassiz about the truth of evolution, by 1860 he would no longer have regarded it as a quasi-science. But this was the belief with which he had matured into scientific adulthood. Indeed, nearly twenty years before, when he had criticized Chambers, he had singled out Agassiz's parallel attack for praise (Gray 1846)! And although he was then turned to evolutionism by Darwin, he saw no reason to lift the belief from the domain of popular science. He debated publicly and his evolutionary articles appeared in popular organs, like the *Atlantic Monthly*.

One concludes, therefore, that Gray, himself a very professional botanist—Agassiz may have founded the National Academy of Sciences, but for many years Gray was president of the Boston-based American Academy of Arts and Sciences, far more prestigious in the eyes of many—regarded evolution much as did the British professionals. I doubt there was significant direct input from Huxley—Gray rather disliked the man, for his perceived religious insensitivity—but for similar reasons and no doubt in harmony with his British (and European) friends, Gray had no desire to make a professional science of evolutionism. To the contrary. In a work which he completed toward the end of his life, *Structural Botany* (1881), Gray did nothing to hide his evolutionism, or even the nature of natural selection and his own disagreements with Darwin. But it was a topic tacked on at the end (pp. 328–331). It was hardly the motor force of a paradigm change, or anything like that.

Even when he could have linked his work with selection, Gray was disinclined. Take phyllotaxy, the arrangement of leaves on the stem. An ardent supporter of Darwin, the pragmatist Chauncey Wright (1871) was able to show mathematically how the arrangements were adaptively valuable. Not only did Gray (as we know) make nothing of the interest the *Naturphilosophen* had taken in the topic, he made even less attempt to bring in a Darwinian perspective! For all that Gray quoted Wright's calculations at length, putative causes went unmentioned (Gray 1879, 119–131). Hence, although Darwin got support in the public arena, he came no closer to building a top-class professional discipline of evolutionism through Gray's efforts in the New World than he did through Huxley's in the Old.

Nor was this result challenged or disturbed at the epistemological level. The pragmatists were enthusiastic about evolution as an idea; but the truth is that the philosophers gave little encouragement to any who would argue that evolution (Darwinism in particular) had a strong potential as a mature science. Peirce (1877) had perceptive things to say about the inherently statistical nature of Darwinism. Yet, apart from Peirce's harsh condemnation of the social (usually Spencerian) implications that people drew from evolutionary thought, as best-quality science, the *Origin* itself was rejected, Agassiz was favored, and toward century's end the judgment became even sterner: "[Darwin's] hypothesis, while without dispute one of the most ingenious and pretty ever devised, and while argued with a wealth of knowledge, a strength of logic, a charm of rhetoric, and above all with a certain magnetic genuineness that was almost irresistible, did not appear, at first, at all near to being proved; and to a sober mind its case looks less hopeful now than it did twenty years ago" (Peirce 1893, 297; cited in Hull 1973, 33–34).

James was kinder in his judgments and elements of Darwinism found their way into his idiosyncratic blend of psychology and philosophy (James 1880). Yet, revealingly, in the course on "Comparative Anatomy and Physiology," which he taught at Harvard in the 1870s, a complaint of a student was that "Darwinism is to be treated metaphysically, that is to say . . . precisely as Darwin and his followers say it should not be treated" (Richards 1987, 425). And James's good friend Chauncey Wright seems to have had a similar opinion as to evolution's status, even though Darwin published a critique penned by Wright against a powerful English critic (St. George Mivart). Darwinism is not to be compared to Newton, selection is not akin "to the principles of physical science as they appear in the natures that are shut in by the empirical resources of the laboratory"; at best the "science of evolution" is up to the standards of political economy or meteorology. We are not even to look for the satisfaction of the epistemic norms of hard science: "Who can tell from these principles [of political economy] what the market will be next week, or account for its prices of last week, even by the most ingenious use of hypotheses to supply the missing evidence" (Wright 1871; reprinted in Hull 1973, 393)?

Wright, in particular, was strong in the belief that mature science does not mix its premises with cultural components. For Gray, given his scientific interests, this would just have reinforced his judgment, mirroring Huxley's, on the proper place of evolution—on the edge of or outside professional science. What is interesting is the way that those after him, even though in respects wanting to make evolution more central to their daily work, likewise preserved evolution's place on the popular side of the sociological spectrum, with related epistemological implications. Hyatt had respect and status. He was a member of the National Academy and the American Academy, as well as a foreign member of the Geological Society of London. He was able to write to Darwin, if not as an equal, then as a fellow worker. Yet one senses that in some respects Hyatt's was not a fully successful academic career. The custodianship apart, he held but minor professorships at M.I.T. and at Boston University. Most of his efforts went into designing museum displays for the Boston Society of Natural History—as well as lecturing to schoolteachers. Under the rule of Agassiz's successor, his son Alexander, Hyatt was allowed to reconnect with Harvard but he was not encouraged to pass on his thoughts and ideas to the young researchers (Winsor 1991). He had one or two followers, but there was certainly no question of founding a school or anything like that.

As significantly, *qua* professional, Hyatt himself seems to have been somewhat diffident about pushing his evolutionism: from the early days, when it was difficult to infer the transmutationism from the content of his papers, to his little book on insects, written for schoolteachers some thirty years later. Evolutionism in his hands was a background belief, not the working tool of a professional. It is true that the insect book was open to evolutionism. But, in addition to the fact he certainly did not promote evolution and its mechanisms as an aid for the biologist, Hyatt warned against exposing young people to the idea.

> We strongly advise teachers not to use this or any theory in teaching immature minds. We give it because we are addressing mature minds, and know that many of them will ask such questions and get no reply. The use of a theory in teaching demands a large knowledge of facts and a capacity to understand and explain numerous exceptions, which bright pupils are very apt to find. Immature minds ought to employ the time wholly in observing, the handling of theory being not only beyond their grasp but injurious, because it leads them to neglect the work which they can do well for a game at speculative guessing. (Hyatt and Arms 1890, 42)

Three decades after the *Origin,* one of the leading evolutionists in North America, a man of proven moral courage, would not have students exposed to evolutionary theorizing. If this does not point to the peculiar status of evolutionism in the world of professional science, I do not know what would. The first thing that Jim Watson did after discovering the double helix was to write a huge textbook (Watson 1965).

I hesitate to say that one who wrote in so foggy a manner as Hyatt was contributing solely to popular science. However, apart from the fact that, when he speculated on such questions as the status of the sexes, Hyatt did go to explicitly popular outlets, by design or by circumstance Hyatt represented the professional marginalization of the evolutionist. Evolutionism as an idea remained more a metaphysical picture, in the sense discussed in earlier chapters. And, if this be true of Hyatt, it is even more so of Cope. His achievements as a paleontologist set him well aside from Spencer, but there are echoes. "Brilliant and erratic" was the way that (the very professional) Poulton described Cope to Osborn (letter, February 11, 1905; Osborn Papers, Department of Vertebrate Paleontology Library, AMNH). No one denied Cope's great abilities or failed to marvel at his fabulous finds but, despite his early training, there was always something of the rich amateur about him. This was true of his publishing outlets, as well, for much of his huge production—up to fifty articles a year—appeared in a journal which he himself owned and supported, the *American Naturalist* (Osborn 1931).

To be fair, Cope was as capable of writing in-the-trenches professional studies as the next person, and much of his straight paleontological work was of precisely this character. (See, for instance, Cope 1883.) When he addressed the topic of evolution, however, it was clear for all to see—not just for us now, but for his contemporaries who read what he wrote—that there was a lot more at stake than straight science:

> How much significance, then, is added to the law uttered by Christ! "Except ye become as little children, ye can not enter the kingdom of heaven." Submission of will, loving trust, confiding faith—these belong to the child: how strange they appear to the executing commanding, reasoning man! Are they so strange to the woman? We all know the answer. Woman is near to the point of departure of that development which outlives time and peoples heave; and if many would find it, he must retrace his steps, regain something he lost in youth, and join to the powers and energies of his character the submission, love and faith which the new birth alone can give. (Cope [1870] 1886, 161–162)

Admittedly, the article from which this passage is drawn appeared originally in a popular magazine. Nevertheless, Cope was prepared to reprint the piece in his definitive collection of essays on evolution.

Predictably, when he did fall on hard times, Cope had great difficulty in landing a job—essentially succeeding only through the kindness of his home town. No doubt this was in part a function of Cope's personal reputation, but it was connected also to the fact that his evolutionizing lay (increasingly) outside what was considered acceptable professional science. Cope could plough his own furrow, regardless of what others might have thought; but the price was that others would refuse to think his (evolutionary) work serious science. Mature science does not flaunt its association with social patterns or cultural ideals. A man who could put down all significant heritable change to "growth energy" or "bathmism" was simply out of touch with professional reality.

Evolution: From University to Museum

I make this last point with some confidence, for we have now entered the period when major changes were occurring in American science, particularly in the professionalization of American biology. With the century entering its final quarter, the really important event was the founding of the new research-oriented university of Johns Hopkins, in Baltimore. This marked a new attitude toward science, a sociological development with epistemological implications. (See Benson 1979; Maienschein 1991;

Rainger et al. 1988.) One of the consequences was a whole new way of doing serious biology. The emphasis increasingly was put on a physico-chemical-based physiology and a microscope-driven morphology, and on newly inspired intermediate subjects, especially on those centering on embryology and development. People became more reductionistic, more interventionist, much more experimental. There was a move from the great outdoors into the laboratory—or the laboratory was taken to the outdoors, especially the seaside. Quantification and measurement became more important. Meeting epistemological standards mattered. (See Allen 1978a; Maienschein 1978, 1987.)

Paradigmatic of the end-of-the-century professional biologist was Edmund B. Wilson, trained at Hopkins and then for many years at Columbia, justly famed for his detailed studies of cell division and development (Maienschein 1991). A master with the microscope, almost by definition he set the standard for what would be considered good professional science. He wrote the classic text (Wilson 1896), he contributed to the new specialized journals, he took a leading role in the increasingly focused societies, he trained the graduate students. And he subscribed to a division between professional and popular science as clear-cut as any to be found in Britain. Most pertinently for our story, increasingly he regarded the study of evolution as beyond professional bounds. I do not mean that Wilson or his fellows had any doubts whatsoever about evolution—they were as committed to the belief as was Huxley; rather, Wilson's science simply presupposed evolution as the metaphysical background, and then he got on with what interested him. In the terminology used earlier in this book, Wilson assumed evolution as fact; he rather looked down upon efforts to discern evolution as path; and he was professionally uninterested in questions of evolution as cause.

Wilson was quite open about all of this. Indeed, he welcomed the way in which the biological spotlight, by the century's end, no longer focused on problems of transformation.

> The relative decline of interest in genealogical questions is partly due, I think, to a healthy reaction against the inflated speculation into which morphologists have too often allowed themselves to fall; but it is also in large measure a result of the growing feeling that the solution of the broader problems of genealogy still lies so far beyond our reach that we would better turn for a time to the study of questions that lie nearer at hand that are, to say the least, of equal interest and importance. (Wilson 1901, 17)

As this passage shows, at least one reason for the flight from evolutionary studies was a reaction against excesses of phylogenetic speculation, which often had been based on altogether inadequate embryological data. America followed Europe in this (His 1967; Nyhart 1986, 1995). Another reason was the desire of biologists in the United States (as in Britain) to get a niche in the medical education curriculum. A case could be made for physiology and experimental embryology, none at all for something like paleontology. The desire was not always satisfied, at least not until American medical education was standardized, in this century (Pauly 1984). But it was a factor.

The most important reason of all for the exclusion of evolutionary studies from the professional domain was the transatlantic influence of Thomas Henry Huxley himself. He spoke at the inauguration of Johns Hopkins, urging his epistemological ideal for biology. With "duly arranged instruction" the student "will come to his medical studies with a comprehension of the great truths of morphology and of physiology, with his hands trained to dissect and his eyes taught to see" (Huxley 1888, 119). Even more significantly, on Huxley's (solicited) recommendation, Hopkins hired H. Newell Martin, student of the English biologist and co-author of his most significant textbook, to teach physiology. Thus, at one remove, Huxley's philosophy of biology was imposed upon (what proved to be) America's leading group of turn-of-the-century biologists (including Wilson). Nor was Martin slow to preach the Master's empiricist creed—a "love of truth must extend to a constant searching and inquisition of the mind, with the perpetual endeavour to keep inferences from observation or experiment unbiased, so far as may be, by natural predilections or favourite theories" (Martin 1876, 183)—at the same time feeding the students a strong diet of the right subjects and the modern techniques (Fye 1985).

Hence, in America by the century's end the official position on evolution was much as it was in Britain. And basically for the same reasons. The kind of speculation in which traditional evolutionists indulged, speculation which was thoroughly P/progressionist, simply was not meshing with the ideals and practices of professional biology. As in Britain, there were those who continued to practice an imported version of (professional) Germanic evolutionary morphology. But as in Britain (and increasingly in Germany), this brand of evolutionism was insecure and apologetic. It was not the best of science. What then of Henry Fairfield Osborn? Biology was being professionalized, but as a consequence evolutionism was being marginalized from the domain of professional science—certainly from the best-quality, forward-looking, proto-mature science. What role was played in this by Osborn, a student and disciple of the key professionalizer (Huxley himself, no less)? Osborn was a thoroughly professional biologist, a skilled administrator, and potential discipline builder, and he openly made a career of evolutionism. Moreover, he was unashamedly Progressionist, indeed could hardly write a word without touching on the topic. He defined evolution as "progress or advance of less perfect to more perfect living organisms" (Osborn 1929a, 350)! How was it possible for Osborn to thrive while rallying behind a theory that was considered "out of bounds" in the new professional era?

The answer is obvious. Osborn is the exception that proves the rule. Or, rather, he is no exception, for he is confirming a rule we have seen exemplified by Lankester and (as you will have noticed) Hyatt. A rule in the establishment of which, even though he was no evolutionist, Louis Agassiz played a major role. (Richard Owen, too, if you think about it.) A rule which, incidentally, was now starting to take effect in Germany, particularly given the problems of evolutionary morphology in the strictly academic community. I refer to the move out of academe and into the museums. Paradigmatically, Osborn shows how one could be a professional and an evolutionist, but as a museum worker—one whose duty to a great extent was to popular science, where one was almost obligated to be a P/progressionist—rather than as a cutting-edge scientist.

Start with the fact that Osborn was a Princeton man, under McCosh. The place was a battlefield, with those favoring science, like McCosh, ranged against those opposing it, like Charles Hodge, principal of the seminary and eminent systematic theologian (Hodge 1872). This was not Johns Hopkins, where the conflict between science and religion was but distant smoke and thunder. At Princeton the relative standing of science was reflected in the kind of teaching Osborn received and was then in turn expected to offer in the classroom. His time in England may have been important, but it was only an interlude. Back at Princeton, there were courses to be given (to undergraduates) on the history of the theory of evolution, and—given the needs of the institution—the context was set in a most un-Huxleyite way. "We have now reached 3 modes in which God is working in living Nature," Osborne noted:

1. direct influence of environment
2. change of s—c following change of environment and causing change of structure

3. Natural selection (most important—because it works with the others.) (Osborn Papers, AMNH)

You would never have had such a reference for (or against) God in lectures in South Kensington, or in Michael Foster's classes in Cambridge or in H. N. Martin's courses at Johns Hopkins. Nor would evolution have figured prominently on examinations, as it did on Osborn's 1883 "Senior Electives in Comparative Anatomy," where four out of the six questions were on that topic. ("6. What are species, how do they originate? Give a brief history of the theory of Natural Selection.")

I am glad to say that Osborn's efforts were appreciated: "I think I owe [my promotion] in great part to Dr. McCloskie who is very much pleased with the assistance I have given him" (letter to father, June 11, 1883; Osborn Papers, NYHS). Probably this encouraged Osborn to make a crucial move in the mid-1880s, when he gave up embryology and turned full-time to paleontology. Almost certainly the major reason here was that Osborn did not need a pragmatic base for his biology—he did not need to link up with the medical profession or with school teaching to justify his existence. He was a rich man—his father gave him $13,000 to build a house at Princeton—and he used his power accordingly. "I am going into Paleontology in earnest now, and upon my return will devote all my time to it and much less to teaching, as you have suggested. I will need a good backing from you" (letter to father, November 29, 1885, from Munich; Osborn Papers, NYHS). He bought off his teaching assignments by using his salary of $1,000 to hire an assistant, and the Princeton trustees had the option of liking it or lumping it.

There are shades of Darwin here: the rich man being able to choose his own path. Unfortunately, as for the older man the freedom backfired somewhat. Osborn was turning away from the best professional biology of the day—and with the move to Columbia the distance increased. Seth Low, then the president of Columbia, was with Osborn a fellow member of the New York elite, and he chose Osborn as much for his value in promoting and developing the college as for his biological expertise (Rainger 1991). It was a wise choice, for Osborn did build a good department. This meant following the lead of Johns Hopkins, and Osborn had no hesitation in hiring people like E. B. Wilson. However, pertinent to our story, we find that Osborn's personal biological enthusiasm and efforts went into raising Columbia's profile in the city—especially through very successful series of public lectures, later published as the very popular *From the Greeks to Darwin.* Osborn was happy with his choice—he liked working in New York, he enjoyed institution building at Columbia, he loved giving lectures to the public, he was glad to turn back his salary to support junior faculty. But the cost was that he was moving, or being excluded, from the forefront of professional science.

It would be a mistake to think purely in negative terms. By about 1890 a reaction was building over the turn in American biology. Critical voices were being raised against its narrow specialization, against its reductionism, against the retreat from nature and into the laboratory. The loudest objections came from Theodore Roosevelt, who sent in a scathing report in 1893 to the Overseers of his old college of Harvard, about the direction taken by the zoology department: "In all our colleges and in Harvard among the number, the modern tendency is to pay attention almost solely to work with the microscope in morphology and embryology, chiefly with regard to the lower organisms. This is a great mistake" (Harvard University 1893, 196). Given that he sympathized deeply with Roosevelt—probably put him up to the critique—one of Osborn's options at this point would have been to take a stand. He could have fought at Columbia, and in that way expanded the argument to the entire professional biological community, for his vision of biology—one with evolution (and paleontology) in a central position. But, putting all of his energies into this option, adopting a strategy which might have meant curbing his overt enthusiasm for progress, was not for Osborn. Lusher fields beckoned. The AMNH also wanted the young scientist, the scion of New York's privileged, and in this institution evolution and paleontology were most welcome, not to mention progress and Progress. An explicit mandate of the Museum was science in its most popular form, and in the words of the President of Harvard (Charles W. Eliot) at the opening of the new building in 1877, "modern science . . . has proved that the development of the universe has been a progress from good to better, . . . a benign advance toward ever higher forms of life with ever greater capacities for ever greater enjoyments" (Eliot 1877).

By the turn of the century, almost all of Osborn's labors were going into the Museum. Although he was certainly still producing numerous scientific articles, his heart was in public displays and exhibitions, books and shorter pieces of the same ilk, and activities which arose out of the Museum work. Osborn had entered fully into the domain of popular science—and he brought his evolution and its progressivist backbone with him. An incredibly successful display of the evolution of the horse became a paradigm for the message of life's upward thrust: from the simple to the complex, from the general to the specialized, from the indifferent to the truly wonderful. Osborn even cadged the skeletons of leading race-horses!

It is in this sense that Osborn reaffirmed his old teacher's beliefs about professional and popular science. Osborn was the scientist of the popular domain. As such, there was no tension in the fact that Osborn was working through the medium of evolution—a belief which for him (and for nigh everybody else) meant progressionism. Yet, there was a price to be paid, and this is the other side to the Huxley dichotomy. By the second decade of this century, Osborn was certainly a professional museum worker; but, for all his stream of papers, his position as a professional biologist was a lot less certain. Osborn himself recognized the difference. Not being at Columbia full time, he was unable to win the battle for his vision of biology: "Natural History is being dropped in our Universities and apparently is centring in institutions like this [AMNH]. I have had great difficulty in securing the perpetuation at Columbia of the kind of courses which I gave on Vertebrate Evolution" (letter to Julian Huxley, April 22, 1919; Osborn Papers, NYHS).

Osborn being Osborn, he fought for his vision at Columbia but, essentially, the world went its own way. Osborn had made his choice, and as the years went by his personal contacts loosened and frayed and his knowledge of the cutting edge of science became increasingly hazy and hostile. In the early 1900s, Osborn's money paid T. H. Morgan's salary. This did not stop Morgan in the next decade from being openly contemptuous of Osborn's work, calling him a "vitalist" (Allen 1978b, 320–321, quoting a letter from Morgan to Osborn, December 26, 1917)—which of course he was, from Morgan's standpoint. (Actually, Osborn himself admitted to being "intensely interested" in the ideas of the arch-vitalist, the father of the *élan vital,* Henri Bergson. See letter to Poulton, May 21, 1912; Osborn Papers, Department of Vertebrate Paleontology Library, AMNH.)

The Boundaries of Professional Biology

A half-century after the *Origin* was published, biology had become a professional science, in Britain and in America. But evolution's share in this triumph was modest. In major respects, evolutionism was *the* science of the popular domain. But its popularity was an indication of its limits, as well as a sign of success. There was a place in academic biology for a German-inspired evolutionary morphology, extending perhaps to areas of paleontology. But progressional though this field was, it was problem-ridden and scorned by the best young researchers.

Every professional biologist was an evolutionist, and it was possible to think professionally about causes, so long as one did not try to look at broader pictures. But Darwin's hopes of a top-quality professional science of evolutionary studies—especially one centering on mechanisms—just did not exist as a general phenomenon. There was no discipline as Darwin had wanted. And progress, reflecting Progress, was a factor in evolutionism's failure to take root in the biology departments of the new research universities. Talk of progress, either biological or cultural, the very essence of evolution as then understood, went against Huxley's idea of a professional science. It belonged rather to secular metaphysics, the domain of popular science. Nor did the shift to the museums alter this fact. Indeed, in major respects it confirmed it, for these were the institutions of popular science with explicit P/progressionist underpinnings.

What had happened was not simply blunt prejudice. There was no *a priori* block to the professional aspirations of evolutionists. Perhaps by its very nature evolution would not have lent itself readily to practical ends, as was the case with physiology, but the sheer interest of tracing the past would have compensated, at least in part. The real trouble was that, epistemically, the work on evolution simply failed to make the mark. Scientific maturity was a chimera. The program of Weldon died with his death, and what remained was second-rate. Evolutionary theory was not sufficiently coherent, comprehensive, consistent, predictive, for the new biology. Progress was a significant item here, both as cause and effect. Epistemic purity was neither achieved nor was it the primary goal. Evolution was a world picture, a metaphysics—indeed a kind of substitute for (or ancillary to) religion. And even for those for whom that is an exaggeration, it was and remained deeply cultural—the culture of Progress.

8

British Evolutionists and Mendelian Genetics

In the middle of this century, evolutionists forged a position known as "neo-Darwinism" or the "synthetic theory of evolution." In this chapter and those following, we turn away from the post-Darwinian era toward the years leading to this event.

If we look just at the history of the idea of evolution, the most important factor in this period was the coming, development, and incorporation of the new theory of heredity. The Moravian monk Gregor Mendel, a contemporary of Darwin's but then unknown, formulated a hypothesis based on the idea of simple factors (what we would now call "genes") being passed entire, from generation to generation. His ideas lay unappreciated until the beginning of this century. Then, it proved possible to link Mendel's insight with the rapidly expanding knowledge of the cell (cytology): the factors were located as entities along the chromosomes, string-like bodies in the central parts, the nuclei, of cells. Thus, in the second decade of this century, T. H. Morgan and his young associates (A. Sturtevant, C. Bridges, and H. J. Muller), at Columbia University in New York City, were able to consolidate these various ideas into what became known as the "classical theory of the gene," with changes in the genes ("mutations") occurring spontaneously without regard to the needs of their possessors (Morgan et al. 1915; Allen 1978b).[1] It was this view of

Edwin S. Goodrich

heredity which ruled until the mid-century, when in 1953 James Watson and Francis Crick unraveled the double helix of the DNA molecule. As the chemistry of DNA became identified with the classical concept of the gene, the molecular biological era was begun (Allen 1978a).

Much has been written on the impact of genetics on evolution, Darwinism in particular. But our concern is not with the history of evolutionary thought *per se*. Rather, it is with evolution and P/progress, especially as this relationship is played out against the backdrop of the dichotomy between professional and popular science. To tell the tale, I shall begin in Britain and move then to America.

Bringing Mendel into Evolution

E. B. Poulton disliked intensely Mendel's British champion, William Bateson. Poulton "did not think that the fundamental qualities of an organism were determined on Mendelian lines, but in a way not yet discovered" (E. B. Ford's answer to questions posed by W. Provine and E. Mayr, 1975; Ford Papers). Yet, by 1908 he had accepted that Mendelism is in some ways true and can be integrated profitably with natural selection (Poulton 1908). Not that he himself wanted to make much of this, unlike Edwin Stephen Goodrich (1868–1946), Linacre Professor of Zoology at Oxford from 1921 to 1945. Goodrich merits our attention, both in his own right and for the influence he had on others (de Beer 1938, 1947).

English-born, Goodrich was raised in France but returned to college in London intending to be an artist. Captivated by the lectures on anatomy by the professor of zoology, E. Ray Lankester, Goodrich switched fields and followed Lankester to Oxford, when the latter accepted his appointment to the Linacre Chair. From then on, Goodrich's was a story of unbroken work marked by similarly unbroken appreciation and success. Beginning with comparative studies of invertebrates, Goodrich reached his apotheosis in 1930 with *Studies on the Structure and Development of Vertebrates*. It tells much about the author that, although coming in at 837 pages, he regretted that it "is not a complete treatise" (p. v).

Although Goodrich defended the evolutionary significance of his professional labors—"The Triumph of the doctrine of Evolution has owed much in the past to the study of the structure and development of the Vertebrates" (p. v)—truly, the chief direct contribution that he made to the story of evolution lay elsewhere, in a little book on the subject that he published in 1912 and updated and expanded some twelve years later. *The Evolution of Living Organisms* had a full and comprehensive discussion of Mendelism as known at that point, and the revision took on board the major discoveries of the past decade by T. H. Morgan and his students. Moreover, Goodrich gave a genuine synthesis. Through vigorous argument, the causal framework of evolution was shown to be natural selection brought on by a struggle for existence. But, as Goodrich stressed, selection demands a theory of heredity. And Goodrich showed that Mendelism, supposing that there are factors passed on uncontaminated from generation to generation, according to fixed laws, provides such a theory. The effects of selection are preserved, and not swamped or blended out of existence by sexual reproduction. Moreover, it is a reasonable assumption that every now and then there appear some new variations. "A truly progressive mutation would be produced when new atoms or compounds become permanently involved in the metabolic cycle of the germ-plasm" (Goodrich 1912, 48–49). Hence, even though these changes would occur entirely without regard to the predicament of their possessors, it is also reasonable to suppose that over time genuine evolution will occur.

What is particularly interesting about Goodrich's treatment of the topic is that he seized on the demonstration by the pure mathematician G. H. Hardy that, in a group with no external influences, Mendelian factors would remain in equilibrium (Hardy 1908; Provine 1971). It is just the principle that evolutionists need, for it shows how Mendelism accounts for background stability. Even the rarest mutation can succeed, if it has a selective advantage:

> The relative scarcity of the mutation at the start does not prevent that a number of individuals interbreeding at random, some with and others without a certain factor, will give rise to a population of impure heterozygotes and pure homozygotes in which the proportion of the three classes will be in equilibrium so soon as the square of the number of heterozygotes equals the number of pure "dominants" multiplied by the number of pure "recessives." If this proportion is not already present at the beginning it will soon become established, and will continue, provided there is no selection to disturb the equilibrium. (Goodrich 1912, 69)

I do not want to overexaggerate the sophistication of Goodrich's thinking. He himself certainly showed no great inclination to explore the consequences, mathematical or otherwise, of what we today call the "Hardy-Weinberg law." Nor did he show inclination to expand on the selection experiments of Weldon, which were mentioned briefly. (Goodrich had been Weldon's student.) But this said, Goodrich grasped the need of a Darwinian selection theory for an adequate understanding of heredity; he saw that Mendelism could fill this gap and how it could fill the gap; and he laid out a blueprint for further investigations by those so inclined.

Emergent Progressionism

What of Goodrich's thinking about biological progress, and how was it embedded in his general views of life? As far as the question of progress *per se* is concerned, Goodrich simply absorbed the idea as part of the essence of evolutionary thought. To him, progress was a basic fact of nature—something which neither Darwinian selection nor the newfound Mendelism challenged or threatened. There is a line (in the animal world) running from the simplest organisms right up to our own species.

Goodrich was fully aware of branching. Indeed, his phylogenies were remarkably bush-like, as opposed to the heaven-reaching tree usually depicted. And, as a pupil of Lankester, Goodrich was hardly unaware of degeneration, though he did not consider it problematical: "Now it is one of the great merits of the doctrine of evolution by natural selection, that it accounts for this simplification as easily as for the development of complexity. For both progressive and retrogressive mutations occur. Variation takes place both in the + and in the − direction, and selection of the one may be as advantageous as selection of the other. Which will be chosen depends on the needs of the organism at the time" (Goodrich 1912, 77).

What Goodrich wanted no part of was anything to do with Germanic idealism. Of course, anyone who worked on morphology necessarily owed a debt to Germany, but Goodrich very much wanted to keep the metaphysics at arm's length. In fact, showing how desperate were morphologists to improve their image, we find him a strong critic of the biogenetic law. He allowed that through similarities of developing forms, embryology can be used to discern phylogenies; but he was firmly in the non-Haeckelian camp: "As far as recapitulation takes place it is a *repetition of ontogenetic stages* of immediate *ancestor* which may be distorted and altered by adaptation at any stage leading to divergence in development and ultimately to violent metamorphosis" (Vertebrate Embryology Notes 1917, Zoology Library Archives, Oxford).

How did Goodrich characterize progress? Like nearly everyone else, at the basic level he thought that complexity functions not as a value in itself but as a flag: "when we speak of higher and lower forms in evolutionary series, we have no desire to make invidious distinctions, but take into account the fact that some organisms are more complex and elaborate in their structure, bodily and mental, than others" (Goodrich 1912, 28). But there had to be something more than this, for at the physical level Goodrich admitted that we humans are hardly much modified from the apes. A significant key to his thinking lay in his enthusiasm for Cope's rule of the evolution of the specialized from the general. Like Cope, Goodrich thought that the secret of success is to stay general, to stay flexible, even (self-predicating thoughts of the writer?) to be a bit old-fashioned. As are we humans: "Extreme specialization may secure temporary triumph, and in very uniform conditions even lasting success, but adaptability is the most precious possession, and it is the creatures most ready to meet new and changeable conditions which have the future before them" (Goodrich 1924a, 170).

Adaptability—that is the key. And immediately and easily Goodrich was able to link adaptability with the large brain of *Homo sapiens*, its associated consciousness and intelligence, and the human ability to respond to new challenges.

> Man has conquered in the struggle for existence not so much because his body is more powerful, his movements quicker, or his senses sharper than those of other animals, but because of his great capacity for retaining the impressions of past responses, and for bringing them to bear on the response to new stimulations. To this he owes his marvellous powers of adaptation to new and varying conditions. And this great development of associative memory has been accompanied by a corresponding enlargement of the brain, especially of the cerebral hemispheres. (Goodrich 1912, 104; repeated verbatim in Goodrich 1924a, 190)

In lectures, Goodrich was given to remarkably Victorian racial speculations. In direct line from Huxley (who, judging from his students' notes, was fascinated by tooth size in Negroes), Goodrich informed his students that the teeth of the "Higher races" are small, with the upper cusps reduced. Those of "lower races" are usually large, with less reduced cusps and more of a tendency to overlap. Apparently these differences can be traced to the propensities of the former alone toward the use of "Implements and cooking" (lecture notes about 1913/14, taken by J. B. S. Haldane; Ms 20576, Haldane Papers, Edinburgh).

Turning next to questions of Progress and of connections with progress, we find that for Goodrich notions of biological progress were simply sucked up with the theory of evolution. For all that he was forward-looking in incorporating Mendelism, we do not see in Goodrich's work any of the wrestling that we find in (say) Darwin's about how a non-directional process like selection can produce progress. Even less do we find wrestling with the implications of the randomness of gene mutation. For Goodrich, progress is part of the picture, and there is an end to it. Not that one should belittle cultural factors: Goodrich was a believer in Social Progress and linked its success with the success of science: "Knowledge is power, and in the long run it is always the most abstruse researches that yield the most practical results" (Goodrich 1922, 85).

Of equal significance, Goodrich—who was a practicing Christian—was keen to see that his thinking meshed with contemporary philosophical thought about how the progress of evolution reflected and manifested a kind of conceptual Progress, a "principle of emergence," as one moves temporally up a hierarchy of complexity: "Water displays certain properties not found in either of its constituents, oxygen and hydrogen; add to a molecule of water an atom of carbon and formaldehyde will be produced, having again quite new properties. So, as step by step the complexity increases, new properties emerge not possessed by, and not predictable from our knowledge of, the lower stages" (Goodrich 1924a, 25). Apparently, these "emergent properties" were key factors in the first steps of evolution, and they remain significant today: "As behaviour becomes more complex and intricate, as the parts of the mechanism become more perfectly integrated, new qualities emerge and on the mental side culminate in consciousness" (Goodrich 1924a, 186).

For Goodrich, an important implication of complex behavior is that it promotes cooperation as well as conflict. Thus, we can see that such cultural entities as morality and religion appeared because of evolution, rather than despite it. Perhaps as a consequence of these ideas, we find Goodrich coming to endorse a kind of Spinozistic identity theory: "we believe that to the continuous physico-chemical series of events there corresponds a continuous series of mental events inevitably connected with it; that the two series are but partial views or abstractions, two aspects of some more complete whole, the one seen from without, the other from within, the one observed, the other felt." What this all means is that mind and body evolved together, with the somewhat striking conclusion that we "have no right to assume that metabolic processes can occur without corresponding mental processes, however simple they may be" (Goodrich 1922, 83–84).[2]

On the matter of biological progress, we have yet to ask our final question. Did Goodrich step beyond his evidence? He would have been surprised and rather hurt at any suggestion that this was a fault (as he would have regarded it) that he might have committed. Indeed, he denied that his linking of progress with complexity had any value connotations, and he was set against those (like the French philosopher Henri Bergson) whose vision of evolution clearly rested on the non-empirical. But there certainly was a move beyond the evidence, especially in the way that Goodrich played fast and loose with the notion of "specialization," in order to fashion (rather than discover) a heroic narrative of evolution, up from the brutes to humankind. Consider a lecture given at the British Association for the Advancement of Science in 1924, in which Goodrich hypothesized about the nature of the first land vertebrates. Apparently, these were early, unspecialized amphibia, which (for him) they would need to be if they were to have great evolutionary potential. "They were clumsy heavily built slow-moving—in shape not unlike Salamander.

Heads large and broad—no neck—body thick—tapering tail. Short clumsy limbs with 5 digits set at right angles to body—probably incapable of lifting it off the ground. More or less covered with *true scales* especially on ventral surface" (notes, Zoology Library Archives, Oxford; also Goodrich 1924b). After all of this, it is somewhat of a relief to learn that they had small brains. No doubt they spoke with Irish accents, and almost certainly had sisters who were nuns.

The point, of course, is that even though Goodrich may have been right in every one of his details, his overall picture of a monster misplaced from a horror movie is tailored explicitly to fit his preconceptions about progress. By his own admission, in the same lecture, such an animal had to be descended directly from a highly specialized fish—one with the ability to survive out of water, as during droughts. Of like forms today, he admitted that they are "All adapted for life in hot regions where rivers apt to dry up and water becomes foul—supplemental gills with air-breathing lungs" (n.d. [about 1920?], Zoological Library Archives, Oxford). This does not sound like an all-purpose general form of fish. So why then judge the amphibia unspecialized? All of the talk of clumsiness is a bit beside the point. By supposition, there were no other land vertebrates. Hence, the early amphibian could flop around all that it liked, without fear of danger. It was not a seal out of water surrounded by tigers. Remember, as Goodrich should have remembered, that natural selection is a relativistic process, and that in its own way our early land ancestor could have been perfectly well adapted.

For all his caution, when it came to progress Goodrich went beyond the given. Hence, our usual pattern is established. Goodrich on progress was Goodrich influenced by cultural values. I shall speak later as to how his work in this area fitted onto the professional/popular scale. For now, I stress that the coming of Mendelism made absolutely no difference whatsoever to the first person seriously and publicly to incorporate the new genetics into his evolutionary thinking. Evolution was progressive to the pre-Mendelian Lankester. Evolution was just as progressive to his student, the post-Mendelian Goodrich.

Fisher—the Darwinian

Ronald Aylmer Fisher (1890–1962) went first to the prestigious English private school Harrow and then on a scholarship to Cambridge (Box 1978). There he studied pure mathematics and mathematical physics.

Although Fisher took no biology courses whatsoever, it is clear that Charles Darwin was ever his hero. This near-worshipful attitude was strongly reinforced by a thirty-year friendship with Major Leonard Darwin, Charles Darwin's second-youngest son, who was to become a real father-figure for Fisher, encouraging, sympathizing, pushing, and (not infrequently) helping with much-needed gifts of cash.

An ardent Darwinian at Cambridge, in the early years of this century, was a man rather alone. However, this was a challenge for Fisher, not a barrier. Having read Bateson's *Mendel's Principles of Heredity,* Fisher at once saw intuitively that Darwinism (meaning natural selection) and Mendelism (meaning particulate inheritance) should properly be considered as co-workers, as he pointed out in a short paper read to an undergraduate society in 1911 (Bennett 1983). To the confident young Fisher, therefore, the task at hand was clear. Someone had to go beyond the biometrical/Mendelian debate and put the Darwinian-Mendelian connection on a truly firm basis—on a firm mathematical basis—and who better than Fisher himself? He grubbed up his biology and, backed by training in statistical mechanics and quantum theory by James Jeans and the theory of errors by F. J. M. Stratton, Fisher created his science wholesale.

Following the first major paper, "The correlation between relatives on the supposition of Mendelian inheritance," which laid the mathematical groundwork for putting heredity and selection into an integrated theory, Fisher—who in 1922 had become resident statistician at the agricultural research station at Rothamstead, where he established a major reputation for the design of experiments—moved on to consider the fate of genes in populations, under various stresses of mutation and selection. Especially important was his demonstration that, under certain conditions, most specifically when a heterozygote is biologically fitter than either homozygote, selection will lead to an equilibrium between alternative genes within a population. As importantly, but in the long run far more controversially, Fisher (1922) investigated the conditions under which a neutral gene might drift in a population, "up" to fixity or "down" to elimination. Not now or ever did Fisher deny the conceptual possibility of this phenomenon—which we (after the American Sewall Wright) call "genetic drift." However, Fisher did show that the effect would be very slow except in small populations, and this brought about—or, rather confirmed—a lifelong belief that, since natural populations are always sufficiently large, accident can never be a significant factor in evolutionary change. It is selection or nothing.

Then, finally, at the end of the decade, in large part as a result of the prodding of Major Darwin—"my wish that you should deal with the whole problem of selection mathematically: you will have a small audience but it will gradually be realized that many of the problems can be attacked in no other way" (Box 1978, 187)—Fisher produced his *magnum opus.* This work, *The Genetical Theory of Natural Selection (GTNS),* is generally considered Britain's greatest contribution to evolutionary thought since Darwin. It is therefore important to stress that, our own species excepted, Fisher was really not much interested in the broader spectrum of evolutionary topics. There was no mention of the fossil record, or of geographical distribution, or of any such set of issues as includes these. Fisher was after a basic mechanism—the spread, maintenance, and decline of units of heredity in populations—and it was on this mechanism and on some immediate implications that he concentrated.

Like Goodrich, Fisher saw the Hardy-Weinberg law as the key foundational principle, as the backdrop for the leading actors on the causal stage. Although he himself had shown how selection can lead to equilibrium, in fact he rather thought such a state to be a minor phenomenon in the real world. What he really wanted to show was that the intensity of selection is in some way a function of the variability of fitness in a group and that, all other things being equal, selection is raising the fitness of the group. This fact Fisher derived as his *fundamental theorem of natural selection: "The rate of increase in fitness of any organism at any time is equal to its genetic variance in fitness at that time"* (Fisher 1930, 35). (See also Price 1972.) Essentially what this rule tells us is that, given some organisms fitter than others, natural selection has an ongoing tendency to push the population up to the summit of fitness (Sober 1984).

Despite the somewhat misleading formulation, the fundamental theorem is very much a law pertaining to *groups* rather than individuals, and you must reckon with the fact that, as things change, this focus on the state of the group could affect internal comparisons of fitness. Moreover, the environment could be "deteriorating" in various ways, so that one step up may be equal to one or more steps down. "Against the action of Natural Selection in constantly increasing the fitness of every organism, at a rate equal to the genetic variance in fitness which that population maintains, is to be set off the very considerable item of the deterioration of its inorganic and organic environment" (Fisher 1930, 42). But, as we shall see, overall the fundamental theorem was to be taken in a positive, creative sense.

With the basic structure in place, much of the *GTNS* was given to tracing quantitative effects on populations of various rates of mutation and degrees of selective severity. Given Fisher's crucial belief that things, taken at the group level, do not happen by chance, he was happy to show that even the lowest powers of mutation will have their effects sooner or later, generally sooner: "in a species in which 1,000,000,000 come in each generation to maturity," given the "familiar mutation rate of 1 in 1,000,000 the whole business would be settled, with a considerable margin to spare, in the first generation" (Fisher 1930, 78). Similar arguments hold for selection, and meshing nicely with these was Fisher's theorizing about the phenomenon of dominance: How and why is it that alleles of one kind mask alleles of other kinds, and why in particular is it that the usually existing form in nature (the "wild type") tends to mask mutants? Fisher argued that dominance is a function of selection: because mutants are often deleterious, organisms develop adaptations which make for the recessiveness of new alleles. One effect of these adaptations is that populations tend to carry a fair number of recessive alleles, and should conditions change and new needs arise and selective pressures change there is an already existing reserve of variation on which to draw.

There is more to the *GTNS*. For now, let it be noted that with the appearance of the book, Fisher's importance as a major scientific theorist was established beyond dispute. Professorships followed, first in London and then in Cambridge. Evolution had a champion of the first rank.

Fisher—the Eugenicist

We turn now to questions of P/progress. First, there is the matter of biological progressionism, *prima facie* an impossible topic since Fisher was simply not interested in the kinds of questions which usually show a commitment (or non-commitment) to the notion of progress.

> If I had had so large an aim as to write an important book on Evolution, I should have had to attempt an account of very much work about which I am not really qualified to give a useful opinion. As it is there is surprisingly little in the whole book that would not stand if the world had been created in 4004 B.C., and my primary job is to try to give an account of what Natural Selection *must* be doing, even if it had never done anything of much account until now. (Fisher to J. S. Huxley, 6 May 1930; in Bennett 1983, 222)

Yet, for all this, the student of Fisher's life and work is left in little doubt about where he stood on the subject. He believed absolutely and completely in biological progress and he thought humans are right at the top. The general case starts to come through in early writings, but it flowers fully in the *GTNS,* especially when Fisher was dealing with his "fundamental theorem." Remember that Fisher had trained as a mathematical physicist, and this early schooling—most surely the influence of the gas theorist Jeans—colored his perspective on the biological world. Genes in populations, more specifically genes in large populations, are to be seen as akin to the inorganic molecules in a physical system: "the whole investigation may be compared to the analytical treatment of the Theory of Gases" (Fisher 1922, 321). It was for this reason that the fundamental theorem was to be regarded as one of the basic laws of biology, analogous to a basic law of physics. Fisher himself went as far as to liken it to the second law of thermodynamics. Just as in any physical system we have this ongoing increase in entropy (presumably to a final state), so in any biological system we have this ongoing increase in fitness (presumably to a final state).

There are differences between entropy and fitness, and it is through these differences that the potential for progress is introduced into the biological model. Whereas entropy seems to run a system down, selection seems to run a system up. There are lots of fits and starts in both cases. To Major Darwin Fisher confided that he thought any species generally to be just about balanced between forces. "I think of the species not as dragged along laboriously by selection like a barge in treacle, but as responding extremely sensitively whenever a perceptible selective difference is established. All simple characters, like body size, must be always very near the optimum" (letter to Leonard Darwin, 7 August 1928; in Bennett 1983, 88). But overall, both in physics and in biology, the forces seem to have a cumulative effect. In the biological world, this leads to progress: "Entropy changes lead to a progressive disorganization of the physical world, at least from the human standpoint of the utilization of energy, while evolutionary changes are generally recognized as producing progressively higher organization in the organic world" (Fisher 1930, 37).

It is tempting to say that Fisher, like his hero Darwin, was leaping from comparative or relativistic progress to absolute Progress. Yet, although later in the *GTNS* Fisher gave examples of comparative progress, he offered no detailed exposition of the notion, certainly not when he introduced the fundamental theorem. The simple fact is that Fisher seems to have been a lot less tense than Darwin about the power of selection to bring about absolute Progress, with us humans at the apex: "I have never had the least doubt as to the importance of the human race, of their mental and moral characteristics, and in particular of human intellect, honour, love, generosity and saintliness, wherever these precious qualities may be recognized" (Fisher 1930, 171).

Turning now to look at Fisher's cultural perspective and the ways in which it might be seen to feed back into his biology, we must now enter a major qualification and introduce a guiding passion in Fisher's life. Fisher saw the general evolutionary picture as one of advance and progress. In the human case he had taken well to heart Darwin's views on the virtues of a division of labor, which he felt necessary for the coming of civilization. And he made clear that he saw civilization as having qualitative as well as quantitative benefits: "It is a matter of experience, which no one thinks of denying, that such an organization does in fact enable a given area to support a much larger population, and that at a higher level of material and intellectual well-being, than the uncivilized peoples who could alternatively occupy the same territory" (p. 175).

Nevertheless, he was worried—desperately worried—about the decline of civilizations. It was the obverse side to progressive change which really gripped Fisher's imagination: "The decay and fall of civilizations, including not only the historic examples of the Graeco-Roman and Islamic civilizations, but also those of prehistoric times, which have been shown to have preceded them, offers to the sociologist a very special and definite problem—so sharply indeed that its existence appears to challenge any claim we dare to make to understand the nature and workings of human society" (p. 174).

Why should this decline occur? Why are societies not forever on the rise? Fisher's reading, as an undergraduate and shortly thereafter, led him to what he perceived as the root of the problem. Unfortunately, as societies advance they become more clearly stratified, and those at the upper levels tend to reproduce less than those at the lower levels. In the words of a major influence, one J. A. Cobb: "Any able man who rises by his ability into a higher social class than that in which he was born will naturally marry into that class, and will be likely to have a less fertile wife and fewer children than his medium brother who remained in the class into which he was born" (Cobb 1912, 380). This sets up very strong selective pressures—much stronger than Fisher thought normally obtain elsewhere in the living world—and before very long, with the worse

R. A. Fisher

Fisher's somewhat dramatic illustration of the causal contribution of the genes to height. (From Fisher 1918b.)

outbreeding the better, the moral and intellectual fiber of a society rots away and a general collapse ensues.

There are shades of Herbert Spencer here, which Cobb somewhat grudgingly and Fisher more graciously acknowledged. But, whoever should get the credit for seeing the problem, to Fisher the solution was clear. One must try to activate some form of human breeding project, devoted to getting the upper levels of society to breed more and the lower levels to breed less. One must be a *eugenicist,* and it is hardly too much to say that on this subject Fisher was a life-long fanatic. At the personal level, Fisher literally lived eugenicism. At times of need, he worked for and was supported by the Eugenics Society—the medium through which he met and became close to Major Darwin, who was its head. He married a naive young girl, just seventeen, who was chosen deliberately because she would be good breeding stock. And then, despite severe financial hardships, he proceeded deliberately to have a large family of eight children, as an expression of his beliefs. At the professional level, again and again it was the dog of eugenics which wagged the tail of biology.

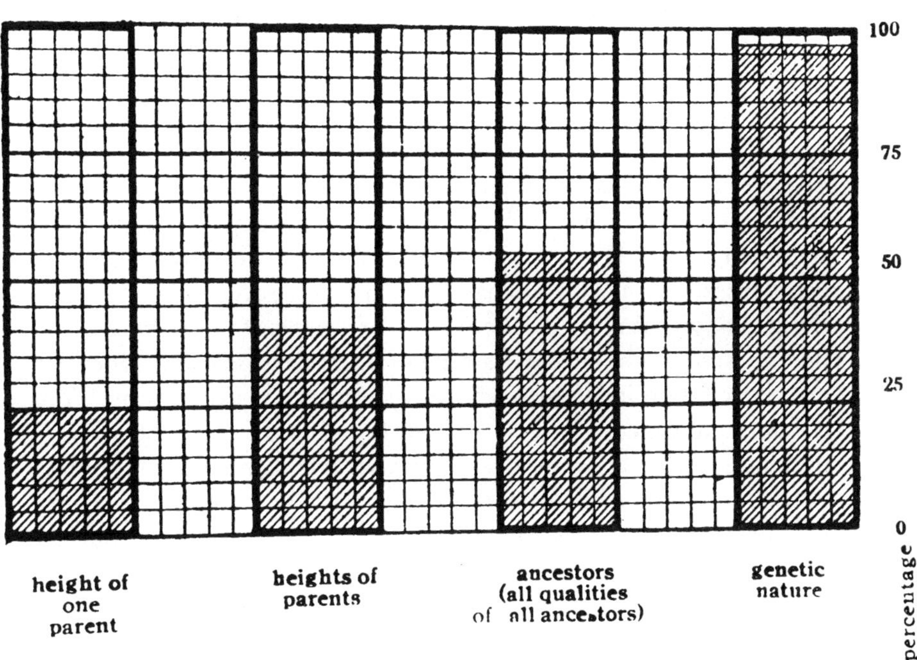

That first undergraduate paper, linking biometry and Mendelism, was read before the newly formed Cambridge University Eugenics Society, and the underlying theme was explicitly eugenical: "the thought of a race of men combining the illustrious qualities of [Shakespeare and Darwin], and breeding true to them, is almost too overwhelming, but such a race will inevitably arise in whatever country first sees the inheritance of mental characters elucidated" (Bennett 1983, 53–54).

Likewise, eugenics was the *leitmotif* of his paper on the correlation between relatives, which forged the mathematical link between Darwinism and Mendelism (Fisher 1918a). Helpfully, for those who would have difficulty following the mathematics, Fisher gave a digest of his position, in words, in the *Eugenics Review* (Fisher 1918b). Having looked at studies of human height, Fisher concluded that the eugenicists are fully vindicated. The variation within our species due to genetic qualities, the amount which is in some sense innate, is over 95 percent. Although Fisher conceded that the other 5 percent "would allow considerable scope for the action of environment in individual cases" (Fisher 1918b, 220), the message and conclusion were unambiguous. Human nature is a function of the genes, and if we are to maintain Progress in our own society then we must, artificially if necessary, do all that we can to maintain progress.

The urgency of Fisher's concerns was transferred straight into the final chapters of the *GTNS,* a work which appeared in the year after the stock market crash, just when the Great Depression was getting under way. Societies seem to rise and then decline. Fisher worried that, having manifested the former phase, British society was now apparently moving into the latter phase. Perhaps reflecting some of the fears of his American counterparts, Fisher was prepared to agree that racial mixing is no good thing. But, for the middle-class Fisher, the real threat was to be found right here at home, with the underfertility of the professional and upper classes and the overfertility of the lower classes. And the answer is clear. We need a family allowance! Not a general one for all—that would be worse than nothing—but an allowance geared toward class and income, so that an upper-level person who decides to have another child is in no way penalized financially: "If this were so, and if, at the same time, preferential promotion of infertile strains from the less prosperous classes were entirely to cease, it seems not impossible that the fertility of the upper classes might be restored, by the differential elimination of the less fertile strains, within no very lengthy historical period" (Fisher 1930, 263).

If ever the biological echoed the social, it is here. So, let us now turn briefly to our third question, about the evidence. Did any of what Fisher had to say about progress or Progress or decline have anything to do with the real world as it was then known, or might reasonably have been known, to Fisher? Note that Fisher's view of evolution made crucial the flow of genes in *large* populations (and that large size was more assumed than proven in the *GTNS*). It was for this reason that minute selective pressures could be effective. Genes enter populations and, if useful, as their effects become felt they eventually become the norm. But in a smaller population the effects of selection change somewhat. The significance of the fundamental theorem is no longer so obvious, and if this theorem no longer applies (or is downplayed) then the background for Fisher's overall progressionism is weakened, at the least. And it is surely weakened even more if one recognizes that the decay of the environment might always equal (on average) the gains through selection. In other words, populations might be running flat out—constantly changing, adding new adaptations—but never "advancing": a phenomenon today known as the "Red Queen effect." This may in fact never happen in nature, but the point is that Fisher did not prove otherwise. (See Chapter 12.)

Also unproven were Fisher's pronouncements on human populations. In early married life, "The Fishers read together in the history of civilizations, noting evidences that the decay of past civilizations had resulted from the pattern of social selection arising in a moneyed economy" (Box 1978, vi). Again and again, however, Fisher leapt from the given to the believed. Most suspect, surely, even on his own terms, was his commitment to the supposedly almost totally innate nature of human abilities. Making this claim just at the time when the social sciences were starting to flower, Fisher should have had some sensitivity to the complexities of human nature. Only then could one start to ferret out those qualities which make for success and failure, and only then could one understand precisely how these qualities might be underpinned by biology. As it is, we have an ideology imposed on the biology. Fisher's conclusions outstripped his data.

The answer to the third part of our question surely supports one overall conclusion: Fisher's thinking about evolution was impregnated with his extra-scientific beliefs about society and its Progress, positive and negative.

Fisher—the Christian

A remarkably coherent world picture is starting to emerge. The biological reflects the social and indeed the two blend into one. Yet our picture is still incomplete, and unless we now pull back for a broader view we shall fail adequately to assess Fisher's place in our story. Adding to Darwinian natural selection and the need of eugenics, there was a third pillar holding Fisher's thought high above ground. This was Christianity (Hodge 1992). For Fisher, a devoted member of the Church of England, God was no ethereal distant Being but a divine artificer actively concerned in His creation. He (for such, I am sure, was Fisher's God's sex) had imposed upon Himself a task, namely that of the creation of life, up to and including humankind. This was genuinely a task, and it was genuinely a task of creation, for God's work had to be done against and despite a background of decay—as postulated by the second law of thermodynamics. Fortunately, thanks to natural selection God has succeeded, and the end result pays testimony to that which is at the center of Christian theology: humans specifically and organisms generally, especially in their intricately and marvelously designed natures—what Fisher referred to as "the high perfection of existing adaptation" (letter to Sewall Wright, 19 January, 1931; in Bennett 1983, 279).

Parallel to God's task, however, is our own, human task. "Man is in process of creation, and the process involves something we can call improvement, in which Man's own co-operation is necessary" (letter to E. W. Barnes, 12 January 1952; in Bennett 1983, 182). Cooperation demands a dimension of freedom, and Fisher stressed that there is an element of indeterminacy to our lives which opens the possibility of choice and creativity. In justification, Fisher drew explicit analogy with the indeterminacy of mutation, and drawing a truly remarkable parallel between putative causes of evolution and essential demands of Christian life, he emphasized that our work has the same theological significance as does God's.

> There is indeed a strand of moral philosophy, which appeals to me as pure gain, which arises in comparing Natural Selection with the Lamarckian group of evolutionary theories. In both of these contrasting hypotheses living things themselves are the chief instruments of the Creative activity. On the Lamarckian view, however, they work their effect by willing and striving only; but, on the Darwinian view, it is by doing or dying. It is not the mere will, but its actual sequel in the real world, its success or failure, that is alone effective.
>
> We come here to a close parallelism with Christian discussions on the merits of Faith and Works. Faith, in the form of right intentions and resolution, is assuredly necessary, but there has, I believe, never been lacking through the centuries the parallel, or complementary, conviction that the service of God requires of us also effective action. If men are to see our good works, it is of course necessary that they should be good, but also and emphatically that they should work, in making the world a better place. (Fisher 1950, 19–20)

It is here, of course, that we find Fisher's Progressionism—the need of human effort. Truly, Fisher's thinking is set against a background of Providentialism, although more appropriately one might speak even of God's contribution as Progressionist rather than Providentialist. Given critical remarks that he made in reviews (in the *Eugenics Review* from 1918 on) of books of a Spenglerian bent, Fisher was in no way drawn to idealists' all-encompassing laws of determined history. On Fisher's theology, God seems to have set Himself genuine work to do—the completion of which He chooses freely, rather than having His works emerge directly and necessarily as a consequence of His Divine Nature. Analogously, tying in eugenics, Fisher emphasized that God expects us to get on with the job that He has started:

> To the traditionally religious man, the essential novelty introduced by the theory of the evolution of organic life, is that creation was not all finished a long while ago, but is still in progress, in the midst of its incredible duration. In the language of Genesis we are living in the sixth day, probably rather early in the morning, and the Divine Artist has not yet stood back from his work, and declared it to be "very good." Perhaps that can be only when God's very imperfect image has become more competent to manage the affairs of the planet of which he is in control. (Fisher 1947, 1001)

Like God, we are faced with a background of decay and decline—the social equivalent of the increase in entropy—and it is our task to try to reverse the process, as God overcomes entropy. Hence: "Instead of being a soulless creed, the possibility of evolution might well be the very centre of our faith and hope" (Fisher 1947, 1001).

The full story of the Fisherian world picture emerges. Biological progress is an absolutely central part of his thinking. Far from Mendelism having expelled progressionism, Fisher thought the new genetics guarantees it! Given that organisms are usually close to their optimum fitness, a large change—a large jump—would take one straight down or over the top and down the other side (Fisher 1930, 38–41; Turner 1987; Leigh 1986, 1987). Large variations would give their own direction to evolution, almost certainly in a deleterious fashion. Very small variations are what we need, for then all of the creative onus can be put on selection. Mendelism therefore gives God and man a free hand to direct evolution as they will. The unique perfect solution exists—*Homo sapiens* for God, upper-middle-class society for Fisher—and it is up to us to get there.

J.B.S.

John Burdon Sanderson Haldane (1892–1964), the son of J. S. Haldane, a noted physiologist, was simply the brightest man that people had ever met (Clark 1968; Sarkar 1992a,b; Maynard Smith 1992). At school and then at Oxford, he triumphed both in mathematics and in classics. Given his family background, Haldane was hardly less proficient in biology. At Oxford, he sat in on Goodrich's course in vertebrate anatomy. His plans to study biology formally were lost, however, when the First World War intervened and Haldane went off to serve his country. Returning to civilian life, Haldane spent ten years at Cambridge, where he did important work in biochemistry, made his contribution to evolutionary theory (generally working on a Saturday morning at an agricultural station, to which he was attached part-time), set out on a career as one of the most talented scientific essayists of all time, and generally scandalized the staid and pompous. "'I am not a prude' Bateson said 'but I don't approve of a man running about the streets like a dog' " (recollections of JBSH by C. D. Darlington, recorded September 30, 1965; Darlington Papers, J68). His father suffered greatly from the son's actions, and one suspects that this may have acted as a stimulus.

By the 1930s, Haldane was a person of some notoriety, invariably good (in the eyes of the popular press) for a pithy quote on anyone or anything. Physically he moved to a professorship in London, and politically he moved from a mild socialism into the Communist Party. He spent time in Spain during the civil war and wrote copiously for the (communist) *Daily Worker.* After the Second World War—during which he inflicted upon himself some horrendous experiences to test the effects of decompression (in submarines)—he broke from the party. His withdrawal was slow, painful, and dragged out to the point of self-degradation, as he—one of this century's champions of modern genetics—defended beyond the limit of reason that travesty of biological thought in the Soviet Union which flourished under the dominion of Lysenko. Finally, Haldane tore loose and, controversial as always, pulled up roots and took off to India.

As an undergraduate, Haldane accepted both Darwinian selectionism and Mendelian heredity as easily and as readily as did Fisher. To a student society, he read papers both on Mendelism and on Darwinism. Having emphasized the intricately adapted nature of the organic world, Haldane wrote: "Darwin's theory of natural selection then, on the whole stands in a better position than it ever did. It is supported by recent work in every branch of biology, and especially on inheritance" (1912 notebook, Ms 20578, 14f., Haldane Papers, Edinburgh). Presumably, Goodrich was an influence here, and indeed it was very much the older man's program that was followed in the postwar technical papers and a synthesizing popular book, *The Causes of Evolution.* Certainly, unlike Fisher, Haldane had no interest in proving background theorems, "fundamental" or other. The Hardy-Weinberg law was all that Haldane needed to build a range of models, introducing extraneous forces into hitherto stable populations, under a variety of conditions like superior heterozygote fitness or sex-linked inheritance, and deducing the consequences (Haldane 1924a,b, 1926, 1927a,b, 1930, 1931a,b).

The results which started to tumble out strongly influenced Haldane's thinking about the overall evolutionary process, especially the conclusion that—Fisher's opinion to the contrary—selection pressures in nature might be very high indeed, and that change could be correspondingly rapid. Even more significant for Haldane was his finding that most new mutations will take many many generations to spread in a population. It would seem that evolution can never get started, except in the most exceptional circumstances—a problem which drove Haldane to posit the necessity of some form of isolation. "This case seems to me very important, because it is probably the basis of progressive evolution of many organs and functions in higher animals, and of the break-up of one species into several" (Haldane 1932a, 102–103). Haldane thought isolation might be very important in the evolutionary process for another, related reason: really significant changes probably require two or more changes occurring at the same time. For example: "A serious improvement in the eye would involve a simultaneous change in many of its specifications" (Haldane 1932a, 103).

Were one looking at Haldane's work for its own sake, there is more one could mention, including some remarkably modern-sounding speculations on the conditions necessary for the spread of biological altruism (Maynard Smith 1992). For our ends, however, we need not review every detail, and so let us turn to our P/progress questions, asking first about Haldane's attitude toward biological progress. As with Fisher, being interested primarily in causes, Haldane the evolutionist was not out to make his mark in the broad sweep of the subject. Yet, Haldane was surprisingly forthcoming on the question. From beginning to end, Haldane nailed his colors to the mast of progress. In another of his undergraduate essays, he wrote: "No one really doubts the facts of selection and variation and the geological record leaves no doubt that animals and plants have steadily been changing, and on the whole improving, throughout geological time." Adding that "the fact that we [humans] have risen so far makes it exceedingly probable that we shall rise higher, especially if we are able to understand the processes which govern our own rise" (Ms 20578, Haldane Papers, Edinburgh). In a similar vein, toward the end of his career in 1953, he said that although "progressive

J. B. S. Haldane

evolution is exceptional, and regressive evolution, leading to loss of organs and capacities, is the rule . . . evolution has been, on the whole, progressive" (lectures on Darwinism given in winter 1953; Box 1, Haldane Papers, London).

In the *Causes of Evolution,* in his conclusion, Haldane gave a very brief history of life, arguing that there has been "fairly steady progress" up from the most primitive forms to humans, and likewise in plants (Haldane 1932a, 151). But this was added more as an afterthought. Haldane's most explicit discussion of biological progress had, in fact, come a year or two earlier (1927), in a book *(Animal Biology)* he co-authored with his long-time friend (they were at school together) and sometime colleague, Julian Huxley. This was a school textbook, and although the style suggests that Huxley was the guiding force overall, most of the evolutionary examples were lifted straight from Haldane's student essays.

Notwithstanding the fact that the authors introduced the topic of biological progress as a variable of complexity, they made it clear that, for them, such structural criteria are but marks of function: "we may say

INDIVIDUATION

	Single cells.	Colonies of cells without division of labour.	Colonies of cells with division of labour.	Metazoan individuals.	Segmented Metazoa.	Colonies of Metazoa without division of labour.	Colonies of Metazoa with division of labour.
Reason, speech, and tradition. Habitual use of tools and fire.						Ancestral Man	Modern Man
Elaboration of intelligence. Evolution of hands, increased reliance on sight.					Higher Primates	Gregarious Primates	
Elaboration of associative memory. Warm blood, increased pre-natal and post-natal care.					Birds, Mammals, Higher Reptiles	Gregarious Birds, Mammals, and Reptiles	
Elaboration of instincts.					Highest Insects, e.g. solitary bees, wasps; Spiders	} Colonial Insects	Social Insects (ants, social bees and wasps, termites)
Terrestrial life fully adopted.					Many Insects and Arachnids, Reptiles		
Terrestrial life, but confined to moist situations, often breeding in water.				Land Molluscs (snails and slugs)	Amphibians, Land Crustacea		
Further elaboration of brain and head.				Many Molluscs (whelks, sea-slugs, &c.)	Fish, higher Crustacea, Lampreys	Gregarious Fish	
Elaboration of locomotor- and sense-organs, primitive head.				Primitive Molluscs (Chiton)	Lower Arthropods (water-fleas, Peripatus, &c.)		
Coelom, elaboration of heart, gills, and feeding mechanism.				Echinoderms, solitary Polyzoa, Bivalve Molluscs, Lamp-shells	Earth-worms, Leeches, marine Annelids, Amphioxus, Simple Ascidians	Compound Ascidians, Colonial Polyzoa	Salps and other colonial pelagic Tunicates
Blood-system.				Nemertine worms			
Anus.				Rotifers, Nematode worms (round-worms)			
Three layers, central nervous system, excretory system.				Liver-fluke, free-living Flat-worms	Tapeworms		
Nerve-ring.				Jelly-fish			Siphonophora (Portuguese man-o'-war, &c.)
Mouth, nerve-net.				Hydra, sea-anemones		Various Corals, &c.	Many Colonial Hydroids
Two layers.				Simple Sponges		Bath sponge and other colonial sponges	
Elaboration of cell organs.	Complex Protozoa, higher Ciliates, e.g. Vorticella	Colonial Vorticellids, e.g. Carchesium	Colonial Vorticellids, e.g. Zoothamnium				
Nucleus.	Simple Protozoa, Amoeba, simple Flagellates, &c.	Pandorina	Volvox				
No formed nucleus.	Bacteria	Colonial bacteria					

AGGREGATION

Haldane/Huxley hierarchical table.

that high and low organisms can be distinguished by the degree of their control over and their independence of environment" (Haldane and Huxley 1927, 232). Apparently, the rise up the chain of life involves a balance between twin (Spencer-like) processes labeled "aggregation" and "individuation": "Individuation is the improvement of the separate unit, as seen, for example, in the series Hydra—Earth-worm—Frog—Man. Aggregation is the joining together of a number of separate units to form a super-unit, as when coral polyps unite to form a colony. This is often followed by division of labour among the various units, which of course is the beginning of individuation for the super-unit, the turning of a mere aggregate into an individual" (Haldane and Huxley 1927, 235). They helpfully provided a table showing not only how these two processes interact in the evolutionary scheme, but who wins and who loses.

One point which the co-authors were particularly keen to emphasize is that one must draw a distinction between short-term specialized adaptive gain and genuine long-term lasting achievement. Consider horses, moles, bats, whales, lions, sloths: "Each of these animals is well adapted for its particular mode of life, but each is by that very adaptation quite cut off from leading the life of any other" (Haldane and Huxley 1927, 240). It is only with genuine progress that we get "greater complexity of organization" and "greater control and independence of the environment" (p. 246). Significantly, the authors tied in this discussion of specialization with one of the clearest statements, since the *Origin,* of the process by which comparative progress is made. However, whereas Darwin saw a slide from comparative progress to absolute Progress, if anything Haldane and Huxley seemed to think that great success in the former jeopardizes genuine success in the latter. Extreme specialization often, if not always, points to a dead end, as organisms are unable to break out to higher levels. The authors instanced the horse—a detailed example in one of Haldane's undergraduate essays—which is perfectly adapted for running on hard ground but which would be quite inadequate on any other terrain. Regretfully, it seems to have no possibility of evolution out from its present state.

Real progress demands that an organism transcend its past state and break through to a new adaptive plateau. Humans are not particularly specialized. If anything, they are rather general organisms. But this is their progressive strength rather than weakness: "The human mind is not merely adapted for solving one or two particular problems: it represents a *method* more efficient than any previously adopted for dealing with any and every problem that may confront an organism" (Haldane 1932b, 248). One might add that as part and parcel of this picture, Haldane and Huxley were keen to stress—as Haldane was always keen to stress in his discussions of biological progress—that the flip side to progress is degeneration and extinction. Specialization often means the loss of hitherto functioning organs, in the name of greater adaptive efficiency in the struggle for existence.

So much for Haldane's views on the pattern and process of evolution. Turn next to cultural factors, his beliefs on Progress, and their possible influence on his thinking about evolution. The most obvious reason for his favorable thoughts on progress—and, I am sure, a true reason—is simply that Haldane absorbed progressionism as he learned about and accepted evolution. The course that he took from Goodrich had biological progress as its backbone—happy metaphor! One went from simple little amphioxus, right up through the mammals, to Western man. Moreover, the appropriate stages of development were located at the appropriate times. Amphioxus, for instance, is primitive because cephalization "has hardly begun" (lecture notes, about 1913/14, Ms 20576, 18f., Haldane Papers, Edinburgh). Likewise pure Goodrich is all of the talk about specialization implying immediate advance, and generalization being required for long-term advance—or, rather, Cope as passed on by Goodrich. Combine this teaching with a friendship with Julian Huxley (see next chapter) and the wonder would be that Haldane were not a progressionist, of the kind that he was.

Did Haldane have personal reasons to make him a progressionist? Religion had little effect, either for or against. Even at school, he showed little enthusiasm for the other-worldly: "Religion, none, but no convictions of atheism. Catholic church very attractive but obviously wrong. Also some attraction to monistic pantheism. Am really a Buddhist-Karma, but with belief in dominance of law, and perhaps illusion of personal identity" (schoolboy notebook, November 24, 1910, Acc12306, Haldane Papers, Edinburgh). This is hardly a total repudiation of anything spiritual. But it is not the Anglican mind of Fisher.

What about Haldane's views on human nature and the relevance of eugenics? Haldane believed that there are innate differences between people, and he agreed with Fisher that the better-quality genes are clustered in the upper strata of societies and tend to be reproduced least: "the people who have the most children are the least desirable socially" (1912 notebook, Ms 20578, Haldane Papers, Edinburgh). Nevertheless, the mature Haldane saw human evolution as a rise up from more primitive types—"Neanderthal man, as shown by his brow ridges and his more rapid development of teeth, was somewhat less foetalised than ourselves" (Haldane 1932b, 150)—and there are strong hints in his writings that Haldane saw some humans as having risen further than others. Perhaps his quite remarkable views that, since Negroes are less sensitive than whites to mustard gas, they should be the shock troops in chemical warfare are not themselves racist (Haldane 1923). But what does one make of the following? "Maybe it will be shown that, even if he is given every educational opportunity, the negro from the tropical rain forest is on the whole a less intelligent being than the European. If so, it may be because an unintelligent cheerfulness is the best possible quality to enable one to survive in what Myres has called the slums of our planet, as it is, perhaps, for survival in our own slums" (Haldane 1934a, 64).

At another level, Haldane being Haldane, he could see the flaws in each of his arguments. Of eugenics, he wrote of its practitioner as "that strange medley of the priest, the policeman and the procurer" (1912 notebook, Ms 20578, 29f., Haldane Papers, Edinburgh). Of Fisher's fears about differential breeding patterns, he wrote that, even though "Mahommedans" should show more business sense than Jews, because their tolerance of polygamy selectively favors the rich, "This is notoriously not the case" (Haldane 1938a, 118). Of the Neanderthals: "I have yet to come across any evidence whatsoever that there has been any advance in the intrinsic factors making for intelligence in Europeans during the last 50,000 years" (Haldane 1934b, 8). And even on the Negro question, Haldane could sound remarkably tolerant: "In most countries the negroes enjoy far worse social and educational advantages than the whites, and no fair comparison can be made" (Baker and Haldane 1933).

It is improbable that Haldane's political philosophy was any more truly significant for his progressionism than his thoughts on the biology of humankind. While it is true that, as Haldane became a Marxist, he did start to interpret evolution in a dialectical sort of way—selection and variation "struggle to produce evolution, and thus manifest the principle of the negation of the negation"—even stressing that "the negation of the negation" was regarded by Marx as "the main source of progress and of novelty" (Haldane 1938b, 31), one should not take this and like comments as more than reinterpretations of positions already held. Significantly, at the time when he was doing his creative work, Haldane was (if anything) critical of Marxism. Explicitly, in *The Causes of Evolu-*

tion, he noted that "with regard to the doctrines of Darwin's great contemporary Marx, it is possible to adopt socialism but not historical materialism" (Haldane 1932a, 2).

Was there nothing that Haldane took seriously, that drove his biological progressionism? In fact, there was one such obsession, that which for him corresponded to the role played by eugenics for Fisher, namely *scientism:* a belief in science as a good thing; a belief in the onward and upward development of science; and a belief that through science will come peace and harmony and all good things to all people. Most revealing was a notorious essay, *Daedalus,* which started life as an undergraduate essay and which appeared in print in somewhat extended form a decade later (1923). In this exercise in scientific futurology—Haldane foresaw the day when power will come from giant windmills and when most babies will be produced by "ectogenesis," outside the womb—he lavished praise on science for itself, on its progressive nature, and on its benefits for society. Ultimately science will lead to the conquest of "the dark and evil elements in [man's] own soul" (p. 82).

Science itself Progresses, and as a consequence it leads to Progress in technology and thence to social and moral Progress. This belief stemmed from Haldane's schooldays: "I believe science to be the only human activity (besides philosophy, mathematics and so on) definitely and continuously making for progress." It holds "the truth, which is able to make you free. So back a winner and help it" (schoolboy diary, November 26, 1910, Acc 12306, Haldane Papers, Edinburgh). But did Haldane see the development of science as a model or metaphor for the development of life? Consider the discussion in *Animal Biology* of comparative progress, which Haldane and Huxley cast in terms of a metaphorical arms race. Organisms are forever trying to out-compete their opponents, whether it be through more lethal methods of attack or perfected methods of defense: "A precisely similar state of affairs is often to be seen in the evolution of the tools and weapons and machines of man. For instance, in naval history, the increase throughout the nineteenth century of the range and piercing power of projectiles on the one hand, of the thickness and resistance of armour-plate on the other, provides a very exact parallel with the simultaneous increase of speed and strength in both carnivores and their prey" (Haldane and Huxley 1927, 237).

Haldane and Huxley continued the metaphor even further, as they used scientific and technological advance to illustrate their crucial distinction between mere specialization and genuine progress. In transportation, there have been many short-term gains, but only the wheel and then the steam engine made for genuine advances: "As a result, steam locomotives became for certain purposes the 'dominant type' of vehicle within an extremely short period of time" (Haldane and Huxley 1927, 249). The authors went on to stress that the evolution of machines in this way is a genuine form of evolution, and since the "struggle" for existence in the animal world is essentially metaphorical, one can just as legitimately speak of "struggle" and "natural selection" in the technological world.

In the *Causes of Evolution,* Haldane touched more on the difficult question of degeneration. Obviously to a prophet of scientism, this raises a problem, since any metaphorical understanding of biological degeneration would seem to demand that, on occasion, science degenerates. The escape hatch apparently lay in the fact that it is the artistic sphere which gives a better sense of decline than does the world of science: "In an evolutionary line rising from simplicity to complexity, then often falling back to an apparently primitive condition before its end, we perceive an artistic unity similar to that of a fugue, or the life work of a painter of great versatile genius like Picasso, who began with severe line drawing, passed through cubism, and is now, in the intervals between still more bizarre experiments, painting somewhat in the manner of Ingres" (Haldane 1932a, 168). The same theme was repeated elsewhere, when Haldane drew the distinction between the primitive with potential (that is, truly progressive) and the specialized but sterile (that is, degenerate).

The case is now sufficiently well made to move on to our third and final query, about the empirical basis for Haldane's thoughts on progress. And the simple answer is that, as he himself admitted candidly, such a basis is thin indeed. There was certainly no attempt to give an objective measure of progress, one through which one might quantify change in the world. In fact, Haldane agreed that everything is a bit subjective: "I have been using such words as 'progress,' 'advance,' and 'degeneration,' as I think one must in such a discussion, but I am well aware that such terminology represents rather a tendency of man to pat himself on the back than any clear scientific thinking" (Haldane 1932a, 153–154).

A similar lack of an evidential base is evident when we look at Haldane's proposals for the cause of (absolute) progress itself. Progress requires an adaptive breakthrough: "evolution has been, on the whole, progressive, because a single species gaining a new faculty such as flight or temperature regulation can become the ancestor of thousands of species which exploit this capacity in different ways" (lectures on Darwinism given in winter 1953; Box 1, Haldane Papers, London). But whether this breakthrough comes through the shuffling and isolation of already existent genes, or the relaxation of selection, or simply through a new mutation, Haldane certainly gave no evidence that any one of these options has actually occurred. Nor, to be honest, is it easy to see how much pertinent evidence could be obtained.

My point is not that Haldane was wrong. He may have been right. My point is that his thinking on progress transcended his evidential base, and this confirms the conclusion that Haldane, after his own fashion, was no less influenced by cultural values in his thoughts on progress than was Fisher, after his fashion.

Edging Toward an Adequate Professionalism

We turn to questions of professionalism, and to do so let us go back for a moment, to the beginning of the new century. What is remarkable is just how many professional scientists preferred to steer clear of evolution in their work, or to make it very clear that the focus of their labors was not on evolution as such. Professionalism lay in other areas of biology, and discipline building was just not a serious project as far as evolutionary theory was concerned. Evolution was not mature science. Take biogeography, deserving of honor by evolutionists if only because of its crucial importance to both Darwin and Wallace. A major work of the time was *Observations of a Naturalist in the Pacific between 1896 and 1899,* by Dr. H. B. Guppy (1906). The author was an evolutionist and progressivist, but this 600-page steady attack on the mechanism of natural selection was prefaced by this disclaimer: "so far as observation of the processes of Nature at present working around us can guide us, each type might well be regarded as eternal. We can never hope to arrive at an explanation of the progressive development of types by studying the differentiating process; and since the last is alone cognisable for us, evolution, as it is usually termed, becomes an article of our faith, and of faith only" (Guppy 1906, ix–x). Hardly a solid endorsement of evolution's professional status!

Among the positive cases, Weldon, with his concern about causes, fits the bill. Yet, professional though he may have been, his labors went nowhere. Weldon was quite incapable of building a discipline. Bateson, his hated rival, was far better at discipline building; but, although his initial impetus for looking at the problems of heredity was the quest for a potentially professional evolutionary science alternative to morphology—revealingly his first major book was titled *Materials for the Study of Variation Treated with Especial Regard to Discontinuity in the Origin of Species*—paradoxically as he and his fellow Mendelians achieved disciplinary status, it became *less* appropriate to consider them front-line workers on evolution. And this is despite the fact that, as their work was established, increasingly evolutionists could make use of it. With the successes of genetics and the development of a professional discipline around it, the problems of genetics became ends in themselves rather than pieces valued for their significance in the overall puzzle of evolution. The major source of funds for early genetics was agriculture—note the histories of both Fisher and Haldane (in 1910 Bateson had become director of the place where Haldane worked, the John Innes Horticultural Institute)—and whatever may have been the private interests of individual geneticists, they were being paid to work on problems of animal and plant breeding, and not on evolution.

So much for the best kind of professional science. Everyone was an evolutionist, but as Huxley had decreed, the idea functioned essentially as a background world picture. It was in the museums that evolutionism, and its attendant progressionism, flourished: as popular science, which was just what it was supposed to be. It is true that one must award comparative morphologists, in or out of the museum, professional status—membership of a discipline even. They had posts, students, journals, and the like. They were evolutionists—their subject was evolution—and they were progressionists. But, as one follows the prospects of evolutionism as mature science from the last century and into this, skies remain overcast. Indeed, for comfortable discussions about evolution and phylogenies, even professional progressionists preferred semi-popular outlets. Not that publishing in these venues was necessarily free of controversy: one happy discussion of arthropod phylogeny in the late 1890s in *Natural Science* evoked so many alternatives that the editors quizzically wondered how one might manage "to reduce these varied views to dogmatic order" (Bernard et al. 1897, 117). What one does sense is that, through the first decades of this century, if anything evolutionary morphology got more and more phenomenal. Public commitments to phylogeny—never a Darwinian science, relying essentially on techniques worked out by non-evolutionists—got slighter and slighter. Even the morphologists realized that evolution and progression were getting them nowhere.

How then do we place Goodrich, if he was as important as I claim in the story of evolution? He fits very easily into our picture, as you might expect of a student of Lankester. Though he was *the* force in British morphology, right through the Second World War, we find him increasingly edgy about theoretical commitments in professional publications. By 1918, in a paper considered to be a triumphant resolution of a decades-old controversy about the vertebrate skull, Goodrich never even so much as mentioned the word *evolution*! Even in his great work on the vertebrates (Goodrich 1930), although he may have defended its evolutionary significance, in truth there was very little between its covers to give offense to the creationist. References to phylogeny tended to be brief and extremely tentative, confirming his admission that many consider the work so insecure that it affords "little trustworthy evidence concerning the process of Evolution" (p. v).

This does not imply that Goodrich himself was other than an ardent evolutionist; but, there was a time and place for everything. And apparently the chief place for evolutionism, even with Mendelism, was in the realm of popular science. The first version of his little book on the subject came out in a general series revealingly entitled "The People's Books," and the preface to the final version expressed the hope that the book "will prove a useful introduction to the study of Organic Evolution not only for the scientific student but also for the general reader" (Goodrich 1924a, 1). Moreover, Goodrich's podia for venturing into the speculative were semi-public occasions. Both the claims about identity theory and the claims about the first land vertebrates came in the course of lectures to the British Association (Goodrich 1922, 1924b). Conversely, whatever "scientific students" may have read *Living Organisms: An Account of Their Origin and Evolution,* let us hope they were not Oxford undergraduates. Judging from the contents of examinations given during the Goodrich era, any such reading would have been a complete waste of time. It is true that progression seeped into his lectures, but their intent was not evolutionary.[3] A former student wrote that, even in 1940, "I went through the whole of Goodrich's enormous course and although, of course, he occasionally mentioned function (although extremely rarely) he never once spoke of natural selection" (letter from Arthur J. Cain to Ernst Mayr, March 1, 1976; Cain Papers).

Move on to Fisher, who was not only a Mendelian and a progressionist but who used the former to underpin the latter. He is an absorbing case. *The Genetical Theory of Natural Selection* is a very peculiar book. It is two-thirds technical theory and one-third eugenics. Given the mathematics, the average evolutionary biologist could not have written it. But, given the structure—specifically the way that, right after the straight biology, Fisher brought in all of the human material, dripping with social values—no average evolutionist, who was also a professional scientist, would have wanted to write it. It cuts straight across the Huxley divide between the professional and the popular. Even today, enthusiasts who write about Fisher's work tend to collapse into embarrassed silence about the final section (Leigh 1986, 1987). It is not just the content but the very form that offends.

Indeed, the "fundamental theorem" itself turns out not to be so very fundamental that any evolutionist ever used it. No-one could ever really decide precisely what form of the theorem Fisher intended. We ourselves have seen how Fisher fails to make crystal clear the exact level (individual or group) at which the theorem operates. Evolutionists are still arguing about its meaning in this respect and others (Price 1972). But part of the reason it has not been adopted was surely that there was something misguided about the theorem, in the sense that it was trying to drive people in a way that they did not want to go, or see any need to go. This point will be set in context as we look at the work of other evolutionists, in Britain and America. But contributing to the unease was the way the theorem so blatantly promotes progress, in a fashion unacceptable to professionalism. The problem is not that Fisher's progressionism is incompatible with Mendelism, but that his progressionism which incorporated Mendelism is incompatible with professionalism.

It would be easy just to end here, simply connecting the sociological with the epistemological. Since Fisher's work so centrally included cultural values, its potential as mature science was negated. However, the story is more complex than this. Fisher's peculiarities clearly stem from a lack of professionalism, but it is important to keep clear in what sense there was a lack. He was a professional scientist, and it seems small-minded to deny that by 1930 he was a professional evolutionist. But, he was not a professionally *trained* biologist. He was trained as a mathematician and physicist and moved into biology as an amateur because of his social interests: Progress and worries of decline. It was for this reason that he trampled so crudely across the landscape planted and tended so carefully by T. H. Huxley and his students. Yet, because Fisher was a professional scientist—and an incredibly good one at that—he had a lasting effect that an unqualified dilettante would not. Not to be unduly cynical, there is nothing that scientists admire (and fear) more than unrestrained mathematical virtuosity—and Fisher showed that, and more. Therefore, he could not be ignored.

Initially, this was the real significance of Fisher. His work functioned at a sociological rather than an epistemological level. People were not falling over themselves to read his book. It took fifteen years to exhaust a printing of 1,500 copies—from the beginning the *Origin* sold 1,000 copies per year (Box 1978; Freeman 1977). Yet, *GTNS* was very important. Would-be professional evolutionists could point skeptics to ferocious mathematics, and not simply to foggy phylogenies. Later, as more people came to biology with mathematical training, Fisher's work could be used (epistemologically) for what it was. But, at first, for all that people would say that in Fisher we have a "reborn Darwinism, this mutated phoenix risen from the ashes of the pyre kindled by [critics]," and while it is true that people could certainly use some of the less mathematical discussions (there was an excellent treatment of mimicry), the real force of *GTNS* was its style rather than its content (Huxley 1942, 28). I noted at the beginning of this book that sometimes professionalization and attendant discipline building depend crucially on the authority of professionals from outside. We are certainly not yet at the point of discipline building for evolutionism, or even of real professionalism. But Fisher made an important start, if not necessarily in the way that people usually pretended he did. I shall return to this point in the next chapter, where, indeed, I shall show also that Fisher had other important social contributions to make.

Finally, we come to Haldane, who begins as the most promising and ends as the most disappointing. As his father's son and as Goodrich's student, he had learned the professional/popular divide, and he knew that progressionism was a major reason why so much in evolution had to stay on the popular side—a ruling with which he seems to have been content. *Animal Biology,* where Haldane and Huxley introduced so full a discussion of progress, was from a series explicitly aiming to raise interest "in the ideas of science" rather than to aid "in assimilating a large quantity of detailed facts" (Haldane and Huxley 1927, v). The *Causes of Evolution* started life as a series of general lectures to Welshmen. Even here, although published in a chatty informal style, Haldane qualified his use of such terms as *progress* and *degeneration* and explicitly he was attuned to the underlying fact/value divide: "And, finally, when we have surveyed the process of evolution we shall have to ask what judgment we can make about it. Is it good or bad, beautiful or ugly, directed or undirected? These are largely value-judgments, and are thus not scientific." Sensitively, he added: "But it is the answer to them which makes evolution interesting to the ordinary educated man and woman" (Haldane 1932a, 32–33). Conversely, there is no whisper of progressionism in a technical appendix to the *Causes of Evolution,* and the same holds true of Haldane's series of (professionally published) papers on population genetics. Here, he truly hewed to the professional science line. A progressionist product like the fundamental theorem could never have had any place in Haldane's work.

By the beginning of the 1930s, therefore, Haldane was poised to break forward from the monolithic evolution-as-popular-science mold. From henceforth there would be a dual approach. On the one side would be a professional evolutionary theory, incorporating Mendelian genetics, non-progressionist, formal. It would of course incorporate Haldane's own particular views on topics like the significance of isolation, but I take it that that would have been open to debate and future effort. In epistemological terms, it could be the foundation for mature science, satisfying epistemic norms—if it was but formal itself, it was formal in the right way, giving hope of an empirically informed, predictive, consistent, fertile (etc.) science. On the other hand, Haldane could continue with a popular evolutionism, likewise Mendelian, but distinctively progressionist. Here, one could do as one wished and no-one could find fault. As it happens, Fisher of all people wrote (but did not publish) a review of the *Causes of Evolution* faulting Haldane for not writing a professional book (Bennett 1983, 289–291). But once the two streams were in place, one could relax, for there would still be a place for popular work.

Admittedly, even had Haldane been generally successful in segregating the professional and popular versions of evolutionary theory, one might still wonder whether there would have been a complete separation between the two. My sense is that, even in Haldane's formal work, he was probing for the causes of things like adaptive breakthroughs, ideas which are the thin end of a large progressionist wedge. As was true for Goodrich, the divide might not have been absolute. But issues like these—whether separation can ever be truly achieved, yielding a progress-free professional evolution—can be discussed in the context of later thinkers. The point now is that what was needed to execute the two-track program to which Haldane pointed was persistence and some discipline-building skills. Ideas had to be set in a context, in a group, with support. And alas, at this point, Haldane proved a failure.

As many have remarked, for all of his brilliance, there seems to have been something lacking about the man. Paradoxically, this may even have been because of the brilliance.

> He would come up with some novel idea (making me feel I was a fool not to have thought of it); he would play this with brilliant verbal prestidigitation, and then sit down, leaving everybody feeling that after that it would be superfluous to intervene with anything more humdrum. I have known a meeting of the Genetical Society to be more or less ruined by Haldane's repeated brilliancies. I don't think anyone doubts that he really was brilliant; and I would think of him myself as a genius who misfired because of personality deficiencies. (Letter from Eliot Slater to C. D. Darlington, December 13, 1968; Darlington Papers)

Haldane seems to have suffered a collapse in creative imagination, and a crisis in personal confidence. Indeed, Haldane's then-colleague C. D. Darlington noted that by the time of the publication of the *Causes of Evolution,* he had run out of steam. "Haldane failed to develop as he and we expected in the years 1926–32. By 1933 he already acknowledged himself to me as a picker-up of unconsidered trifles" (letter to S. C. Harland, October 23, 1968; Darlington Papers). Whether or not this be entirely true—in addition to the noted interest in altruism he worked on problems of genetics, particularly human genetics, as well as on issues in the negative costs of natural selection (Maynard Smith 1992)—his slowdown in the area of evolutionary theorizing was certainly not compensated for by discipline building. His brilliant touch at writing for the common man, the deep loyalty and affection he inspired in his friends, the sacrifices he willingly made for his country—these parts of his personality were marred by an explosive temper, a readiness to behave ferociously rudely to others (especially underlings), and a congenital inability to request grant money through normal (or abnormal) channels. There is no surprise that he responded to the joys of communism and the thrill of the Spanish civil war. If he had not discovered these enthusiasms, there would have been something else. Whatever the reasons—and an element of self-hatred may have been involved ("In my opinion he was one of the few genuine masochists I have come across. If you reflect on a) the type of auto-physiological experiments in which he indulged b) the two wives he married I think you may agree"; letter to Darlington from Miriam Lane, December 2, 1968; Darlington Papers)—Haldane was simply not the man to build a professional evolutionary discipline, with or without progress at its heart.

Overall, therefore, our conclusion has to be that the first wave of evolutionists in Britain in this century were more successful at incorporating Mendelism than in upgrading evolution fully to the status of high-quality professional science. In major part, Goodrich did not want to do this, and Fisher and Haldane were mixed in motives and achievements. But, this having been said, from the vantage of history, the prospect of professionalism is becoming clearer. Maturity is a possibility. Let us therefore move straight on to the next wave of thinkers and workers.

9

Discipline Building in Britain

We have crossed the Mendelian divide. By the 1930s, the forward-looking evolutionists in Britain embraced both the selectionism of Darwin and that particular theory of heredity started by Mendel. The mathematicians had dealt formally with the subject, and the biologists could focus their efforts on discipline building. The foreground is, as always, taken by questions of progress, Progress, and the interrelations between them. I shall look at the work and influence of three men and through their labors bring our story in Britain up to 1959, the one hundredth anniversary of Darwin's *Origin.*

The Evolution of the Nucleus

Cyril Dean Darlington (1903–1981) ended his academic training with a pass degree in agriculture from London University. Noteworthy was that the future Sherardian Professor of botany at Oxford University managed to fail his first attempt at the intermediate examination in botany. He joined the John Innes Horticultural Institution—Bateson's institute and, as such, a place for research into plant breeding—and through the years he rose from junior unpaid worker to director, a post he relinquished only in 1953, when he left to go to Oxford. Darlington's labors centered on plant cytology, more strictly plant karyology (study of the nucleus), a field in which he blossomed and was, for many years, a world leader (Lewis 1983; Olby 1990).

His most significant scientific work was published in the 1930s. In *The Evolution of Genetic Systems* (1939)—anticipated in Darlington (1932)—Darlington attempted to make his own contribution to the story of evolution, reflexively turning the Darwin-Mendel approach to evolution back on the very mechanisms of evolution themselves: "In the present sketch I have attempted to show genetics as the study of systems of heredity and variation, systems which rest on a basis of the chromosomes and are related to one another by processes of natural selection" (p. v). Taking the reader from first principles, Darlington introduced the basic premises of genetics and linked them to the physical basis of heredity within the nucleus. Then, having traveled through a range of pertinent topics, as a kind of climax Darlington was ready to look at the evolution of the reproductive system. Extrapolating from the variety of extant organisms, he inferred that this phenomenon happened in a sequential manner, with genes interacting in ever more complex ways: "There are three important levels of organisation within this series. The lowest has no differentiation between genes. The second has a differentiation of genes but no sexual reproduction. The third has sexual reproduction with an alternation of haploid and diploid phases" (Darlington 1939, 123–124).

Apparently, the move from the first to the third level required two important innovations. "The first step was the differentiation of the genes which are still undifferentiated in viruses and bacteria. It is made possible by the invention of mitosis and the arrangement of the genes on linear threads. This means, as we have seen, that a chemical equilibrium was superseded by a mechanical one" (p. 124). Then: "The second step was the adoption of sexual reproduction. It depended on the invention of the meiosis as a modification of mitosis" (p. 125). It should be noted that, for all that Darlington portrayed himself as an ardent Darwinian cherishing the worth of natural selection, at this crucial second step abruptly he abandoned the system: "The origin of meiosis and sexual reproduction . . . shows the most violent discontinuity in the whole of evolution. It demands not merely a sudden change but a revolution. It is impossible to imagine it as the result of a gradual accumulation of changes each one of which had a value as an adaptation either Lamarckian or Darwinian" (p. 125).

To Darlington, the chief point at issue was not how such an event as this occurred, nor even its magnitude, but that its occurrence was not a function of need. Rather, the change occurred simply for mechanical reasons and then, because it conferred benefit on the species (note the group-selectionist perspective), it survived, reproduced (itself), and throve. And the same holds true of the variants which are built upon the sexual system. A case in point is apomixis, which is an escape from sterility through self-fertilization: "The elaborate genetic processes of self-sterility and the endless devices securing cross-pollination can yield no reward except in the qualities of the progeny. All these changes anticipate not merely the act of selection but the generation in which selection occurs. They all of them therefore put out of court any assumption of direct adaptation or the inheritance of acquired characters" (Darlington 1939, 131–132).

What of biological progress in all of this? Through the first ten years he was at the John Innes, Haldane was Darlington's part-time colleague. The somewhat uneducated young cytologist was much impressed and influenced by the brilliant geneticist, then right at the height of his powers. ("As a young man I . . . had a great admiration for JBSH: he was my idea of an educated man, equally willing to talk and to listen"; answer to psychology questionnaire, 15 June 1964, Darlington Papers, A12.) One would therefore expect some flavor of progressionism about Darlington's evolutionary work, which indeed we have already had with the talk of "three important levels of organisation" and "important inventions" fueling changes from one to another. Elsewhere, also, in *The Evolution of Genetic Systems,* the reasonable interpretation is that Darlington was working in a progressionist mode. For instance, the whole discussion of apomixis was framed in a contest of nature trying, ultimately unsuccessfully, to combat the forces of decline and decay: "Thus we see apomixis saves what can be saved when sexual fertility has been lost. Sexual reproduction provides in recombination the basis for the adaptation of all its posterity. Apomixis provides for its immediate progeny." Although, alas: "Apomixis is an escape from sterility, but it is an escape into a blind alley of evolution" (Darlington 1939, 113).

Traces or hints about progress were all that one did get in these cytological speculations. Presumably full-blooded sexuality was being prized because it does give more scope for evolutionary novelty, advantage, and so forth—it brings together valuable genes or combinations into one organism or group—but Darlington did not say so explicitly.

Certainly, he made no attempt to tie in his cellular theories with humankind. For that, one had to wait for Darlington's later writings. When these came—reflecting concerns which went back many years—they more than compensated for the delay by their blunt candor. In line with then accepted thought, Darlington claimed that we humans had evolved away from the apes about 20 million years ago and that the chief change was the explosive growth of our brain, as we descended from the trees to live on the plains.

The increase in brain size set up a kind of feedback process, as the success of one adaptation triggered a circular chain of improvement in related adaptations: "Clearly one invention led to another. Each development in the association of hand and eye by the brain opened a new field of activity which made further associations of hand and eye more useful. And indeed more necessary; for with each advance man's intelligence displaced his former methods of defence by tooth and claw and made him more dependent on his intelligence. Thus a process of directed evolution of the brain was set in motion" (Darlington 1969, 25). As when he wrote about cytology, Darlington's approach and language showed that he thought in terms of an overall upward climb: "improvement" and "advance"; he gave no hint of relativism.

Darlington's picture of the endpoint to human evolution makes Fisher look like a sociologist. Not only are we humans (in today's terminology) "genetically determined," but even the very professions that we follow seem to be a function of the genes: "Individuals and populations cannot be shifted from one place or occupation to another after an appropriate period of training to fit the convenience of some master planner, any more than hill farmers can be turned into deep-sea fishermen or habitual criminals can be turned into good citizens" (Darlington 1953, 304). Of course, being different does not mean being better or worse, but Darlington made clear that some conventional Victorian assessments obtain: "The greatest achievements of paleolithic man are shown us by his tools, his art, and the evidence of his prowess in hunting . . . They are paralleled but not equalled by what we can still see of the surviving paleolithic peoples, the Eskimo, the Bushman, the Ainu of Japan and the Australian Aborigines. For these are all peripheral people and peripheral people are usually backward" (Darlington 1969, 29).

There was, however, more to Darlington's thinking than just a straight labeling of European man as the "superior." Just as in his cytological writings he credited the development of sexuality with bringing diverse elements creatively together, Darlington argued that the key to human progress lies in the hybridization of diverse human groups. It is true that, in a sense, any particular group or type is as good as any other. "The superiority of the hunter-collector over the civilized man in what concerns his own survival is therefore unquestionable. It is a genetic superiority for which no training can compensate" (Darlington 1969, 30). In another sense, a mixture is better, even a mixture of the civilized and the primitive. Think of America: "From the point of view of supporting a population of human beings there can be no doubt that the white-plus-negro society is more efficient than the Indian was or ever could be" (Darlington 1953, 292–293).

American society, as he depicted it, would seem to be essentially a cultural mix, although given Darlington's position it is necessarily a genetic mix also. It is not, however, the epitome of civilized advance, for that requires individual diversity: "It is therefore not surprising that the Indo-European area with relatively free communications but bounded at both ends by an impassable ocean, preventing until recently a loss by migration, has preserved the greatest human diversity and has become the chief seat of cultural advance" (Darlington 1953, 413–415).

Adding a spin of uncertainty to all of this is the fact that although outbreeding between groups can lead to defectives, it can also lead to the creative genius: "He is the inventor, the artist or the hero." Moreover, although the genius probably cannot prevent ultimate extinction for the human species, "on a shorter view, where the events of a few hundred years or the fate of only a few nations seem to matter, he may alter it" (Darlington 1953, 678–679). With views like these, it is small wonder that to many the name of Darlington was anathema—especially to liberal-minded Americans, and even more especially to liberal-minded Americans who themselves had inclinations toward genetic determinism. (See, for instance, Dobzhansky 1954; Dobzhansky and Penrose 1954.)

What led Darlington to take his position on progress? What guided him—other than an undoubted element of self-predication, for he prided himself on the fact that he was a product of outbreeding and drew the appropriate implications (Lewis 1983, 113)? As far as progressionism *per se* is concerned, remember Haldane. Notwithstanding his later views on humankind, the younger Darlington was caught by Haldane's interests in Marxism—"In the period from 1928 to 1934 I encountered Haldane, discussing Marx and Freud with him" (from his autobiography, written 1 November 1980; this is a fragment scribbled on scrap paper, Darlington Papers)—and, by the 1940s, was likewise describing himself as a Marxist. "Pretends to be a Marxist nor-nor-west./ Left school at 16 (and never regretted it)" (letter from Darlington to E. J. Holmyard, March 27, 1942, describing himself; Darlington Papers). "His philosophy and politics are tinged with Marxism which in his *Conflict of Science and Society* (1948) however deviates from the orthodox colour" (a self-description, from *Endeavour,* April 1949.)

These influences certainly explain the kind of progressionism that we find in the *Evolution of Genetic Systems,* with its emphasis on levels of organization and humans moving up the hierarchy. The philosophy of the book was materialistic—the very aim was to show that heredity is based on cellular parts. At the same time, there was a stress on organization and on integration, showing that the whole is more than just the sum of the parts—the result being the natural evolution of the complex from the simple. Yet, although his early exposure to Marxist theory was significant, we should probably not make too much of Marxism as an influence. By his own admission, Darlington was at best a Marxist of a strange variety. The very important non-gradualist, non-adaptive element in his thought owed little to dialectical materialism and more to the fact that he was *de facto* a student of Bateson, an ardent saltationist and non-adaptationist. Likewise, Darlington's group-selectionist yearnings, which persisted through his later writings, were primarily attributable to the general assumption of evolutionists of the 1930s that species evolved through group rather than individual selection. Indeed, far from Marxism being a blanket causal explanation, much of Darlington's genetic determinism of later years came in reaction to Marxism—especially to Marxism as manifested by Haldane's support of Lysenko. In the 1920s, Darlington had grown close to the Soviet agricultural geneticist Nikolai Vavilov. When Vavilov fell before the Stalinist regime, Darlington became a prominent critic of Lysenko, strenuously fighting the disinclination of the British establishment to say anything offensive to the Soviet regime.

Not only did Darlington come to deny that he was Progressionist in a Marxist sense, his emphasis on the indeterminacy brought on by the acts of "Great Men" in making history was intended to underline the point. As a related matter, Darlington turned from skepticism toward eugenics to enthusiasm. Yet, although this change in his thinking led to a break with Haldane—compounded by the fact that Darlington became director of the Innes, a post Haldane coveted—Haldane's influence persisted, for it was a variant of Haldane's scientism that truly gave Darlington his

C. D. Darlington Julian Huxley

philosophy of Progress. Take the lecture given in the late 1940s, on "The conflict of science and society." There certainly was little Marxism of an "orthodox colour." Darlington explicitly saw intellectual advance as a Darwinian process of struggle and selection: it is the outcome of the conflict between the creative advances of the innovator and the natural conservativism of society, including scientific society. But the solution to this conflict does not lie in giving science free reign, for both elements are essential: "On the one hand, the need for continued research and discovery, if we are to survive, is no longer to be doubted. On the other hand, even the most revolutionary thinker must need intellectual security and stability in the intervals between his revolutionary thoughts. Otherwise a coherent individual would be as impossible as an established society" (Darlington 1948, 46). The trick lies in combining the new with the old: "In research, as well as in education, [new] developments give us the chance of correcting the maladjustments of the last century and establishing a new harmony." Here, as always for Darlington, the key to success lies less in dialectical materialism and more in some kind of hybridism. Different elements make something altogether better in combination than they could be individually.

Darlington's views on cultural Progress were at one with his views on biological progress, and the former is a likely influence on the latter. This conclusion is backed by the evidence—or, rather, by the non-evidence, for certainly none is offered to support a claim that the sexual organism is better than the non-sexual organism or the modern-day American than the Native American. *Prima facie,* if the non-sexual can do as well as the sexual, and in many cases it seems to do precisely that, or better, then that is enough for Darwinism. Likewise, as far as humans are concerned, one wonders what Darlington would make of the present-day success of Southeast Asia. Allow that he could accommodate the recent turn of events within his theory, and that he would be forgiven for not recognizing at the time he was writing that the North American lust for automobiles and electronic equipment would point to a world order that reverses the defeats of the last world war. Surely, nevertheless, more proof is needed to establish the inherent superiority of Western civilization. No doubt Darlington preferred his own life to that of an Australian aborigine—as do I. But that is another matter. Preference is not proof. Here, as throughout, the argument slips from wishing to justifying.

The Modern Synthesizer

Julian Sorell Huxley (1887–1975), the oldest grandson of Thomas Henry Huxley, was by his own admission born with great advantages and grave handicaps (Clark 1960; Baker 1978; Huxley 1970, 1973; Waters and van Helden 1992). Nature and nurture combined to give him a brilliant start to life—not only was he a descendant of "Darwin's bulldog" but also of Thomas Arnold of Rugby, the historian and the greatest influence on secondary education in Victorian Britain. His uncle was Matthew Arnold, one of the century's finest poets and essayists, and his aunt was Mrs. Humphry Ward, the authoress of *Robert Elsmere,* the paradigmatic nineteenth-century novel of religious doubt and the search for spiritual and moral meaning. Expected to perform brilliantly, young Huxley did, first at Eton and then up at Oxford, where he gained his first in biology and won the Newdigate Prize for poetry. Thus launched, he soon became professor at the Rice Institute in Texas. This proved to be a comparatively brief interlude abroad, and already Huxley was showing the down side of his inheritance, for like his grandfather he was subject to periodic episodes of depression: "nervous breakdowns." This unstable disposition was in major part responsible for the fact that he was never in his life to hold a long-term post, although by his own admission he disliked the routine of teaching and hated the drudgery of university administration.

In the 1920s, after some work on the effects of hormones on development, he became a co-author, with the novelist H. G. Wells and his son, of a massive semi-popular account of the science of biology, thereby gaining always valued experience of trying to reach a general audience (Wells, Huxley, and Wells 1929–30). Almost naturally he moved next to become secretary and *de facto* director of the London Zoo. This was a post for which he would have seemed perfectly fitted, given both his biological expertise and his now explicit determination to address people beyond the narrow confines of academia. Nevertheless, Huxley managed so to irritate his governing board they fired him! The Second World War was in progress, and Huxley by now had other irons in the fire—not the least of which was participation in a highly popular radio (later television) quiz show, the "Brains Trust." With the founding of the United Nations after the war, Huxley became the first director-general of UNESCO. During this time he traveled extensively, and when the term was over the pace increased to an almost frenetic level, as Huxley slipped comfortably into the role as *the* elder statesman for science.

For all his surface brilliance, as a creative scientist Julian Huxley was not of the first rank. The root cause was a fundamental insecurity about his own abilities. In a searing letter written to a former student, Huxley told of his earliest work as "completely in vain" and of his own sense of being "horribly dissatisfied at having had to fall back on 'second-hand work' " (letter to Alistair Hardy, May 12, 1921; J. Huxley Papers). Nor did an escape seem easy: "I knew that I was very ignorant outside one or two special subjects. Physiology was a closed book to me; I had tried to read some at Oxford, but had decided I did not understand enough mathematics, physics, and chemistry." Hence, the flight to Texas: "There for 3 years I led the life of an intellectual semi-invalid—never daring to work hard, only doing my routine of lectures and laboratory, still feeling that I should never be able to concentrate enough to do real research."

Yet, Huxley's ambitious drive equalled his sense of insecurity. As the great (and deserved) success of his collaboration with the Wellses *(Science of Life)* shows fully, he turned his weakness into a strength. If he could not compete in the conventional way, then he would compete in a way of his own making—and for this reason, Huxley is perhaps more important to our story than a more conventional, more creative scientist. By his own admission, Huxley was a generalist, a synthesizer of ideas, rather than a specialist. This means he was eminently able to pull together a variety of ideas and present the "big picture"—painting in broad, vivid strokes what his contemporaries saw but dimly. Most important, it was Huxley's role to articulate and to put the final seal of approval on the synthesis between Darwinian selection and Mendelian genetics. He made himself the spokesman for the twentieth-century evolutionary edifice. We find him preparing for this task as early as 1915, in a series of public lectures he gave at Rice. (Huxley was not as idle in Texas as he later implied.) But the place where Huxley really made his definitive case was the book which he rightly regarded as his *magnum opus, Evolution: The Modern Synthesis* (1942).

Let me enter a word of warning. In *EMS,* as elsewhere, Huxley rather gives the impression that—the mathematical geneticists having given the theoretical structure, the logical skeleton—he and those upon whom he was reporting were now simply adding the empirical support, the evidential flesh as it were. But this is not so. Not only did Huxley eschew formalisms, and not only did he avoid all those parts of the theorists' work which did not appeal (there was no mention of Fisher's fundamental theorem, for instance), his very approach was simply not that of the mathematicians. Remember, this was the man who dropped physiology because of the sums. Huxley's true intellectual mentor was in fact the most obvious candidate, once one thinks about it. Like his good friend Jack Haldane, it was his teacher back at Oxford, E. S. Goodrich. Huxley's whole approach to evolution was Goodrichian in style. Rather than tight formal arguments the reader got much more of a discursive, inductive approach. Huxley surveyed the pertinent literature, amassing examples to prove and buttress his main points: namely, that natural selection is a powerful force in nature, and that the raw material on which it works is provided by the Mendelian theory of heredity.

Huxley was Goodrichian in scope also, as compared with the likes of Fisher or Darlington. For him, evolution covered the whole of biology. Thus in *EMS* one got a careful and full discussion of problems of speciation, of taxonomy, and of biogeography. All of the classic problems about the significance of islands and of barriers and of ecology and of climate and of competition and of much more were raised and discussed and evaluated and catalogued and—with overwhelming detail—illustrated.

Likewise, problems of paleontology were given full attention. Addressing what was then a hot topic among paleontologists, namely the supposed existence of trends in the fossil record, Huxley suggested that what often seems to be a non-adaptive trend in the fossil record, "orthogenesis," can be readily explained as a consequence of interconnected "allometric" growth (p. 535). Some body parts grow as a logarithmic function of the whole organism, and thus a trait that is quite disadvantageous in itself can, as it were, piggyback in on a trend for a trait much favored by selection.

Huxley's intentions were synthetic, and synthesizing is precisely where he succeeded, as never before. Yet, as I turn to analysis, let me leave you with a niggling worry—certainly one which Haldane felt after he had read *EMS*. For all that Huxley was portraying himself as the champion of modern evolutionary thought, of modern *Darwinian* evolutionary thought, in respects there was something of his grandfather about his approach to evolution. There was a part of Huxley that was never that deeply committed to adaptationism. Without wanting to say something ridiculously paradoxical, as that the great synthesizer of neo-Darwinism did not accept natural selection—he did—one does sense that there was not the deep internalization of selection that one finds in Weldon and Fisher (to name two fanatics). In fact, in the case of orthogenesis Huxley did not exclude absolutely the possibility of non-selective causal forces. Ostensibly, he wanted no truck with mystical non-scientific forces, but a space was left in his theory for mechanisms of an unknown kind.

All of which leads one to suspect that along with the orthodox tunes, Huxley was hearing melodies of another kind. We must see if they become louder, as we turn now to the question of P/progress.

Religion without Revelation

Julian Huxley was ever an evolutionary progressionist. The metaphor of the arms race, which was presented in the textbook jointly authored with Haldane, predates the First World War and can be found in Huxley's first published book: "The correlated evolution of weapons of offence and defence in naval warfare is closely similar, though simpler far" than the course of correlated biological evolution. "Each advance in attack has brought forth, as if by magic, a corresponding advance in defense" (Huxley 1912, 114–115). And down through the years his writing was seeded with progressionism—not just the comparative progressionism of arms races—until it all flowered in *EMS,* which revealingly started life as an address to the British Association for the Advancement of Science (in 1936) with the title "Natural selection and evolutionary progress." Of course, progress does not occur all of the time and in every case, throughout the organic evolutionary process. No-one denies that there are reversals and extinctions and sometimes, like the brachiopod *Lingula,* long periods of stagnation. In fact, the rule is extinction and the exception is advance. But overall there is progress, and the endpoint is our own species: *Homo sapiens.*

Wherein lies the essence of progress? Huxley responded with warmth to the notion of increased complexity. "High types *are* on the whole more complex than low" (Huxley 1942, 559). Indeed, in one of his essays ("Higher and Lower," from *Essays of a Humanist*), Huxley went so far as to say that Spencer's speculations about homogeneity giving way to heterogeneity contain "a great deal of truth" (Huxley 1964, 35). However, this approach does not really capture the nub of progress. There have been complex organisms which have gone extinct. Rather, what Huxley favored were *control* and *independence,* more ideas for which Spencer got praise as a forerunner. Those organisms that can better control their environment, and those organisms that are more independent of their surroundings than other forms of life, are higher up the ladder of evolutionary progress. So, for instance, the mammals with their methods of maintaining a constant body temperature are higher than the brutes without such methods.

Clearly, with its capacity for control and with its great independence, the human species is right up at the top. More than any other, it has "increased control over and independence of the environment"; or, as one might put it alternatively, it has raised "the upper level of all-round functional efficiency and of harmony of internal adjustment" (Huxley 1942, 564–565). Later, Huxley added *variability* and consequent *adaptability* to his list of the marks of biological improvement (Huxley and Huxley 1947, 128, 182). But he had always stressed that "specialization," as is caused by competitive trends, can rarely if ever lead to true progress. With specialization you get instant advance but rarely genuine "improvement." Going too far down the path of specific adaptive complexity apparently precludes that generality which is needed for genuine evolutionary advance. A significant point about humankind is that in many respects, physically and mentally, we are generalists—indeed our strength is that we have *not* taken the path of specialized "advances" and have as a consequence remained highly adaptable to change.

What is the causal force behind true progress? One of the virtues of an informal, discursive, and lengthy style is that you can readily contradict in one place what you have said in another. Thus, at one point, Huxley denied a role for selection. "All that natural selection can ensure is survival. It does not ensure progress, or maximum advantage, or any other ideal state of affairs" (Huxley 1942, 466). Then, happily, at another point, Huxley gave it an important role: "It should be clear that if natural selection can account for adaptation and for long-range trends of specialization, it can account for progress too. Progressive changes have obviously given their owners advantages which have enabled them to become dominant." Admittedly: "Sometimes it may have needed a climatic revolution to give the progressive change full play." But this is hardly a great surprise, for we know that: "It seems to be a general characteristic of evolution that in each epoch a minority of stocks give rise to the majority in the next phase, while conversely the majority of the rest become extinguished or are reduced in numbers" (Huxley 1942, 568).

Obviously, in separating the competitive drive to specialization from genuine improvement, Huxley had barred Darwin's slide from comparative to absolute progress. He seemed never to get on top of an alternative way to derive progress from selection, which in any case was probably never going to be adequate alone. Not that he wanted a general mechanism for progress, for what was crucial to Huxley's thinking was the belief that, finally, only one type of organism would reach the pinnacle of evolution. And, to a certain extent, as it ascended it pulled the ladder up after itself. Thus, we humans have evolved to the highest point, and at the same time we have made it impossible that any other organism should evolve to become our equals. Indeed, at this point, Huxley became almost nasty about natural selection. We humans have, or should, transcend any such vulgar mechanism: "True human progress consists in increases of aesthetic, intellectual, and spiritual experience and satisfaction" (Huxley 1942, 575–576).

If you sense a note of necessity, of predestination, about this whole process, you are right. "One somewhat curious fact emerges from a survey of biological progress as culminating for the evolutionary moment in the dominance of *Homo sapiens*. It could apparently have pursued no other general course than that which it has historically followed" (Huxley 1942, 569). The channels seem to be laid down in advance, so much so in fact that one might fear a shift from progress to Providence. But, in word (and, to his great credit, in personal deed), Huxley did stress the need for human effort. In his later writings, perhaps as a consequence of his time spent working for UNESCO, Huxley picked up in earnest the theme of earlier evolutionists about the threat of decline. Huxley had long been a eugenicist. Now, more ardently than ever, he preached the cause of family planning and population control, warning that all of humankind's advances will be lost unless we stem the horrendous explosion of human numbers: "Birth-control is . . . necessary, on a world scale and as soon as possible" (Huxley 1957, 212).

Where did this ultra-progressionism all come from and why did it drive Huxley so? At one level, having traced his roots back to his university training, we have the answer already (Swetlitz 1991). Huxley's denial that specialization can lead to genuine progress is simply Cope's rule of the evolution of the unspecialized: a non-Darwinian rule from a non-Darwinian thinker. It was the basis of Goodrich's thinking about the fossil record, and if more reinforcement were needed it came directly from H. F. Osborn. During his early American sojourn, Huxley was welcomed by the older man—small return for the latter's having bored everyone for years with tales of the grandfather's teaching—and Huxley was given the party line (that is, the Cope line) on progress.

Significantly, by the 1950s Cope's rule was under heavy attack. G. G. Simpson, the paleontologist, argued strenuously that there is little reason to think that the early mammals were unspecialized reptiles—and he told Huxley this, at length (correspondence found in both Simpson Papers and J. Huxley Papers). At the same time, Alistair Hardy (Huxley's former student) was pushing a critique of the rule—even in a collection co-edited by Huxley himself (Huxley et al. 1954)! Toward the end of his career, therefore, we find Huxley downplaying the rule; but, as ardent a progressionist as ever, he now promoted the virtues of variation, arguing that the truly progressive organism is not (as previously) the generalized one, but the organism or group with or resulting from variation. Expectedly, *Homo sapiens* scores well (Huxley 1954a, 1955).

Yet, there is still the question as to why these influences struck so deeply. Here, we look for motives. Could it have been eugenics? "I regard it as wholly probable that true negroes have a slightly lower average intelligence than the whites or yellows" (Huxley 1941, 53). The same holds true of the Irish in Ireland, not to mention country folk in every industrial society: "The difference between the southern Irish in America and in Ireland strikes every observer: we can hardly doubt that is due in part (though doubtless not entirely) to a sifting of more from less adventurous types" (Huxley 1941, 56–57). Yet, apart from the fact that eugenical concerns are more a response to worries about the decline of progress than a reason for the initial belief in progressionism, this kind of thinking was not the driving force it was for Fisher. Basically, Huxley was simply showing the colors of his middle-class professional status; his male status ("the inherited constitution of the whole [female] sex, has made their possibility of achievement not only different from, but less than man's"; lecture 3, p. 40, J. Huxley Papers); and his status as one who knowingly agreed to teach at a segregated college ("a negro buck with Reckitt's blue trousers, Coleman's mustard shirt, oxblood tan shoes and a face like a polished grate is a grand sight"; Huxley 1924).

Could Huxley have been a scientific Progressionist and simply have read this into biology—as did Haldane? Apart from the talk about arms races, from his earliest writings Huxley promoted an analogy between biology and culture, specifically between evolutionary progress and scientific Progress. In 1912 we find him extolling the virtues of a "division of labour," both among humans and among biological parts. "As with men, so with cells—a jack-of-all-trades cannot advance in any, and the same lesson of sacrifice has to be learnt before the colony can become an individual organism" (Huxley 1912, 111–112). In the first published full discussion on progress, the analogy occurs again: "What could be more striking than the parallel between the rise of the mammals to dominance over the reptiles, and the rise of the motor vehicle to dominance over that drawn by horses" (Huxley 1923, 36).

Yet, passages such as these are more icing on the cake than the main ingredient in the concoction. Huxley was not so much using the analogy to inspire the theme of biological progress as to promote and explain it. We must continue our search for the grounds of Huxley's life-long devotion to biological progress. Fortunately this is no very difficult task, for he was really quite open on the subject. Apart from a significant reading of *Robert Elsmere,* the influence of a much-beloved mother, whose death just when he was on the verge of manhood scarred Huxley psychically for life, was crucial. He spoke of her "sense of wonder" which "I have inherited, and though I cannot believe in the essential goodness of things as a whole, I have stressed the fact that evolution, including that of our own species, is essentially *progressive*" (Huxley 1970, 20).

Together, his two legacies—from T. H. Huxley and from the Arnolds—drove Julian Huxley to his faith in biological progress. Through this notion, Huxley thought he could stay sincere to science, all the while extracting that sense of meaning and value which is the epitome of religion: "It is immaterial whether the human mind comes to have these values *because* they make for progress in evolution, or whether things which make for evolutionary progress become significant *because* they happen to be considered as valuable to human mind, for both are in their degree true" (Huxley 1923, 59–60). Thus Huxley developed his "evolutionary humanism," or—to use the title of one of his books—"religion without revelation."

This is not to say that Huxley's thinking remained absolutely static. Apart from the family influence, he was certainly primed by a neo-Hegelian haze which permeated his Oxford college (Balliol) when he was an undergraduate. To this must be added Germanic elements in his scientific training. For instance, the early Huxley fully accepted the biogenetic law. In addition, his first (1912) discussion of arms races was introduced in explicitly Spencerian terms: "A state of equilibrium may for a time exist, but every balanced organism is as it were pressing against every other, and a change in one means a rearrangement of them all" (Huxley 1912, 114–115). At the time that Huxley began his career, in England an open profession of Spencerianism was hardly calculated to engender respect—but perhaps there was always a family enthusiasm for such thinking.

All of this past experience pales beside the next major outside influence on Huxley—the most profound of all and that which was to stay with him forever: *Creative Evolution* by the French philosopher Henri Bergson. Huxley's first book explicitly acknowledged the debt—"It will easily be seen how much I owe to M. Bergson, who, whether one agrees or not with his views, has given a stimulus (most valuable gift of all) to Biology and Philosophy alike" (Huxley 1912, vii–viii). Paying tribute was only proper, not only because Bergson gave Huxley faith in the overall idea of progress but because a particular feature that Huxley seized on as the mark of progress was pure Bergson: "Civilized man is the most independent, in our [i.e., Bergson's] sense, of any animals: this he owes partly to his comparatively large size, more to his purely mechanical complexity of body and brain, giving him the possibility of many precise and separate actions, and most to the unique machinery of part of his brain which enables him to use his size and the smoothly-working machine-actions of his body in the most varied way" (Huxley 1912, 6–7). There is no surprise that Bergson liked Huxley's book very much (letter to Huxley, December 14, 1912; J. Huxley Papers).

Bergson was a vitalist, and by the time he got to Rice Huxley had so far followed Bergson that he was himself an overt vitalist: "When I was last in New York, I went for a walk, leaving Fifth Avenue and the Business section behind me, into the crowded streets near the Bowery. And while I was there, I had a sudden feeling of relief and confidence. There was Bergson's élan vital—there was assimilation causing life to exert as much pressure, though embodied here in the shape of men, as it had ever done in the earliest year of evolution:—there was the driving force of progress" (lecture 1, n.p., J. Huxley Papers).

This was a vitalism that Huxley was trying to link to some notion of a deity, through the medium of evolution. The same things that the man of higher civilization "recognizes as having value for himself he finds by purely objective study, to have had value in the whole of evolution. In other words, he finds in evolution *something outside of himself which gives him a rational basis for his idea of God*" (lecture 6, pp. 36, 38, J. Huxley Papers; Huxley's italics).

Back in England, Huxley restrained himself. Ontologically he moved toward a position that simply had to be described as "atheistic" (Baker 1976). Tactically he came to realize that discretion is demanded: "Bergson's *élan vital* can serve as a symbolic description of the thrust of life during its evolution, but not as a scientific explanation. To read *L'Evolution Créatice* is to realize that Bergson was a writer of great vision but with little biological understanding, a good poet but a bad scientist" (Huxley 1942, 457–458). Yet Huxley never relinquished this philosophy. He always saw a foreordained upward rise to the evolutionary process. And this belief was backed by a reading of several idealistic thinkers of the 1920s: A. N. Whitehead; J. S. Haldane (J.B.S.'s father, much concerned to reconcile science and religion); and (again picking up the Goodrich theme) C. Lloyd Morgan, author of a neo-Spinozistic theory of "emergent evolution": "At every upward stage of emergent evolution there is an increasing richness in stuff or substance. With the advent of each new kind of relatedness the observed manner of go in events is different. In a naturalistic sense each level transcends that which lies below it" (Morgan 1927, 206).

Morgan was a student of T. H. Huxley, so in a sense (the emphasis on non-adaptive saltations) there was continuity in the commitment, over the generations, to non-selective evolutionary processes: "Increase in organization is for the most part gradual, but now and again there is a sudden rapid passage to a totally new and more comprehensive type or order of organization, with quite new emergent properties, and involving quite new methods of further evolution" (Huxley and Huxley 1947, 120). There was continuity also in that Julian, like his grandfather, never expected change to occur without effort and struggle. The difference was that whereas T.H. thought Progress is achieved by combatting evolution, Julian thought that it is achieved by ensuring evolution: "From the evolutionary point of view, the destiny of man may be summed up very simply: it is to realize the maximum progress in the minimum time. That is why the philosophy of Unesco must have an evolutionary background, and why the concept of progress cannot but occupy a central position in that philosophy" (Huxley 1948, 11).[1]

Huxley's biological progressionism was embedded in his evolutionary humanism, and the link between the two was a phenomenon of culture rather than of straight science. This is a conclusion justified further when we ask about the empirical evidence. As commentator after commentator has pointed out, much of Huxley's argumentation for the importance of progress skated on very thin evidential ice indeed (Greene 1990; Waters and van Helden 1992). It is hard to see why a flying insect or a bird should be denied a level of independence or control beyond that commanded by humans. I cannot soar like an eagle, nor can I dive like a whale. Of course, in respects, we humans have more control over our environment than other species have, but in other respects we have less. And, in any case, proof is needed to show the inevitability of the progressive process of evolution. Perhaps with humans around, it would be difficult for other intelligent beings to evolve. We would kill them all first. But why it was determined that humans would evolve in the first place is quite another matter. At the very least, we would like some sort of theory showing how our ancestors were channelled down a certain path. This Huxley did not offer.

Obviously, we have the conclusion that Huxley's progressionism was a function of his cultural commitments. Toward the end of his life, Huxley became a (non-Believing) enthusiast for the ideas of Father Pierre Teilhard de Chardin, the Jesuit paleontologist who saw the whole of reality as undergoing constant change and moving upward to Christ, the "Omega Point." In a move which shocked his orthodox fellow biologists, it was Huxley who wrote the introduction to the translation of Teilhard's masterwork, *The Phenomenon of Man* (Huxley 1959). Given Teilhard's deep debt to the Bergsonian philosophy, and given what we now know of Huxley, for us the shock would have come had he been indifferent to Teilhard's work.

E. B. ("Henry") Ford

E. B. Ford and the British School of Ecological Genetics

Edmund Brisco Ford (1901–1988) spent his whole career at the University of Oxford, rising from undergraduate through lecturer and reader to the (especially created) post of professor of ecological genetics. A bachelor of the very old school, stepping straight from the pages of an Evelyn Waugh novel, he was legendary for expressing his dislikes (especially of things North American). Chicago: "I think this the bottomless pit of a town: vast, black, sordid and cheerless." Minneapolis: "Rather a dreadful place." Columbus, Ohio: "nearer Hell than anything I have seen." Charlottetown, Prince Edward Island: "it is frightful, . . . it is such a ghastly place." It is somewhat of a surprise to learn that: "England is the most unpleasant country I have ever visited." (Quotes from letters to Philip Sheppard, November 22, 1959; March 19, 1966; May 5, 1966; April 6, 1969; November 21, 1967; Sheppard Papers.) Teeming with prejudices though he may have been, "Henry" Ford proved also to have a powerful vision: he had determined on a professional, discipline-based, evolutionary biology.

Oxford in the mid-1920s offered very little to one interested in evolution, such as he. "I remember more than one person speaking in almost the same terms (J. Z. Young was one) and saying 'surely you are not now going on with *Genetics* when we are seeing what can be done in experimental embryology' " (reply to questions posed by W. Provine and E. Mayr, 1975; Ford Papers, A9). Analogously: "one must remember, too, that those who had made a great name in zoology then (e.g. Goodrich) were almost wholly non-experimental, being chiefly concerned with comparative anatomy in its most observational aspects" (answer to Mayr's questionnaire, Ford Papers).[2] Yet, if not a practicing evolutionist, Goodrich was creating an environment very favorable to the Darwinian evolutionist, and this was reinforced for Ford through a friendship with Poulton—a friendship grounded in the fact that, from boyhood, Ford (like so many English schoolchildren) had been an ardent butterfly collector. But the real influence, one cemented by a close, forty-year friendship, came from Fisher, who, on Julian Huxley's suggestion, sought out Ford and drew the young man (still an undergraduate) into his orbit (Box 1978). Thanks to Fisher—his friendship, his encouragement, his collaboration, his ideas, both about evolution *per se* and about (statistical) techniques of experimentation—Ford was able to articulate and actualize his own program of evolutionary biology: "We were most intimate friends from 1923 to his death in 1962. His encouragement was magnificent . . . It would not have been possible to work on the synthesis we are considering without Fisher's *Statistical Methods for Research Workers.* His *Genetical Theory of Natural Selection* was its great early landmark" (Mayr's questionnaire, Ford Papers).

Ford, whose earliest training was as a classicist, adapted well to biology. He proved a talented student and a gifted writer. Well-crafted books flowed from his pen for the next half-century, beginning with a little book (written in the Goodrichian style) showing the pertinence of Mendelism to evolution (Ford 1931). His definitive tome, *Ecological Genetics* (1964), summarized over thirty years' work by himself and his students: "ecological genetics deals with the adjustments and adaptations of wild populations to their environment. It is thus . . . essentially evolutionary in outlook" (Ford 1964, 2–3). Taking a strong selectionist line, Ford constantly reiterated the theme that the world of organisms shows ubiquitous biological adaptive advantage. The most striking example of this notion came from Ford's protégés, Philip Sheppard and Arthur Cain, who showed that the colors and bandings of the shells of snails are brought about through selective predation by birds: the evolution of these shell patterns involves adaptive camouflaging responses to the background hues of the snails' respective ecological niches (Cain and Sheppard 1950, 1952, 1954).

Like any good scientist, Ford liked an opponent against whom to define and articulate his own position. With Fisher, Ford found his rival in the person of the American population geneticist Sewall Wright. Ford argued first against Wright's theory that in certain circumstances gene ratios can "drift" in a population randomly, and then against a challenge Wright mounted against Fisher's theory of the evolution of dominance (Fisher 1927, 1928, 1930; Wright 1934b). Fisher's theory depended on the modification of heterozygotes by selection until they no longer reveal deleterious alleles, and it demanded that there be heterozygotes in sufficient numbers in populations for selection to be effective. Ford made a life-long career of demonstrating in nature what Fisher had shown mathematically, namely that selection can keep different forms "balanced" in populations, with all of the consequent heterozygosity that that entails (Ford 1931, 54–55). The classic study of balanced "polymorphism" was done by Ford (with Fisher's help, especially in the analysis) on the scarlet tiger moth, *Panaxia dominula.* They demonstrated a balance between two genes that have significantly different effects on the insects' wing markings (Fisher and Ford 1947).

Fisher was the theoretician. Ford was the naturalist, the empiricist. What then of Ford's beliefs in progress, especially given Fisher's very strong views on the subject? There are some whiffs of progressionist thinking in Ford's work. Ford certainly showed an interest in humankind that transcended the disinterestedly objective. A book he wrote toward the end of his career, distilling all of his thoughts about polymorphism, deliberately and distinctively devoted a whole separate chapter to our species (Ford 1965). Elsewhere in his writing Ford likewise gave *Homo sapiens* special attention, most obviously in a book on medical genetics which contained some familiar sentiments: "A clever man should seek in marriage a partner whose mental powers are good, for . . . on the average, they will produce intelligent children" (Ford 1942, 139). Perhaps connected to this "proposition" were some very old-fashioned views held by Ford on the subject of women. Notoriously it was he who, when once faced with a class filled only with female students, turned and left, remarking: "Since there is no one here, there will be no lecture." When, in 1978 (1978!), the suggestion was made that women might be permitted to lunch in public at Ford's Oxford college, All Souls, he countered with vigor: "profound physiological and genetic differences separate men and women. The distinctions between them are greater than those separating distinct *species,* and not very closely related species either" (Ford Papers, B7). He followed up this proclamation with a memorandum relaying this information: "Women pass out of the maximum learning period, as they enter it, earlier than men. It is far from true that outstanding women undergraduates, compared with men, make brilliant research workers" (Ford Papers, B8). Shades of the biogenetic law!

Of course, much of this was "Henry doing his thing"—performing in public because it was expected of him. But, there was an underlying belief, and there are other hints that Ford in some respects saw "advancement" in evolutionism (with man the most advanced). For instance, at the urging of Leonard Huxley, Ford joined the London Eugenical Society, showing that even if he was dubious about progress he was certainly worried about decline! Yet, ultimately, whiffs and hints are all one can find. Although Ford openly and genuinely professed his allegiance to the Fisherian perspective on evolution, essentially as a professional biologist he wanted nothing to do with biological progress. Most revealing was the first book that Ford wrote, *Mendelism and Evolution.* The preface was ardent in its praise of Fisher. Indeed, of the several works published on the bearing of genetics on evolution: "Outstanding among these is Dr. R. A. Fisher's valuable and illuminating treatise on *The Genetical Theory of Natural Selection*" (Ford 1931, vii–viii). And the praise was genuine, for he included detailed discussions of a number of topics taken from a Fisherian perspective. Fisher was referred to explicitly on sixteen pages of text, Darwin on two, Mendel on three, Haldane on seven, and Wright on one. Yet there was absolutely no mention of the fundamental theorem, the cornerstone of Fisher's progressionism. That aspect of Fisher's work was simply gutted right out, and so it was always to be.

As an explicit topic of discussion, Ford never once raised the question of progress. He always dealt with changes of gene ratios in limited time spans, with no thought for the long term. Why was there this total silence? One might argue that the essence of Ford's work excluded thoughts of progress. Balanced polymorphisms are, by definition, to do with organisms held in a stable situation. There is therefore no place for change, especially progressive change. However, although the point draws attention to the fact that Ford's focus was on *micro*-changes (or non-changes), nothing in his work on polymorphism precluded an additional interest in macro-changes, which could well be progressive. Another argument might be that Ford wanted to exclude *all* social or value considerations from his professional science, in line with the traditional approach (as taken by T. H. Huxley). But, even if true, this prohibition would hardly exclude some surreptitious backsliding, or at least some nod toward Fisher's enthusiasms.

The real answer may just have been that Ford was simply lukewarm to progress, as he undoubtedly was to many things considered Progressive (especially if they were American). His early classical training, together with related reading, may have been crucial here. To a student enthused by Progress, Ford recommended the antidote of Gibbon's *Decline and Fall of the Roman Empire:* "I suggest that a close reading of Gibbon will make you a little less enthusiastic about claims of real progress" (letter to author, from Paul Handford, November 21, 1995). This skepticism about Progress may have converted into skepticism about progress. Toward the end of his life, on being asked in general terms about biological progress, Ford replied: "I am not even clear what 'progress' consists in (e.g. in a coelenterate and a mammal). It has been said that it is decreasing the dependence of the organism upon the effects of the environment. (In Man, perhaps increasing ability for conceptual thought?)" (Mayr's questionnaire, Ford Papers). But even this puzzlement does not really explain Ford's indifference. Why did Ford not use his discussions of human genetics to engage in Fisher-like discussions of genetic decay and remedies for its counter? In no way did he attempt such a treatment of the topic. Ford was much more interested in (what today is called) "genetic counseling," a focus on the elimination or avoidance of bad genes rather than on wholesale societal upgrading through selective breeding. He offered real practical advice to real practical people (Ford 1942).

And this offers a clue as to why Ford really wanted no truck with notions like progress in his professional evolutionary biology. What drove Henry Ford had little or nothing to do with grand metaphysical schemes; nor was he a Dickensian reactionary standing in the way of Progress. He had other ends in view. He wanted his science to be *useful,* for then it would attract support and students. In other words, then it would have a base for discipline building. Ford was not the dreamer so many of the earlier evolutionists were. He was, rather, the consummate professional, with precisely the aims and skills that Darwin had sought a hundred years before. To see this aspect of Ford's story in full context, we need to step back and look at the social side to English science, as it evolved in the years after the mathematical geneticists had done their work.

Making a Discipline

There is one thing on which we can agree. Whatever else the significance of the population geneticists, they did not offer a simple theoretical foundation on which the next wave of evolutionists based their efforts. The new generation—Ford particularly—used some of their ideas, but most biologists of the mid-century took flight from the formalisms of the mathematicians—a point, incidentally, that was appreciated explicitly by Darlington: "I think that Fisher, Haldane and Sewall Wright in talking about isolated genes and isolated organisms were not talking about evolution as I understand it: they were just doing very interesting exercises" (notes on evolutionary synthesis, February 21, 1975, following Ernst Mayr's Conference, 1974; Darlington Papers, H163).

Yet the formalists were far from irrelevant. On the one hand, their work gave the new crop of researchers a sense of confidence that a Mendelian-reinforced selectionism—bolstered by various bits and pieces—would work. Even if one could not follow the main moves of the mathematical theory of evolution by selection, let alone the details, one knew (or "knew") that one's general approach was correct. On the other hand, as noted in the last chapter, those evolutionists who forged links with the population geneticists got the sideways support of professionals from other fields raising the status of the contributions of one's own field—and this support was reinforced by the gloss that mathematics brings to any scientific enterprise. I have acknowledged Fisher in this respect, and even though I have been more negative about Haldane, he should not be discounted entirely. He had worked significantly in biochemistry, as well as in his forays into evolutionary formalism (Sarkar 1992a). So he too helps establish the point that, whatever it was that drove them personally into the field of evolution, the would-be evolutionist in the 1930s and 1940s had an opportunity to work as a professional within the context of a discipline as had never before existed.

A discipline demands discipline builders, and Darlington had aspirations in that direction. He was not employed by a university until later in life, so he did not have a ready supply of students for much of his career. Indeed, for all that he later became an Oxford professor, there was an element of contempt for major academic institutions in his attitude. "Being largely home-taught, research-taught and travel-taught he does not attach the highest value to formal education" (autobiographical notes, June 3, 1969; Darlington Papers, A6). But it is clear that many came to the John Innes expressly to work with Darlington. They absorbed his visions as well as his techniques, and they fanned out to great effect in their subsequent careers. "During the period of his Directorship [1939–1953], among the staff, workers and students, eleven were elected F.R.S., eight became university professors, three directors of Research Institutes and one a Vice-Chancellor" (Lewis 1983, 123). Apart from running the John Innes for nearly fifteen years, Darlington was also an energetic president of the Genetical Society (1943–1946), where "he influenced the course of genetics in Britain when he transformed the year book by including a list of relevant journals, research institutes and university departments concerned with genetics and cytology, a miniature of Darlingtonian synthesis" (Lewis 1983, 145).

Most crucially, there was the printed word. Darlington was an indefatigable producer of textbooks (for instance, Darlington and Mather 1949). Even more important was his decisive move into journal editing—an area which gave him freedom to pursue his own agenda, bound only by the loyalty of his readership; an area where his energies and abilities could be used in crafting and correcting the work of those who wanted or could be coerced to share his aspirations. He did not take this step by chance. In the 1930s, Darlington experienced difficulties in placing his papers. There was no established niche into which his articles on cytology could fit comfortably—the geneticists wanted the results of breeding experiments rather than work that focused directly on the workings of the cell. Fisher was consulted for advice, and eventually (after the Second World War) the two men decided to found their own journal, with the intention always that Darlington would be the real, active editor and that Fisher would lend his support and name (together with an initial financial guarantee). In return, Fisher gained an open avenue for publishing his work and that of his research group: "It would be enough for me to do the work and for him to be able to publish the papers of his department in the journal" ("Origin and Development of *Heredity,*" unpublished notes, April 10, 1970; Darlington Papers, G53). As so it proved. *Heredity,* as the journal was called, was very much the product of Darlington's unaided labors, as he prodded and pushed and revised contributions. "I am filled with admiration by the trouble you take with manuscripts" (letter from Kenneth Mather, January 11, 1958; Darlington Papers, G37). His efforts were amply rewarded, however, for the journal rapidly gained a reputation and influence as one of the leading outlets in the life sciences in Britain.

Now, what influence did Darlington display as a discipline builder? And why in particular should it be of interest to us in our inquiry into biological progress? The crucial point is that, in line with his initial discussions with Fisher—who also had had his troubles with placing papers, especially in the biometricians' house journal, *Biometrika*—Darlington always intended a broad domain for his publication. As against existing journals, Darlington wrote to the publishers that he and Fisher "agree, I think in wishing to pursue a somewhat more adventurous policy designed to advance the subject (as well as the journal)" (letter to Oliver and Boyd, May 30, 1946; Darlington Papers, G28). The last thing Darlington wanted was to publish a journal narrowly devoted to issues in cytology. Rather, he hoped to provide a far-reaching umbrella—an umbrella of genetics, broadly construed, which could thus encompass not only cytology but also pertinent parts of biochemistry, statistics, systematics, agricultural breeding, and more. In short, *Heredity* was founded on "the view of genetics as a many-sided science in process of becoming the causal framework of biology" (memo, April 10, 1970; Darlington Papers, G6).

Where did evolutionary studies fit into all of this? It is certainly not the case that Darlington thought of evolution *per se* (fact, path, or theory) as the overall, arching synthesizer. "I had explained in 1933 that I regarded the Origin of Species as a mixture of problems, not a subject for synthesis" (letter to W. Provine, January 23, 1975; Darlington Papers, H161). Rather, as comes through clearly in the information that Darlington prepared for would-be contributors to his journal, work on evolution itself was a part, but only a part, of the total picture: "*Heredity* publishes quarterly original articles of interest to research workers, teachers and students in the fields of experimental breeding, cytology, statistical and biochemical genetics and evolutionary theory" (prospectus, Darlington Papers, G32). Evolution *per se* was not the glue holding together the framework of Darlington's vision; genetics, in some broadly construed sense, would unite the new discipline:

> [Genetics] is not only rapidly advancing but it is also rapidly broadening its scope. Indeed it is beginning to cover every aspect of biology and every form of life. It is beginning to weave the threads of many different sciences and techniques into a single pattern. Darwinism and biometry, Mendelism and cytology are being joined together and the needs of medicine and agriculture are being met by the establishment of a single fundamental discipline. (Darlington 1947b)

Note how the shrewd Darlington was pitching his claim to the practical value of his would-be discipline—in medicine and agriculture. Why indeed should he have regarded evolution in any broad sense as central?

Darlington was forging a professional discipline, with (speaking now epistemologically) genetics in some sense as the intellectual glue. Genetics (especially as one infers from the contents of the publication) offered the epistemic virtues of good-quality (proto-mature) science: consistency, predictive potential, unificatory power, and so forth. Hence, by looking at the nature of Darlington's nascent discipline and its science, we are getting close to extracting the implications of Darlington's discipline building for the status of claims about progress. But, let us just ask one more question—a question about Darlington's reading of evolutionism. What sort of evolutionism is it that would be done by professional biologists who wanted to publish their work in *Heredity,* where it would be but one of several related subjects? Most obviously one would look for a narrow, technical reading of evolution. There is no place for paleontology and biogeography in Darlington's professional journal. Rather than broad-sweeping generalizations about the nature (and meaning) of life, one would expect experimental or detailed observational studies about short-scale changes and issues of adaptation. In other words, one would anticipate the kind of work that Weldon initiated.

And so, indeed, it proved. For all that Fisher was a co-editor of *Heredity,* Darlington had little time for purely formal discussions—indeed, two decades after the journal's founding we find him criticizing the editor at that time for permitting too much mathematics (letter to Dan Lewis, May 5, 1969; Darlington Papers, G61). But other options for publishable work were made available, specifically research that used mathematics but went beyond it. Most particularly, in the early years of *Heredity* the work of Ford and his students in "ecological genetics" was coming to fruition. "I was hoping you would particularly arrange with Ford about the work he is doing partly with you. Can you do anything on these lines?" (letter to Fisher, 7 September 1946; Darlington Papers, G25). Darlington solicited this work for his journal, and indeed published what are now considered to be the classic papers (Cain and Sheppard 1950, 1952; Kettlewell 1955, 1956). He thus defined his vision of evolution through his choice of articles to publish.

Yet, interestingly, Darlington's attitude to this work showed also how his respect for evolution's status was bound by limits—Darlington certainly did not consider the material from the ecological geneticists to be the only or even the most central of the contributions to *Heredity*. Indeed, after a few years, he fell out with the leading junior members of the school—ostensibly purely over trivial matters of style—and refused to publish more of their work: "I hope you understand that I receive many papers which require no alteration at all, and if papers were usually submitted to me in the condition in which I received yours I should have to give up most of my other duties. If you, therefore, prefer to send your future contributions to some less exigent editor I shall be very happy" (letter to A. J. Cain, November 21, 1951; Darlington Papers, G43). Darlington marked the carbon to this paper: "No more papers from Cain and Sheppard." And so it proved.

Perhaps, in line with his own work, Darlington's attitude at this point reflected a gut feeling that ultimately heredity is more significant than selection and adaptation. Certainly, American critics thought Darlington's thinking insufficiently "integrated" with "selectionist thinking" (H. Carson, 1973, reply to Mayr's questionnaire; Ford Papers). Be this as it may, the implications of all of this for thoughts of biological progress are obvious. Grant that the editors of and (perhaps some of the) contributors to *Heredity* believed in biological progress. Grant also such beliefs were triggered by thoughts of cultural Progress. The fact remains that the kind of discipline Darlington proposed through *Heredity* effectively excluded the expression of progressionist beliefs from the domain of professional scientific discourse. Evolutionism was part of a broader discipline of genetics, and there was simply no place within it for talk of trends, whether from bad to good, or simple to complex, or whatever. Biological progressionism could be considered only in work directed at the broader public: "The result [of my inquiry] will mostly, I believe, make sense for the general reader" (Huxley 1942, 9).

Darlington had a discipline to forge, although it is a moot question whether he approached this project with the explicit exclusion of progress in mind. On the one hand, one might reasonably argue that progress was unintentially overlooked. Darlington grew up under the tutelage of Bateson and Haldane and others, and like those around him, he thought there was a place for professional work and a place for popular work, and that was that. Evolution was just not a topic that moved Darlington that easily. On the other hand, a little book (1959) that Darlington wrote to mark the centenary of the *Origin,* suggests that he knew just what he was about. Darlington's attitude to Darwin was condescending to say the least. He claimed that, because of his refusal to experiment and to work from nature, Darwin never grasped the truths of heredity. The result was that a door was left wide open for those who would interpret evolution in grandiose, progressivist ways; hence, the scene was set for years of bad science. The unstated moral was that some—no names need be mentioned—had learned the lesson of history. Certainly, by his own admission, Darlington did not start presenting his progressive evolutionism in the public forum until after he had got his fellowship in the Royal Society (autobiography, November 1, 1980; Darlington Papers, A3).

Darlington proposed a science unfriendly, both sociologically and epistemologically, to progressionism. Of course, there could still be feedback, including a feedback of progressionism, from the popular realm to the professional. This message is to be found in his writings on cytology if one looks carefully. In the same vein, if we look at the contents of early volumes of *Heredity,* we find that Darlington was not above publishing work which would later be used as support for his speculations about humankind (Darlington 1947a). But the progressionism had to infiltrate the professional science surreptitiously. Although a back scullery window may have been left slightly ajar, Darlington's biological vision barred progression from using the front door. Progressive notions of advance and improvement were illicit in his professional, mature science.

Grandson versus Grandfather

Given our history to this point, either there was something naively and radically wrong about *Evolution: The Modern Synthesis,* or Huxley was deliberately trying to break the mold—the mold which his grandfather had set. The rules were that one spoke not at all of evolution and progress in the same breath, or did so only in the realm of popular science. Admittedly, there was nothing exceptional about what proved to be the embryonic form of Huxley's book, for it was written as an address to the British Association for the Advancement of Science, a forum in which traditionally biologists were allowed to speculate. But what of *EMS* itself? It was serious and sober—yet led to open talk of progress.

Of course the strategy was deliberate. It could not be said of Julian Huxley, as one might argue for Fisher, that the author was not a professionally trained biologist and had to be excused because his book was too innovative to ignore. It is obvious that Huxley, the most skilled writer in the field, knew full well what he was about. In his preface to *EMS,* he acknowledged the general reader but made an open pitch to the professional: "the layman interested in biology will, I hope, find the book suited to his needs, though I hope that it will appeal mainly to professional biologists interested in the more general aspects of their subjects" (Huxley 1942, 8). Moreover, if we look backward and forward, we see a pattern which governed all of Huxley's thinking about evolution. He wanted his work taken seriously by professional biologists, and for him that meant not only an acceptance of his synthetic vision of the field but also the inclusion of progress in the professional domain. In this respect, he wanted to fracture his grandfather's distinction between professional and popular science.

We see this agenda right from the beginning, first in the little book Huxley published before he left England and then in public lectures at Rice, where he laid out his vision. Through the 1920s, it was the goal behind Huxley's evolutionary efforts. The textbook series, which he edited, explicitly made progress the conceptual framework for biology, as of course did his own contribution (jointly authored with Haldane). If one can set people off into their careers thinking that evolution equals progress, then the job is half-done—especially if the collateral reading for students, parents, and (probably) teachers is Wells, Huxley, and Wells. After *EMS,* Huxley kept up the pressure. Literally every opportunity to address biologists was used to promote progress.

> As regards lecturing I would prefer latish in the term. What about March 1st? I would like to talk (not lecture formally) on "Evolutionary Progress and Related Problems", but would like to make it rather short, with a view to getting a good discussion. I am hoping you will get in one or two palaeontologists if there are any good ones about. Le Gros Clerk might be interested. If Zukermann or J.Z. Young or Medawar would be willing to revisit their old haunts for the purpose of the discussion I would welcome this very much. (Letter to Hardy, November 7, 1950; J. Huxley Papers)

It is perhaps a little odd to be the editor of one's own *Festschrift;* but, in celebration of his sixtieth birthday, that is essentially what Huxley became. At least, performing the honors oneself does mean that one can ensure that the volume's introduction promotes one's own views. Consequently, at the head of what was in fact a report on the state of the (professional) art in the early 1950s, Huxley once again aired all of the old themes: "evolution is on the verge of becoming internalized, conscious, and self-directing" (Huxley 1954a, 13). And so on and so forth.

Evolution equals progress, and that is a message that should be carried to all quarters.

Huxley's synthesis was likewise part of his game plan. Unlike Darlington, he simply *had* to bring in paleontology, because it was the definitive testing ground for all claims about specialization, improvement, and the like. More than this, paleontology apparently gave the greatest support to progress: only the fossil hunter may trace the record from monad to man. As a matter of fact, much of the attention that Huxley gave in *EMS* to taxonomy and biogeography was primarily a function of a recent collaboration with a number of full-time taxonomists (through the newly founded "Systematics Association") on precisely such questions (Huxley 1940). But these subjects also provided grist for the mill, not the least for the claim that humans represent a separate kingdom.

Grant that Huxley wanted to bring progress into professional evolutionary biology. Why should he have been driven this way and felt it possible? The answer to the first part of this question centers on his vitalism, and the answer to the second is at least related to it. Simply speaking, Julian Huxley did not accept the dichotomy drawn by David Hume between fact and value—that dichotomy on which his grandfather's whole strategy was based. For Julian, the world of objective things was also the world of subjective values. They were merely different ways of looking at matters: "We saw mental states as the inner aspect of the same reality of which the outer aspect consisted of physical happenings in the brain. In just the same way we can think of evolutionary progress as a whole. Objectively considered, it resides in the increase of control. But from the other side it is seen, with equal truth, to be an increase in knowledge, love, will, and joy. When these increase we know that we are in the stream of progress" (lecture 2, pp. 38–39, J. Huxley Papers).

For Julian Huxley, therefore, it would have been a wasted opportunity, if not a downright mistake, had progress not been introduced into all aspects of evolutionary thought. Put in epistemological terms: he saw no essential difference between epistemic norms and cultural norms, and hence he saw no theoretical reason why a mature science should not (in principle) be molded by a cultural norm.

What success did Huxley have for his efforts? He had many resources. He had authority in the (professional) field. He was no outsider but the grandson of the great and much respected T. H. Huxley. He had been educated in the right places, had produced work to his credit, and by the late 1930s had been elected as a Fellow of the Royal Society (*the* mark of approval). He had organizing and promotional abilities. Additionally, there were Huxley's very great talents as a writer and as a synthesizer. *EMS* is an impressive work. It is beautifully written and a masterpiece of organization, with attention paid to details and broader issues alike—not to mention the fake gloss of mathematical competency. Huxley knew all the tricks about promoting one's own ideas. His tutelage under H. G. Wells paid dividends.

Yet, Huxley's armor was not without its chinks. Everybody knew that he was really not that significant a scientist. From his early years, Huxley had been getting negative responses from referees because of the slapdash nature of his work. (See, for instance, letter from E. J. Allen to Huxley, February 28, 1922, J. Huxley Papers.) The F.R.S. was long in coming, and there were hints that he got it primarily on the basis of his secretaryship of the zoo. He had given up his academic institutional base and so had no steady flow of students. Perhaps also the very success of Huxley's intentions for *EMS* spoke against it. As a work, it was less a call for action than a massive testimonial to past achievements—more of value to those facing comprehensive examinations than for those seeking a thesis topic. (See Baker, 1976, for hints that this was so.) Then, as an organizer, his strengths—enthusiasm, vision—were matched by flaws—impatience, lack of tact. Witness the zoo dismissal. In any case, after *EMS* was published, Huxley was having to earn a living through such time-and-attention-consuming work as the directorship of UNESCO.

But what really tipped the scales against Huxley's attempts to install progress in the science of professionalized evolution was the fact that he was not beginning with a clean slate. Lacking his vitalism, lacking his readiness to dismiss the fact/value distinction, his fellow scientists accepted the dichotomy between professional and popular science, and they had already banished progress to the popular side. Revealing on this point was Huxley's Bateson Lecture at the John Innes. Invited to speak for 1953, Huxley was told by Darlington that he should feel free to expand and break from the formal mode. Apparently he did just this, and more so! By now it was crucial to the progressionist cause to promote variability. Cope's law was under attack and the case for progress needed reformulating: "Then, one great moral—one great implication—of this evolutionary view of the world, is that variety is the spice of life—but not merely the spice of life: variety is the seed of progress. In order to progress, in order to realize our evolutionary possibilities to the full, society must have people who are artists, who are scientists, who are engineers, who are willing to devote themselves to education"—etc., etc., etc., etc. (Huxley 1954b). Huxley used the occasion to explore the undoubted extent of natural variation, even—or especially—in *Homo sapiens*. It seems that he also used the occasion to draw the consequence that variation is the fuel of progress, and Darlington was horrified that this might end up in his journal: "I have been through your manuscript as you will see from the comments. The first fifty pages with no alteration beyond adding a closing paragraph would make a satisfactory article for *Heredity*" (letter to Huxley, July 13, 1953; Darlington Papers). So much for the final pages!

Huxley fought back, pushing Darlington as hard as he could: "I have now been over my material carefully and would like to suggest the following arrangement—that you would publish the lecture in two halves, of which I would send you the first now and the second as soon as possible after my return" (letter from Huxley to Darlington, August 24, 1953; Darlington Papers). He was not even above a little bribery. Darlington was not a member of a special dining club associated with the Royal Society.

> What about the publication of my Bateson lecture? I've just got back from my eight months' trip, and should be ready to get on with the revision of the second half of the lecture for publication as soon as I have cleared my desk.
>
> Shall I see you at the Royal Society Club on the thirteenth? (Letter from Huxley to Darlington, May 8, 1954; Darlington Papers)

> Now that you are back you might like to consider again whether this material is not more suitable for a book than for Heredity. If you did think so, I should be very glad to publish a summary or the first 23 pages of the present manuscript in Heredity, . . . (Letter from Darlington to Huxley, May 10, 1954; Darlington Papers)

> We must certainly try and get you elected to the R.S. Club. Meanwhile could you dine with me there on Thursday May 27th? about 6.15 at the Athenaeum. There we could talk over the problem of my Bateson Lecture and its publication: offhand I still feel the publication of pages 1–23 only would be inadequate . . . (Letter from Huxley to Darlington, May 17, 1954; Darlington Papers)

But essentially in the end Huxley could claim no more than a draw. He got his paper published (Huxley 1955); the case was made for widespread variation; yet all of the overt connections between evolution and progress, especially the key mediating role of variation, were forced to go unmentioned. Darlington's journal remained true to its editor's professional criteria and excluded Julian Huxley's progressionism. In the end, it was the grandfather that triumphed.

The Ford School

But what of evolution? The grandfather had wanted to keep that out of professional science also. Weldon had shown one way to get it in—drop everything that might hint of progress—but he left only an example and not a discipline. Darlington had followed a related tack, but evolution was only a side interest in the genetics-focused discipline he sought. It was E. B. Ford, in England, who—partly by using the labors of others, like Darlington, and partly by pursuing precisely his own battle plan—finally secured the beachhead and moved on inland. At last there was an academic niche for the professional evolutionary biologist.

Ford's starting point was that, in Britain in the 1920s, no such niche yet existed for evolutionary biology, nor even one tailor-made waiting to be taken down and used.

> The Honours School of Zoology at Oxford, which I read, covered the whole field of zoology but with special reference to comparative anatomy and embryology. Genetics was the subject least dealt with, and there was little in the lectures on it that I was not getting out of the literature. I myself was concerned with the study of evolution by means of observation and experiment: . . . No such idea was considered in the course at Oxford. (Mayr's questionnaire; Ford Papers)

Ford had to make a place for himself in the academic sunshine, and it was not always easy. He had to struggle to get and keep a job at Oxford, certainly to move up the ladder. In the early years, because his talents and interests were not needed, he was even unable to get a college fellowship. Colleges wanted tutors to coach in physiology and anatomy, not genetics.

> My own reappointment for the next four years has just gone through Convocation all right. This is for both my posts. Reader in Genetics, and University Demonstrator . . .
>
> After the war, I trust I may look forward to having a Department of my own: and perhaps I may reasonably be thought qualified to do so. I do not want to end my days in bloody Oxford. I have a great affection for *this Department,* but not for the University. And how could I? I have

> not been particularly well treated here. I am not even a Fellow of a College! I was not made a Reader here until five years after I was offered a Readership in London and refused it in order to remain here. (Letter to Julian Huxley, June 6, 1942; J. Huxley Papers)

Through networking and hard work, Ford did get ahead. Sometimes the helping hand came from others. It was in Fisher's interest that his collaborator be seen as a respected biologist. Not only did he help Ford through a spot of that bother that tends to trouble bachelors in Evelyn Waugh novels, but soon after the Second World War there was the coveted fellowship in the Royal Society. (In a letter to Huxley, January 18, 1941, Fisher pushed Ford's entry above Darlington's; J. Huxley Papers.) Sometimes the helping hand came from home. Ford himself was a master at promoting his own ends. Consider, for instance, the following letter written to a prospective student.

> There is one thing that I always insist from people who come here. It is that they should have some problem, or problems, which they themselves are burning to solve. The one thing I do not like is anybody to arrive merely just wanting to work here. We have, of course, a good many problems always on hand which there is no one to work upon, and it sometimes turns out that a man's own ideas need readjusting or, in fact, prove to be impracticable, relative to our facilities or for other reasons. In those circumstances, one is always happy to put other possibilities forward, provided that an applicant has come with clear ideas of his own that he is really keen to solve. (Letter to D. A. Jones, March 25, 1957; Ford Papers, F48)

Who, especially if they had real ability, could resist a challenge like that? There is work to be done. One will be part of a team. Yet one will be required to keep one's own individuality. Certainly, it was no chance that Ford gathered in such brilliant young men as H. B. D. Kettlewell, Sheppard, and Cain. And once they came into the group, they were encouraged and pushed and promoted. "Paul Handford . . . sent in his 2 highly original, and to me most important, papers on electrophoresis in *Maniola* to *Heredity*. They came back with a number of alterations in his conclusions and deductions: all in an anti-selectionist direction: by the person to whom the papers had been submitted . . . He therefore withdrew them from *Heredity*." Later: "Paul Handford's 2 papers on electrophoresis in *Maniola jurtina* are well on the way to production, in the *Proceedings of the Royal Society*." Openly, Ford admitted to Sheppard that he had been "a good deal involved," and when the papers appeared he concluded that "They are really very satisfactory" (letters to Sheppard, November 30, 1972; March 31, 1973; April 6, 1973; June 22, 1973; Sheppard Papers). "He really was on a mission, and he wanted ammunition. In my Maniola case, he wanted to get the stuff out, together with all the selectionist flavour intact" (letter to author from Paul Handford, November 21, 1995).

When one adds, to all of this, happy camping trips to the Scilly Isles to collect butterflies, "love" is not too strong a term for what bound teacher and pupils in the "Ford school." The correspondence in the 1970s, between Ford and his most brilliant student, Philip Sheppard, as the latter lay dying of leukemia, makes heartbreaking reading. These were Ford's men, and once trained they fanned out through the universities to spread the message—or, for those who could not learn at first-hand, through a series of wonderfully accessible texts (Cain 1954; Sheppard 1958).

Research needs support, and here again Ford triumphed. The nineteenth-century biologists had found backing for physiology in the medical profession and for anatomy in the teaching profession. So, likewise, Ford sought and—through the right contacts—found backing for his program in evolutionary biology. This came courtesy of the good graces of the Nuffield Foundation. During the Second War, Lord Nuffield, the (real) Henry Ford of Britain, set up a foundation to support, with other items, medical research and "the advancement of social well-being . . . in particular . . . by scientific research" (Nuffield Foundation 1946, 9). By the early 1950s, because of biology's sad state in Britain, the Foundation was deliberately seeking to support work in the subject, and through Ford's department chair, the Linacre professor Alistair Hardy, application for funds was made successfully: "I put up a programme of research to the Nuffield Foundation on which they have awarded me £10,000 spread over five years, and promise powerfully to support application for another £4000. This latter is still under way, but the £10,000 is an accomplished fact. There is much to arrange in getting Kettlewell from Africa to work with me etc. But-it-*is* rather nice" (letter to Huxley, March 2, 1951; J. Huxley Papers).

Thus, at the beginning of the decade, there was money for research in ecological genetics. More money was given through the decade, and so more work was done and more researchers supported. Then, toward the end of the 1950s, a part of the Oxford group budded off, moving across country to promotions at Liverpool University, where they continued in the Ford tradition. Funds from the Nuffield Foundation went too, finally climaxing in 1963 with the largest grant the Foundation had ever given (a third of a million pounds) (Clark 1972, 187–188). Ford, who was openly acknowledged as the key factor behind this massive award, had succeeded splendidly in his efforts. Coincidentally he published his *magnum opus,* defining and summarizing his evolutionism (Ford 1964), and was awarded his Oxford professorship.

But for all of this support there was a price to be paid. Most particularly, Ford and his people had to strive to show that their work advanced "Social well-being," as much as possible. In this respect, Fisher again showed the way, for since becoming a professor at London he had interested himself in human genetics and its possible value to our species. Ford readily took up the theme—witness his book for medical students, originating in lectures given to undergraduates. And, in a master stroke, he linked his insect work with medicine. He argued that since balanced polymorphisms in nature show evidence of selection, we should investigate such polymorphisms in humans for evidence of biological effects (Ford 1942). Moreover, having shown (in part through examining traits in chimpanzees at the London Zoo, courtesy of Huxley) that there are indeed stable polymorphisms in humans (Fisher, Ford, and Huxley 1939), Ford began forecasting that human variability as manifested by such things as blood groups will prove to be associated significantly with various diseases. In other words, Ford argued that there is a direct chain of inference from work on butterflies to the prediction of patterns of illness in *Homo sapiens!*

Famously, some years after his forecast Ford was shown to have been correct (Aird, Bentall, and Fraser Roberts 1953). There are significant links between blood groups and susceptibilities to illness. Naturally enough, almost all of the Nuffield grants paid explicit attention to the possible handsome medical dividends for any work in ecological genetics. In fact, the initial approach to the Foundation, by Hardy, made this point: "I do not wish to suggest that Dr. Ford's work will be of *direct* medical value, but rather that it is part of the essential groundwork paving the way to a more direct application of genetics to medicine" (Clark 1972, 187). And this theme was continued, as is shown in the following typical example drawn from the Foundation's report for 1959:

> Dr. Sheppard and Dr. Cain have already provided one of the clearest and simplest demonstrations of the operation of natural selection, showing

> that birds act as predators upon snails and choose those colour-patterns which are the least well adapted to local conditions. The breeding results carried out as part of the programme are of considerable importance in human as well as animal genetics. They are proving the evolution of linkage, a concept of direct application to the human blood groups. (Nuffield Foundation 1959, 19)

Not that, in the best tradition of grantsmanship, the Ford school was incapable of diversifying and multiplying the supposed practical virtues of its efforts. Having found that organisms show sensitivity to environmental changes by taking widely different genetic strategies, "Dr. Ford and his colleagues" drew some obvious conclusions: "This may well throw light upon the way in which crop plants and domestic animals respond when transferred to a district new to them" (Nuffield Foundation, 1963, 31). There is no surprise that the massive grant to Liverpool was directly and explicitly given to fund a department of medical genetics—staffed and run by Ford's men, who as part of their duties continued work on non-human evolutionary studies.

Stories like these will have the ring of authenticity to anyone who has submitted a grant application successfully. There is no outright dishonesty or even subservience, but there is certainly a shaping of one's project to meet the donor's interests. And it hardly needs saying how antithetical was Ford's strategy, from the earliest pre-grant days right down to the time of full success, to thoughts of culture—the very mark of the non-professional approach to evolutionary biology. "Even now, my dear Henry, I have not the faintest idea of what your religious views (if any) are; never once in all these years have you mentioned them, even in social conversation, let alone in scientific" (letter to Ford from Cain, April 2, 1985; Cain Papers). The last thing that Ford needed at any time was to be infected with a reputation for foggy metaphysics, be it Christianity or be it Progress. Like Weldon, he wanted science without the scientists and was very uneasy about contributing memories to a biography of Fisher: "It is in his work and thought, rather than in himself, that true interest lies" (letter to Sheppard, June 14, 1968; Sheppard Papers).

Ford needed to be considered a hard empirical scientist, a researcher whose work had the prospect of practical payoffs. Epistemologically, he needed to present his science as having all those lovely grant-worthy values like predictive ability. And this need was a part of an ongoing battle, for there were lots of other coarse feeders out there, ready to grab every tasty morsel thrown into the funding pool. Experimental embryology may have collapsed in on itself, but now the molecular biologists were flexing their muscles. The precarious state of his new discipline was brought home brutally when, on Ford's retirement, his professorship was taken from the evolutionists. Despite frenetic campaigning on behalf of Sheppard, the selection committee listened to Peter Medawar—Nobel Prize–winner and no friend of Darwinism—and awarded the post to the molecular geneticist Walter Bodmer. (See especially letters from Ford to Sheppard, August 17, 1969, and September 8, 1969; Sheppard Papers.)

In any case, if proof of the wisdom of the policy of excluding progress from professional evolutionism were needed, there was the example of Huxley. Just at the time Ford was applying to the Nuffield Foundation, Huxley was applying to the Ford Foundation (in America) to support a project of his own, a think tank with Progress as a major unifying theme: "Exploration of the nature of man, notably the exploration of human possibilities (of course taking account of human evolution and history, physiology and psychology and the relations of man with external nature)" (letter from JSH to Robert Hutchins, March 2, 1951; J. Huxley Papers). Not only was Huxley rejected but, just at that very moment, the Rockefeller Foundation was giving large sums for work on orthodox genetics to Huxley's sometime Rice assistant, H. J. Muller (letter to Huxley from Warren Weaver, April 13, 1951; J. Huxley Papers). (Muller was getting $200,000 over five years.)

The "writing on the wall" was there for all to see—literally, in fact, for it was Medawar who wrote the most public and most savage critique of Huxley's infatuation with Teilhard (Medawar [1961] 1967, 71–81). And so we find Ford defining his vision of evolution (the very term *ecological genetics* sounds forward-looking, non-threatening, and socially useful) to exclude topics like paleontology, which his contact with Goodrich and Huxley must have shown could only lead to trouble. He wrote, for example, in response to Mayr's questionnaire (Ford Papers): Q. "What fields contributed most/least to synthesis?" A. "Obviously some of these are only marginally relevant *e.g.* palaeontology compared with classical genetics." What makes this exclusion particularly ironic is that Ford's attitude certainly did not stem, as one might surmise, from the fact that his interest in biology arose from butterfly collecting rather than, say, fossil hunting. His hobby was archeology, where he was given to tracing past history with great enthusiasm (Ford 1953; Clark, Ford, and Thomas 1957). But not in evolutionary studies!

As Ford excluded the dangerous subjects, as he strove for scientific credibility and promise of practical worth, as he forged his discipline, the evolutionary synthesis endorsed by him and his school included and stressed areas of biology in which thoughts of progress would not naturally arise. A fitting epigraph is given by the description provided by his students for their "Msc Advanced Course in Evolution, Variation and Taxonomy" at Liverpool University:

> Considerable stress will be placed on modern advances in understanding the origin and fate of variation in both animal and plant species. It will, therefore, be appropriate to anyone who wants to appreciate the fundamental importance of evolutionary processes in living organisms, whatever they are ultimately destined to become, for instance plant or animal breeder, entomologist involved in the control of insect vectors by insecticides, evolutionist, taxonomist, conservationist or ecologist. (Ford Papers, F76)

Everything is there, except a place for progress. Paradoxically, given how the Nuffield Foundation was forever stressing its role in promoting scientific advance, the aim for cultural Progress pushed aside thoughts of biological progress.

The British side to our history is completed. The Mendelian divide has been crossed, but we see now how wrong it would be to say that those who made the journey—the self-named "ecological geneticists," the taxonomists, and others—simply clothed the population geneticists' formalisms with empirical findings. There was certainly continuity between the older mathematicians and the younger empiricists (ignoring the fact that Huxley was older than anyone!), but the connections were much more complex than many suspect. And the same is true of progress, biological and cultural. Post-Mendelians like Darlington accepted the traditional prohibition established by T. H. Huxley: keep your professional biology value-free and put your metaphysical evolutionary speculations, which will almost undoubtedly be progressionist, into your popular writings. Julian Huxley, an ardent progressionist, wanted to have his professionalism and his progress too. He tried to create an academic discipline of evolutionary studies that would be firmly progressionist—one that would reflect the Progressionism of its practitioners. Alas, his aims were not to be realized. E. B. Ford was the first successfully to create a secure school of professional British evolutionists. The price of success, however, was the total exclusion of all hints of progressionism. Charles Darwin had at last got his professional discipline, and at its core it was indeed strongly Darwinian. Whether it was quite what Darwin had in mind is another matter. There was no room for progress, Darwinian or otherwise. But then, Darwin never had to go to the Nuffield Foundation for his money.

10

The Genetics of Populations

In 1913, Osborn's sometime student and assistant William K. Gregory started an article on the evolution of the mammals by admitting to the low status of the kind of work in which he was engaged. General opinion was that "our theories of phylogeny and morphology are too much the product of the undecided imagination, which seizes gladly upon favorable evidence but which fails to seek the unfavorable." Little wonder that "zoologists have in great numbers turned away from vertebrate comparative anatomy as a thankless task and have come to regard its labyrinths as leading nowhere" (Gregory 1913, 1). The Huxley philosophy had conquered in the New World as firmly as it had in the Old, a point well demonstrated by Edwin Grant Conklin, graduate of Hopkins and student of H. N. Martin, and one of the most influential American biologists in the first half of this century (Maienschein 1991). *Qua* professional, Conklin churned out desperately serious papers on invertebrate development, making the contribution of the cytoplasm his special area of concern and research (Conklin 1897, 1902). And *qua* popularizer, he wrote book after book on evolution, unabashedly enthusiastic for a P/progressionist picture of all reality (e.g., Conklin 1921).

As in England, the talk here is of status and not interest. Around the turn of the century in America, decades before and decades after, the interest in matters evolutionary was intense. This is shown vividly by a major controversy about evolutionary causes which exploded into the journal *Science* in 1906. All sorts of topics—isolation, mutation, and variation—were raised and thrashed, nigh to death. People really cared. Yet, in those days *Science* made an explicit pitch to the general reader as well as to the expert: "Our journal . . . will occupy itself mostly with those broader aspects of thought and culture which are of interest not only to scientific investigators, but to educated men of every profession." All of the argument about evolution and its causes—much of which, typically, was sparked by general addresses to scientific societies—was kept firmly in those sections of the journal intended to have the broadest appeal. (See Abrams 1905; Allen 1905, 1906; Berry 1906; Casey 1906; Cook 1906a,b; Davenport 1906; Gulick 1906; Jordan 1905a,b, 1906; Lane 1906; Lankester 1906; Lloyd 1905; Merriman 1906; Metcalf 1906; Ortmann 1906a–f; Robertson 1906; Stejneger 1906; Vaughan 1906.)

There were distinctive American twists to the work of people like Conklin and others. Most noticeable was the influence of Herbert Spencer. Although his stock had sunk dreadfully low in Britain, in America Spencer still had considerable authority (Russett 1966, 1976; Richards 1987). His standing was due less to his identification as the father of the notorious Social Darwinian movement and more to his reputation as the prophet of P/progress, especially inasmuch as readers were captivated by his thoughts on equilibrium. For remember, in Spencer's eyes evolution was not an anchor holding firm beneath static balance; rather, it was ever-changing and forward-moving, in the ontogeny of the individual and in the progressive evolution of the group.

Spencer's "dynamic equilibrium" had transplanted very happily to the United States, where it was taken up by proselytizers like the sociologist Frank Lester Ward (Russett 1966). By the beginning of this century, Spencerian Progressive change—an ever-present tendency to equilibrium between conflicting forces being complemented by outside (or other) disruptions, thus triggering advance to a higher plain of being—became virtual orthodoxy in many circles. People of the left (including Marxists!) were as enthusiastic as people of the right (Pittenger 1993). Expectedly, therefore, American evolutionism was often deeply Spencerian.

> Life itself, as well as evolution, is a continual adjustment of internal to external conditions, a balance between constructive and destructive processes, a combination of differentiation and integration, of variation and inheritance, a compromise between the needs of the individual and those of the species. And in addition to these conflicting relations we find in man the opposition of instinct and intelligence, emotion and reason,

> selfishness and altruism, individual freedom and social obligation. Progress is the product of the harmonious correlation of organism and environment, specialization and co-operation, instinct and intelligence, liberty and duty. (Conklin 1921, 87)

With all of the organic metaphors, with the balances between positive and negative forces, with the move to specialization and cooperative harmony, with everything pointing to Progressive advance, the old man himself might still have been at work.

Was there no-one comparable to Weldon in America? The answer is both "yes" and "no." There were people who were committed evolutionists, professional scientists, believers in selection (natural and artificial), and ready to tell their students about such things. But these were not researchers in the fashion of Weldon, working on evolution for its own sake. The American evolutionists were located at schools of agriculture, practicing (as they had long practiced) methods of selection to improve crops—a matter of increasing importance in America at this time, as rural areas began a trend of depopulation while the urban centers experienced explosive growth (Kimmelman 1987). At Cornell, for instance, we find Liberty Hyde Bailey, willing to tell his students that he was an evolutionist and that natural selection is important; but Bailey made this claim simply as a prolegomenon to his main pedagogical intent, which centered on detailed discussions of artificial selection for practical ends. Those were the foci of his own professional interests, and totally and utterly the source of his support (Bailey 1896, 1897, 1898a,b, 1901, 1904, 1911a,b).

The same was true of others in comparable situations. Bailey's colleague, the Cornell University entomologist John Henry Comstock, wrote a well-known essay arguing that one must interpret classifications in an evolutionary manner, and that the underlying causes of change are selective, no less (Comstock 1893). He himself was particularly interested in the evolution of insect wings (Comstock and Needham 1898–99). But Comstock and his students were certainly not professional evolutionists in the sense desired by Darwin and exemplified by Weldon. Like Bailey and his students, they made their living off the back of agriculture, for their expertise (and most of their time) lay in the economic implications and control of insect pests (Henson 1990, 144–146). Although it is true that they had a genuine interest in evolution, it was all very much a sideline.

In any case, apart from the fact that Comstock's thinking about the past showed a far deeper debt to Agassiz and German thought generally than it ever did to Darwin (Comstock and Needham 1899)—hardly a surprise since Comstock's major teacher was Burt Green Wilder, an Agassiz student (Comstock 1925)—Comstock himself kept up the Huxley divide between popular evolution and professional biology, just what one might expect from someone who explicitly modeled his own course and first textbook on Huxley and Martin (Henson 1990, 289; see also Comstock 1875, 1888). He confined his evolutionizing to such occasions as a *Festschrift* volume—a place where, notoriously, one can say what one likes, especially when (as in this case) one is both editor *and* publisher! When Comstock came to the presentation of actual taxonomic results, his work read just like that of an ideal morphologist, one who referred not to ancestors but to "hypothetical types" and whose major interest was in homology. Evolution and selection went unmentioned in Comstock's professional work. And the same is true of overt progress, for Comstock warned his students to avoid such inflammatory terms as *low* and *high* and to stick to such safer words as *general* and *specialized* (Comstock and Needham 1898–99; see also Comstock 1888, 1918, 1924; Comstock and Comstock 1895; Comstock and Kellogg 1895).

What about the people who, like Bateson, were channeling their energies into problems of heredity? There were American equivalents. Revealingly, much of their work was funded by the Carnegie Institute for Studies in *Experimental Evolution* (my italics). But, as in Britain, genetics took on a life of its own, independent of evolutionary studies *per se*. Take the work of W. E. Castle (1867–1962), professor of zoology at Harvard's Bussey (Agricultural) Institution and, through his long career, one of the most influential teachers of this century (Dunn 1965; Provine 1986). Like nigh everyone else, Castle started on the morphological track, specializing as an invertebrate embryologist; as soon as Mendel's laws were rediscovered, however, he stopped, retooled, and emerged as a student of heredity. A committed Mendelian, Castle was impressed by the degree of variation one finds in the features of small mammals. Therefore, he began a series of now-classic selection experiments on rats, in which he tried to achieve maximum morphological divergence in once-identical forms and to tease out underlying causes (Castle 1911, 1916, 1917).

Castle succeeded wonderfully in his labors. Constrained only by the bounds of physical possibility, he showed that there seems to be no limit to the effects of selection. Moreover, although cessation of selection undoubtedly leads to a certain degree of backsliding ("regression"), the

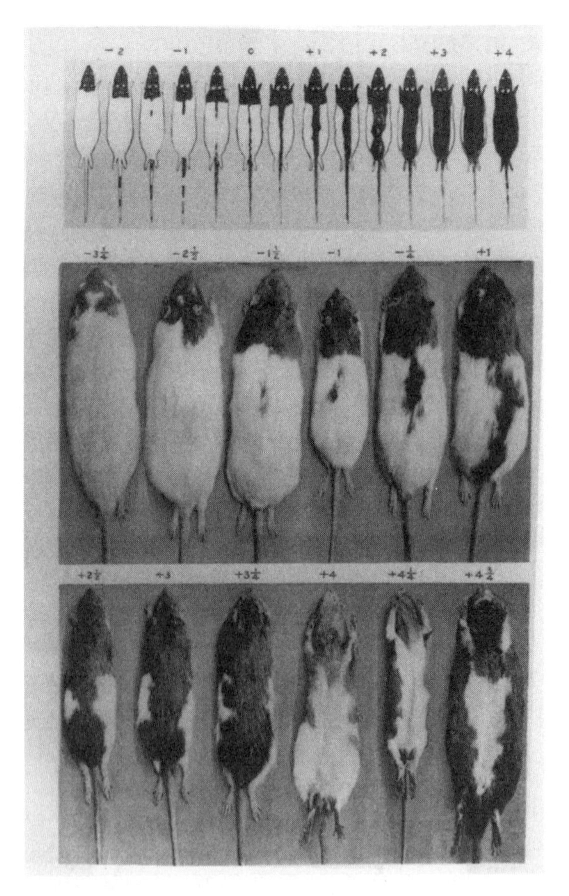

The effects of selection on coat color in rats. *Top row:* Arbitrary grades used for classification. *Middle row:* Selection for white coat. *Bottom row:* Selection for dark coat (the animals with grades 4, 4¼, and 4¾ were entirely dark from above). (From Castle and Phillips 1914.)

succeeding generations do not revert back to the original form. "The results of such plus or minus selections are permanent, for return selection is not more effective than the original selection, and during return selection regression occurs *away from* the original mode, that is, toward the mode established by selection" (Castle and Phillips 1914, 22). Finally, and significantly, selection seems to have no memory, in that by ceasing and resuming the selection pressure one can with equal ease select backward or forward.

However, the crucial point is that the role of natural selection for Castle was not that of selection in the *Origin.* Nowhere in his seminal writings did Castle direct his selection experiments to problems of adaptation—the very starting point for Weldon. One is as far from British natural theology as one could be. Castle's interests were (as with Bateson) the mechanisms of heredity. Specifically, he wanted to know whether or not the Mendelian units get retransmitted entirely from generation to generation, or whether they are in some way themselves modified by selection. Eventually, Castle had to admit that his experiments disproved his own predilection, that selection can in itself alter the units of heredity (Provine 1971, 109–114); but this fact, though important, is no part of our tale. What is important to us is that, although Castle did show the significance of selection, we must not pretend that he did more than that. He does not shake an overall conclusion similar to that drawn by British researchers. Professional evolutionary biology did not stand at the cutting edge of the life sciences. The best people had turned their attentions elsewhere.

Yet, within a very few years, certainly by mid-century, evolution in America was as vital and professional a field as anywhere in the world. A discipline had been created and, epistemically, the science had been upgraded. To start the story going again, I turn to two of the most significant figures in the history of evolutionary thought.

From Animal Breeding to Theoretical Population Genetics

Fisher and Haldane were difficult men. Their behavior could be, and frequently was, absolutely appalling. The third of the great pioneers in theoretical population genetics was very different. Sewall Green Wright (1898–1988)—he dropped the middle name, giving the English endless urges to hyphenate him—was the nicest of human beings. Alas, he was also a rather dull man who would mumble into symbols on the blackboard for two hours when asked to give a one-hour lecture. His idea of a fun night out was a discussion with his university chums at the local faculty club. The son of a college professor, who was teaching at a small Universalist college in the Midwest, Wright was steered toward biology in his senior year by an inspiring teacher: Wilhelmine Marie Entemann (Key), one of the first women to graduate with a doctoral degree from the University of Chicago. For graduate work he went to Harvard, where he studied under Castle, and this led to a job with the United States Department of Agriculture. Here, he worked intensely on problems of heredity, most particularly on the physiological effects of the genes: the ways in which they express themselves and how they interact in systems. Called to the University of Chicago's biology department in the mid-1920s,

Sewall Wright with guinea pig, his favored experimental organism.

Wright's two-dimensional adaptive landscape, seen from on high. The plus signs represent peaks, and the minus signs valleys. (From Wright [1932] 1986.)

Wright was to remain there for thirty years, until his retirement and a move to Wisconsin (Provine 1986).

Wright's enthusiasm for evolution was life-long. Four things in particular motivated him in his work and led to the distinctive form his ideas were to take (Wright 1923, 1977). First, there was the time spent at Harvard with Castle, who was just then engaged on the experiments selecting for coat color in rats. The success of this work convinced Wright utterly of the power of selection, although the attendant risks of infertility made him wary of the assumption that selection is usually simply working its powers in large groups. Second, there was Wright's own thesis topic, concerned with interactions of genes in guinea pigs, again primarily coat color.

> From my studies of gene combination, . . . I recognized that an organism must never be looked upon as a mere mosaic of "unit characters," each determined by a single gene, but rather as a vast network of interaction systems. The indirectness of the relations of genes to characters insures that gene substitutions often have very different effects in different combinations and also multiple (pleiotropic) effects in any given combi-

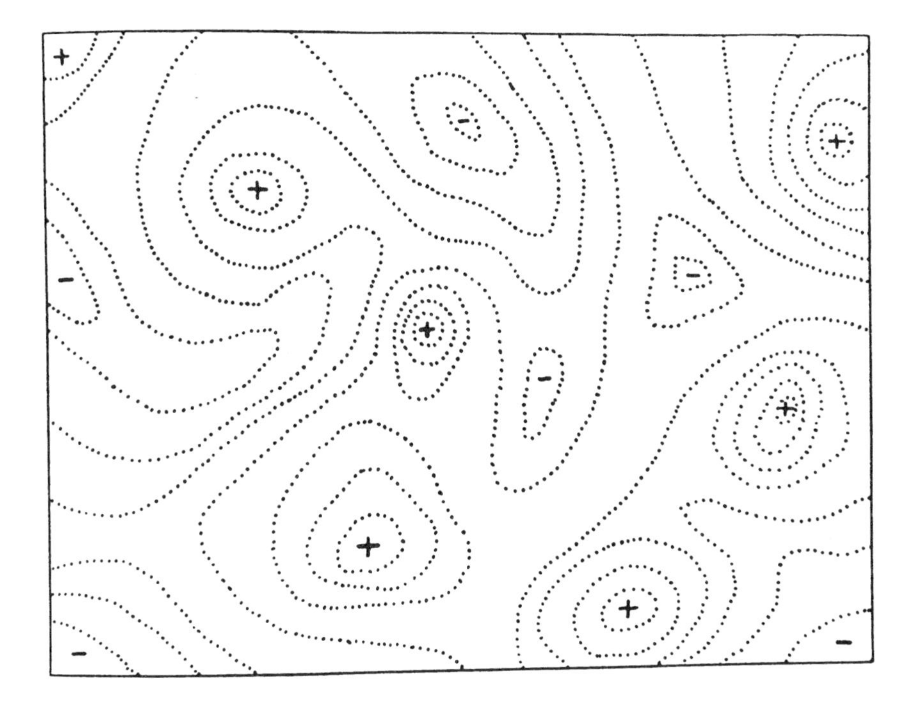

nation. The latter consideration gives another reason for the tendency of mass selection to lead to deterioration. (Wright 1977, 9)

Third, from his work on guinea pigs Wright came to see in a dramatic way how great is the range of heritable variation. Building on this insight, he determined that isolating a small number of organisms and inbreeding can "fix" randomly a particular sub-section of the available variation, thus leading to a group of animals with their own very distinctive characteristics. Fourth, and related in a sense, from a major study he made of the breeding history of shorthorn cattle, which he performed as part of his duties for the Department of Agriculture, Wright came to see that success in breeding stems from the isolation of a small number of animals with highly desirable properties and the very heavy use of them to upgrade the total stock. One does not try for a general improvement of one's herds. One selects drastically and breeds from a very favored few (Wright 1923).

Although his mathematical training was minimal, Wright showed an early flair for the subject. He applied his talents brilliantly to problems of inbreeding—in the course of which he developed his method of "path coefficients" for dissecting combined causal components (Wright 1921, 1922). Attracted by the related work of Fisher, Wright tried his own hand at population-genetical theorizing, and around the time of the shift to Chicago he put together his ideas in a massive paper—one which, however, went unpublished until the year after Fisher published (Wright 1931).

Despite their techniques being very different, their conclusions were formally compatible. And, naturally enough, particularly in the beginning parts of their theorizings, there were some topics discussed by both men (and by Haldane also). For instance, Wright considered the question of superior heterozygote fitness and its consequences for balanced proportions of genes within populations. However, showing his own influences, Wright then went on, on a track quite unlike Fisher's, to examine in detail the fate of genes in small populations—specifically those where the overall numbers might not be great enough to counteract the effects of chance, taken down through a finite number of generations. This led to an appreciation of "genetic drift": Wright showed mathematically that, in small populations, certain genes might move randomly up to total fixation or down to extinction, purely because of "sampling effects." And this change in a gene's proportion would occur even though forces of selection, and mutation, were working in the opposite direction.

The formal quantitative work completed, Wright was now ready to turn to qualitative theory and to present his vision of evolution's mechanisms—the "shifting balance theory of evolution": "Evolution as a process of cumulative change depends on a proper balance of the conditions, which, at each level of organization—gene, chromosome, cell individual, local race—make for genetic homogeneity or genetic heterogeneity of the species . . . The type and rate of evolution in such a system depend on the balance among the evolutionary pressures considered here" (Wright 1931, 158).

What did Wright have in mind here? Not the process of Fisher, who argued that one selective force works uniformly on a united population. At best, Wright thought this mechanism would yield very slow change, if that. Rather, drawing on his experiences with guinea pigs and shorthorns, Wright thought change occurred in small sub-populations, in which drift could be a real factor and could create altogether new gene combinations/ratios, and from which the rest of the total population could then be seeded and upgraded.

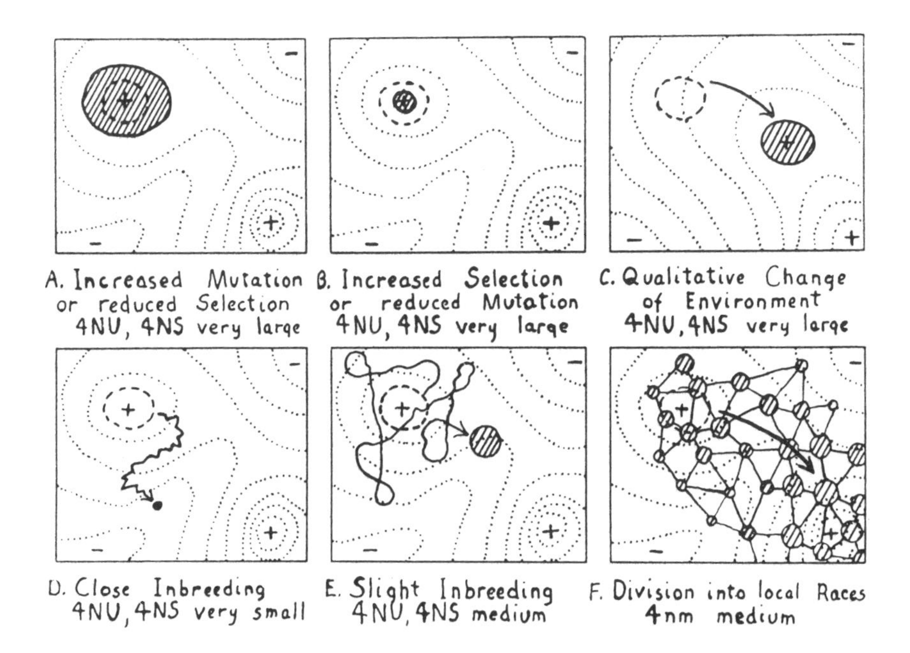

Various possibilities for adaptive landscapes with variation in selection *(S)*, mutation rate *(U)*, population size *(N, n)*, and population exchange with rest of species *(m)*. Note that *(D)* a small inbred population is liable to go extinct, but that *(E)* with isolation and moderate inbreeding will lead to genetic drift. The fastest and most efficient evolution occurs *(F)* when there is some gene exchange among populations of a fragmented species. (From Wright [1932] 1986.)

> Finally in a large population, divided and subdivided into partially isolated local races of small size, there is a continually shifting differentiation among the latter . . . which inevitably brings about an indefinitely continuing, irreversible, adaptive, and much more rapid evolution of the species. Complete isolation in this case, and more slowly in the preceding, originates new species differing for the most part in nonadaptive respects but is capable of initiating an adaptive radiation as well as of parallel orthogenetic lines, in accordance with the conditions. (Wright 1931, 159)

A picture is worth a thousand words. In the eyes of evolutionists sixty years ago, a picture was worth a million symbols. In 1932, a year after his paper appeared, Wright was asked to give a short, clear presentation at a congress. For once in his life Wright obliged: he explained his theory without the mathematics and introduced the problems and results associated with small groups in terms of an inspired biological metaphor, that of an "adaptive landscape" (Wright 1932).

Think of a gene or a set of genes, which might occur in combination with other genes and which might be found in organisms in a variety of circumstances; these are two dimensions of a description of a gene pool. The location of that gene or set of genes in the "adaptive landscape" will be associated with a certain biological fitness or degree of adaptedness. Fitness can be represented by peaks or troughs in a third dimension, thus giving rise to a contour map or landscape—moving around the landscape takes you to other genes or gene combinations. Organisms (or, rather, gene holders) will generally cluster around peaks. If they are not on top of the world, then selection will act to drive them up there. But note that a group of organisms (say a species) at the top of one peak will not generally be able to move to another peak, even though the second may be higher. The valley between peaks, representing low fitness, will prevent such lateral moves. With some slight fluctuations, the adaptive landscape must stay as it is.

In terms of this pictorial metaphor, Wright could now present his theory to the mathematic-phobic. The most obvious way in which change might occur is through a change in the environment. An adaptive landscape is not literally embedded in rock. If a valley is removed, then a species might move sideways and onto another peak. Apparently, however, this cause of change is not all that significant—the impression given is that although the landscape may not be rock, it is pretty treacly. Far more important movements result from a population being broken into small sub-groups. Here, as theory shows and as the results of breeding confirm, when selection is sufficiently slight the genes tend to drift to fixation or elimination. But, although one small group on its own will probably not effect major change—most likely it will suffer from such extreme inbreeding that it will go extinct—a population divided into a number of small groups, with some slight genetic interchange between them, will simultaneously be able to form new genetic combinations under drift and prevent the sterility consequent on inbreeding.

Thus, the way is paved for a situation like the shorthorn cattle, where the whole group is dragged across a valley and up a higher peak by virtue of a small number of superior individuals.

> Let us consider the case of a large species which is subdivided into many small local races, each breeding largely within itself but occasionally

> crossbreeding. The field of gene combinations occupied by each of these local races shifts continually in a nonadaptive fashion . . . With many local races, each spreading over a considerable field and moving relatively rapidly in the more general field about the controlling peak, the chances are good that one at least will come under the influence of another peak. The average adaptiveness of the species thus advances under intergroup selection. (Wright 1932, 168)

When Wright used the terms *intragroup* and *intergroup selection,* he was probably not drawing the distinction between individual and group selection. Rather, he was thinking more in terms of individual selection either working solely within the group or between members of different groups.

The shifting balance theory was presented in the early 1930s. What is quite remarkable is the way Wright held to it for the next sixty years, even using the same pictures in his publications! (See Wright 1982.) It is true that sufficient empirical evidence was uncovered to push Wright to a somewhat more selective stance than he took initially. At first he inclined to think that interspecific differences tend not to be of great adaptive significance. Apparently drift works at various levels, from populations to species. Later, he changed his mind somewhat on this point and gave more work to natural selection. But, this adjustment aside, what you got at first was what you got at last: populations breaking into small subgroups, non-adaptive drift resulting in new gene combinations, and then selective response as the whole population would be dragged up another adaptive peak.

Genetic Drift and the Problem of Progress

Let us see where Sewall Wright stood on the issue of biological progress, and let us start with the easy part. Wright clearly accepted a form of limited progress—a kind of relativistic or comparative progress. That notion lies at the very center of the balance theory, especially as it is described by the landscape metaphor. Organisms move from one adaptive peak to another, and it is clearly understood that the second peak is higher than the first. Logically, obviously, there is no need for the second peak to be higher—certainly, once achieved it might sink—but generally this is not what Wright had in mind. Basically, Wright's metaphor was of a fixed landscape, not of the ocean's surface or a water bed in which there is constant motion: "I recognized

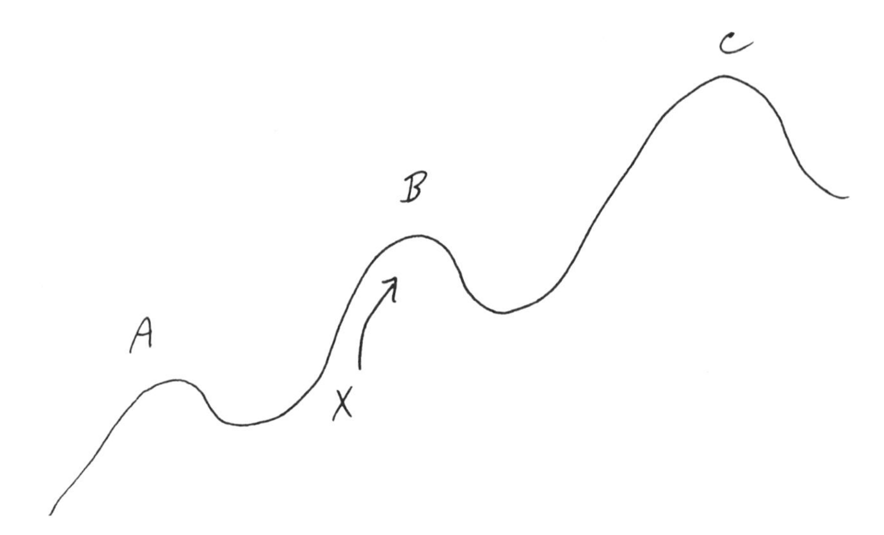

Sewall Wright's first illustration of an adaptive landscape, showing (presumably unintentionally) the ease with which the metaphor takes on a progressionist reading. (Redrawn from a letter to Fisher, February 3, 1931.)

that there were an infinite number of peaks, so you wouldn't get simply a reversal—you just oscillate between peaks. That wouldn't happen with an infinite number of peaks. The chance of going back to the original one with a second change of conditions is very small, so that the population would be kept continually on the move. The general tendency would be to get higher and higher" (interview with W. Provine, May 18, 1978; Wright Papers).

In this context it is instructive to look at his first informal, two-dimensional picture of his theory. The upward direction is explicit and there is no talk of future sinking. (See letter to Fisher, February 3, 1931, in Provine 1986, 272; also Wright 1939.) Elsewhere, Wright made it clear that he thought that this is the kind of progress which leads to specialization, perhaps even to quasi-directed evolution. For instance: "General control by natural selection seems an adequate explanation of such cases as the somewhat orthogenetic sequence of genera of the horses from the Eocene to the present" (Wright 1948, 536).

What then of the move to a more absolute kind of progress? With many other evolutionists, Wright repeatedly asserted that on some few rare occasions there are, as it were, evolutionary breakthroughs, by which whole new forms of life "emerge." Sometimes these events are a function of the organism: "A form, in the course of its gradual evolution, may acquire a character or character complex that happens to be of general rather than merely special significance and which thereby opens up the possibility of a relatively unexploited way of life" (Wright 1949, 568). Sometimes the driving force seems to be external: "A form which reaches relatively unoccupied territory also has before it a major ecological opportunity independently of any character differences from its parent form" (Wright 1949, 568). Note that whatever the situation, there is the implication of real advance.

Does all this convert into real, absolute progress? As one concerned essentially with mechanisms, this question was not pressed on Wright in the way that it was (say) on Osborn. But Wright did sometimes address broader issues, for instance in encyclopedia articles. If the measure of progress is complexity, perhaps related to size, then—although Wright noted that there are many exceptions—the answer seems to be affirmative: "Evolution, presumably starting from self-duplicating organic molecules, proceeding through prokaryotes, unicellular eukaryotes, and simple multicellular forms as higher plants, insects, and cephalopods, birds, and mammals, has been characterized in the main by increasing complexity of organization" (Wright 1968–79, 4, 492). Moreover: "Another rather general tendency has been that toward greater size along each lineage (Cope's rule) . . . Most of the well-established phyletic lines of mammals followed Cope's rule" (ibid.).

What of the crucial question about ourselves? What of *Homo sapiens?* Compared with almost all other evolutionists, Wright was very reserved in explicit discussion of the status of humankind—so reserved, in fact, that I shall return to the point. Nevertheless, overall the strong feeling one has is that Wright (*qua* evolutionist) did care very much about and favor humans. He built his balance theory thinking of humans: "I had human evolution very much in mind in formulating my ideas on the roles in evolution of local random drift, intrademic selections and interdemic selection in 1931, and later, although I touched on it only very briefly" (letter to A. M. Brues, October 30, 1963; Wright Papers). In a four-volume *opus* Wright wrote in retirement, our species gets detailed treatment (more than for any other organism), and significantly this discussion comes right at the end of the work (Wright 1968–79, vol. 4). Moreover, Wright saw some fairly standard orderings within the species: "I lean toward a primarily genetic interpretation of the high productivity of the Jews in Science and the Arts. They have undoubtedly been subjected to exceptionally strong selection for intellectual ability for at least 3000 years" (letter to H. Mehlberg, August 3, 1977; Wright Papers). Analogously: "The capacity to anticipate and plan for the future is a mental attribute which would be favoured under northern conditions and selected for insofar as it has a genetic basis. This would presumably come to be more advanced in the temperate zone than in the tropics" (Wright 1968–79, 4, 456). Here one is hearing echoes of the terror of a small boy who was once set upon by blacks on his way home from school (unpublished autobiography, Wright Papers).

Wright was indeed a progressionist, both comparatively and absolutely. In fact, both in its own right and in the light of its subsequent history, the landscape metaphor is deeply progressionist, whatever the strict logic of the case may be. But Wright is somewhat hazy over the precise endpoint of absolute progress. If one wanted to argue that humans are uniquely or jointly the endpoints of progressive evolution, one certainly could use the balance theory for support. But, for all the hints he threw out, in Wright's writings you were not going to find explicit claims to that effect. Their absence makes all the more pressing the second part of our inquiry, namely into Sewall Wright's cultural beliefs. Where did he stand on Progress, and is there reason to think that this stand had any effect on his biological position?

Neither Christianity nor extreme scientism seems to have been a factor. Wright worshipped as a Unitarian, and nowhere in his writings do we find Haldane's fondness for science and technology in itself. Indeed, let me emphasize just how distant Wright was from Fisher and Haldane, especially the former (Hodge 1992). Although he certainly believed selection was important, in respects he was as anemic about adaptation as was Castle. Selection meant one organism or set of organisms winning out over another. Even putting questions of drift aside, there was in Wright's theory simply no feel whatsoever for the design-like nature of organisms. This was not a denial but an attitude. Wright's "comparative" progressionism was far from the metaphor-laden natural history of the British. At Chicago, for example, Wright had started a collection of spiders but then dropped it as too time-consuming, whereas Fisher always loved butterfly collecting, despite his dreadful eyesight. Significantly, the factor which sparked what was to be the life-long bitter dispute between Wright and Fisher centered on the true evolutionary cause of gene dominance, Fisher claiming it was adaptive and Wright countering that it was but a by-product of the mechanics of the Mendelian process (Fisher 1930, 1934; Wright 1934a,b).

Do we have in Wright an exception proving the rule—namely, a man who was a little dubious about cultural Progress, and hence a little wary of biological progress? Certainly, he shared some commonly held concerns about societal decline. For many years Wright was on the board of directors of the American Eugenics Society—a role backed in both his scientific writings and his private correspondence by some fairly standard sentiments: "Those who are still social parasites after a fair chance should be eliminated eugenically" (letter to brother Quincy, October 17, 1915; Q. Wright Papers). But eugenics never played a huge role in his thoughts for or against progress. Wright was only too aware of the relativism of most eugenical programs, as he was likewise cautious about the underlying notion of a direct analogy between cultural Progress and biological progress. He wrote that: "The mode of evolution of culture is analogous to that of genetic system. Invention is the analog of gene flow. Cultural variation is continually subject to selection on the basis of utility. There is random cultural drift, exemplified by the breaking up of languages into dialects." However, for all that he was prepared to refer to the "evolutionary advance since the beginnings of agriculture" as the "last phase of the shifting balance process" (Wright 1968–79, 4, 455, 454), Wright was no great enthusiast for Progress of this kind. He noted that the great civilizations of the past have collapsed and there seems no guarantee that ours is uniquely better or more stable.

By this point, one might be inclined to argue that the search has gone far enough. Biological progress is at best a hazy notion for Wright, and he lacks an interest in cultural Progress. But, as I have said, in its way the landscape metaphor is deeply progressionist. The problem is that we are still missing the essential key to Wright's thought. The animal breeding and so forth is crucially important, but it is not the whole answer.

Dynamic Equilibrium as a World Philosophy

Right in the concluding sections of his classic exposition of the balance theory, Wright inserted this: "The present discussion has dealt with the problem of evolution as one depending wholly on mechanism and chance. In recent years, there has been some tendency to revert to more or less mystical conceptions revolving about such phrases as 'emergent evolution' and 'creative evolution.' The writer must confess to a certain sympathy with such viewpoints philosophically" (Wright 1931, 155). He said no more at this point, but let us ask first about the "mystical" views to which he referred. Precisely what views did he have in mind? In later years, Wright expressed strong sympathy for the philosophy of Whitehead. But earlier, when Wright was first moving into a career in science (in 1912), as for Julian Huxley the key influence was the French metaphysician Henri Bergson and his *Creative Evolution* (Wright 1953, 14; 1964, 281). Bergson had argued that a blind, mechanistic evolutionism is inadequate and that a "vital impulse" or *élan vital* somehow gives a sense of order to inert matter. It was this impulse that leads to the emergence of organic complexity: "an *original impetus* of life, passing from one generation of germs to the following generations of germs through the developed organisms which bridge the interval between the generations" (Bergson 1911, 92).

Even if we worked only from Bergson, we would be on the way to understanding Wright and the kind of evolutionism he favored. For instance, one of the main reasons Wright proposed and supported genetic drift was to explain the origin of new features. It was just this problem that Bergson highlighted as insoluble on the basis of pure Darwinism and for which he invoked the *élan vital.* Again, in the matter of Wright's fuzzy attitude to the nature and aim of progress, we find that his views are a reflection of Bergsonian thinking. The Frenchman certainly saw an upward trend to evolution and wanted to give humans a favored status, but he had to admit that although we might be thought of as "the reason for the existence of the entire organization of life on our planet," one must add the qualification that this "would be only a manner of speaking" (Bergson 1911, 195).[1]

Bergson apart, another significant influence on Wright was a reading of Karl Pearson's *Grammar of Science,* which led to a life-long conviction that truly only minds are real and that material objects are therefore one set of minds as they appear to others. In fact, as Wright realized many years later, this doctrine—"panpsychic monism"—was not held by Pearson himself but by his friend, W. K. Clifford (1879). But whatever the true origin, it fitted nicely into Wright's metaphysics and explained to him how minds (which are of a form logically different from matter) could evolve. It is not so much a question of something new appearing as of the already-present making itself manifest to other minds. And once again we see in Wright's thinking belief in a particular kind of Progress, backing progress of a Wrightian kind. He supposed less a qualitative change than a quantitative change through evolutionary history; it is, however, a change which can be all-embracing, going beyond the merely human.

There were other formative factors, including a youthful enthusiasm for Osborn: an avid reading of the *Age of Mammals* (1910) and other works as they appeared in print. Little wonder that even in Wright's seminal paper we find talk of "adaptive radiation" and "parallel orthogenetic lines," albeit with a Mendelian background alien to Osborn. Yet, there is still more, and to find it let us ask a simple question about the name of Wright's theory, something which often puzzles people. Why "balance"? As we have seen, balance supposedly occurs at all levels: "gene, chromosome, cell, individual, local race" (Wright [1931] 1986, 158). It is a balance between forces leading to uniformity (homogeneity) and forces leading to diversity (heterogeneity). Moreover—a point which one must stress—this notion of balance was absolutely crucial to Wright. Even though his theory became more selectionist through time, he always maintained that it was unchanged: as it was, with respect to balance. Conversely, Wright never had much time for any thinkers who wanted to explain everything in terms of drift. His objection came not so much from Wright's being an ardent selectionist as from his fear that balance would be lost.

With the talk of the "balance" between "homogeneity" and "heterogeneity" one starts to think of our old friend Herbert Spencer. Even if we gloss over the very great debt owed by the young Bergson to Spencer, at the direct level such thoughts are not misdirected. Sewall's brother Quincy, just then writing his Ph.D. thesis on international law, knew all about Spencer, and we find the Wright brothers corresponding (briefly but knowledgeably) about Spencer (letter from Sewall to Quincy, December 14, 1915; Q. Wright Papers)—apparently Sewall had a picture of Spencer (as well as Darwin) on his wall. At the very least, Wright would have been exposed to such ideas around the family dinner table and at college, for his father (and teacher), Philip Wright, was an enthusiast for sociology and economics and a radical thinker drawn to some form of Progressionism: "For hundreds of thousands of years since men have been on the planet there has been war and violence and cruelty, yet, miraculously, as the water lily grows out of the mud individual tenderness has blossomed here and there and a growing vision of the brotherhood of man. The eternal creative forces will continue with the persistency of sunlight after this tempest of madness [World War I]" (letter from Philip to Quincy Wright, April 21, 1916, Q. Wright Papers; a letter of January 4, 1923, denies a belief in a guiding Providence).

This selfsame father moved to the faculty of Harvard, to teach economics, and remained there the whole time that his son was a graduate student. Sewall himself used to go to talks on economics. Moreover, the move from the socio-economic to the biological would have been so natural for Wright as to have been unremarkable, given that his biology teacher, Mrs. Key—who later went to work full-time for the eugenics movement—was a graduate of the North American home of neo-Spencerian progressivism and had been supervised by the chair of the biology department, Charles Otis Whitman, an explicit enthusiast for Spencer's equilibrium thinking (Pauly 1988, 131).

Wright could not have been unmoved. Not that we should suppose that a young evolutionist in America around 1915 would be so very exceptional in such inclinations. Think back to the beginning of this chapter, to someone like Conklin, for an example. Conklin was explicit not only in his P/progressivism but in his belief that biological advance is a kind of movement up nature's hierarchies, a result of a constant striving for balance between opposing forces. (This view had at least a resemblance to Samuel Alexander's emergent evolution—see Alexander 1920—another, somewhat later Wright enthusiasm.)

L. J. Henderson

We can forge the link to a neo-Spencerian P/progressivism more strongly yet. Primed as he was, Wright was ripe for the teaching of his chemistry professor at Harvard, the archetypal proponent of dynamic equilibrium thinking, L. J. Henderson (Parascandola 1971). In class Wright got all of the ideas of Henderson's classic book, *The Fitness of the Environment,* and of related writings (Henderson 1913). There were organismic analogies, likening groups and societies to individuals; there were analogies and metaphors about balance and regulation; there was much talk about how everything (inorganic as well as organic) must be adapted to everything else; and there was an overall teleology leading nature upward, through hierarchies, as a dynamic consequence of balancing mechanisms at lower levels. There was also lots of praise for the Synthetic Philosopher. "Spencer's belief in the tendency toward dynamic equilibrium in all things is of course fully justified" (Henderson 1917, 138).

And then, if this were not enough, everything was reinforced by a reading of *The Origin and Nature of Life,* by Benjamin Moore, professor of biochemistry at the University of Liverpool:

> It may then be summed up as a general law universal in its application to all matter, although varying in intensity in different types of matter, and holding throughout all space as generally as the law of gravitation—a law which might be called the *Law of Complexity*—that matter so far as its energy environment will permit tends to assume more and more complex forms in labile equilibrium. Atoms, molecules, colloids, and living organisms, arise as the result of the operations of this law, and in the higher regions of complexity it induces organic evolution and all of the many thousands of living forms. At still higher levels, it forms the basis of social evolution and leads to that intellectual development in individuals and community which surmounts the whole and is ever building upwards. (Moore 1913, 188–189)

Wright always acknowledged these influences—"I was very much impressed with Henderson's ideas" (interview with William B. Provine, June 4, 1976; Wright Papers)—and they are confirmed by letters written, at this time, to his brother Quincy. There was the charisma of Henderson: "I found him a very stimulating lecturer and got a lot of ideas from him, 'condition of dynamic equilibrium' etc." (letter of January 10, 1916; Q. Wright Papers). There was the organismic analogy: "Thus the body is not an absolute monarchy in which the bulk of the cells are mere mechanisms, directed in every action by a central unit. It is a democracy or perhaps better is limited monarchy. In the main each part knows what to do and does it of its own accord, as occasion arises. Regulation from outside comes rather from suggestions from numerous peers, not in a single command from above" (letter of December 14, 1915; Q. Wright Papers). And, above all, there was a hierarchical view of nature, explicitly linking equilibrium and evolution:

> My original idea was to classify all sciences by the unit of organization—electron, atom, animal etc.—with which they deal subdividing on a fourfold basis—
> A. Condition in equilibrium
> 1. Description of organization
> 2. Mechanism of maintenance of equil.
> B. Change of equilibrium (Evolution)
> 3. Description of changes (history)
> 4. Mechanism of change

However, Wright was not happy with his first efforts. He therefore proposed various hierarchical classifications—as, incidentally, he was to continue to do through his long life—but never committed to a single one. The following passage, which again reinforces the equilibrium/evolution link, spells out the problem:

> The difficulty of classification is well illustrated by my own science, genetics—from one point of view it deals with the organization of the cell and has very close relations with cytology, then it deals with the mechanism of individual development—the mode in which developmental factors are represented in the one cell stage,—and finally it deals with both the maintenance of equilibrium in the species (heredity) but also the mechanisms of change in this equilibrium (variation by recombination of factors and otherwise). (Letter of February 27, 1916; Q. Wright Papers)

This all fits with an earlier remarkable letter, pregnant with anticipation, in which Wright expounded on his beliefs about balance in societies and the way forward:

> Darwinists would hold that the most rapid evolution would follow from a happy mean between conditions which permit the existence of a wide range of variations,—many of them more or less injurious—which can recombine in all possible ways—and conditions which tend to eliminate the more unfit. To use a human analogy, we do not expect civilization to advance most rapidly either in the arctic zone where existence depends on following one very definite mode of life or in the tropics where conditions of life are too easy . . . The greatest progress should result in a society which is neither crystallized into a caste system nor so fluid that individuals of a family, which has produced favorable variations and done much for progress in the past, receive no advantage over inferior families. The problem of statesmanship is to adjust laws so that there is just the degree of viscosity in all respects which gives the maximum progress. It is a problem of maxima and minima and therefore much more difficult than progress toward an absolute democratic or absolute aristocratic ideal. (Letter of October 17, 1915; Q. Wright Papers)

This is not yet Wright's shifting balance theory of biological evolution; but, with all of the emphasis on achieving a happy mean between opposing ends, it is hard to imagine ground better prepared.

We can see now why Wright simply had to be a P/progressionist, for his philosophy was P/progressionist through and through. Indeed, there are passages in the writings of those influencing Wright which seem almost to have been translated literally into the language of population genetics:

> It is only necessary for the atomic basis to our chemistry to realize that the atom, just like the chemical molecule at the different stage, or the fixed organic species of the biologist, is a point of the stable equilibrium in upward evolution. Between each two such points there lies a region of unstable equilibrium, and as matter becomes more charged with energy, surgings and transformations occur, and in the greater number of cases when the cycle is complete, the matter drops back again to its stable point. But occasionally when a supply of energy at high-potential, or concentration, is available, there is a huge wave of uplifting which carries the matter involved over a hill crest into a higher hollow of stable equilibrium, and a new type of matter becomes evolved at the expense of kinetic energy passing over into latent energy or potentia. (Moore 1913, 40)

Yet, at the same time, these excerpts from Wright's correspondence indicate a major part of the reason why the writings of the public Wright are so slippery when it comes to notions of progress. For him, progress (and Progress) did not really involve a simple march up the Chain of Being to humankind. It was, rather, a question of increase in complexity and integration at higher levels up the hierarchy. Remember how often we have seen Wright gliding from an individual-type thinking to a group-type thinking. And, as one gets the movement, one gets a fresh level of mental awareness.

> The greatest difficulty is in appreciating the possibility of the integration of many largely isolated minds into a higher unitary field of consciousness such as must necessarily occur under this viewpoint in the organism in relation to its cells; in these in relation to their molecules and in these in relation to their molecules and these in relation to more ultimate entities. The observable hierarchy of physical organization must be the external aspect of a hierarchy of mind. (Letter to J. T. McNeill, November 12, 1943; Wright Papers)

The conclusion, therefore, seems to be that the whole world will eventually be integrated into one universal mind. Hence, everything—all of nature—is involved in P/progress. Humans, and especially human societies, are obviously important forerunners of this univeral mind, and they will no doubt play a significant role in the final outcome. But the implication does seem to be that we lack any ontologically unique status. There are potential minds at levels higher than any of us.

Adaptive Landscapes and the Hard Evidence

After Darwin's natural selection, Wright's landscape metaphor proved the most powerful in the history of evolutionary thought. Its success was well earned. Whatever one may think personally about dynamic equilibrium ideas, one must acknowledge Wright's genius in meshing his underlying social philosophy and metaphysics with his theoretical knowledge of practical agriculture. Nothing said in the last section challenges Wright's own recollections of his route to the shifting balance theory—but he was juggling (balancing!) far more factors and concerns than hitherto acknowledged.

Yet, turning finally and briefly to the evidence, again we find confirmation of the overall picture being sketched. Wright's conceptual innovations are not all that they seem—or, rather, they are somewhat more than they seem. Despite all of Wright's interests in gene interaction, his formal models could deal only with alleles at one locus (Mayr 1959). There was no mathematical guarantee that traits controlled by multi-locus genes could evolve by genetic drift. And if it be objected that the guinea-pig story shows that drift can actually work when several loci are involved, it can be counter-objected that in nature such novel combinations might be destroyed by selection. And these objections are quite apart from the fact that Wright had no knowledge that populations in nature are ever sufficiently fragmented for drift, or that migration might not swamp such fragmentation.

Problems multiply. Even if the model works perfectly up to this point, who then is to say that new combinations would not be broken down as soon as they re-entered the general population? By this stage of the argument, it hardly needs to be added that conceptually there is something very strange about the adaptive landscape model itself—not that this should be a matter of any great surprise, given how much information is being crammed together and simplified to get it to work. If nothing else, the landscape is hardly the continuous surface that most take it to be (Wright 1982; Provine 1986). It is rather a series of points, representing individual hypothetical genotypes, crammed together. But can one put such information side by side in this way? Is a one-gene-at-a-time change going to make for a smooth trip up or down a fitness surface? Might not there be radical discontinuities? Adding to confusion, sometimes Wright spoke of the landscape as a two-dimensional picture of relative gene ratios—pointing, at a minimum, to ambiguities in his thought between individuals and groups.

Wright, especially the Wright of 1931 and 1932, was doing more than simply extrapolating from his knowledge of agriculture. Something else, something deeper, was driving him on. Some of Wright's concerns were straightforwardly epistemic. He wanted to produce a theory with which people could work—one which laid the way open for prediction and control and so forth. But he also wanted more, and it is in his larger concerns that we find room for his cultural influences. Just as Fisher's theorizing was quintessentially British in its reflection of a Christianity which makes central a design-based natural theology, so Wright's theorizing was quintessentially American in its reflection of an ordered science-cum-philosophy which makes central an ever-dynamic inorganic/biological/social equilibrium. Fisher and Wright shared debts both to Darwin and to Spencer but, whereas ultimately Fisher in his heart was the purest of Darwinians, Wright in his heart was the purest of Spencerians.

Wright will prove as interesting on the professionalism front as he is on the progressionism issue. But now, I want to turn to the man who took Wright's ideas and made them into a compelling evolutionary theory.

Genetics and the Origin of Species

Theodosius Gregorievitch Dobzhansky (1900–1975) was born and educated in Russia. The early 1920s found him teaching at the Polytechnic Institute of Kiev, collecting and studying ladybird beetles *(Coccinellidae)*. The little fruit fly *Drosophila* was the hot organism, however. Dobzhansky moved to Petrograd (immediately to be named Leningrad) to work on its biology, under leadership of one of his country's leading geneticists, Iurii Filipchenko. In 1927 the opportunity came to study in Morgan's lab, and naturally Dobzhansky jumped at the chance. He was never to return to his homeland. (The key source for Dobzhansky's early life is his "oral memoir" (1962b), deposited at Columbia University. See also Lewontin, Moore, Provine, and Wallace 1981.)

Dobzhansky learned quickly. Uniquely, he was able to bring together his Russian background and his newly acquired American knowledge of genetics and cytology. This synthesis lay behind the lectures on evolutionary theory he delivered at Columbia University in the fall of 1936, published as a book the next year. And with this, Dobzhansky's reputation was made, for *Genetics and the Origin of Species* has fair claim to being the most influential book on evolution since Darwin's great work. Not that Dobzhansky stopped here. For the next forty years he contributed mightily to our empirical understanding of the evolutionary process. Keenly aware always of his own mathematical limitations, Dobzhansky carefully allied himself with those possessing talents he lacked. A most fruitful collaboration in the early years was with Sewall Wright, who provided full technical analysis of Dobzhansky's findings (Dobzhansky, Wright, and Hovanitz 1942; Dobzhansky and Wright 1946). Much other work was produced with his many students, relations with whom tended to be intense in a very non-Anglo-Saxon manner. With good reason, Dobzhansky spoke of them as his "children" (diary, September 19, 1969; Dobzhansky Papers). And they responded in kind—with deep affection, with protectiveness, and with occasional irritation at not being allowed full independence, not to mention a good measure of sibling rivalry: "I hate him, I hate him, I hate him, . . . and I don't care who knows."

We shall start with the first edition of *Genetics and the Origin of Species* and move then briefly to cover later developments. For Dobzhansky, always, the key to the very largest evolutionary changes lay in the small details: "Experience seems to show, . . . that there is no way toward an understanding of the mechanisms of macro-evolutionary changes, which require time on a geological scale, other than through a full comprehension of the micro-evolutionary processes observable within the span of a human lifetime and often controlled by man's will" (Dobzhansky 1937, 11–12). Focusing therefore on the micro-changes of evolution, and borrowing a metaphor from the physical sciences, Dobzhansky invited us to consider two phases to the problem before us.

The first half of *Genetics and the Origin of Species* covered what Dobzhansky called the "statics" of evolution. Here was discussion of the factors lying beneath evolution, particularly those pertaining to variation and to its causes. A major place was given to work on which Dobzhansky had collaborated, experiments that used chromosomal changes to infer the phylogenies of widespread and distinct races of *Drosophila pseudoobscura*. In the second half of *Genetics and the Origin of Species*, the emphasis was on evolutionary "dynamics," the various factors interacting and causing biological change. For Dobzhansky, it was ever axiomatic that our knowledge of gene mutation and of chromosomal variation is not some academic, laboratory-justified theory. It is a reflection of nature's reality. Populations in the wild contain masses of variation. The notion of an "ideal type" of any group (like a species) is chimerical, because every individual is different: "Despite their external uniformity, the free living populations of Drosophila carry a great mass of chromosomal variations as well as of recessive mutant genes." Fortunately, although many of these variations and mutants apparently bode ill for their possessors, "the species is, according to the succinct metaphor of Tschetwerikoff (1926), 'like a sponge,' absorbing both the mutations and the chromosomal changes that arise and gradually accumulating a great store of variability, mostly concealed in heterozygous condition" (Dobzhansky 1937, 125–126).

This fact of variation raises two questions. First, a question which Dobzhansky himself admitted is only partially scientific: if there is so much variation, and much of it is of a kind which seems positively deleterious (producing ill effects in the homozygous state), then is this not a major danger for each and every species, especially our own? Apparently not: "prophets of doom seem to be unaware of the fact that wild species in the state of nature fare in this respect no better than man does with all the artificiality of his surroundings, and yet life has not come to an end on this planet" (Dobzhansky 1937, 126).

But if the variation is not all that dangerous, can we turn the question around and, at the risk of sounding somewhat teleological, ask whether the variation has some positive value? Apparently so:

> Looked at from another angle, the accumulation of germinal changes in the population genotypes is, in the long run, a necessity if the species is to preserve its evolutionary plasticity. The process of adaptation can be understood only as a continuous series of conflicts between the organism and its environment . . . But nature has not been kind enough to endow the organism with ability to react purposefully to the needs of the changing environment by producing only beneficial mutations where and when needed. Mutations are random changes. Hence the necessity for the species to possess at all times a store of concealed, potential, variability. (Dobzhansky 1937, 126–127)

The second question centers on the way(s) natural variation is kept in equilibrium. Why does the variation, originating with mutation, persist in populations? One might think that the time had come now to turn for help to natural selection, and indeed Dobzhansky did launch into a detailed discussion of selection. But, for all that his book's title echoed Darwin's, one must not assume that the rest of the book does, too: "Some modern biologists seem to believe that the word 'adaptation' has teleological connotations, and should therefore be expunged from the scientific lexicon. With this we must emphatically disagree. That adaptations exist is so evident as to be almost a truism" (Dobzhansky 1937, 150). However: "Whether the theory of natural selection explains not only adaptation but evolution as well is quite another matter" (p. 150).

For Dobzhansky—that is to say, the Dobzhansky of the first edition of *Genetics and the Origin of Species*—aside from its somewhat peripheral adaptation-producing role, selection was primarily a negative force, threatening (and making necessary) variation rather than promoting it. Although he admitted that selection did operate, Dobzhansky stated flatly that: "The inadequacy of the experimental foundations of the theory of natural selection must be admitted, I believe, by its followers as well as by its opponents" (p. 176). But, if not pure selection, then what? Here, Wright's shifting balance theory came to the rescue. Dobzhansky had encountered Wright's paper at the congress in 1932, and in his own words "fell in love with" Wright immediately (Provine 1981, 56). Consummating the passion, Dobzhansky put the balance theory up front in *Genetics and the Origin of Species* and used it to generate and justify natural variation:

> If the entire field of possible gene combinations is graded with respect to the adaptive value, we may find numerous "adaptive peaks" separated by "valleys." The evolutionary possibilities are twofold. First, a change in the environment may make the old genotypes less fit than they were before . . . The second type of evolution is for a species to find its way from one of the adaptive peaks to the others in the available field, which may be conceived as remaining relatively constant in its general relief. (P. 187)

Taken at face value, Dobzhansky's description of the adaptive landscape was a little bit misleading. Dobzhansky's problem was not really Wright's problem. For all his doubts, Wright always thought selection a significant causal factor. Wright invoked the adaptive landscape, together with the idea of drift, to isolate populations from each other, make new gene complexes (with corresponding new organic features), and then bring them back into the general group. Dobzhansky, much more ready to discount adaptive underpinnings, used the landscape to isolate populations, thereby creating masses of new variation (through drift), and thus feeding his general view of species.

> The present writer is impressed by the fact that this scheme is best able to explain the old and familiar observation that races and species frequently differ in characteristics to which it is very hard to ascribe any adaptive value. Since in a semi-isolated colony of a species the fixation or loss of genes is to a certain degree independent of their adaptive values (owing to the restriction of the population size), a colony may become different from others simultaneously in several characters. (Dobzhansky 1937, 190–191)

After all of this, the discussion of the ostensible topic of the book, speciation, was somewhat anti-climactic. Most significantly, to the end Dobzhansky continued to regard selection as somewhat of a bit player in the evolutionary drama.

Taking his text as programmatic rather than definitive, Dobzhansky and collaborators plunged into the empirical testing and analysis of the mechanisms of change, both as they occur in nature and then as they can be replicated in the laboratory. From this effort stemmed an ongoing series, *The Genetics of Natural Populations* (Lewontin et al. 1981), as Dobzhansky was forced into a substantial revision of his position. Hitherto paradigmatic examples of drift were now seen to be instances of tight control by selection. Dobzhansky was able to reproduce his results artificially, showing that the successes of strains are undoubtedly linked to such ecological factors as food supplies and temperature (Dobzhansky [1943] 1981; Dobzhansky and Spassky [1944] 1981; Wright and Dobzhansky 1946). These findings, together with continual adaptationist pressure from England, led Dobzhansky (and other American evolutionists) to start downplaying the significance of drift and to stress that even apparently non-adaptive features of organisms may have selective virtues. Thus occurred what Stephen J. Gould has referred to as the "hardening" of the synthesis (on evolution), as adaptationism rose in strength (Gould 1983).

Converted to selectionism, Dobzhansky had now to search for causes other than drift for intra- and inter-populational variation. Selection works on the variation, but what maintains the variation? In particular, what guarantees does one have that selection will not simply wipe out a whole set of variations? Pushed into looking for ways in which selection itself could do the work of maintaining variation in populations,

Theodosius Dobzhansky

Dobzhansky's "balance" hypothesis, shown by pairs of corresponding chromosomes of fellow species members. Any particular locus, with few exceptions (*D* in this example), is filled by a range of alternate alleles.

$$\frac{A_3\ B_2\ C_2\ D\ E_5\ \ldots\ Z_2}{A_1\ B_7\ C_2\ D\ E_2\ \ldots\ Z_3} \qquad \frac{A_2\ B_4\ C_1\ D\ E_2\ \ldots\ Z_1}{A_3\ B_5\ C_2\ D\ E_3\ \ldots\ Z_1}$$

Dobzhansky turned to superior heterozygote fitness (heterosis) for an answer. Thus, as we move from the 1940s and into the 1950s, he looked for evidence of heterosis and came to favor a suggestion by his friend, I. M. Lerner (1954), that heterozygotes will possess virtues of both homozygotes and thus will be fitter. Hence, heterozygosity "provides a mechanism for maintaining genetic reserves and potential plasticity" for the evolution of populations and species, and also "it permits a large proportion of individuals to exhibit combinations of phenotypic properties near the optimum" (Dobzhansky 1955, 797).

Thus, Dobzhansky formulated what he called his "balance" theory of evolution, the scientific position of his maturity: Any gene locus in a population (generally) has a whole range of different alleles which can occupy it. These alleles are held in place by various mechanisms, prominent among which is superior heterozygote fitness. There will be fluctuations in gene ratios over time, but the crucial point is that the variability has no innate tendency to disappear. As new selective demands are made, by nature, appropriate variation is generally there for the using. Extinction may be the ultimate fate of any line, no matter how successful; but, in the meantime, there is never need to wait passively for just the right random mutation. Capacity for response is a fact of biological life (Lewontin 1974; Beatty 1987).

Dobzhansky and Biological Progress

Turning now to the culture questions, we begin by asking about Dobzhansky and biological progress. Although there is a little bit on macro-evolution at the end of the third edition of *Genetics and the Origin of Species,* Dobzhansky as a populational geneticist was concerned primarily with micro-evolution. Hence, questions of biological progress tended not to get raised overtly in Dobzhansky's strictly scientific writings. There is nothing particularly progressionist about the chromosomal phylogenies he traced. But, this said, in personal conversation and correspondence Dobzhansky was quite open in his belief in such progress.

> I refuse to abstain from talking about progress, improvement, and creativity. Why should I? Some extreme "scientists" would eliminate expressions such as that the eye is built so that the animal can see things. Perhaps extremes converge, and you would join these mechanistic purists. In evolution some organisms progressed and improved and stayed alive, others failed to do so and became extinct. Some adaptations are better than others—for the organisms having them; they are better for survival rather than for death. Yes, life is a value and a success, death is valueless and a failure. So, some evolutionary changes are better than others. (Letter to John Greene, November 23, 1961; Dobzhansky Papers)

Moreover, in the last twenty years of his life, Dobzhansky wrote several works on and around our own species in which he made it unambiguously clear that he saw our own species standing on the highest "pinnacle" (his term). The first words of *Mankind Evolving* are: "The universe, inanimate as well as animate matter, human bodily frame as well as man's psyche, the structure of human societies and man's ideas—all have had a history and all are in the process of change at present. Moreover, the changes so far have been on the whole, though not always, progressive, tending toward what we men regard as betterment" (Dobzhansky 1962a, 1). Elsewhere, analogously, he stated that: "Judged by any reasonable criteria, man represents the highest, most progressive, and most successful product of organic evolution. The really strange thing is that so obvious an appraisal has been over and over again challenged by some biologists" (Dobzhansky 1956, 86).

What was Dobzhansky's criterion of progress and how did he think that evolution had been so powerfully progressive? He presumed some fairly conventional attributes, like intelligence and so forth. Specifically for humans, it seems better to be tall and handsome, like Cro-Magnon man, rather than "short and squat" with large teeth, like the Neanderthals (Dobzhansky 1962a, 180). But, beyond all else, for Dobzhansky the key progressive attribute, that which we humans possess in abundance, is flexibility or *adaptability.*

> Less than two thousand years ago, the ancestors of most modern Americans and Europeans were barbarians eking out a rough and precarious existence in the forests and swamps of northern Europe. But these barbarians responded magnificently when they were given an opportunity to borrow foreign cultures developed by the rather different peoples who inhabited the lands around the eastern part of the Mediterranean Sea. Successful cultural reconditioning can now be observed any day in the large universities in such centers as New York, London, Paris, or Moscow, which attract students from all over the world. (Dobzhansky 1956, 47)

Adaptability is a function of the genes reaching up into culture: "The transition from the adaptive zone of a pre-human primate to the human adaptive zone was brought about by the development of the biological basis for the ability to use symbolic thought, language, to profit by experience, to learn, in short by the development of educability" (Dobzhansky 1956, 121).

Dobzhansky saw adaptability as being crucial at the group level as well as that of the individual. Indeed, by the third edition of *Genetics and the Origin of Species,* he was speaking of a species as "not merely a group and a category of classification. It is also a supra-individual biological entity, which, in principle, can be arrived at regardless of the possession of common morphological characteristics" (Dobzhansky 1953, 6). The species that is richly endowed with genetic variability, much no doubt concealed, stands ready to respond to nature's demands in a way that a genetically uniform group would not. Fortunately, such variability is the norm (this is the implication of the balance hypothesis, as it was earlier of the more drift-inspired position), and in all biological respects humans fit the pattern more than adequately.

Does not this range of variation detract from human perfection? Does it not mean that we have a "genetic load," referring now to the deleterious mutants that we will all carry, even if usually recessively? For all of the cavalier attitude ("prophets of doom") that Dobzhansky expressed in the first edition of *Genetics and the Origin of Species,* he worried about the point a great deal. At least, he wrote on it incessantly. "If the fitness of a species depends to any appreciable extent on the presence of heterotic gene alleles . . . there must be a source of supply of new alleles to replace those that become lost by chance or otherwise from the gene pool. Mutation is, then, not only the price for evolutionary plasticity; it is also the tax levied in order to preserve the status quo" (Dobzhansky and Wallace 1959, 165).

Luckily, there are some mitigating factors. First, not all variability, even that which is bad from a biological viewpoint, is necessarily socially or humanly bad. Slightly reduced fertility might be no dreadful thing. Second, genetic variation usually leads to heterozygosity, which may well involve increased fitness and well-being. Third, one must never confuse biological desiderata with moral worth. Human genetic variability may be the price we pay for success, but—far from despair—this could be the foundation of a rich and well-functioning society: "It would be naive to claim that the discovery of this biological uniqueness constitutes a scientific proof of every person's existential singularity, but this view is at least consistent with the fact of biological singularity" (Dobzhansky 1962a, 219). Putting matters another way, one might say that a society where some of us are hewers of wood and drawers of water is better than one in which we are all fruit-fly geneticists.

How has progress come about? The notion of adaptability, particularly at the group level, was the very heart of Dobzhansky's evolutionary thought. But this hardly explains why the process of evolution should have produced humans. Even after he became more selectionist, Wright's shifting balance theory was at the center of Dobzhansky's evolutionizing. Hence, clearly inasmuch as the theory could be said to support progress of various kinds—and Wright certainly thought it could—Dobzhansky had a mechanism there. But other than the implicit support from Wright, the general assumption seemed to be that evolution is a kind of cumulative process and that therefore "obviously" there will be a rise in degree of the most important attributes. As Dobzhansky wrote, tangentially: "Organic diversity arose in evolution because life had to continuously produce new forms able to cope with the progressively more complex and diversified environments" (Dobzhansky 1956, 70–71).

Not that Dobzhansky saw evolution as progressing necessarily and smoothly upward. Like everyone else, he was aware that the full picture is much more complex: "evolution is primarily a groping and only secondarily a directional process; evolution is not infrequently regressive rather than progressive" (Dobzhansky 1967, 36). The point is, however, that humans did evolve and they are better than anything else. Indeed, Dobzhansky was prepared to go so far as to say that the evolution of humans represents a "quantum" leap above other organisms, analogous to the degree that such other organisms are above the inorganic: "Inorganic, organic, and human evolutions occur in different dimensions, or on different levels, of the evolutionary development of the universe" (Dobzhansky 1967, 44). In support of his position here, Dobzhansky appealed to the hierarchical thinking of the dialectical materialists—although, to cover himself, he appealed also to the ideas of the American theologian Paul Tillich!

The Biology of Ultimate Concern

The biggest mistake that one could make in thinking about Dobzhansky would be to regard him merely as an American with a funny accent, a kind of Slavonic Spencerian. His American identity, for which he was deeply grateful, was a veneer on his first identity, whose core was as Russian as that of his (distant) cousin Dostoevsky. And these earliest influences extended to Dobzhansky's science. Although the theoretical superstructure of *Genetics and the Origin of Species* is Wright's balance theory, as soon as one looks at the references one realizes that the underlying empirical basis was in major part Russian. (Remember the reference to "Tschetwerikoff," the Russian population geneticist Sergie Chetverikov.) This is significant because, at the macrolevel, Russian evolutionism was a tradition which started shortly after the publication of Darwin's *Origin,* and—like so much Russian science, being deeply grounded in Germanic ideas—the tradition rather downplayed Darwinian mechanisms and made (embryologic-like) progressionism central.

This was a tradition which flourished when Dobzhansky was a young man and which survived the Revolution and into the first years of the Marxist era. Pertinently, it was Dobzhansky's tradition for, moving on from a youthful enthusiasm for Bergson, he was by the Leningrad years entirely captivated by the speculations of L. S. Berg, Russian author of one of the most famous of non-Darwinian evolutionary tracts, *Nomogenesis* ([1926] 1969). Berg put forth a version of the biogenetic law, that "the laws of the organic world are the same, *whether we are dealing with the development of an individual (ontogeny) or that of a palaeotological series (phylogeny)*. Neither in the one nor in the other is there room for chance" (Berg [1926] 1969, 134; his italics). Berg underlined his pre-formation progressionism by making H. F. Osborn—a writer for whom Dobzhansky later expressed admiration—his second-most referenced author after Darwin. There is hardly any surprise that Berg saw the evolutionary process as culminating in our own species.

Dobzhansky broke from Berg's influence but, just as his cool attitude toward selection would have had Russian roots, biological progressionism for Dobzhansky simply came packaged with macro-evolution. He had little doubt of the eventual triumph of humankind because his youthful mind was set by embryological pre-formism. This is not to explain fully progressionism's persistence in Dobzhansky's thought. Nor is it to explain the particular form that it took in his thinking. It is to say that the presumption is that Dobzhansky would be a progressionist—as he was.

What then of cultural Progress? With respect to scientific Progress, Dobzhansky was quite explicit: "Science is cumulative knowledge. This makes scientific theories relatively impermanent, especially during the epochs when knowledge piles up in something like geometric progression" (Dobzhansky 1962a, xii). Moreover, the optimism seemed to extend to society, for Dobzhansky thought that we face challenges and that they are in principle soluble. Most particularly, we must confront our biological and cultural heritages and acknowledge the fact that ultimately they are one: "To deal with this challenge successfully, knowledge and understanding of evolution in general, and of the unique aspects of human evolution in particular, are essential" (Dobzhansky 1962a, 22).

In this context, where did Dobzhansky stand on the thorny question of eugenics? The answer is mixed. In general, Dobzhansky thought biological variation a good thing. Society needs different types, both to function properly now and to hold in reserve the means with which to meet any new challenges it may face. Hence, he adamantly opposed proposals for positive eugenics, aimed at producing a race of supermen. Nevertheless, Dobzhansky certainly did not think that all of the variants that might reside in the human gene pool were that desirable. Indeed, at times he was even given to some Fisher-like speculations: "In technologically advanced societies the business of propagation seems to be entrusted largely to people with mediocre to inferior qualifications for parenthood." Although some of the direst predictions are unwarranted, beware of complacency: "Genetic consequences cannot, however, be ignored . . . It cannot be gainsaid that there is a predicament here which should cause concern" (Dobzhansky 1962a, 312–313).

Move on to the interaction between biology and culture. Here, we must turn the master key to Dobzhansky's life and thought—by his own explicit admission and confirmed by repeated writings. This is *religion,* especially in the broad sense of concern about humans and their spiritual destiny: "I would say that of all the scientists I have known and admired, Doby came closest to being a really religious man" (G. L. Stebbins, interview with author, May 25, 1988). It was this faith which attracted Dobzhansky to evolutionism and which drove him on through all of his work. Fruit flies were just a step toward understanding our own species and our place in the divine scheme.

> It is hardly surprising that both during the pre-Revolutionary days in Russia and during the post-Revolutionary days, to biologists [the] philosophical-humanistic implications of evolutionism were in the center of attention.
>
> I think it is not an exaggeration to say that probably this interest is what made me, if not a biologist, at least an evolutionist. (Dobzhansky 1962b, 56, 351–352)

Dobzhansky's own background was Christian, and he was prepared to make an explicit avowal of faith (letter to Sophie Dobzhansky, May 16, 1957; Dobzhansky Papers), even though he preferred not to be too dogmatic: "Modern man . . . needs nothing less than a religious synthesis. This synthesis cannot be simply a revival of any one of the existing religions, and it need not be a new religion. The synthesis may be grounded in one of the world's great religions, or in all of them together"

(Dobzhansky 1967, 109). The important thing is that this synthesis must be, in essence, evolutionary. Fortunately: "Christianity is basically evolutionistic. It affirms that the meaning of history lies in the progression from Creation, through Redemption, to the City of God" (Dobzhansky 1967, 112). And, apparently, we humans are part of the process: "the world is not a 'devils vaudeville' (Dostoevsky's words), but is meaningful. Evolution (cosmic and biological and human) is going towards something, we hope some City of God" (letter to J. Greene, November 23, 1961; Dobzhansky Papers).

If we grant that Christianity is in some sense developmental or evolutionary, do we not now run afoul of the distinction between Progress and Providence? Is not Christianity, strictly speaking, Providential, meaning that the course of history is in God's hands and not ours? There are certainly hints that Dobzhansky could see this implication, although he realized that accepting it raises other problems: "I see no escape from thinking that God acts not in fits of miraculous interventions, but in all significant and insignificant spectacular and humdrum events. Pantheism, you may say? I do not think so, but if so then there is this much truth in pantheism" (letter to J. Greene, November 23, 1961; Dobzhansky Papers). Undoubtedly, it was through his pondering on such issues as these that Dobzhansky was led to strong sympathy, a sympathy which shocked his more orthodox scientific friends, for the thinking of Teilhard de Chardin: "Perhaps Teilhard had a hint, very obscurely expressed" (letter to J. Greene, November 23, 1961; Dobzhansky Papers). Possibly the early enthusiasm for Berg's directed evolutionism prepared the way here.

Yet, even if we accept that the distinction between Progress and Providence need not be clear-cut, we have still not yet dug down to the heart of Dobzhansky's Progressionism—to see, in particular, why he so prized adaptability both at the cultural and at the biological level. More than for any other figure in this book, it is the personality of the man that is crucial. Absolutely central to Dobzhansky's very being was the classical problem of reconciling a good God with the all-too-apparent fact of evil. In the Revolutionary period, his father had died of syphilis; his mother had choked on a piece of dry, ersatz bread; and he had starved. "[These years] gave experiences which in the United States few people had and few understand. Starvation and undernourishment are more than physiological states, they leave longlasting psychological scars. So does insecurity and anxiety, inevitable in a police state" (unpublished autobiography, circa 1970, 2; Dobzhansky Papers).

Dobzhansky was filled with a terror of pain and of death. When his thyroid malfunctioned in the early 1950s, he was convinced that he was going to die, typically interpreting his ordeal in religious terms: "I am going through strong spiritual crisis, and hope that its results will be beneficial. The starting point is, of course, a fear of death, and a fear that death may be close. I felt the need of an intimacy, and Natashe [his wife] and myself spent some time together, and we cried several times. I feel a terrific desire to love and all my soul rebels against death" (diary, February 16, 1952; Dobzhansky Papers). Combined with all of this was overwhelming guilt that he had escaped the Russian terror: "In the evening listened to the Beethoven's Ninth. Ode to Joy—what a great wisdom! What else but joy is the good in life. And we had our portion of joy, in fact far more than any one of the friends of our youth, although some of these were no less worthy than we. If we did not have more joy than we did, it was our own stupidity, for life has not refused us much which it refuses to so many. These Russian boys and girls who were our playmates and our friends in youth are mostly dead, or else leading a mere existence. What a luck, so far!" (diary, February 17, 1952; Dobzhansky Papers).

One effect of the insecurity of his early life was to make him appallingly overprotective of his daughter, his only child. Sensibly, she fled the nest to go to an out-of-town university and did not send home for parental examination the requested carbons of all her term papers. Another effect was to drive him to seek theological meaning to life's dreadful terrors and unfairnesses. Like Dostoevsky—no doubt following Dostoevsky in respects—Dobzhansky found the answer in the fact of freedom. Evil exists, but it is in part a function of human freedom and in part a challenge to freedom. The battle against evil is the task God has set us, and it is in this sense that Christianity incorporates hope of Progress, as we free beings try to conquer evil. "Life is a rather tragic business, and wisdom consists not in ignoring this fact but in accepting it and still going on with living" (letter to Sophie Dobzhansky, December 6, 1952; Dobzhansky Papers).

The existential significance of freedom was clearly illustrated for Dobzhansky in 1953, when he was faced with a job offer from Texas. At one level, the unsympathetic observer will see an indecisive man: "never before have I felt more keenly the limitations of one's intellect in making a rational decision of even one's own problems. How can one foresee the future? And without this, how can one know if a decision is right or wrong? May be when one will know it will be too late to change the decision. Oh, there is no 'may be' about it—some decisions are irreversible" (letter to L. C. Dunn, December 31, 1953; Dobzhansky Papers). But at another level—a very real level to Dobzhansky—is the cutting edge of human existence: "There has been some pleasure in this otherwise painful ordeal. This is the feeling that one is *free* to decide between two possible courses of action. The very fact that the merits of the case are not evident means that one can make a free choice" (letter to Sophie, November 15, 1953; Dobzhansky Papers).

At a more general level, Dobzhansky's religious concerns led to a crisis when they were juxtaposed against science: "Some sort of reconciliation or harmonization seemed necessary. The urgency of finding a meaning of life grew in the bloody tumult of the Russian Revolution, when life became most insecure and its sense least intelligible" (Dobzhansky 1967, 1). Fortunately, reconciliation came in a Progressivist reading of reality: in Dobzhansky's vision organic evolution is itself progressive, culminating in the organism with the greatest adaptability or freedom, humankind. This cannot be a hard-line Providentialist reading. "Any doctrine which regards evolution as predetermined or guided collides head-on with the ineluctable fact of the existence of evil" (Dobzhansky 1967, 120). Rather, it must be genuinely Progressivist, allowing in a Teilhardian sense for "the replacement of creation." Most particularly: "On the human level, freedom necessarily entails the ability to do evil as well as good. If we can do only the good, or act in only one way, we are not free. We are slaves of necessity. The evolution of the universe must be conceived as having been in some sense a struggle for a gradual emergence of freedom" (Dobzhansky 1967, 120).

We start to see now most fully why the supposition of natural variation, whether the product of drift or of selection, was so important for Dobzhansky. With the reserve genetic variation held at the ready, a population is maximally free to respond to nature's challenges. One might complain that this reasoning rather blurs the distinction between individual freedom and group freedom; but, apart from the fact that Dobzhansky came to think that heterosis promotes both individual excellence and freedom as well as group excellence and freedom, we know that Dobzhansky himself was rather inclined to blur the distinction between the individual and the group. At some ultimate level, organic adaptability and human adaptability, group adaptability and individual adaptability, blend into one freedom, as all of life is striving upward toward the final blissful state where we are all part of the divine supraorganism (Dobzhansky 1967, 131): "Evolution, human and biological and cosmic, is not simply a lot of whirl and flutter going nowhere in particular. It is, at least in its general trend, progressive . . . Man's individual life is a component part of the evolution of the universe; man's ultimate concern, and his individual meaning and dignity are atoms of the meaning of the whole cosmos" (Dobzhansky 1967, 116).

The Question of Evidence

Dobzhansky was a biological progressionist because this position harmonized with and led into his religious or spiritual Progressionism. His spiritual Progressionism depended crucially on a belief in freedom, and that in turn made biological freedom imperative. This latter came through the flexibility conferred by the genes, both at the individual and the group level. That much of this line of reasoning was without firm empirical support—leading to the conclusion that culture was indeed the dog wagging the tail of science—hardly needs further argument. Most centrally there is Dobzhansky's belief in the genetic variability of populations and in the power of either drift or selection to maintain and even promote this variability (Beatty 1987). In the 1950s, when the balance position was at its height, it was powerfully opposed by H. J. Muller, the former associate of T. H. Morgan (Muller 1949; Dobzhansky 1955). Supporting the "classical" theory, Muller argued that organisms are fundamentally standardized. There is indeed a type for a group like a species, and all who fail to fit this type are generally below the adaptive optimum. Normally, selection works to weed out the inadequate, the variant. But, because mutation is forever emptying new forms into a population, and especially because most new mutant alleles tend to be recessive, there will always be a certain genetic background interference to the perfection of any group, however strong the purifying forces. That variations were generally undesirable seemed obvious to Muller, when judged against the highly deleterious effects he usually produced through subjecting organisms to mutagen processes like X-rays.

As it happens, in the mid-1960s much light was thrown on the Dobzhansky/Muller dispute by the use of powerful molecular techniques (Lewontin 1974) (although whether these proved everything everybody thought they would is perhaps another matter). What is crucial for us here is that, until that point—that is, for virtually the whole time that

Dobzhansky was working as an evolutionist—there was really no firm way of measuring the variation in species. Everyone knew there was some, but how much was another matter. So too was the question of whether such variation was increasing or decreasing or just held stable. It is true that, even if we disregard the contentious issue of the evidence for drift, there were some undoubted cases of heterosis, most notably (and repeatedly mentioned) the case of sickle-cell anemia in humans (Ruse 1982, 141–142, 266). But, as Aristotle said, one swallow does not make a summer.

The idea of variation was important, really important, for Dobzhansky. Indeed, in the 1950s, as part of his fight with Muller, he got embroiled in the controversy over atomic radiation, especially over the danger of the fallout from bomb testing (Beatty 1987). As a decent human being, Dobzhansky hated the idea of such testing, and as one whose love for the country of his birth never diminished, he feared terribly were such bombs ever to be used in earnest. Yet, whereas for Muller the issues were simple—testing simply increases our genetic load and that is always an unequivocally bad thing—for Dobzhansky there were powerful forces pulling him the other way. It is true that he (and his student Bruce Wallace) produced evidence supporting his position, but then so did Muller (and his student Raphael Falk) produce evidence supporting the contrary position (Dobzhansky and Wallace 1953; Wallace 1952, 1963; Muller and Falk 1961; Falk 1961). Ultimately, what counted was the prior commitment, and so he had to argue that the threat posed by fallout is not really that great. After all, species "soak up variation like a sponge" (Dobzhansky 1937, 126).

I am not now arguing that this most significant and most influential evolutionist was indifferent to empirical evidence. Dobzhansky worked hard, all of his life, to find such evidence. And, as his swing to selection shows, he could be moved by it. But I do conclude that the question of evidence supports the conclusion that Dobzhansky read his progressionism into nature, rather than from it, and that his science was shaped in such a way as to support such a reading.

Sewall Wright: The Cautious Professional

Wright came to scientific maturity when men like Conklin were pushing a progressionist evolutionism, comfortably and openly, but only in the popular realm. They made no effort to integrate their evolutionism into their professional science. They wanted to keep up the Huxley divide between professional and popular, with evolution essentially on the popular side. Wright differed radically—at least, he did in a very important surface sense. He wanted his evolutionary biology to be well within the bounds of his professional biology. And indeed, when Wright wrote on matters evolutionary, anything less of a writer of popular science it would be hard to imagine. But it was not simply that Wright tended to be a dense thinker that made his work professional—actually, as a writer he was reasonably lucid. Nor was it the fact that he used mathematics that made his work professional—although, as in the case of Fisher, given the dreadful awe felt by most biologists at the sight of a symbol, his mathematics certainly did not hurt. Rather, Wright set out deliberately—far more self-consciously than either Fisher or Haldane—to be a professional scientist. This ambition extended to his evolutionism, thinking now not just sociologically but epistemologically also, and it framed his treatment of progress.

First and foremost, the reason for his attitude was that, as a geneticist in the early part of this century, Wright had entered a young and rather insecure area of biology. It was not yet (as today) at the forefront of the field—it was something of a fringe topic. I have emphasized the significance for professional discipline building of ready financial support. Early geneticists found their money at eugenics congresses and agricultural stations—remember Bailey and Castle, not to mention Bateson, Fisher, and Haldane—neither of which had great prestige (as, say, medicine). And Wright himself was tarred with the same low-caste brush, for he did not start his working life as an academic but as a civil servant for the Federal Department of Agriculture.

It therefore behooved him, especially, to persuade the outer world, within and without biology, that he was a legitimate professional scientist (not a mere breeder) in a legitimate professional science. And this pressure was kept up even after—especially after—Wright moved to Chicago. He was the only geneticist among a crowd of embryologists and others, and he was looked upon with some suspicion. "Genetics to them was a side issue to the fundamental problem of development" (Provine 1986, 168). It is not insignificant in this context that Wright had taken developmental aspects of genetics as the main focus of his research, thereby putting his career path as close to the mainstream as possible.

It is not insignificant either that, at Chicago, developmental genetics remained Wright's basic life work. Indeed, Wright was so far concerned to be a regular professional that—refusing to respond to risky new developments—he became positively anachronistic. By the time he retired, his kind of work on guinea pigs, laboriously working out gene interactions, had been quite overtaken by the new molecular genetics. And this kind of caution extended to Wright's evolutionism. He was incredibly careful about its presentation. It always came in "top quality," respectable journals. It was never contaminated with metaphysical speculations, however dear they might have been to Wright's heart. It is true that in his main paper he mentioned his yearnings, but immediately he qualified with "they can have no place in an attempt at scientific analysis of the problem" (Wright 1931, 155). Nor did they have any place in the very cautious graduate course on evolution that Wright gave for twenty-five years at Chicago, which started by being exclusively on his own theoretical ideas and only slowly was extended to the work of the next generation of evolutionists. Even in 1940, the course included virtually nothing on the work of others, despite *Genetics and the Origin of Species* being used as a text. (Information from class notes of W. D. Burbanck; in possession of D. Parker, Aarhus University, Denmark.) Later in life, Wright let out his personal views but sparingly and in (what were traditionally taken to be) appropriate contexts: he expounded on his view of life's history only through the outlet of an encyclopedia article; he exposed his metaphysics directly to his biological colleagues only on the occasion of his presidential address to the American Society of Naturalists.

In all of this care of presentation, Wright was totally self-conscious. He had learned from Pearson that science deals only with appearances, even if only science deals with appearances.

> One may recognize that the only reality directly experienced is that of mind, including choice, that mechanism is merely a term for regular behavior, and that there can be no ultimate explanation in terms of mechanism—merely an analytic description. Such a description, however, is the essential task of science and because of these very considerations, objective and subjective terms cannot be used in the same description without danger of something like 100 percent duplication. (Wright 1931, 155)

Furthermore, Wright had learned from Henderson that you can be as metaphysical as you like—it just so happened that both Wright and Henderson liked very much to be metaphysical in a deistic/agnostic sort of way—as long as your metaphysics remains parallel to your science and not part of it. "The man of science is not even obliged to have an opinion concerning its [teleology's] reality, for it dwells in another world where he as a scientist can never enter" (Henderson 1917, 311). As Wright wrote about his teacher to his brother: "He took a very rigidly mechanistic view and was very severe with Driesch and Bergson whom he looked upon as the most dangerous enemies of scientific progress" (January 10, 1916; Q. Wright Papers, box 18, folder 5). It was for slopping over these bounds that Henderson criticized the vitalism of Bergson. The Frenchman's vision may be right—in my opinion it would be difficult to be more teleological than Henderson—but it has no place pretending to be science: "To postulate such a tendency is, however, in itself rather a philosophical than a scientific act" (Wright 1931, 281).

Given these pressures, external and internal, then apart from the fact that it was quite natural that Wright, as a theorist of mechanisms, should avoid talk of long-term effects, there is little surprise that Wright never forced grand Spencerian discussions of progress into his straight science. Epistemologically, given his education and given his location in an insecure branch of science, the last thing that Wright wanted was a cultural value intruding in on his biology. This is the major reason that in public Wright was so restrained about our own species. Wright was a progressionist, his Progressionism informed his science, his views were there to be seen by those who would see—that was an important part of Wright's monism—but he was certainly not going to offer progress as a consequence of his science. To do so would be absolutely to destroy his aims for himself as a professional and his aims for evolutionism as mature science. Even as it was, his graduate students were appalled at his presidential address—"We really thought he had lost it" (letter from Janice Spofford, Wright's last doctoral student, to author, August 10, 1995).

Yet—and here I pick up on an earlier comment about Wright's functioning in a "surface sense"—*qua* evolutionism these aims were at heart deeply troubling. Wright simply had to go beyond the likes of Conklin and push right into professional, causal speculations. He had done the work and, with the passing of time, there were the examples of Fisher and Haldane to emulate—and better. What is more, as a student of Henderson Wright felt the need to complement one's metaphysics with hard science. Wright could not bring his progressionism over into his professional science; but, as a progressionist, he had to provide the professional science to mirror it. He had to come up with the epistemologically tough theoretical product.

Some of the tension following from his various aims may have lain behind the fact that Wright, America's leading theoretician, did virtually nothing in his own right to found a professional discipline of evolutionary studies—he was dragged in only as others set to work. Although the main reason was undoubtedly his personality: "He was both shy (of students) and preoccupied, and could go for several terms without doing more than saying hello . . . He definitely did not want to talk science with his graduate students en masse. So far as I know, he never gave a seminar course in which current literature was read and discussed" (letter from Spofford to author).

Add to this the fact that the 1930s especially were poor times for graduating students to find research jobs (only six out of twenty of Wright's students went this way), and that the typical biology student back then simply did not have the background or ability to handle the mathematics at Wright's level. Not that this is so significant. Mathematics itself may have impressed, but Wright's sort of work was just not encouraged at Chicago: "there was a strong bias in the department toward awarding degrees for original research and this meant laboratory or field observation-type research. One could use any amount of statistics or mathematical modeling in relating one's results to theory, but pure theory development was for the ridiculed Committee on Mathematical Biology" (letter from Spofford to author).

In sum, Wright's failure as a discipline builder becomes nigh overdetermined. But perhaps there was also a psychological reason at work: that deep down inside the cautious Wright agreed that the Huxleyites—not to mention his Chicago colleagues—were right. A real discipline does not lie in this direction. Evolutionism, at least of a whole-organism, empirical kind, can never truly be a professional, mature science, certainly not in our lifetime and before more conventional matters of development and the like are solved. "Wright obviously thought that understanding the nature of the gene and how it operated was one of the great questions in biology, and expended most of his time on that" (letter from Spofford to author).

An extreme conclusion; but learn that, of Wright's twenty graduate students, only two worked directly on problems of evolution. Calculate how Wright had been at Chicago a full five years before he published his major paper—by which time he was virtually forced into print because Fisher had published his book and Haldane much of his work. Discover that, although it was Dobzhansky who made much of Wright's work and pushed evolutionism forward through a most fruitful collaboration, Wright was always a reluctant partner. As soon as he could, pleading the pressure of other work, he pulled away. Appreciate that "Wright regarded his physiological studies on guinea pigs as perhaps his major work" and that he shared his colleagues' views on students and mathematics: "He didn't think biology students should write theoretical theses" (letter to author from James F. Crow, August 14, 1995). And conclude with the memory that Wright deferred the publication of his major work on the topic of evolution until he retired—and that only because there was no room for guinea pigs in Wisconsin.

I see tension and balance in Wright's own life—the bland exterior, the metaphysical interior; the son of a socialist, the conservative about Vietnam; the descendant of abolitionists, the fear of blacks—as well as in his science. The tension spilled into his treatment of evolutionism, torn as he was by his twin urges to progressionism and professionalism. He could provide a blueprint for others to ground a professional science but, emotionally, he could at best be a camp follower in the building of a discipline. For himself, it was enough that he had provided the formal structure for a picture he found metaphysically satisfying. Nothing else really mattered, and, significantly, believing that the cream had been skimmed, "he took rather little interest in the heavy mathematical developments of the 60s and after" (letter to author from Crow). If ever there was a reluctant revolutionary, it was Sewall Wright.

Micro-professional, Macro-populist

Dobzhansky is a fascinating contrast in just about every way, beginning with his attitude toward the professional/popular divide and his placing of the correct position of P/progressivism. When he was working on the second edition of *Genetics and the Origin of Species* in 1940, he wrote to his friend, the geneticist M. Demerec, that: "I certainly do not intend it to be a textbook, and it is not one in the proper sense of the word. There is a lot of creative work involved in it, of a semi-philosophical kind of language and I enjoy it greatly. It is useful occasionally to leave one's immediate fold of interest, and to look over the broader aspects of one's science" (March 26, 1940; Dobzhansky Papers). Then, a few years later, to his colleague L. C. Dunn he wrote (about the views of the anarchist Russian, Prince Peter Kropotkin, that selection promotes group harmony): "I do not believe that biology and philosophy are quite so separate as you seem to imply, in fact if they were I would rather lose interest in biology. Surely it is a legitimate biological problem to ask whether natural selection favors combative or cooperative behavior, and it is a real change in biology that in the past the former and at present the latter is emphasized" (November 23, 1955; Dobzhansky Papers).

And yet later, and especially about progress, he wrote to the historian John C. Greene, who wanted to put a barrier between science and religion, that:

> I do not doubt that at some level that evolution, like everything in the world, is a manifestation of God's activity. All that I say is that *as a scientist* I do not observe anything that would prove this. In short, as scientists Laplace and myself "have no need of this hypothesis," but as a human being I do need this hypothesis! But I cannot follow your advice and put these things in water-tight compartments, and see only "change" and no "progress", only change and no "trial and error". For as a scientist I observe that evolution is on the whole progressive, its "creativeness" is increasing, and these findings I find fitting nicely into my general thinking, in which your "creative ground" is perfectly acceptable. (Letter to J. Greene, November 23, 1961; Dobzhansky Papers)

By this stage, one might be wondering (fearing?) whether the most influential evolutionist of this century was so far blurring the professional/popular divide that he was simply dragging all of Sewall Wright's good efforts down into the public arena. Or was Dobzhansky merely trying to run on a track parallel to that of Julian Huxley, hoping to create a professional science with progressionism at its core? In fact, neither of these suppositions is correct—quite apart from Dobzhansky's total contempt for Huxley's work. (He advised Mayr not to review *Evolution: The Modern Synthesis* (Mayr to Dobzhansky, June 11, 1943; Dobzhansky Papers), and he loathed Huxley's atheistic Teilhardphilia.) The key to Dobzhansky's strategy was the dichotomy between micro-and macroevolution, a belief in the importance of which he inherited from his teacher in Leningrad, Filipchenko (1929).

Dobzhansky needed no urging to keep his work, *qua* micro-evolution, professional. By the time he came to write *Genetics and the Origin of Species,* Dobzhansky had been ten years in Morgan's laboratory, a home which was notoriously hostile to religion or to any "soft" thinking: "his idea was, religion feeds on mystery. The way to combat religion is to combat mystery, hence to show that the biological phenomena are not mysterious, but they are scientifically explicable" (Dobzhansky 1962b, 255). In his work *qua* macro-evolution, however, there was, to be honest, something of a sleight of hand. In his professional writings, that is to say his writings about micro-evolution, Dobzhansky claimed continuity between micro and macro. Hence, even though in such places he himself did not talk much at all about macro-evolution, it was open to the reader to—it was indeed implied that the reader should—read the course and process of life without adding value-laden interpretations. Dobzhansky then sat down and wrote a series of books in which he read precisely what he felt like reading into life's history. Hence, because Dobzhansky wanted it to, macro-evolutionary progress thrust ever higher.

The escape clause, naturally, was that these progressionist books were for the general reader. Thus, although Dobzhansky had a metaphysics, unlike Weldon and Ford, for example, and although he did not stay silent about it, as Wright generally did, he could not be faulted as was Julian Huxley for mixing professional and popular science. He went beyond Goodrich and Conklin in offering a professional micro-evolutionism—something epistemologically careful and ostensibly culture-free—but mirrored them in giving a popular macro-evolutionism. And, in all of this he was reasonably successful. The more serious of Dobzhansky's professional colleagues, even the progressionists, found his religious enthusiasm a trifle embarrassing. "So your Biology of Ultimate Concern got published after all! Many thanks for sending it to me. I read it with interest and needless to say I would agree with almost everything you say. The exception of course is chapter 6. To me Teilhard is a step in the wrong direction" (letter from Mayr, June 12, 1967; Dobzhansky Papers). But, in a sense, these publications were a bit of fine-tuning. Those who liked the religion could keep it, and those who did not could drop it.

The point is that socially, as well as intellectually, Dobzhansky built on the work that Wright had started. The native-born American gave evolutionists a firm foundation on which to raise a professional, mature science, one which was—and had to be—ostensibly progress-free. Dobzhansky carried this work forward, raising this professional, mature science, one which was—and had to be—ostensibly progress-free. But Dobzhansky went beyond Wright in providing also a non-professional, popular science. In the public realm one could be as openly progressionist as one wished. Thus, thanks to this dual approach, one could have the status one coveted and yet acknowledge the motives that drove one.

We are almost, but not quite, there. Apart from the fact that the macro-evolutionist, no less than the micro-evolutionist, might yearn for the possibility of professional status, we still have no professional *discipline.* Could Dobzhansky and those influenced by him do what Wright could not? It is to this question, and to the disposition (in America) of progress, that we turn next.

11

The Synthesis

Genetics and the Origin of Species provided a conceptual framework for research that motivated people to go forth and extend its message in new fields. In the next decade (the 1940s), three people in particular in America seized on Dobzhansky's program and produced very influential books within and beyond their own fields of biology. Since in many respects they thought of themselves as working together, they may be said to have forged a New World equivalent of Julian Huxley's evolutionary "synthesis."

From Lamarckian Naturalist to Darwinian Systematist

Ernst Mayr was born in Bavaria in 1904, the son of a judge, a man of liberal thought who was virtually the only male member of his family who was not a doctor. The young Ernst was also encouraged to take up medicine, but the pull of natural history soon proved too strong. Mayr graduated from Berlin University, having covered himself with great distinction in all of the biology courses. Thanks to the limited advancement opportunities under the German professorial system, Mayr soon looked westward. From childhood Mayr had been an evolutionist but, like his teachers and colleagues, until he came to America he had downplayed the significance of natural selection and had favored Lamarckian causal processes. These beliefs faded rapidly, and, especially under the

Ernst Mayr

spell of Dobzhansky, Mayr became a committed Darwinian. Much inspired, in 1942 he produced his best-known book, *Systematics and the Origin of Species, from the Viewpoint of a Zoologist.* Fame and security followed. From being a yearly appointment on soft money (and an enemy alien, to boot!), within a few years Mayr had become a Harvard faculty member (1953), rising to the directorship of Agassiz's Museum of Comparative Zoology (1961). The "immigrant boy" had made good (Dobzhansky, letter to E. Mayr, June 19, 1974, on the occasion of Mayr's 70th birthday; T.D.–E.M. Correspondence). (A special issue of *Biology and Philosophy,* 1994: 9/3, celebrating Mayr's ninetieth birthday, contains much information about his life and work.)

In turning to *Systematics and the Origin of Species,* we must take care. We must not assume automatically that, as Dobzhansky had given a (Russian) geneticist's reading of Sewall Wright, because he had been spurred to action by Dobzhansky Mayr simply gave a zoologist's reading of *Genetics and the Origin of Species:* "It is altogether misleading to say that 'the great achievement of the synthesis was [to show] that all biological data . . . could be reduced to the neo-Darwinian principles,' if this implies reductionist population genetics" (Mayr 1986, 4). Indeed, as far as genetics was concerned, considered from a positive and formal perspective, Mayr was virtually silent. References to population genetics occur in only 68 of the 11,324 lines of this volume (Mayr 1992, 6).

As with Dobzhansky, to understand Mayr's work properly we have to go back to his non-American origins. The battle for Mayr in Germany was not the battle for evolution. That had been long since won. Nor was it the battle for causes, in terms of heredity. *The geneticists were the enemy!*

> It is convenient, even though it is a vast oversimplification, to say that in the 1920s the evolutionists fell clearly into two camps: the naturalists-systematists were anti-Mendelians; they believed in gradual evolution, and the origin of organic diversity was one of their chief interests. The geneticists still showed their origin from Mendelism as characterized by De Vries, Bateson, and Johannsen. Their major interest was in the phenomenon of mutation, the change of a given (usually closed) population, and in gene physiology. (Mayr 1992, 1–2)

For Mayr, the battle was species and speciation. The first wave of geneticists, thinking that the key mutations were large, had embraced saltationism. The fact of natural variation making possible gradual change was the crucial phenomenon or problem facing the naturalist, and it was precisely this that the traditional geneticist denied.

In America Mayr learned from Dobzhansky that one could be a gradualist and a Mendelian, in the updated (post-Morgan) sense of a genetics which supposed key micro-mutations. Relatedly, therefore, one could accept lots of natural variation. But Mayr still saw a very live threat to proper thinking coming from the early geneticists' claim that new species come from instant macro-mutational steps. "I wasn't interested in the principles of evolutionary genetics, what I was interested in was to prove that the attacks on Darwin's gradualism by all the saltationists were refuted by showing the gradual origin of species through geographic speciation" (Mayr 1986, 7). These worries were intensified in 1940, when the German-born geneticist Richard B. Goldschmidt published a major book, *The Material Basis of Evolution,* reiterating the claim that speciation is a one-step event. Mayr had all the stimulation that he needed, and it is against this background that *Systematics and the Origin of Species* should be considered: less as an extension of Wright-Dobzhansky, and more as the manifesto of a naturalist fighting (old-fashioned) genetics.

Mirroring Dobzhansky's division of the statics and dynamics of genetics, Mayr's book fell into two sequential parts: the statics and dynamics of speciation. Thus, first Mayr looked at systematics—the problems of classification—at the level of the species, in the light of modern biology. The key problem-setter and organizing principle was the all-important variation that occurs throughout nature. This is variation which comes at all levels, from the lowest to the highest, and taxonomic groups absolutely must reflect the phenomenon. Most specifically, species taxa simply cannot and must not be defined or characterized in terms of abstracted physical, morphological features. Mayr allowed that it is true that one uses morphology to delimit species and to assign specimens to species; but, he maintained, given the variation, morphology cannot be part of the definition of what it is to be a species. Citing Dobzhansky, who had defined species as "that stage of the evolutionary process at which the once actually or potentially interbreeding array of forms becomes segregated into two or more separate arrays which are physiologically incapable of interbreeding" (Mayr 1942, 119), Mayr therefore argued that: "Species are groups of actually or potentially interbreeding natural populations, which are reproductively isolated from other such groups" (p. 120). This definition avoids reference to morphology (as in "species are groups of similar-looking organisms"), and as it stresses biological processes it allows for internal variation.[1]

In the second section of his book, Mayr's main concern was with the nature and causes of the formation of species, which are now to be understood as ever-fluid breeding groups rather than as immobile Platonic essences. The key is the gradualness of evolutionary change, for one-off instant switches simply would not do the job. Hence, Mayr's case was based on an extension of his systematics—a geographical, not a temporal, extension. As we find variation in any group, so we find all degrees of difference between groups, most especially between those we would rank as "species." Differences range from the imperceptible, from almost total morphological identity and full reproductive abilities, to complete morphological difference and total reproductive isolation. Most dramatic are rings of subspecies, where the links of the chain can interbreed but the end populations, although overlapping, are isolated (Mayr 1942, 183). This is gradualness caught in action.

Again and again we find Goldschmidt was used as a foil. Most particular scorn was reserved for the one-step species formation hypothesis: "To believe that this could actually happen, as Dobzhansky has said in review of Goldschmidt's work, is equivalent to 'a belief in miracles' " (Mayr 1942, 155). So far opposed was Mayr to this kind of thinking that he argued strenuously that (in the animal world at least) only under the most exceptional cases could speciation occur straight from one group to another, without some form of separate physical geographical separation. There is no reproductive isolation—the Mayrian mark of species-hood—without a physical break for a period of time. Otherwise, genes are simply passed back and forth between populations, and no barriers to interbreeding can be put in place and maintained. Technically, virtually all speciation is "allopatric"—"sympatric" speciation is nearly impossible.

In *Systematics and the Origin of Species,* Mayr's problem was not really that of adaptation. "My major objective was to solve the major problem of evolution not solved by the geneticists, the origin of discontinuities; or to put it in different words, the origin of organic diversity" (Mayr 1992, 6). Yet, Mayr's naturalist background had always inclined him to take seriously the reality of adaptation. Whether he was a Lamarckian or a Darwinian, his impulse was to look for function. Indeed, in 1939 (before the so-called hardening) we find Mayr challenging Dobzhansky about two bordering populations of *Drosophila.* "I am always somewhat suspicious of accepting the principle of simple random variation without a selective environmental influence. This is related to some of the problems which I thought were not sufficiently explicit in your book" (letter to Dobzhansky, February 28, 1939; T.D.–E.M. Correspondence). Since the conventional wisdom around 1940 was that many species differences are non-adaptive, it was hardly surprising that Mayr did bend to the wind, somewhat: "Many combinations of color patterns, spots, and bands, as well as extra bristles and wing veins, are probably largely accidental" (Mayr 1942, 86).

Moreover, like everyone else, Mayr invoked the authority of Sewall Wright and genetic drift, although he claims now that he was tense about this: "What always intrigues me is the inconsistency of just about everybody, including myself. There, in my 1942 book, in order to be 'modern', I quote Sewall Wright copiously. However, in my actual thinking and working I was very much opposed to him. And I fought Dobzhansky all along when he wanted to believe in the neutrality of the human blood group genes and the Drosophila gene arrangements" (letter to author, November 20, 1991).

We must interpret this recollection with care. Mayr was not much of a "drifter," but he certainly took in the landscape metaphor. In fact, it is no exaggeration to say that the metaphor was behind Mayr's causal thinking in *Systematics and the Origin of Species,* as well as in subsequent thinking. It is all very well to refer to the gradualness of evolutionary change, but there must in turn be a reason for this change itself. In Mayr's case it was that organisms go up and down Wrightian hills, for Darwinian (plus Mendelian) reasons. Thus, again and again, the contour imagery was brought smoothly into Mayr's discussion, as in the case of Darwin's finches on the Galápagos: "The absence of competitors, to use Sewall Wright's language, apparently facilitates the crossing of valleys between one adaptive peak and another" (Mayr 1942, 271).

One can truly say that, social factors apart, it was this metaphor—if we understand it in the enriched, Dobzhanskian sense of reflecting natural variability and the clustering of organisms into isolated units—that tied Mayr into the synthesis. His work was in no sense a derivation from the theory of the *Genetics and the Origin of Species.* More honestly, it was a broadside against the tradition of the genetical opponents of his youth. But, Dobzhansky (using Wright) made available an intellectual niche—a niche using the best of modern genetics—within which Mayr could work. Conversely, Mayr provided evidence which tangentially supported the kind of picture Dobzhansky was articulating. Everybody benefited.

Even into this, his tenth decade, Mayr has worked without cease, but for all that he has trimmed and shaped his thinking, with the 1942 book the essential picture was in place. Thus, *Animal Species and Evolution,* a Wagnerian 1963 summation of his science, states flatly that, if an organic type ("phenotype") is put under a selective pressure and responds, and if the pressure is then relaxed but the type fails to revert entirely to its old form, it is likely "that an alternative adaptive peak had been climbed. This alternative peak is equivalent to the original genotype as far as general fitness is concerned but superior with regard to the specific phenotypic character that had been under selection pressure" (Mayr 1963, 288–289).

By the time Mayr wrote this, he and Wright had fallen out, somewhat, so there was certainly no gracious acknowledgment of intellectual debts. But the ideas were unchanged. So, let us now move on to analysis.

Mayr and Progress

In Mayr's writing of recent years, he has been reflecting on a life's work as an active scientist. He has been crossing the *t*'s and dotting the *i*'s and adding the umlauts, as it were. And he has been making crystal clear his commitment to progress: "On almost any measure one can think of a squid, a social bee, or a primate, is more progressive than a prokaryote" (Mayr 1982, 532). In more detail:

> who can deny that overall there is an advance from the prokaryotes that dominated the living world more than three billion years ago to the eukaryotes with their well organized nucleus and chromosomes as well as cytoplasmic organelles; from the single-celled eukaryotes to metaphytes and metazoans with a strict division of labor among their highly specialized organ systems; within the metazoans from ectotherms that are at the mercy of the climate to the warm-blooded endotherms, and within the endotherms from types with a small brain and low social organization to those with a very large central nervous system, highly developed parental care, and the capacity to transmit information from generation to generation? (Mayr 1988, 251–252)

What of one's criterion of progress? Complexity is a poor guide, since ancient forms (like trilobites) were more complex even possibly than humans. However, there are two reliable criteria of progress: "One of these is parental care (made possible by internal fertilization), which provides the potential for transferring information nongenetically from one generation to the next." The other, linked to the first, is culture, "together with speech setting man quite aside from all other living organisms" (pp. 252–253). It is important to stress that, however defined, advance comes about through the normal processes of natural selection. There is no need to postulate unseen forces or causes, or any similar thing. Indeed, the branching and irregular nature of the fossil record quite precludes any such unorthodox mechanisms.

Are there, then, no deeper causes underlying Mayr's beliefs in biological progressionism and the possible connections with beliefs about cultural Progress? At the basic level, evolutionary theory is a package deal for Mayr. If you believe in evolution, then you believe in progress. It is as simple as that. Mayr grew up in Germany, reading and accepting a tradition much influenced by Haeckel, a tradition which just equated evolutionism and progressionism. As a schoolboy Mayr read Haeckel's *Riddle of the Universe* with enthusiasm. He remembers primarily the anti-Christian message, which helped convert him from childhood Lutheranism to life-long skepticism; the very point, however, was that evolutionary progressionism was being offered to take the place of conventional belief. Mayr also read Wilhelm Boelsche, another evolutionary progressionist, and even more popular than Haeckel at the beginning of this century in Germany. And, most important, with particular reference to social Progress, there were family influences: "In my youth, I was deliberately educated by my parents to believe in the possibility of a better world and my duty to help toward the creation of such a better world" (interview with author, March 30, 1988).

Mayr worked for twenty years in the American Museum of Natural History, the temple of progressionism. Even though he personally disliked Osborn, he would have been caught in the general philosophy, both social (Progress) and biological (progress). The idea of Progress in general has gone undimmed from that day to this: "In your treatment, about half the time you talk about Progressionism with a sneer, always illustrated with some detestable examples of racism or male chauvinism. I think much of your writing would be improved if you would admit that much of Progressionism was a rather noble philosophy. In fact a very good case could be made for the claim that the current mess is result of the loss of this philosophy" (taped commentary on an earlier version of this manuscript, Spring 1993).

In fact, social issues are not the driving passion. Although Mayr shares the eugenical concerns of evolutionists of his generation, he is less concerned with manipulating the human gene pool and more concerned with seeing that social identity is not imposed in the name of social equality. As for any good German, it is the idea through the ages which truly counts. In science, as a massive Mayr-authored history of biology shows well—only too well, in the opinion of many professional historians—he believes strongly that real advances are made, and as they are truth emerges and error is strongly eliminated. Indeed, so firmly does Mayr believe this that he has trouble dealing with that which is so obviously wrong: "It would be quite justifiable to ignore Spencer totally in a history of biological ideas because his positive contributions were nil" (Mayr 1982, 386).

Mayr claims to be "greatly attracted by the Hegelian dialectic of thesis-antithesis-synthesis" (Mayr 1982, 9; sentiment repeated in letter to author, July 20, 1994). This is certainly plausible, for the emphasis in Mayr's thinking is more on the conceptual than the empirical: "Scientific progress consists in the development of new concepts, like selection or biological species, and the repeated refinement of definitions by which these concepts are articulated" (Mayr 1982, 43). For Mayr's present-day picture of the path of science, however, Germanic-type influences function now more as background. More recently significant was the influence of Mayr's fellow ornithologist Julian Huxley (before he got caught up with

Teilhardphilia), whose writings—read and digested in the 1930s—gave Mayr just the secular Darwinized world philosophy he had long sought. Although Christianity may be inadequate, a philosophy of P/progress can substitute. "To be frank with you, Huxley's *Religion without Revelation* very much appealed to me. I am of the opinion that you can have a deep religion, even being an atheist" (comments on manuscript, Spring 1993).

In respects, of more significance than the Progress of science is the Progress of technology. This is tied tightly to the progress of organisms.

> I feel, for instance, that a modern motor car is better than an old Model T Ford, or something like that and at the time when the Model T Ford was built neither Mr. Ford or anybody else had a picture of the car of 1988 in his mind. It developed gradually by each car manufacturer, constantly each year according to the American principle that only that is good which is new, built in all sorts of new things into their cars. Some were highly successful, some were total flops. But as a result there was a progress to better and better cars. Without any teleological component in it. There wasn't anything. And that's exactly the way it is in nature. I mean, there the natural resources were limited. There was constant competition for them, and every time an organism made an invention, it either helped it to utilize the resources better or it made a mistake and it didn't. (Interview with author, March 30, 1988)

This gives us a kind of Darwinian comparative progress. What about absolute progress? Would one want to say, for instance, that the motor car is intrinsically better than (say) the bicycle?

> Well, under many circumstances they are clearly superior. As a boy, I bicycled through large parts of Germany and it took me 6, 7, 8 days to get from one place to the other. It was part of the time raining and I got very wet, I had a head wind on one of my trips all of the way and it was extremely strenuous. I would say to have now a car available to cover this distance is a considerable advantage. (Interview with author, March 30, 1988)

This is a metaphor and a conclusion that Mayr repeats in print and in correspondence. It is heartfelt. The Progress of technology—specifically the Progress of automobile technology—impresses Mayr at a significant level.

What about the factual basis to Mayr's beliefs? The answer surely supports the conclusion thus far established. If one so wishes, one can certainly endorse a Mayrian view of progress. But, apart from any doubts one might have about the archetypical Germanic belief that the automobile is the epitome of technological Progress, there is no reason to think that nature—or biological theory—makes this progress mandatory. Domestically, there is something rather touching about Mayr's belief that parental care represents advance, especially when one combines it with another belief he holds about the essentially monogamous nature of humans; but biologically there is nothing to support any of this. Evolutionists today distinguish between two reproductive strategies, one that involves parental care (*K*-strategy) and one that does not (*r*-strategy). Sometimes the investment strategy pays and sometimes it does not. It is all a question of the availability and reliability of resources. Where food and shelter and the like are of irregular supply, especially where one has sometimes famine and sometimes glut, the *r*-strategy can make a lot more sense biologically than the *K*-strategy. At least, it can make a lot more sense in the only language evolutionary biology understands, namely increased survival and reproduction (Clutton-Brock and Godfray 1991).

In short, everything points to the conclusion that Mayr's thoughts on progress are a function of the biological milieu in which he grew up in Germany and that which he entered in America, spliced with the influence of Julian Huxley and with his present cultural convictions, especially those about scientific and technological Progress.

Synthesizing Genetics and Paleontology

George Gaylord Simpson (1902–1984), described by his sometime supervisor as the "greatest paleontologist since Cuvier" (Simpson 1987, ix), was raised in Denver, Colorado. After study at Yale and time in London, Simpson joined the American Museum of Natural History in New York, which was to be his academic home for three decades. The 1930s saw Simpson establishing himself in the front ranks of the world's students of fossil mammals. Increasingly, however, his interests turned to causal questions. It would be unfair to suggest that Simpson unaided brought paleontological theory into the twentieth century. Indeed, two of Osborn's own students and co-workers, Gregory and W. D. Matthew, were pushing the subject toward a more selectionist perspective (Rainger 1991). But it was Simpson who reached out to the rest of the biological world, thanks in major part to a facility with mathematics: "When I was definitely working toward a synthesis in my own mind, I read the works that became among the sources of the synthetic theory, notably those of

George Gaylord Simpson

Facing page: Simpson's reconstruction of the evolution of the horse, using his modified version of Wright's landscape metaphor. (From Simpson 1944.)

Fisher and Haldane. I was a bit late in getting to Sewall Wright, but did so in time" (Mayr and Provine 1980, 456). Yet, as with Mayr, it seems to have been the influence of Dobzhansky that was really crucial. *Genetics and the Origin of Species* "profoundly changed my whole outlook and started me thinking more definitely along the lines of an explanatory (causal) synthesis and less exclusively along lines more nearly traditional in palaeontology" (Mayr and Provine 1980, 456).

Firmly committed to a Dobzhansky-like approach, Simpson strove to bring paleontology under this new perspective. This he did in 1944 with his most important work, *Tempo and Mode in Evolution,* which moved him to the front of his field, ready to influence a whole new generation: an influence extended through a 1953 revision, retitled *The Major Features of Evolution.* Simpson's tenure at the AMNH was followed by a decade at Harvard, and then in his last years Simpson moved to Arizona. Honor and fame were showered upon him, yet one senses that there was not a great deal of love. Other than to his family, and one or two special friends, Simpson was not a man of great personal warmth and sociability.

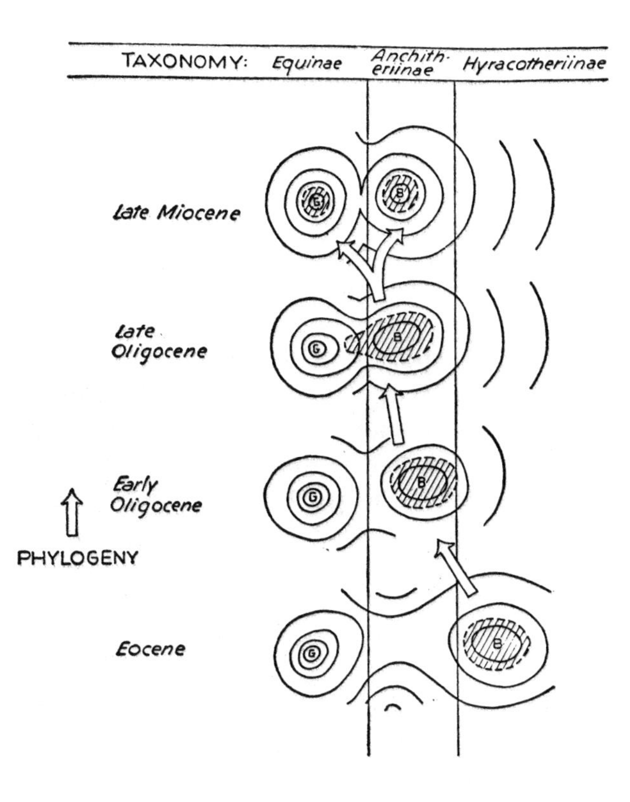

Whatever his personal failings, however, Simpson was a man of ideals. He had a deep sense of patriotism, something which led to unpleasant and dangerous Army service in the Second World War, at an age when he could readily have escaped—not to mention horrendous conflict with General George Patton, who tried (unsuccessfully) to get Simpson to shave his beard.

In *Tempo and Mode in Evolution,* after a prologue surveying rates of evolution ("tempo"), Simpson addressed his crucial task ("mode"): a detailed presentation of a Dobzhansky-like view of the nature and spread of genes in populations and of the various effects of natural selection. Like Mayr, Simpson often used opponents as whipping boys to make his own points. Goldschmidt was once again given a full critical treatment. But, technically, Simpson's discussion was significantly more detailed than anything offered by Mayr. Mutation, effective population number, selection, and more, got fair and detailed coverage. As was to be expected, since he was writing at the time before the idea came under heavy attack, Simpson was quite sympathetic to the notion of non-adaptive drift. By the time Simpson restated his thesis in *The Major Features of Evolution,* this sympathy had declined—although one should note that Simpson was ever essentially an adaptationist, believing in the prime importance of natural selection.

From Wright, Simpson also absorbed the metaphor of an adaptive landscape. As a paleontologist, Simpson had to modify the picture to his own ends—to think of phenotypes, not genes, and of much longer time scales—but the image was there: "the field of possible structural variation is pictured as a landscape with hills and valleys, and the extent and directions of variation in a population can be represented by outlining an area and a shape on the field" (Simpson 1944, 89). Simpson was aware that the appropriate metaphor is more like a waterbed than a landscape, but he acknowledged that he thought in absolute terms rather than in dynamic, relativistic terms (p. 90). One consequence of this way of thinking was that Simpson was inclined to suppose that there are ecological niches waiting to be occupied as soon as a group has climbed a particular adaptive peak. In other words, adaptive success is not just a relativistic phenomenon; it really represents something "out there," in nature.

Consider, by way of illustration, Simpson's discussion of the evolution of the horse. At the time of the Eocene, two adaptive peaks were available for the horses (the Equidae): one for browsing and one for grazing. However, "only the browsing peak was occupied by members of this family" (Simpson 1944, 91). As time went by, the animals grew in size. Hence, "the browsing peak moved toward the grazing peak, because some of the secondary adaptations to large size . . . were incidentally in the direction of grazing adaptation." Eventually, some animals sat in the middle, between both peaks. This is not stable and some were pushed hard toward grazing. "This slope is steeper than those of the browsing peak, and the grazing peak is higher (involves greater and more specific, less easily reversible or branching specialization to a particular mode of life)" (Simpson 1944, 93). Meanwhile, the browsers scrambled back up their hill, and those in the middle got wiped out. Now there were two separate groups, and finally (at the end of the Tertiary) the browsers went extinct, leaving only the grazers alive today. Note that not only was the grazing peak waiting there to be occupied, but also that in some sense the new peak was "higher" than the old one.

With the populational theory of the geneticists expounded, Simpson was now free to turn directly to the problems of his own science. One major problem for a neo-Darwinian was the question of evolutionary "trends," the raw material of all theories of orthogenesis. Simpson tackled this issue both negatively and positively. On the one hand, he argued that the fossil record is nowhere like as trendy as many enthusiasts claim. Many supposedly paradigmatic examples of evolutionary trends, like the horses, contain stops, starts, and frequent reversals. Often, major features are not involved in trends at all: "*Hyracotherium* was already a vertebrate, a mammal, a placental, an ungulate, a perissodactyl, a hippomorph, and an equid, which is a classificatory way of saying that the vast majority of its multitude of morphological characters were already the same as those preserved in *Equus* and in all equids as well as in many other more or less related animals" (Simpson 1944, 158–159).

On the other hand, Simpson argued that where trends do exist, they are readily explicable by selection working on random mutation. For instance, like most vertebrate groups, the horses grew in overall physical size. However: "Gradual increase in size is one of the most obviously advantageous specializations under most, but not all, conditions" (p. 159). Likewise with brain size: "A broad trend toward greater intelligence, disregarding fluctuating details of brain structure, is obviously of selective value to animals such as the horses under any ecological conditions" (p. 160). Even the overshooting of adaptation into a non-adaptive mode can be related to modern theory. Explicitly following Haldane, Simpson argued that selection for desirable features in youth may lead to the persistence of undesirable features in old age: "Thus, in the example of the Irish deer, early development of large antlers, for instance, by a high degree of positive heterogony, may be advantageous to the young bucks and may be favored by selection among them, but may result in an excessive maximum antler size in older bucks" (pp. 173–174).

Completing his book, Simpson stressed that life in the past followed no one set pattern, and that hence only the Darwin-Mendel synthesis is sufficiently powerful and flexible to deal with all that we find in the fossil record. The time has come to bring paleontology into the contemporary biological fold.

The Meaning of Evolution

Simpson was a good friend of Julian Huxley. They talked and wrote at length about the nature and significance of biological progress. Yet, whereas Huxley brought thoughts of progress explicitly into virtually all of his writings, Simpson took meticulous care not to do this. *Tempo and Mode in Evolution* does not end with a final chapter on "The Meaning of It All," nor does *Major Features of Evolution* carry extended discussion of the proper meanings of "higher" and "lower." In fact, Simpson's first major published exercise in paleontology was an attack on the traditional Chain-of-Being interpretation of mammalian history, from primitive monotremes, through less-primitive marsupials, to advanced placentals (Simpson 1928).

Is there nevertheless hope of some sort of upward rise or progress, within the Darwinian schema? At least until the writing of *Tempo and Mode,* criticisms notwithstanding, Simpson seems simply to have absorbed and endorsed a general progressionism as a matter of unquestionable fact. But, increasingly, perhaps as he was struck by the full implications of theoretical Darwinism and of his own work in paleontology, he came to feel that more is needed than mere assertion, and that a notion like progress is truly something we impute to the fossil record rather than find from it. Not that this in any sense changed Simpson's ultimate convictions. In the *Meaning of Evolution,* produced between the successive editions of this main work, Simpson was univocal. There is progress. Moreover: "Man *is* the highest animal" (Simpson 1949, 285, his italics). The problem then is to prove the fact, or rather to prove it without getting into the circle of defining "progress" as "humanlike" and then simply awarding ourselves the prize! What we must do is look at other, independent criteria of progress and see where humankind comes out on the scale.

To this end, Simpson ran critically through a list of candidates. On the standard most generally applicable, *tendency for life to expand,* humans do quite well, but whether one would think of such expansion as necessarily progressive is another matter. The problem with Julian Huxley's criterion, *dominance,* is whether one can truly say (for instance) that the aquatic vertebrates are less dominant than the terrestrial vertebrates. On the other hand, it cannot be denied that within a particular group judgments can be made. Progress with respect to *adaptation* is clearly important, and we humans do well with respect to that—except, interestingly, we do badly with respect to the related *increase in specialization.* We also do well on the criterion of *independence from the environment,* and even better on *ability to control the environment.* With respect to that old war-horse *complexity,* Simpson was not so sure of our status. Like Mayr, however, he was not that struck with complexity as a mark of advance, anyway. A recent crustacean like a crab seems simultaneously more specialized and less complex than an ancient crustacean like a trilobite. *Parental care* gets a mention: like Mayr, Simpson took this seriously. Then, Simpson started to get really excited about the suggestion of one C. J. Herrick, that progress entails "change in the direction of increase in the range and variety of adjustments of the organism to its environment" (Herrick 1946, 469, quoted in Simpson 1949, 258). There are two consequences to this type of change, on both of which humans score at the top. First, there is the general *awareness of the environment,* the ability to think about it and then to act on judgments. Second, there is *individualization,* "a prerequisite for the human type of socialization and finds in the latter opportunity for its greatest possible development" (Simpson 1949, 261).

All in all, concluded Simpson, although we humans are not necessarily highest on every scale, and although certainly he would not want to claim that there is any necessity about this, the notion of evolutionary progress is meaningful and our species sits alone on the peak. But, we then ask, do we sit on a peak with the ropes pulled up around us? For all of his friendship with Huxley, Simpson had little time for this viewpoint. He may have been a progressionist, but he was smashing nearly a hundred years of tradition that included Cope's law. To Huxley he wrote:

> A difficulty is that "generalization" in a full sense does not really exist in nature, and that "specialization" is often judged *a posteriori.* Viviparity was evidently more "specialized", oviparity more "generalized," so we have, as everyone knows, specializations that are progressive and that broaden opportunities rather than restricting them. Other specializations preceded extinction, so we judge them bad for progress, but how hard it is to achieve a rigid criterion that could possibly have predictive value! (Letter to Huxley, August 2, 1950; Simpson Papers)

As Simpson explained in a later letter, he did not want to deny that, in some sense, evolution may be funnelling itself into a channel: "There is, to be sure, strong evidence that evolution is running down in one sense of the words. The physical, structural and functional differences involved have tended to become less and less radical" (letter to Huxley, December 19, 1950; Simpson Papers). But Simpson did not want to preclude further significant progressive advance. It is true that the horse cannot get less than one toe, but who is to say that something as specialized as the nervous system has exhausted its options? "I cannot feel any confidence in prediction that the nervous system is now at its limit, or even that quite other possible channels of progress, in the nervous system or in other systems, are really non-existent" (letter to Huxley, August 2, 1950; Simpson Papers). Indeed, Simpson did not want to preclude the re-evolution of a human-like form, perhaps from an opossum. It is not likely: "Yet it remains true that manlike intelligence and individual adaptability have high selective value in evolution and that other animals have a conceivable basis for similar development" (Simpson 1949, 327). Proofs of humans' necessary uniqueness sailed too close to the orthodoxly theological for Simpson's taste.

What of Simpson's beliefs about Progress of some kind or another, and what of their possible connections to his thoughts on biological progress? A clear picture emerges. Simpson did believe, very strongly, in Progress. Moreover, this belief was reflected in his biology. Start with Simpson's religious belief, or more precisely, with his lack of it, in a formal way. Though he was raised a strict Presbyterian, a fundamentalist no less, Simpson's faith faded as he grew to adolescence. It was never to be replaced, yet at the same time he felt a void, the need for something to take its place: "I want desperately to believe in something more comforting, but how can one? Inner Faith?—a gift which cannot be sought, and probably a false gift since it is common to all forms of religion and since faith often has been placed in things which are obviously false. Miracles?—none has ever been authenticated for which a natural explanation was not conceivable. Divine consolation?—Many people are consoled by delusions and errors" (letter to his sister Martha, March 5, 1931; in Simpson 1987, 156).

Gradually, science began to take the place of his childhood beliefs—science as a way of life, as something which gives meaning, and as something to which one can make a commitment. To his parents, looking after his children on one of his collecting absences, he wrote in thanks: "I do not know how I can repay you, but I deeply appreciate it, I love you very dearly, and what I can accomplish for science and for the richness of human knowledge is due as much to you as to anything or anyone" (October 1933; in Simpson 1987, 185). But what gave science its meaning and its importance was the fact that it, above all other human accomplishments, is Progressive—it is the epitome of forward-moving human knowledge. Contrasting it with art, Simpson argued that "little as I know about art, I know that for me it would be vastly less satisfying than science for (please keep your seat) science progresses, it *does* find out new things that were absolutely never known before. Art does *not* progress and can't" (ibid.).

The difference, apparently, is between the unsatisfactory static and the more pleasing dynamic: "Beauty, relative as it is, & personal as it is, is stationary & cannot be expanded. Scientific knowledge can be & is expanded. That is what I mean by art being (always from my own personal point of view, in ordering my own life)—being impossible of progress. When I find & describe a new pantothere I am making a positive & permanent advance over all that the human race ever knew before, and that is a fine thing, even tho it be inconsequential & futile, as fine things so often are" (letter to Martha, April 20, 1927; in Simpson 1987, 82).

Despite a lack of formal belief, emotionally, Simpson (like the Huxleys) was a deeply religious man: "I respect religion, any religion in which men believe honestly" (letter to Martha, March 5, 1931; in Simpson 1987, 156). If he himself could not accept dogma, he got as close as he could by worshipping with the Unitarians for many years. The idea of Progress, therefore, bound as it was to the act of science, was central to Simpson's very being. This is not the robust scientism of Haldane, who displayed almost a love of technology for technology's sake, but science as a sacrament, scientific discovery as the ultimate miracle. Although he could not accept the science/religion synthesis of his good friend Teilhard de Chardin, Simpson's affection for Teilhard was because of the priest's faith, rather than despite it. Where the Presbyterian disapproval entered was over Teilhard's womanizing—or, rather, over the duplicity of one who voluntarily renounced sex in the name of his Lord and then traveled, openly, with his mistress: "I do not myself consider poverty, chastity, and obedience to be virtues, but anyone who voluntarily took what he recognized as sacred vows to observe them and then egregiously and constantly broke all three vows throughout his life without ever renouncing them—such a man must be considered a hypocrite" (letter to D. A. Hooijer, July 11, 1972; also letters to S. J. Gould, April 14, 1976; J. Hemleben, May 28, 1965; L. Laporte, April 7, 1984; Simpson Papers).

A view of science as sacred points toward connections between biological progress and scientific Progress, for evolution in itself was ever at the heart of Simpsonian theology. As he developed his philosophy of life (especially in *The Meaning of Evolution*), the links were forged even more strongly: "In the basic diagnosis of *Homo sapiens* the most important features are probably interrelated factors of intelligence, flexibility, individualization, and socialization." Apparently, these features occur (and mark progress) throughout the animal kingdom. However: "In man all four are carried to a degree incomparably greater than in any other sort of animal" (Simpson 1949, 284). Simpson argued further that not only is knowledge, especially scientific knowledge, essentially Progressive, but that we have a positive moral obligation to see that this Progress continues: "promotion of knowledge is essentially good" (Simpson 1949, 311). He meant this primarily at the cultural level, although (incidentally) he did drop hints that he would not be entirely opposed to biological schemes aimed at raising overall human thinking ability.

Yet, with ability comes responsibility—and to have responsibility, we must have freedom to choose and to sift ideas. This freedom is linked with individuality. From the fact that humans are high on the scale of individualization "arises the ethical judgment that it is good, right, and moral to recognize the integrity and dignity of the individual and to promote the realization or fulfilment of individual capacities. It is bad, wrong, and immoral to fail in such recognition or to impede such fulfilment" (Simpson 1949, 315). From here, Simpson slipped easily into a condemnation of totalitarianism and a praise of democracy. Writing in the late 1940s, Simpson was obviously reflecting a loathing of the defeated fascist states. In his wartime letters home to his wife, Simpson had emphasized his hatred not just of the political systems of the enemy but of the very German and Italian peoples. Obviously, as the century came to its midpoint and the Cold War settled into its long winter, Simpson had also in mind those societies then perceived as America's "peacetime" opponents. This was the time of the Lysenko affair.

Tight indeed is the connection between Simpson's views on biological progress and social Progress. Can we push the connection, perhaps going so far as to see a direct analogy between biological and cultural evolution? We are well on the way to making this claim, inasmuch as the well-functioning biological society prized individualization, as also does the well-functioning social society. But we can go further, if we judge by an (unpublished) address to the American Academy of Arts and Sciences in 1960. Biologically: "Orientation in evolution is adaptive, i.e. it tends toward an adjustment between the whole organism and the whole of its environment that is favorable to survival of the population of organisms under existing circumstances" (Simpson Papers, File 31, v). This is brought about by the sequence "prolificness → struggle for existence → survival of the fittest." Analogously in culture: "Orientation in social evolution is usually, but probably not always, adaptive in the same sense as biological adaptation. At least it tends in any one culture in a direction of increased personal values as values are accepted by a majority of the population. Whether such values realistically favor biological survival is a moot point. There is certainly, as a special case of this general tendency, a 'struggle for existence' among cultures."

In Simpson's case, it is unnecessary to go on to ask our usual third question, about the extent to which Simpson's views on biological progress outstripped the empirically given. By his own admission, he was imposing his views on his biology, rather than reading from it: "In sober enquiry, we have no real reason to assume, without other standards, that evolution, over-all or in any particular case, has been either for better or for worse" (Simpson 1949, 241). The conclusion that the cultural value controlled the biological claim is a conclusion that Simpson would have been happy to share with us.

Botany and the Synthesis

George Ledyard Stebbins Jr. (1906–), author of *Variation and Evolution in Plants,* the botanical contribution to the synthesis, was born of a successful New York lawyer and a mother high on the social scale. Sent to Harvard, Stebbins rather drifted into botany, as much because other options failed to interest him. Soon he was hooked, however, and Stebbins launched into graduate work and a full-time career as a botanist. Even as an undergraduate, he had been an enthusiastic Darwinian, although without much sympathy from his elders and supposed betters. Of a second-year biology course, taught in part by the zoologist G. H. Parker, Stebbins wrote: "His lectures on Charles Darwin and Natural Selection set me on fire. Even during the last moments of his first lecture, I had begun to think about the problem." Unfortunately, Parker himself belittled the significance of selection. "I became like a young lover who had just fallen in love with a glorious vision, only to be spurned and cast aside. However, I was not shaken in my belief, and during my career as an evolutionist have welcomed the revival of natural selection as of prime importance for major evolutionary changes" (Smocovitis 1988, 20–21).

As for Mayr and Simpson, the true catalyst, the really powerful causal

George Ledyard Stebbins

force, was Dobzhansky. Although Stebbins had heard Wright's 1932 presentation, he was still mystified by his model. "I never really understood him. I don't know that I understand him even now. My mathematical background has always been poor. I've never been adept at mathematics" (interview with author, May 25, 1988). The adaptive landscape metaphor struck home, however. Unforgettable was "the poster demonstration by Sewall Wright in which all of us saw for the first time all those diagrams in his 1932 paper that Dobzhansky used in the first edition of *Genetics and the Origin of Species.*" It took Dobzhansky to turn Wright's idea into a fully fledged theory, and it took Dobzhansky also to persuade Stebbins of the full significance of the new theory, to the point of pushing him into writing a botanical component to the synthetic theory:

> Q. Did you consciously write *Variation and Evolution in Plants* as the botanical equivalent to the books of Dobzhansky, Mayr, and Simpson?
>
> A. That's exactly what Doby asked me to do. He invited me to give the Jesup lectures in 1946 . . . While I was there, [in New York] he would have it no other way but that I live in the same apartment as him. We ate our meals together. We walked back and forth to the campus together. I just got loaded with hints. He said: "Ledyard, I just want you to see what you can do with population concepts in plants—with adding plants to the synthetic theory." This was a commission from Dobzhansky. Really paying him back for all of the wonderful things he did for me. (Interview with author, May 25, 1988)[2]

Finally, fifty years after the rediscovery of Mendel, the synthetic theory was complete.

After the writings of the animal biologists, there is much in *Variation and Evolution in Plants* that strikes an unfamiliar note. There is, for instance, a detailed discussion of the phenomenon of hybridization. This occurs in the animal world, but most authorities denied that it is a significant factor in evolutionary change. In the plant world, however, hybridization cannot be so quickly dismissed. It is often a dead-end step; nevertheless: "the selective advantage of occasional hybridization between species, that is, the ability to produce radically new adaptations to new environmental conditions which may arise, may outweigh the disadvantage incurred by the sterility of such hybrids" (Stebbins 1950, 252–253). Another phenomenon, essentially alien to the zoologist but likewise given major treatment by Stebbins, is that of polyploidy (the multiplication of chromosome sets during reproduction). In the so-called higher plants it is widespread, and indeed is characteristic of a great many valuable crop plants: wheat, oats, cotton, tobacco, potato, banana, coffee, sugar cane (p. 299). It too can allow rapid adaptation to fast-moving environmental changes.

There were other unfamiliar topics which got detailed treatment from Stebbins—for instance, asexuality. Yet, as one reads through his pages, one starts to realize that the underlying structure of Stebbins's view of the plant world was precisely the synthetic theory put forward for animal life, with the peculiarities merely variations on the top. Most particularly, in the world of plants as in the world of animals, Stebbins saw the neo-Darwinian mechanism of natural selection working on random variation as the central force of evolutionary change. There is variation in plants, just as there is variation in animals. This is the basis of differential reproduction, change, and adaptation. Backing this claim was a massive amount of empirical evidence on plants. Then the conclusion was drawn:

> The material presented . . . is intended to show that individual variation, in the form of mutation (in the broadest sense) and gene recombination, exists in all populations: and that the moulding of this raw material into variation on the level of populations, by means of natural selection, fluctuation in population size, random fixation, and isolation is sufficient to account for all the differences, both adaptive and nonadaptive, which exist between related races and species. (Stebbins 1950, 152)

This is the message that Stebbins has kept repeating in books and articles, for the full four decades since *Variation and Evolution in Plants* appeared.

Progress and the Plant World

It was Asa Gray who opined that, given there is no obvious candidate for the zenith of the plant world as there is for animals, the whole notion of botanical progress is in a sense unneeded, if not simply inapplicable. A quick look at *Variation and Evolution in Plants* is not exactly calculated to raise one's hopes. Apart from anything else, the book focuses more on mechanisms than on tracing out paths of evolution. Stebbins was just not about to make the sort of inquiry which leads to thoughts of progress.

If we have learned anything by now, however, it is that answers are there for those willing to seek. So, let us start with one certain fact, namely that in addition to his regular scientific works, Stebbins—like other major architects of the synthetic theory—has been given to writing books directed to a more popular audience. Moreover—again like the other major architects of the synthetic theory—in those other writings Stebbins has shown himself an explicit and enthusiastic biological progressionist (see table). Indeed, he has even gone so far as to write one whole book exclusively on the subject (Stebbins 1969), as well as label a textbook *Darwin to DNA: Molecules to Humanity.*

The key notion to understanding Stebbins's views on progress is that of "hierarchy." He believes that the living world is organized in a layer-like fashion, with (eight) ever-increasing degrees of complexity. These degrees are not given any formal characterization but are taken as intuitively

Hierarchy of life, from *The Basis of Progressive Evolution* (1969), by George Ledyard Stebbins

Level	Examples	Years ago when first appeared
8. Dominance of tool using and conscious planning	Man	50,000
7. Homoiothermic metabolism (warm blood)	Mammals, birds	150,000,000
6. Organized central nervous system, well-developed sense organs, limbs	Arthropods, vertebrates	600,000,000 450,000,000
5. Differentiated systems of organs and tissues	Coelenterates, flatworms, higher plants	1,000,000,000 ? 400,000,000
4. Multicellular organisms with some cellular differentiation	Sponges, algae, fungi	2,000,000,000 ?
3. Division of labor between nucleus, cytoplasm organelles	Flagellates, other protozoa (eukaryotes)	??????
2. Surrounding cell membrane with selective permeability and active transport of metabolites	Bacteria, blue-green algae (prokaryotes)	3,000,000,000 ?
1. Earliest self-reproducing organic systems	Free-living viroids (none still living)	??????

obvious or illustratable by example. Thus, for instance, working down from macro to micro within an animal muscle we find fibers, and then single fibrils, filaments, and—at the bottom—myosin molecules.

> In the long run, organisms repeatedly have evolved new ways of exploiting the same or similar environments. In doing so, their bodies have from time to time evolved new levels in the hierarchy of complexity from macromolecule to organelle, cell, tissue, organ, and organ system. Achieving these levels required the accumulation of new genetic information, concerned largely with the integration of development and metabolism and with regulating the translation of genetic information into form and function. It seems to me that from the standpoint of life as

> a whole, the achievement of these new levels of complexity can be designated as progress, wherever they have arisen. (Stebbins 1969, 29–30)

There is a striking difference in approach here between Stebbins and his fellow synthetic theorist, G. G. Simpson. For the latter, there is no unique mark of progress and essentially it is all a matter of interpretation. For the former, although he sometimes gets a little self-conscious about his position ("Progress is not an intrinsic quality, that exists independently of human thought"), essentially there seems to be some sort of objective—or at least quasi-objective—mark of advance, which is signaled by achievement of a certain level of (albeit unexplicated) complexity. Not that Stebbins, any more than Simpson, favors an inevitable teleological rise up the chain of progress. Rather, from Stebbins we get an image of organisms feeling their ways onward, driven or propelled by the dictates of adaptive necessity. They are forever probing according to selective demands.

In other words, complexity is not of value in itself—is not self-manifesting as it were—but the material basis for efficient adaptive powers. In general, the better the adaptation—especially, the better the adaptation leading to some sort of breakthrough—the more the complexity.

> Q. How would you define progress?
>
> A. The phenomena that in biology analogous to what we call progress in humans are any one of various successions of evolutionary changes that bring a particular line of populations of organisms toward some particular adaptation that later turns out to be very widespread.
>
> Let's compare, among amphibians, salamanders with frogs. There I would say frogs are more progressive than salamanders, because the frog structure not only enables jumping but once the jumping habit of the ranid type is perfected, there was a radiation all over—tree frogs, toads, all sorts of variants on the frog pattern—whereas if anything salamanders lost a little, they remained very sluggish and so on. And while they still exist, they certainly don't dominate the field to any extent.
>
> There is I would say, progress for frogs. No progress, or little progress if any, for salamanders. (Interview with author, May 25, 1988)

But why should there be an upward movement, through the levels of complexity? At this point, Stebbins introduces his own innovation. Like everyone else, he starts with Sewall Wright's (1932) landscape picture of organisms making their ways up, or sitting on the tops of, peaks. For Sewall Wright, the height of the peak would be linked to adaptive superiority—and so it is also for Stebbins. For Stebbins, however, in addition the peak represents complexity—this follows from his linking of greater adaptive powers with greater complexity. This connection is crucial, because once a species acquires complexity, it is in a very real sense impossible to lose it. There is a kind of "ratchet" effect, which entails that complexity keeps building on complexity, without falling or slipping back.

Indeed, synthetic theorist though he is, Stebbins sees this process—which is based on a principle he calls the "conservation of organization"—as coming about through a kind of modern equivalent of the biogenetic law. What happens in individual (ontogenetic) development is that organisms build on what has gone before, compressing the past down the hierarchy of growth. A level of complexity in development is achieved that cannot be wrecked, else the adult organism would cease to function: "Once a unit of action has been assembled at a lower level of the hierarchy of organization and performs an essential function in the development of organization at higher levels, mutations that might interfere with the activity of this unit are so strongly disadvantageous that they are rejected at the cellular level and never appear in the adult individual in which they occur" (Stebbins 1969, 105). As evolution proceeds, if a species is to jump up in adaptive worth—in some real quantitative sense—it must build on what has been achieved, rather than take away. Hence, through phylogeny we get the development of greater and greater complexity.

One final question. Where do humans stand on this upward ladder of progress? Sometimes Stebbins writes, and talks, as if his prime concern is with a comparative progress of the Darwinian kind. "I think that progress, let's say of the anteater line towards the aardvark for efficient subsistence on termite nests, is just as much progress as the evolution of humans to society and supposedly greater intelligence." However, Stebbins really thinks that we humans are rather special and do have a favored status at the top.

> The really distinctive feature of humans as compared to any other animals is the transmission of principles of ethics and so on by imitation and learning and the fact that our life is based on that. No other species has combined the trait of working with another for little more than recognition of the other person's worth. That I think is the crux of humanity. (Interview with author, May 25, 1988)

In Stebbins's opinion, all of the great religions of the world are built on this insight. "We are not just another animal."

Obviously, we are now at the edge of, if not already into, our next question, namely about Stebbins's thoughts on cultural Progress. Nor will there be any surprise to find that Stebbins is an ardent champion of human Progress. Through our science and technology we humans have pushed ever onward and upward. "In respect to cultural evolution, we see [Progress] most clearly in the invention and perfection of increasingly complex and efficient methods of travel and communication" (Stebbins 1969, 133–134). Moreover, the same causal processes are at work whenever Progress is an issue: "in both organic and cultural evolution, the ultimate progress is successive levels of greater complexity can be attributed in large part to the principle of the conservation or organization." Thus: "changes leading to a higher order of organization, particularly when they make efficient use of the lower order as units in the higher order, may from time to time increase the potentialities for dominance of either a species of animal or human society" (Stebbins 1969, 134–135).

Fleshing out his discussion, Stebbins gives carefully drawn reference to the biologically evolutionary virtues of small isolated populations of highly atypical individuals: "Is it not likely that the same principle holds for sociocultural evolution? For populations of exceptional genotypes substitute communities of people having exceptional combinations of ideas, inventions, and customs. Unless such communities are allowed to flourish and pursue their own course around the fringes of society, we may be held down forever by the dead weight of tradition that binds so many of us to old, often obsolete ways of doing things" (Stebbins 1969, 142–143). Dare one inform the reader that Stebbins, writing in the late 1960s, was a faculty member on a Californian campus?

Is it proper to conclude that Stebbins's beliefs in cultural Progress fed directly back into his science, thus making him a biological progressionist? He has certainly asserted this: "Progress is not an intrinsic quality, that exists independently of human thought. It is a metaphor, originally used to characterize certain properties of individual humans, i.e. Pilgrim's Progress, or societies, e.g. progress toward democracy or toward a more perfect union of our United States. Its application to theoretical biology is a secondary extrapolation" (written comments on a talk by the author, April 25, 1988). Yet, in a way, rather than seeing a straight analogy between biology and culture, it is more appropriate that we interpret Stebbins's thoughts on biological progress and cultural Progress as part of one overall world picture. It would be too slick simply to refer to Stebbins as Spencerian, but there are parallels—for both men, complexity (heterogeneity) emerges through evolution because of ontogenetic constraints. This line of thought ties in with influences closer to home: Stebbins was at Harvard in the 1920s, and, as he moved across the spectrum in search of an adequate world view, he would have got a full blast of the same philosophy that moved Wright: "I became a unitarian, a liberal, when I was at Harvard. I shed my episcopalian religion and my republican political affinity. I was a democratic unitarian when I got mixed up with the liberal club at Harvard" (interview with author, May 25, 1988). No wonder he has a fondness for hierarchies and took so readily to the landscape metaphor.

Although Stebbins has pulled back somewhat—at one point he was an ardent socialist—his overall philosophy has long been that of a liberal humanist. He stresses that he is "not militant atheist," but it is many years since he was able to believe in any kind of personal God. His commitment has always been to the best in humankind, especially with regard to care and cooperation: "and the nearest I can come to it is that it is a synthesis of human morals, thoughts, aspirations which is handed on down and is something strong enough for many people to live by." Into this world view, a strong commitment to P/progress, biological progress leading to humans and cultural Progress taking up the development of humans, fits naturally.

> When I read about England in the early nineteenth century reading *David Copperfield* say, or *Pickwick Papers,* I say certainly for the middle classes things are better now than they were at that time in England. Although they were better in England at that time than they were in most other countries. So that is Progress. When I think when I was in college, as far as racism is concerned, the word "Kike" for a Jew, everybody used it! Supposedly intelligent Harvard students used it. And the word "nigger" was equally common. And, the great heroes of sports, like Babe Ruth in the 1920s were wildly racist, both against blacks and against Jews. These things aren't happening now so there are certainly changes. (Interview with author, May 25, 1988)

For a man who thinks like this, a belief in biological progress is well-nigh inevitable.

Statements like these make a little redundant any queries about Stebbins's science and its measure against his progressionist beliefs. Clearly his biological progressionism does outstrip the empirically given. The claims about adaptation and complexity, and progress upward, are hardly read directly from nature. One need only mention the kinds of criticisms Simpson brings against complexity as a mark of progress to see this point. If more is needed, one might add that Stebbins's failure to give a quantitative analysis of his notion of complexity renders his account so loose in respects as to be compatible with almost any (at least, far too many) states of nature.

Here again, therefore, we must conclude that the culture and the biology hybridized into one living whole.

Forging the Synthesis

The synthetic theorists were proud, ambitious, hard-working men. They loved their subject and they wanted respect for their labors—respect for their labors *as evolutionists.* "When I was at Harvard during the 1920's, physiology, experimental embryology and chromosome cytology attracted the best minds, and evolution was regarded by many biologists as a sort of fuzzy-minded natural history, of a lower grade" (G. L. Stebbins, Mayr's questionnaire; Darlington Papers). The evolutionists needed to turn their subject into a professionally organized and recognized discipline. Just as there were professional morphologists and professional physiologists, they wanted recognition as professional evolutionists. It is significant to learn that when he wrote his doctoral thesis, back in Germany in 1926, Mayr had actually shrunk from evolutionary speculations, precisely because at that time they were not thought to have the status of legitimate professional science. It was only as he absorbed more fully the culture of the museum, the one place where evolutionism throve as part of one's profession, that it began to infuse his work. And this new emphasis on evolutionary speculations made pressing the need for a discipline (Beatty 1994).

It is to the forging of the synthesis, as a professional discipline, that I turn now—ever mindful of the implications for progressionism. *Genetics and the Origin of Species* was the guiding light, the intellectual inspiration, the epistemological foundation, and through Dobzhansky's encouragement of Mayr, Simpson, and Stebbins individually, a basis for a synthesis. In addition, there were pragmatic reasons for cooperation (Cain 1992). By about 1940, Simpson as a paleontologist was finding it particularly difficult to get grant money. This applied especially to sources which traditionally funded professional biology: "I had a *talk* today with Dr. Hanson, of the Rockefeller Foundation, in the hope that his organization might support the proposed research on variation and speciation. He was very dubious about this, especially on the grounds that the study does not in fact have much bearing on genetics or the problems of experimental biology" (letter to Dobzhansky, January 22, 1941; Simpson Papers). Self-interest was a fuel for the synthesis, as it is a fuel for so many other human activities.

Thanks to prodding by Julian Huxley, formal organization of evolutionary studies began around 1940. Building on the intellectual ferment and personal interests, Dobzhansky and Mayr and others formed a small Society for the Study of Speciation. Because of the war, this group seems not to have developed very quickly, but by 1943 a Committee on Common Problems of Genetics and Palaeontology (with "Systematics" added shortly thereafter) was circulating stenciled bulletins (edited by Mayr). Very much in the spirit of synthesis, it encouraged evolutionists from different areas to talk openly to each other. Supported as the evolutionists were by the (U.S.) National Research Council, they were fortunate recipients of a then-general move in science organization toward the fostering of collaborative work. The example of U.S. industry and (especially) the war effort was apparently a major influence here. Presumably, the synthetic theorists themselves were stimulated directly by these factors.

At the end of the war, things moved rapidly. In 1946, on the ashes of the moribund Society for the Study of Speciation, a new, better-organized body was formed. This group, the Society for the Study of Evolution (President: G. G. Simpson; Secretary: E. Mayr; Council Members: S. Wright and Th. Dobzhansky), had synthesizing as its explicit mandate: "The Society is a common meeting ground for representatives of all fields of science concerned with organic evolution, including genetics, palaeontology (vertebrate, invertebrate, plant), taxonomy (animal, plant), ecology, anthropology, and others" (E. Mayr, "History of SSE," unpublished; *Evolution* Papers).

In 1947, as part of its bicentennial celebrations, Princeton University hosted a major conference on "Genetics, Palaeontology and Evolution" (Jepsen, Mayr, and Simpson 1949). The self-congratulation following in its wake suggests that it served as much to inspire as to inform. Then, everyone's dream came true. Funds were obtained and a journal was launched (Cain 1994). The first editor of *Evolution* was Ernst Mayr, who strove mightily (albeit at times with patchy success) to see that all branches of the synthesis would be represented. Paleontology and botany were not to be pushed out by *Drosophology!*

At the same time this group activity was occurring, the evolutionists were moving back from the museums into the university departments. After the war, Simpson got a functioning adjunct professorship at Columbia University. Mayr began teaching, and shortly after the half century moved to Harvard—part of the Museum of Comparative Zoology (MCZ) but also as part of the biology faculty. At the end of that decade, Simpson joined him there. Previously, Castle had taught one course focusing on evolution (in 1908–09) and E. M. East another (in 1937–38) (Smocovitis 1992). Now, Mayr and Simpson team-taught a course in evolution, and they plotted graduate programs in evolutionism.

> *Introduction:*
> (1) Scope of Evolutionary Biology. All biology influenced by concept of evolution but it usually goes no further than phylogenetic interpretations of specific phenomena. Research in evolutionary biology is predicated on the principles of evolution and in turn contributes to it. Included here: systematics, palaeontology, population genetics, ethology, animal behavior, and comparative aspects of morphology and physiology.
> (2) Recent trends of research in Evolutionary Biology. Mention ecological genetics, population ecology, behavior.
> (3) Need for extensive training in *biology* in contrast to needs in molecular biology. (From a proposed prospectus, in 1961, for "Graduate Study in Evolutionary Biology at Harvard University"; Simpson Papers)

Major efforts were made to get evolutionists into high-status scientific posts and organizations, like the National Academy of Sciences. Dobzhansky was a great help here, especially in later years as his own students came to maturity. First there was Richard Lewontin. "I never voted for any candidate with greater enthusiasm. My paternal pride is immensely swollen!" (letter of April 24, 1968; Dobzhansky Papers). "I am particularly happy to think that something that I have done has added to your own life since I owe so much to you" (reply by Lewontin, May 3, 1968; Dobzhansky Papers). Then came Bruce Wallace—a difficult candidate, given his involvement in the classical/balance controversy. Then, right at the end of his life, Dobzhansky was politicking away on behalf of Francisco Ayala.

All of this sounds like a success story, and of course in a major way it was. Yet, as in England, ever-present constraints threatened to hold back the evolutionists in postwar America. From above, there were the already-established—or soon-to-be-established—biologists. They had no respect whatsoever for evolutionary studies: certainly not for evolutionary studies considered as professional science. When Simpson and Mayr went cap-in-hand to the American Philosophical Society, in 1947, to ask for $5,000 to start *Evolution,* the one person on the committee who was opposed was the representative biologist, Edwin Grant Conklin. For him, prolific writer on the subject though he was, evolution was just not professional science (Mayr, interview with author, March 30, 1988).

Some years later, in 1955—when the Society for the Study of Evolution and the journal *Evolution* were well established, but also when molecular biology was starting its dizzy climb upward—the lowly status of their subject was brought home brutally to the evolutionists at a National Research Council conference on scientific concepts. Asked to identify "the *few* [4 to 8, as it turned out] concepts and principles *most* basic for biology," virtually everyone in the group—biophysicists, biochemists, physiologists, embryologists, and Mayr, Simpson, and Wright—named some form of organization and equilibrium. Almost no-one other than the evolutionists named natural selection (Cain 1992).

The secret worries of the evolutionists were confirmed. Even that eternal optimist Dobzhansky wrote a rather sad review, in 1960, of a book on taxonomy by Simpson (1962). Explicitly, Dobzhansky acknowledged the world's ordering of things. He defiantly (and revealingly) argued: "It is really the presumption that being at the top of the present peck order somehow confers a high intellectual status, together with claims for juicier research grants and academic preferments, that must be disallowed." He then went on to say that although "much of the work in molecular biology is exciting and thought-provoking, some of it is intellectually shallow." But, in the end, he had to admit that: "G. G. Simpson's book is about taxonomy—near the bottom of the peck order" (Dobzhansky Papers). After this confession, it would be cruel to repeat what Jim Watson, Mayr's sometime colleague at Harvard, said about traditional biology in his *Double Helix* (1968).

The synthetic theorists were threatened from beneath, also. In the 1940s and 50s, there were still lots of people around (as there are today) who were happy to parade evolutionism—their version of it, that is—right through the public arena. They wanted to make (or keep) a popular science of evolutionism, usually because they saw a way thereby to introduce and promote their metaphysical or religious beliefs. As the paleontologist of the group, Simpson was particularly sensitive to the messages of these people, because often they wanted to stake their case on the historical record—precisely that on which Simpson was building his case for a professionalized Darwinism.

In the late 1940s, Simpson's millstone was the French scientist Lecomte du Noüy, author of the best-selling *Human Destiny.* Endorsed, Simpson noted glumly, by clergy, journalists, and even some scientists, not to mention the *Reader's Digest,* the work pushed just the view of evolution that was anathema to the synthetic theorists—especially to the theorists as founders of a professional discipline. It postulated a special line of ascent to our own species, pointed to a metaphysical urge ("telefinalism") driving this line, and offered "proof" that life could not emerge through behind chance. And God's will hovered over everything. What made things even worse was the fact that, although the author presumed to speak with authority on things evolutionary, his background was in physiology and biophysics. In the 1950s, some people, like the retired plant scientist, Edmund W. Sinnott (1955), were pushing a neo-vitalistic evolutionism—centering on a "Principle of Organization," which appears remarkably like God ("a liberal American Protestant one," remarked Simpson sourly). And, of course, swamping everything there then came the writings of Teilhard de Chardin (1955): learned and popular, and (horror of horrors) accepted enthusiastically by J. S. Huxley and Dobzhansky. As they were being belittled from above by the established and successful biologists, the synthetic theorists were being threatened from below by popular writers, whose works suggested that the established biologists might have a good point!

The twin sets of pressures did not mean the failure of the discipline-building enterprise.[3] But the pressures did mean that the synthetic theorists had their work cut out for them, and that they had to accept disappointments with triumphs. Take, for instance, the matter of cash, as much a question of concern for the Americans as for the British. A promising possibility was an application by Mayr, Simpson, and Dobzhansky to the Rockefeller Foundation. In the early 1950s, under the direction of Warren Weaver, this institution was trying explicitly to rejuvenate biology through the concepts and methodology of the physical sciences. In the end, however, this philosophy proved too strong, and proposals in evolutionary studies met with a flat refusal (Beatty 1994). The same negativism existed among private funding agencies, and for essentially the same reasons, that Simpson had encountered a decade earlier.

Fortunately, the government counterpart, the National Science Foundation, was newly founded and started to give out money in 1952 (Appel 1992). The Foundation did much to encourage biology generally, but the evolutionists had to fight for their share. Initially, whereas such subjects as comparative physiology and experimental embryology were included, evolution was left out of the designated grantable areas, leaving it to slip in under such areas as marine biology or genetics. Later, the division was made so as explicitly to recognize "systematic zoology," which obviously did make a real place for evolution, and although there were failures (turned down: a project to consider the evolution of insect resistance to DDT) as well as successes (funded: "Origin and evolution of caste behavior among certain bees"), overall the evolutionists did comparatively well.

Not that any of their gains came by chance: the evolutionists had to push hard for their place in the sunshine of professional science. They did this in part in time-honored ways, for instance by encouraging their friends and harming their enemies and critics. The referee's reports written by Simpson to the National Science Foundation are particularly insightful. In 1955, for example, we learn of a proposal by the paleontologist Everett C. Olson, then-editor of *Evolution,* that: "This application is first-class in every respect." In 1965, however, of an application by one of the leaders of the school of numerical taxonomy—a group trying to expel undue evolutionary hypothesizing from systematics—we find that: "His approach is narrow-minded and shows consistent lack of insight into biological, as distinct from strictly mathematical, aspects of the problems considered." It is hardly surprising that the application was rated "questionable."

Another generous source of funds, especially in the 1950s, was the Atomic Energy Commission. This body was worried—desperately worried—about the effects on humans of radioactive materials, especially from the fallout produced by the testing of atomic and hydrogen bombs. Hence, thinking that one might find some answers in population genetics, the explicit aim of which is to discern the nature and rates of mutations, and their consequent effects, the Commission backed much of the seminal work of the decade. Major beneficiaries were Dobzhansky's students,

STUDIES ON IRRADIATED POPULATIONS OF *DROSOPHILA MELANOGASTER**

By BRUCE WALLACE

Biological Laboratory, Cold Spring Harbor, N.Y.

(With Five Text-figures)

(*Received* 28 *July* 1955)

Introduction

It is well known that the widespread use of ionizing radiations, because of their genetic effects, poses a problem regarding future generations. These radiations induce gene mutations. The vast majority of mutations have deleterious effects on individuals carrying them. Under the pressure of continued mutation, these deleterious mutations will accumulate in populations. Therefore, an irradiated population will, on the average, be harmed—have its 'fitness' reduced—by a continual exposure to irradiation.

The present article summarizes observations made on irradiated populations of *Drosophila melanogaster*. Some of the material presented here has been published previously (Wallace, 1950, 1951; Wallace & King, 1951, 1952). This summary, however, will introduce new material in addition to extending the original observations.

Material and methods

The experimental populations. The experimental populations of *D. melanogaster* are kept in lucite and screen cages. The original flies were obtained from an Oregon-R strain kept by mass transfer for many years. Fourteen lethal- and semi-lethal-free second chromosomes were extracted from this strain through the use of a series of matings identical to those described later (Fig. 2). Flies carrying these second chromosomes and mixtures of Oregon-R and 'marked stock' chromosomes other than the second were the parental flies of the populations.

Brief descriptions of the populations are given in Table 1. The left-hand column gives the identifying number for each population. The second column indicates the origin of the population. 'Stocks' indicates populations whose original flies carried lethal-free second chromosomes of Oregon-R derivation. Three more recent populations are subpopulations of populations 5 and 6; the designation in the table gives the parental population and the generation during which eggs were removed to start the new populations. The third column indicates the number of adults in the population cages: 'large' refers to populations of about 10,000 individuals, 'small' to populations frequently with fewer than 1000 individuals. The last three columns of Table 1 give the type of exposure, the dose, and the date the population was started for each population. Chronic exposure refers to continuous exposure to radium 'bombs'. No exposure in the case of popula-

* This work was done under Contract No. AT-(30-1)-557, U.S. Atomic Energy Commission.

An article on fruit flies by Bruce Wallace, a student of Dobzhansky. (Note the sponsor: the U.S. Atomic Energy Commission.)

who were happy to persuade the Commission that results of experiments on fruit flies have relevance to the fate of humans. Shades of butterflies and the Nuffield Foundation!

Funding aside, right through this period we find that the synthetic theorists tailored their attack to the specific situation. They worked hard at the epistemological side to their science, to achieve their sociological ends. They tried to make their work consistent with other sciences, to unify beneath overarching hypotheses, and so forth (Smocovitis 1992). By his own admission, a major reason Mayr took up philosophy in the 1960s was to find anti-reductionist arguments, to counter the threat of molecular biology.[4] He made much of the distinction between "proximate" and "ultimate" causes (the latter, uniquely, the domain of the evolutionist), and in parallel he looked for acceptable notions of organization and teleology (which he found and called "teleonomy") to drive back the forces of physics and chemistry as they moved into biology (Beatty 1994).

> I am not aware of a single biological discovery that was due to the procedure of putting components at the lower level of integration together to achieve novel insight at a higher level of integration. No molecular biologist has ever found it particularly helpful to work with elementary particles.
>
> In other words, it is futile to argue whether reductionism is wrong or right. But this one can say, that it is heuristically a very poor approach. Contrary to the claims of its devotees, it rarely leads to new insights at higher levels of integration and is just about the worst conceivable approach to an understanding of complex systems. It is a vacuous method of explanation. (Mayr 1969, 128)

Overall, it was personality which counted: "When one is not stationed at a university (as I was in the 23 years at the American Museum) and is almost consistently ignored in the awarding of honors, etc., and when one furthermore works in a Cinderella field like systematics, one perhaps owes it to one's self not to be too modest" (letter from Mayr to author, January 22, 1991).

The tensions and constraints on the synthetic theorists come through very plainly in the early years of the editing of *Evolution*. There was considerable soul-searching about the true focus of the professional evolutionist—and hence the proper content of the journal. One had to think in terms of causes, and it would be fatal to put everything in terms of strictly genetic causes. Nevertheless, one must certainly stay away from neo-Haeckelianism.

> I would like to discuss one or two thoughts with you that came to my mind while reading your paper. I think you might say that there are four major evolutionary problems:
> (1) The fact of evolution
> (2) The material of evolution
> (3) The origin of discontinuities
> (4) The course of evolution
> As far as (1) is concerned, it requires no integration with genetics and is the proper field of phylogenetic palaeontology, comparative embryology, and comparative anatomy. It is (2) that is the proper domain of genetics. It is (3) that is the principal domain of taxonomy since the origin of species is what you really meant by the origin of discontinuities. As far as (4) is concerned, it deals with the factors and causes of evolutionary change and it is here where the various fields concur and where the field of ecology enters the picture. Since there is so much agreement nowadays on (1), (2), and (3), the trend of evolutionary research has been very strongly toward (4). (Letter from Mayr to G. G. Ferris, March 29, 1948; *Evolution* Papers)

The emphasis of a professional evolutionism had to be on experimentation, or epistemic activities with the virtues of experimentation: hard evidence, the exercise of control, and the use of measurement.

> After all, even though there may be considerable agreement about the major outlines of the evolutionary field, there is still a great deal of disagreement as to many concrete problems. Furthermore, the field has reached a point where quantitative work is badly needed. Also, evolutionary research, as you realize, has shifted almost completely from the phylogenetic interest (proving evolution) to an ecological interest evaluating the factors of evolution. (Letter from Mayr to W. A. Gosline, March 5, 1948; *Evolution* Papers)

Papers which just went in for path tracing were rejected.

> Your manuscripts have been scrutinized by two readers and both of them report that they consider them unsuitable for publication in EVOLUTION. I have tried to get some detailed criticism for you (as you asked) but there seems to be nobody in this country now who is interested in phylogenetic speculations. (Letter from Mayr to F. Raw, August 2, 1949; *Evolution* Papers)

Above all, one had to avoid "philosophy." This was just not acceptable in a professional discipline.

> It has so far been the editorial policy of EVOLUTION to present concrete facts in every paper followed by the conclusions to be drawn from these facts. This policy was adopted deliberately because the prestige of evolutionary research has suffered in the past because of too much philosophy and speculation. (Letter from Mayr to G. G. Ferris, March 29, 1948; *Evolution* Papers)

Given these concerns, you can well imagine that certain ideas were discouraged or rejected by Mayr as editor of the journal. Above all, anything with even the slightest whiff of real teleology or special (upward) channeling was unwelcome. The break had to be made with the flabby, popular past.

> Your manuscript, "Orthogenesis in Evolution," has been studied by two readers of the Editorial Board who have reported back to the Editor that they do not consider the manuscript suitable for EVOLUTION. (Letter from Mayr to A. Cronquist, November 14, 1949; *Evolution* Papers)

Indeed, even the language of "orthogenesis" was taboo.

> It might be well to abstain from use of the word "orthogenesis" (harmless as it is, in my opinion), since so many of the geneticists seem to be of the opinion that the use of the term implies some supernatural force. (Letter from Mayr to R. H. Flower, January 23, 1948; *Evolution* Papers)

The simple fact of the matter is that the synthetic theorists were absolutely terrified of anything hinting of built-in direction, especially anything which interpreted such direction in terms of upward progress. Twenty-five years later, Mayr was still showing that the bogeyman lurked close to the surface of his thinking. In a questionnaire he sent to various synthetic theorists, about their memories of the movement, we get more information about Mayr's worries than about those of his respondents.

> Q. The 1930's were 75 years after 1859. To what do you attribute the long delay of the synthesis? Which misconceptions or deficiencies of factual information do you consider as particularly crucial?
>
> In what order would you rank these negative factors:
>
> . . . (7) A belief in a built-in capacity for improvement (orthogenesis, aristogenesis, etc.).

All of the old-fashioned talk about progress was the legacy of popular evolutionism. There was no place for it in the work of the professional evolutionist. To put it in the language of this book, in order to achieve disciplinary status it was essential that evolutionary science—the theory or paradigm in and inspired by Dobzhansky's *Genetics and the Origin of Species*—be put on the road to maturity. The emphasis must always be on epistemic values. Note how Mayr extols experimentation and (ironically so, given his and Dobzhansky's and Stebbins's discomfort with mathematics) the need for quantification and the like. As in Britain, these things count, and even if one cannot do them oneself, it is important to make sure that there are those in one's movement who can. And above all, the culture of professional science demands that one expel the culture of social belief.

Yet, as we know too well, all of these synthetic theorists were biological progressionists! Here we have the distinctive contrast with the most successful, discipline-building English evolutionists. For the Ford school of ecological genetics, the long-term historical dimension was missing and, given the tradition of Weldon, the question of progress did not arise. For the American evolutionists—partly for personal reasons, partly because the paleontological record was always a central component of the American synthesis—the question of progress did arise.

The problem had to be solved. How was one to reconcile the urge to progressionism with the opposing urge to professionalism? A number of strategies were tried, starting with one which we have seen already, namely to bring progressionism into line with the synthetic theory. Old notions like directed mutations really did get discarded. *Evolution* did not just drop Osborn-like language. It dropped Osborn-like ideas. And such ideas found no sympathy elsewhere—even in the most private correspondence. Leaving on one side for the moment the professional/popular divide, thoughts of progress were re-interpreted in terms of adaptive landscapes, and so forth. The theorists may have gone beyond the evidence—we know that they did—but they stayed within bounds of professionally accepted mechanisms.

Another strategy for handling the professional/progress problem was to restrict very carefully what one would allow—or, more precisely, what one thought one would allow—in one's professional science. Sanitized or not, the very idea and the language of progress were forbidden. Even speculation about phylogenies and the like was discouraged. Not that this latter could be an absolute prohibition—then or now. As an example of evolutionary change, Simpson talked about the evolution of the horse in both *Tempo and Mode in Evolution* and *Major Features of Evolution.* Significantly, however, his main work on the horse was a popular book (Simpson 1951).

The line to be trodden here was very fine indeed. If one's professional concerns banished thoughts of progress from one's work, then when one did want to think about progress, one would have no scientific base at all from which to start. Hence, what we do find among the synthetic theorists is a great interest in such topics as evolutionary *trends.* We have seen this already in *Tempo and Mode in Evolution,* and—thanks to Simpson's urging—it was one of the major themes of the 1947 Princeton conference. Although there was a fair amount of progressionist bashing—du Noüy's *Human Destiny* "exhibits an amazing ignorance of both genetics and palaeontology" (Romer 1949, 105)—there was also a fair amount of showing how "orthogenesis" and like phenomena can be derived on Darwinian principles: "A special case of 'orthogenesis' is the very common trend toward increasing body size seen in group after group. So often do we see this occurring in the fossil record that one gets the impression that this trend was universal . . . There are obvious advantages to size, in greater efficiency in energy conservation, greater safety from attack, and other features" (Romer 1949, 109). Qualifications were generally added—they were in this case—but a foundation for the progressionism was certainly well dug.

The third and perhaps most significant strategy was to extend that taken by Dobzhansky. It was to write a second book (or set of books) paralleling one's first book (or set). The difference would be that the first books were explicitly professional, whereas the second books were popular. As long as one kept the two apart—and did not blend them, as J. S. Huxley tried to—then the feeling was that the rules of professionalism were being obeyed. After all, this was precisely what everybody from T. H. Huxley down through Conklin had been doing for nearly a century. The break with tradition was that the evolutionists wanted both their professional and their popular books to be on evolution. So first they wrote their professional books. Then they revised them, gutted them of the mathematics (not much work here!) and the technical language, added a couple of chapters on progress, labeled the work in the preface "for the general reader," and the popular product was complete.

Simpson, who was the brightest and most self-conscious of the theorists, is the paradigm. *Tempo and Mode in Evolution,* finished in 1942 and published in 1944, contains not an explicit whisper of progressionism. *The Meaning of Evolution,* published in 1949, a work which Simpson stressed meant much to him (it was not a quickie, to make money), has all of the biology of the past, none of the population genetics, and concluding chapters on progress. Moreover, "This presentation is . . . written in the common language, not in the abbreviated technical jargon of the professionals, and it is not addressed primarily to the professionals" (p. 6). Then in 1953, with *The Major Features of Evolution,* an updating of *Tempo and Mode,* we are back in the land of the professional, and progress has vanished again.

Of course, there are variants on this pattern. Mayr's non-professional biological work tends to be explicitly philosophical or historical. He would think of it as being professional for those audiences. And one must give the other synthetic theorists credit for going beyond Dobzhansky, who wanted to restrict macro-evolution to the popular realm. Simpson was clearly having nothing of this. But, overall, both were following the same general formula.

Value-Laden Professional Science?

Grant that the synthetic theorists were successful in their drive to found a professional discipline of evolutionary studies. They truly worked and succeeded in the spirit of Darwin, overcoming (while accepting) the professional/popular division of Thomas Huxley. Yet, was there a possibility that culture—beliefs in P/progress, specifically—seeped back into the professional science? After all, the idea of Progress had been expelled only for the sake of achieving professional status, not because people had ceased to believe in it. The answer, surely, is that there was some two-way traffic between science and culture, with notions of biological progress being hauled into the professional science. Above all, there is the centrality to all the synthetic theorists' work of Wright's adaptive landscape, a deeply progressionist notion. Not that the metaphor entails progress logically: that it does not is one of its major selling points. If there were a necessary connection, no-one would touch it. It would be on a par with aristogenes. But, like an old coat, the metaphor slips very easily and comfortably onto a progressionist form. That is another major selling point. Certainly, it was a major attraction for the men of this chapter.[5]

Start with Mayr. As with Dobzhansky, it is true that Mayr's primary interests as a working scientist have never been in those aspects of evolution which lead most readily to thoughts, favorable or otherwise, about progress. He is a neontologist—a student of life today—not a paleontologist—a student of life in the past. Nevertheless, Mayr is nothing if not comprehensive, especially when it comes to evolution. And in his formal scientific writings, as well as in his more reflective thought about overall patterns, he declared quite positively for progressionism. In *Systematics and the Origin of Species* the discussion was brief and cursory. Nevertheless, Mayr made it clear that he saw movement upward as some species hop from one adaptive peak to another, higher one. More particularly, sometimes a real breakthrough is made, as an organism finds itself up in a whole new ecological zone.

> Wherever there is strong competition, specialization undoubtedly gives an advantage, and there is always the possibility that such specialization will lead to a new "adaptive plateau," with unsuspected evolutionary possibilities, such as were discovered by the mammalian and avian branches of reptiles while most other reptilian lines came to an end owing to overspecialization along a less-promising line. (Mayr 1942, 294)

One's sense that the focus here is on underlining real progress is the fact that the automobile analogy makes an early appearance.

Ideas of this nature were repeated through the years, until by the time of *Animal Species and Evolution* Mayr was ready to declare openly for progression, indeed tying the idea to (what is for him) that supremely significant event when organisms break into groups, most particularly when they break into species.

> I feel that it is the very process of creating so many species which leads to evolutionary progress. Species, in the sense of evolution, are quite comparable to mutations. They also are a necessity for evolutionary progress, even though only one out of many mutations leads to a significant improvement of the genotype. Since each coadapted gene complex has different properties and since these properties are, so to speak, not predictable, it requires the creation of a large number of such gene complexes before one is achieved that will lead to real evolutionary advance. (Mayr 1963, 621)

But where does all of this progress lead? Is it all just relative—comparative in the Darwinian sense—or is it absolute? It was obvious that the kinds of changes Mayr referred to added up to more than mere relative change, for Mayr topped off his long book with one chapter uniquely devoted to a single species, *Homo sapiens,* and his language betrayed his (absolute) scale of values: "Primitive characters" of *Homo erectus* include "its low forehead, the lowness of the skull as a whole, the high line of attachment of the occipital bones, the strong supraorbital ridges, and the heaviness of the skull." The brain is small, even though "this volume represents an enormous advance over the Australopithecines" (Mayr 1963, 632). Significantly, Mayr feared that progress has come to an end, and may—alas!—be on the decline. He noted the failure of brain size to increase over the past 100,000 years; he felt (as he still feels) that modern society confuses social equality with biological identity; and he agreed with Julian Huxley on the differential rates of human reproduction: "man's genetic nature degenerated and is still doing so . . . There is also the fact that modern industrial civilization favors the differential decrease of the genes concerned with intelligence . . . If this process were to continue, the results would be extremely grave" (Mayr 1963, 658–659, quoting Huxley 1953).

Simpson was a subtle thinker. You would certainly not catch him with overt progressionist claims in his professional writings. I have made this point repeatedly about *Tempo and Mode* and *The Major Features of Evolution.* Yet, the metaphor was there, and it did likewise structure *his* vision. Simpson's view of the past was one which could take a progressivist interpretation, and it could do so because of factors which, in the needed fashion, went beyond the empirically given. The key, as always, was Wright's metaphor, or rather Simpson's modification of it, and the way he thought that organisms occupy peaks and hop from one to another. Moreover, the hopping tends to occur under the force of selection, with Simpsonian progressive-type features becoming ever more manifested.

Life's changes are not just simply a matter of chance in Simpson's works, nor is a progressionist reading inappropriate. The niches are there, and note how (essential for an absolutist progressionist reading) it is part of Simpson's professional science that these niches or zones have a distinct objective ontology: "Ichthyosaurs became extinct millions of years before their mammalian ecological analogues among the Cetacea appeared. Pterodactyls were long absent before bats occupied a similar or overlapping zone. Dinosaurs became extinct before the larger terrestrial mammals so quickly radiated into much the same spheres" (Simpson 1944, 212). The niches exist and so do the pressures on organisms to fill them, as well as the ways organisms may find to escape upward. What has happened once was not pure chance. Simpson noted that there are remarkable parallels in evolution, as between the mammals of North and South America. Moreover, remember that he thought that the human niche is sufficiently stable and attractive that, were we to disappear, it might be filled by other intelligent beings. Biological progress has its reasons.

Hence, although Simpson did not boldly drag progress or Progress into his strict science, the framework of his science was in important respects rooted in the imagination—the imagination of convenience. Given the indebtedness to Sewall Wright's metaphor of an adaptive landscape, Simpson's professional science and Simpson's culture are not quite as distinct as one gleans from the surface impression he himself strove to maintain. Simpson wanted to get from evolution what he criticized his predecessors for reading in. Perhaps he was not so far from the tradition as he implied.

Finally, we have Stebbins. Is there reason to think that his P/progressionism fed back from his popular writings into his straight science? The answer is that although there can be no suggestion of progress up to humans in his science, which was botany, *Variation and Evolution in Plants* carried all of the elements of his theory of progress—and the same holds true of later writings. Consider modes of reproduction.

> The three levels recognized in [my] hierarchy of reproduction are as follows. The lowest is that containing those organisms which rely solely on the large number of gametes and zygotes they produce. A large number of aquatic plants, particularly among the algae, belong here, and these have tended to retain simple reproductive structures and reproductive cycles . . .
>
> The middle position in the hierarchy of reproduction is occupied by many plants, ranging from the algae to the flowering plants, which produce large, heavily coated, highly resistant resting spores or seeds . . . The highest position in this hierarchy is occupied by those angiosperms which are cross-pollinated by insects, which have fruits adapted to dispersal by animals, or which rely on animals for both pollination and seed dispersal. (Stebbins 1950, 559–560)

The plants mentioned last are, in some general sense, "higher" than the plants mentioned earlier—and their highness seems to be a function of some kind of adaptive advance that leads to widespread success: "many more adaptive gene combinations are possible in plants which rely on animals for their vital reproductive functions."

Backing claims like these, Stebbins gave an explicit discussion of the kind of theoretical reasoning behind his speculations about progress. In discussing the flowering plants, Stebbins picked out eight diagnostic characteristics which show whether particular specimens are "primitive" or "advanced" (his terms). He then went on to point out that not all possible combinations actually exist: "the total number of combinations realized is only 86, or 34 percent of the 256 possible, and 37 of these are represented by only one or two groups . . . Thus, the eight characters studied are far from being combined at random in the different families and genera of angiosperms" (Stebbins 1950, 503).

Why should this be so? Sewall Wright's adaptive landscape was brought out as an explanation, and we learn that only certain combinations occupy the tops of peaks—those which actually obtain, or (at least) obtain more often. Some peaks, we learn, are difficult to climb. It might be noted that among those families that have climbed such peaks, Stebbins listed Orchidoceae, Gramineae, Leguminosae, Malvaceae, Compositae—virtually exactly those he listed at the top of the reproductive scale! Finally, advance was tied in with adaptive specialization, and the reader learns that when a certain degree of specialization is achieved it is difficult or impossible to go back and try moves open to more primitive plants (in other words, it is easier to get up a peak than to get back down it). This is the ratchet theory of one-way advance by another name, or rather by no name at all.

Thinking like this as part of the straight science of evolution was not a one-off aberration. We find elements of it even in Stebbins's most recent work, where he has been suggesting that major adaptive moves, leading to significant evolutionary advance, are a combination of fortuitous preadaptations (adaptations for other needs) combined with open ecological opportunities. But, once the moves have been made, the "economy of evolution" dictates that the essential new features remain forever ingrained, subject to "tinkering": "the most readily available and smallest amount of change needed to form the adaptation that meets the immediate evolutionary challenge" (Stebbins and Hartl 1988, 5145). Another peak is conquered, in the upward progress of life, and there can be no return to the simpler life of the valleys below.

The case is complete. To put the point in evolutionary terms, the synthetic theorists had evolved beyond the stage where, because of its progressionism, their field was necessarily in the popular realm. But, as always in evolution, the marks of history remained ingrained in the most modern of functioning entities. I doubt that, as a trait of evolutionary theory, progressionism was always all that vestigial, even. In epistemological terms, the science had certainly been purified of the most overt elements of culture. But these things are never simply black or white, in or out. Science, at least this science, is deeply metaphoric: first the tree, now the landscape. For these would-be professionals, the landscape could be justified again and again for its *epistemic* virtues. Thanks to Wright's contribution, evolutionary thought could be made more formal and—desirable trappings apart—hugely more elegant, more predictive, more fertile, and so forth. The catch, although frankly I am not sure that the synthetic theorists would have thought it a catch, is that this self-same powerhouse of epistemic integrity is precisely what opens the door for the reentry of *culture!* As so often in life, the complete picture—the landscape metaphor—was a package deal, in which the epistemic and the cultural were fused together.

12

Professional Evolutionism

I turn now to the post-synthesis period, the contemporary evolutionary scene. I do not provide here a complete survey of recent evolutionary science—that would be impossible and unnecessary. Instead I focus on some of the really good work and some of the major controversies and new ideas. I shall bring together the British and the American strands of our story. In fact, the synthesizers themselves were often separated by more than the Atlantic. Darlington returned the critical opinion of his work by not reading Wright or Dobzhansky (Mayr questionnaire; Darlington Papers). And we know Dobzhansky's views on *Evolution: The Modern Synthesis*. But starting with Cain and Sheppard, who were postdoctoral students with Mayr and Dobzhansky, respectively, barriers decayed. They are not gone. England is more Darwinian and America more Spencerian, and the preponderance of the British in this chapter and the Americans in the next is not entirely artifactual. In this age of cheap transatlantic travel and of e-mail, however, it is fair to consider the community of evolutionary scholars in Britain and the United States as one discipline.

Rather more significant than nationality is the fact that one effect of professionalism, at least as we know it in the late twentieth century, is that it is considerably less easy to pick out "Great Men" than it was previously. Science, and this includes evolutionary science, has become much more of a collaborative effort, whether this means scientists actually working together or criticizing and reacting against a position. A Charles Darwin, acting in secret and delaying publication for twenty years, would simply be impossible today. Hence, although my emphasis will still be on individuals, I shall have to move more strongly to a structure based on areas or problems. The contemporary culture of science demands such a shift. Yet, as always, my eye is on questions of P/progress.

Adaptationism

Let us start with natural selection and adaptation. In the 1950s, the cutting edge of adaptationism was that found in Ford's school of ecological genetics—Ford's (1964) own work on balance, Kettlewell's (1961, 1973) studies of industrial melanism, and above all the Cain and Sheppard (1950, 1952, 1954) papers on snail variation. But science moves quickly on, and there is none more cruel than the next generation: "I remember a streetside conversation (outside the Bodleian) with [a younger researcher who said] something to the effect of: 'Yes, well, we're all just waiting for Henry to pop off and then we can get in and find out what is really going on with Maniola wing spotting polymorphisms. It is clear that there really is something interesting going on, but I don't think Henry's 30 years of data have told us much about what it is' " (letter to author from Paul Handford, November 21, 1995). In the decades since, in the evolutionary field as a whole, there have been two major moves or refinements: a rethinking of the nature and working of natural selection, and a move from what organisms are to what organisms do.

First, today there is renewed concern with an issue dividing Darwin and Wallace, namely the *level* at which natural selection operates: the precise "unit of selection" (Brandon and Burian 1984). The population geneticists generally favored Darwin's viewpoint, that selection must favor the individual, but by mid-century almost all evolutionists would have been with Wallace, happy with a theory that could work at nigh any level: "I have always been wondering whether there isn't in species other than man a selective advantage of groups and populations as a whole and not only of individuals" (Ernst Mayr to Theodosius Dobzhansky, January 10, 1955; T.D.–E.M. Correspondence). In the 1960s, adaptationists returned sharply to a Darwinian perspective. Above all it was a theory from the English William Hamilton (then a graduate student) which symbolized this move to "individual selectionism." In one clever stroke,

William Hamilton

Nicholas Davies

Hamilton (1964a,b) solved a problem which had bedeviled evolutionists since Darwin: namely, why is it that, if evolution puts such a premium on reproductive ability, social insects have sterile worker castes? Seizing on the fact that, in the Hymenoptera (ants, bees, and wasps) females have both mothers and fathers and hence one chromosome set from each ("diploid") whereas males have only mothers and hence but one chromosome set ("haploid"), Hamilton argued that the sterility of workers acts in their own reproductive interests! In particular, females (the sex of workers) are more closely related to sisters (that is, they share 75 percent of the same genes) than they are to daughters (50 percent). Hence, by proxy as it were, they improve their biological contribution (judged in terms of gene copies transmitted to the next generation) by raising the offspring of their fertile mothers more than by raising their own offspring.

The elegance and predictive success of Hamilton's hypothesis (now labeled a special case of "kin selection"), together with work in a similar vein, has been at least part of the reason behind the second major shift in emphasis since the 1950s, namely toward an appreciation of the significance of *behavior*. Again, this idea is to be found in Darwin; but, for a number of reasons—some obvious, like the difficulty in quantifying behavior, and some less obvious, like the rise of the social sciences and the drive of social scientists to bring behavior beneath *their* domain—it seems fair to say that evolutionists tended to ignore or downplay that aspect of the organic world. Given the power of conceptual tools like kin selection, together with a willingness to make the empirical effort, adaptationists today think as readily about the importance of what animals do as about what they are.

From the many studies that illustrate both of the just-mentioned fresh emphases, let me refer to the work of one of today's active evolutionists, a man who is only now into mid-career but who attracts general praise. The ornithologist Nicholas Davies (at Cambridge) has recently completed a ten-year study of dunnocks, so-called hedge sparrows (Davies 1990, 1992). His main interest lies in their mating patterns, for truly they evince a sexual diversity that would demand the attention even of Hugh Hefner. For all their dull, rather retiring appearance, hidden beneath the hedgerows they show enough different mating combinations to keep busy a whole department of cultural anthropologists. In one small flock, there were solitary males, there were polyandrous unions (two or three males with one female), there were monogamous unions, there were polygynandrous unions (the polite phrase for "group sex"), and there were polygynous groups (one male, two females).

What are the reasons for and consequences of this catholic sexual system? By making genetic fingerprints of dunnock blood samples—illustrating, incidentally, how in respects molecular biology has become the handmaiden of the evolutionist—Davies showed that virtually all and only those male birds who had sexual access to females take part in raising the young. Dunnocks are prepared to make an effort in this life, but apparently only if and when they are promulgating their own genes. Conversely, everyone tries to get the maximum from the system that they possibly can.

> Where a female is able to gain her optimum at the expense of males, we observe cooperative polyandry. Where a male is able to gain his optimum at the expense of females, we observe polygyny. In monogamy the female has not been able to gain another male and the male has not been able to gain another female. Polygynandry (e.g. two males with two females) can be viewed as a kind of "stalemate", the alpha male is unable to drive the beta male off and hence claim both females for

> himself, and neither female is able to evict the other and claim both males for herself. (Davies 1990, 476)

But, granting that this and many similar examples make the point about the new focus on individual selection and on behavior, what has any of this to do with progress? Apparently in evolutionary research *qua* adaptationist studies, whatever may be the personal proclivities of individual evolutionists, thoughts of cultural Progress are irrelevant and thoughts of biological progress are gone. These ideas simply do not arise in the kind of work in which someone like Davies is engaged. Moreover, if you ask the members of his group about progress, they confirm the conclusion of non-relevance: "Me and my mates just choose a little problem to work on and bother away at that" (interview with author, November 1, 1990). The broader metaphysical speculations are not part of biology as they know it.

Science as Ultimate Concern

That thoughts of P/progress are irrelevant to the new professional evolutionist is too important just to be left resting on but one example. Let us dig a little more deeply into the question, and to do this I shall turn to the work and thoughts of a man who is today generally regarded as Britain's most brilliant middle-level evolutionist—a man who was elected to the Royal Society in his mid-forties, even though his education and situation are decidedly more humble than those of his competitors. Geoffrey Parker was both undergraduate and postgraduate at the University of Bristol (1962–68), after which he moved to a position at Liverpool (in Cain's department), where he is still today. It cannot be said that he had much training as an evolutionist: there were virtually no lectures on the topic for the B.Sc., and his Ph.D. supervisor had the remarkable educational philosophy that he should not read his students' theses lest he make too much input of his own. Perhaps the ideas were in the air, however, for Parker and his fellow students were able to work out the significance of an individual-centered approach to natural selection. Applying this theory to organisms, especially to problems of behavior, he was launched on a career as a professional evolutionist.

Able to think symbolically with the ready ease that Parker showed in studying nature, I shall refer here to a much-praised series of papers that Parker produced from his thesis work. These papers center on one of the less prepossessing members of nature's creation, the dung fly *Scatophaga stercovaria,* properly so called because its courting and egg-laying activities take place on freshly deposited pats of cow feces. As soon as the mammal expels its waste, large numbers of male flies flock to the deposit. Then come the females (outnumbered by males four to one). They are grasped by a male, who copulates and then remains on the female until she has deposited her fertilized eggs. At this point the insects disconnect and fly off, leaving the eggs to hatch and the larvae to use the dung as their first home. (See Parker 1969, 1970a–g, 1974 a,b.)

There are here a large number of points of biological interest. A good example of Parker at work is his analysis of copula duration in his flies (Parker 1970c). On average, the flies copulate for about 35 minutes. This is not generally sufficient time to fertilize all of the female's eggs. Why then, given the fact that the male has to go to the trouble of finding a female, does he not stay on the job long enough to fertilize the lot? Parker's answer, showing his commitment to an individual perspective, is that it is simply not in the male's reproductive interests to prolong the action. On the one hand, time spent copulating is time lost searching for new partners. On the other hand, since females are receptive to multiple matings and it is always the *last* mating which rates most success, time spent in copulation increases the risk of being overwhelmed by an attacking rival. Better, therefore, to cut the time short, even if it means reducing one's immediate expectations.

This example in particular, and Parker's work in general—backed as it always is by quantitative discussion—is a paradigm of how the modern professional evolutionary biologist advances today. And one can see readily how this kind of work would bring joy to the hearts of the great synthetic theorists, not to mention Darwin himself. It exhibits in abundance all of the right epistemic values: an experiment that controls and explains the phenomena at issue in a mathematized manner quite beyond the abilities of the founders themselves. This is mature science, produced by a professional working in his discipline—even if he had to find the discipline on his own.

Is there nothing of a cultural nature in all of this careful work? By now, the sensitized reader will know how to read the text, and thus read one comes away with a full haul. In particular, the whole study reeks of the sexual revolution of the 1960s, with males competing for females who share their favors far and wide. Unrestrained copulation by both sexes is the only game in town—or, rather, in the cow field. Another clue is

Parker's (1970g) openly anthropomorphic language: he talks of the females of some species as being "promiscuous" (he puts this in quotes) and of the occasional "rape" (he does not put this in quotes). Some might indeed claim that his language is heterosexist, for in his discussion of intersexes we get all sorts of talk of "normality" and the inference that time spent with them is "wasted" and "fruitless" (Parker 1969).

Most striking of all is the fact that Parker's work tracks the growing social sensitivity to females. The early papers portray the female flies as basically passive, directed partners in the whole sexual enterprise. They seem to be archetypical Victorian maidens awaiting the wills of their masters. Then, as we move to the mid-seventies, the female flies come much more toward center stage, even if it is only as triggers for male responses. Thus: "The manner in which females arrive at the dropping will be an important determinant of male search strategy" (Parker 1974a, 102).

I should say, incidentally, that Parker himself is remarkably candid about his extra-scientific concerns.

> Q. The 1960s was the generation of sexual freedom. Does this tie in with interest in sexual selection? Is this a crazy idea?
>
> A. No, I don't think it's crazy. It was a time of liberation. It was time of sexual freedom. I think it is impossible if you are interested in biology, and if you are interested in behaviour, not to think about human behaviour. And we certainly did. We would talk about what would be in male interests and what would be in female interests, and this sort of thing . . . There is no question about that sort of link. (Interview with author, Spring 1991; following quotes from same interview)

Indeed, Parker is quite willing to go much further than this, to tie in all of his subsequent work as an evolutionist to personal concerns.

> I started working on sexual selection, as we talked about. And that led to sperm competition and the evolution of two sexes. The next thing I worked on then was the fighting behaviour, and that led on to the interest in arms races. The thing I started working on quite a lot in the late seventies was parent-offspring conflict. It's almost a history of my life! [Laughs] I was interested in sexual selection when I was young and interested in sex, I suppose more than perhaps now. And then fighting behaviour at a time when I was at the bottom of a hierarchy in a university department and wanted to be able to understand why the rules of the game were such. Then, sexual conflict appeared briefly, and parent-offspring conflict I was well into in the late seventies. I suppose

my children by 1979, Claire would have been six and Ronald would have been three.

Yes, it wasn't totally *in vacuo* I suppose. [Q.: Senescence next?] [Laughs] Senescence will be next, and maybe God, although I don't think I've quite got to that stage.

Actually, Parker's recent interests are in clutch-size problems. One might have predicted that he would have become an expert bantam fowl breeder: with 250 birds in his back garden, he is a repeated prizewinner and a valued show judge.

What is significant for our story is that there is nothing overtly related to progress, nor does there seem to be much which is covert, in Parker's studies.

Q. Do you think very much about the overall process of evolution or are you mainly interested in problems?

A. Well, I'm more interested in problems really. It's a very interesting point that, because there are some people who really want to be able to explain the world, and everything there in it. I can only face one problem at a time, really.

It is true that if you squeeze Parker, you can get some fractured admissions on complexity. But the note is hardly enthusiastic.

Q. Complexity?

A. I would certainly believe that you can decrease complexity—I mean there is lots of evidence that many parasites have reduced complexity. But one thing I am fairly sure of is that to get to high complexity you have to start from a point of low complexity. So, complexity has to build up, I think.

With a man who does tie his science so frankly to his personal life, one might well think that there has to be some connection between his basic lack of interest in progress and his personal philosophy. In fact, it is true that (for once) we have a scientist not absolutely convinced of his subject's Progress. As a post-Kuhnian (by whose major book he was much impressed), Parker inclines to a more cyclical (and cynical!) than linear view of science. "I do think that if there is a theory which is appealing, the empiricists are under tremendous pressure to find evidence to support it." (He made this comment, during our interview, in the light of William Hamilton's theory of sexuality—of which more in a moment.) But the drive to exclude progress and like ideas from science is more than a negative reaction. It has a positive, moral force. Having lost his Christian faith at sixteen, Parker turned more and more to science.

> Q. Interest in philosophy?
> A. Yes, I think I have been interested in those sort of things. But I'm not a very metaphysical sort, really. I think I do believe in rationality and logic and so on. I don't believe in ghosts, fairies, a god as such at all, in any way, shape or form. And I suppose that for me science is something slightly more than just science. Only a little bit more, if you like. It's not a religion at all. But it is to some slight extent a way of life.
> I suppose as a child I can remember being incredibly frightened of the Inquisition and the like—the horrors that religion could impose upon man. But not only that. The way that superstition in its various forms could grab the human mind and make life a misery. And it seems to me that the way out of this was to believe that which one has good grounds to believe—and never to spend one's life throwing salt over one's shoulder and avoiding green and all things of that sort, which does seem to impose an immense burden on many people. I think that if science does have a message it is probably best to be rational, logical about things.
> What I want to get out of science is a freedom from these forces [e.g., Islam] on the human mind. I think science is a form of freedom, that shouldn't be lightly cast away. I just feel that science and logic and liberal thought, if you like, allow us to live our lives—not selfishly I hasten to say, because that's not a conclusion I think you would come to from science and the like—but to live in a much freer way. In a mental freedom, and in the end in a physical freedom, because I think there are horrors with believing in things for which there is no rational evidence.

I have taken Parker as my model of a modern professional evolutionist. I am certainly not going to impose on him also the burden of being my sole model for what contemporary evolutionists believe about progress. In virtually the same words, however, he does confirm the attitude of my Cambridge evolutionist, namely that there is today a place for the professional who wants no truck with progress. Both confirm, furthermore, that there is a place for the professional who wants positively to get away from the "woolly" metaphysics of the past.

Arms Races

Actually, as we well know, there has long been such a place—at least, there has been such a place in England, from the time of Weldon. Forget

Geoffrey Parker

Richard Dawkins

long time spans. Think only of mechanisms and micro-problems, and progress does not really arise. But what if you do start to stretch out the calendar? What if you start, working now still very much as a hardline Darwinian adaptationist, to inquire about comparative progress? Do we find echoes of this idea in contemporary evolutionary studies, and do they tend to slide into modern-day absolute progress? And, if so, why?

When last encountered in a major way, Darwinian comparative progress was being interpreted in military metaphors, as competitions between rival warring factions. This notion of progress as the result of an "arms race" has figured in the (British) literature right up to the present. (See, for instance, Cott 1940, xii.) Indeed, recently it has received considerable attention, both theoretical and empirical. The reader may have noticed a pertinent reference, just above, by Parker. A major overview, stimulating much interest, appeared in an article published in 1979 by John Krebs, now a Royal Society professor, and Richard Dawkins, author of some of today's most successful popular books on evolutionary theory (Dawkins 1976, 1986, 1995). Dawkins and Krebs argued that competition between members of groups ("lineages") is common in nature and that this conflict leads to a "kind of evolutionary escalation of ever more refined mutual counter-adaptations" (p. 490), suggesting "progressive rather than random . . . trends in evolution" (p. 492). Of course, it is individuals who fight and fight back, but "it is lineages that evolve, and lineages that exhibit progressive trends in response to the selection pressures set up by the progressive improvements in other lineages" (p. 492).

How has modern professional evolutionism been able to go beyond the theoretical and speculative? Among others, Nicholas Davies has worked on this very topic, in studies of the relationship of the English cuckoo, *Cuculus canorus,* with the various birds that it parasitizes, especially meadow pipits, reed warblers, dunnocks, and pied wagtails (Davies and Brooke 1988, 1989a,b; Davies, Brooke, and Bourke 1989). Different brands (gentes) of cuckoo parasitize different species, and it has been Davies's aim to look at the coevolution between hosts and parasite. Analyzing the evolutionary moves made in terms of the military metaphor, which he does with some vigor, Davies feels that he can show a profitable interaction between theory and observation. On the one hand, he finds that the cuckoos have various adaptations enabling them to parasitize nests. Thus, for instance, a cuckoo can lay eggs very quickly, and the eggs have distinctive camouflage markings, those of one particular brand (gens) matching the eggs of its particular host (there is one exception to the rule). On the other hand, hosts have adaptations to detect and avoid exploitation by cuckoos. For instance, they tend to be very sensitive to oddly colored or marked eggs.

Darwin's relativistic kind of progress is alive and well in contemporary evolutionary circles. The forces which lead to a comparative highness are recognized and studied—although the result is not absolute progress. Like Parker, Davies and other of his Cambridge colleagues seem to share a casual feeling that there is an overall upward pattern to evolution. Davies himself admits to a belief in a hierarchy of levels, with evolution having moved living things up the hierarchy. And another member of the group, T. H. Clutton-Brock, author of a much-praised study of the evolutionary ecology of the red deer (Clutton-Brock et al. 1982), seems almost to have a Chain-of-Being view of life. A volume he edited on reproductive success takes us from insects, via the amphibians and birds, up through the mammals to humans (Clutton-Brock 1988). But there is no strong interest in these ideas among these scientists, nor the desire to put them on a firmer footing. As the already-quoted spokesman said: "I'm not very good at thinking about the broader things." Likewise Parker:

> I think your perspective on arms races is quite different if you ask a question if what if at any point in time, might we expect to see? And I would expect to see some sort of equilibrium, in which the rules of the game are prescribed at that time, and then we can . . . answer what's happening at that time. The perfectly valid alternative question would be—forget your micro problems of temporary equilibrium or temporary rules of the game, or what have you—what happens over long, long periods of time when there is change in the rules of the different games you were playing? When there are global modifications or complete changes in the organisms, meaning that which was interesting as an explanation at a particular geological time is no longer or barely relevant at another geological time. That's a different sort of perspective, and that's macro-evolution, if you like, rather than micro-evolution. And I've been interested in micro-evolution mainly. In problems, if you like, on a micro-evolutionary scale. (Interview with author, Spring 1991)

My hunch is that for Parker the idea of "equilibrium" has a strong moral force, above any epistemic virtues. "Balance" implies moderation or decency. It surely did for Wright.

On-board Computers and Absolute Progress

Darwin and his successors had a tendency to move from comparative progress to absolute progress. Has this tendency died completely? Although Cambridge and Liverpool may disappoint us in our quest, Oxford is more hopeful. In his highly successful overview of evolutionary thought, *The Blind Watchmaker,* Dawkins (1986) returns to the question of arms races and of their relevance to progress. In his original coauthored paper, he had hinted at rather more than relative progress. That article admitted that long-term, progressive change might occur: "it does at first sight seem to be an expectation of the arms race idea that modern predators might massacre Eocene prey. And Eocene predators chasing modern prey might be in the same position as a Spitfire chasing a jet" (Dawkins and Krebs 1979, 506). At times there is even a hint of something stronger yet: "Directionalist common sense surely wins on the very long time scale: once there was only blue-green slime and now there are sharp-eyed metazoa" (p. 508).

However, now Dawkins speaks sternly against the fallacious view (an "earlier prejudice") that there is something "inherently progressive about evolution" (Dawkins 1986, 178). All sense of progression, therefore, comes from the relativistic process of an arms race, and we are warned off "a Victorian idea of the inexorability of progress, each generation better, finer and braver than its parents" (p. 181). In truth: "The reality in nature is nothing like that." Yet, even here, Dawkins cannot help concluding that: "Nevertheless, when all this is said, the arms-race idea remains by far the most satisfactory explanation for the existence of the advanced and complex machinery that animals and plants possess" (p. 181). Moreover, he backs up this claim with an interesting discussion about the way in which modern military arms races have gone heavily into electronic competition, a development that is paralleled in the organic world by the evolution of the brain. Relying on some controversial speculations by the American brain scientist Harry Jerison (1985), who has tried to devise a general measure of intelligence—the "encephalization quotient" (the EQ)—for all animals, Dawkins drops some suggestive hints about who are the winners of this particular race. We are warned that "The fact that humans have an EQ of 7 and hippos an EQ of 0.3 may not literally mean that humans are 23 times as clever as hippos!" (p. 189). But, as he himself concludes, it does tell us "something."

By now that "something" will be starting to have a familiar ring to us, and the odor of absolute progressionism grows yet stronger when we learn that paralleling Dawkins's enthusiasm for the evolution of bigger and better on-board animal computers is a long-term affair with human-made computers. Indeed, the great strength of *The Blind Watchmaker* is the way Dawkins brings his talents in this latter direction to bear on so many of the public misconceptions about the workings and powers of natural selection. Biological values and cultural values seem to be melding into one. And in some of Dawkins's more recent writings and statements he has made, they have become truly one. He speculates now that there may be such a thing as the "evolution of evolvability," an opening, as it were, to new dimensions of evolution. "What if there are certain major water-shed events in evolution which, whether or not they lead to any massive improvement in the adaptedness of the individual organisms that show them, nevertheless open the floodgates to future evolution? I think of something like segmentation, the invention of segmentation, either in the—call it vertebrate lineage—or in the arthropod lineage" (transcript of conference discussion, July 1989, Melbu, Norway, in possession of author).

These earliest new forms may not be particularly fit, but they paved the way for others to come: "Suppose we rank embryologies in order of evolutionary potential. Then as evolution proceeds and adaptive radiations give way to adaptive radiations, there is presumably a kind of ratchet such that changes in embryology that happen to be relatively fertile, evolutionarily speaking, tend to be still with us" (Dawkins 1988, 218). Again the analogy is drawn with the progressive development of computers.

> Computer evolution in human technology is enormously rapid and unmistakably progressive. It comes about through at least partly a kind of hardware/software coevolution. Advances in hardware are in step with advances in software. There is also software/software coevolution. Advances in software make possible not only improvements in short-term computational efficiency—although they certainly do that—they also make possible further advances in the evolution of the software. So the first point is just the sheer adaptedness of the advances of software make for efficient computing. The second point is the progressive thing. The advances of software open the door—again I wouldn't mind using the word "floodgates" in some instances—open the floodgates to further advances in software. (Transcript in possession of author)

Dawkins specifies that he has in mind the invention of a new compiler or a high-level language.

Finally, we can note that ultimately Dawkins's (1988) vision of evolution is cumulative: it has "the power to build new progress on the shoulders of earlier generations of progress, and hence the power to build up the formidable complexity that is diagnostic of life" (p. 219). When questioned, he allowed that: "I was trying to suggest, by my analogy with software/software coevolution, in brain evolution that these may have been advances that will come under the heading of the evolution of evolvability in [the] evolution of intelligence" (transcript in possession of author).

We have seen ideas like these before, of course, most particularly in the joint work of Huxley and Haldane. There is a denial of any kind of Germanic-type absolute progress but endorsement of a kind of absolute progress through adaptive breakthroughs. Darwin would have felt right at home with the new version of progressive development—although no doubt he would have noted and regretted the fact that by this point the discussion has slid from the professional to the popular. With Dawkins, we start in the *Proceedings of the Royal Society*, very respectably scientific, move out through a book for the general reader to essays from a free-wheeling conference, and end with off-the-cuff remarks at another conference.

It would seem, therefore, that thoughts of absolute progress may well be with us still—but they are where they have always been, or ought to have been: in the popular domain. Truly, they have no place in today's professional science. The university-based, full-time evolutionist has simply moved on beyond such thoughts—at least, when he or she is doing what the job calls for. Yet, we know too well to stop here. There may well be thoughts of absolute progress, beneath the surface, guiding and influencing evolutionists in their work. We must persist a little longer before we throw in the towel, defeated in the last bout.

The Problem of Sex

Going right to the top, let me turn to what Parker characterizes as "probably still the fundamental problem we've got to solve": the origin and maintenance of sexuality. What is the biological point of sex? This may seem a ridiculous question, for the answer is surely obvious. Asexual reproduction limits the rate of genetic change in a population. With sexual reproduction, however, together with shuffling ("crossing over") at the chromosomal level, genetic material is passed along to the next generation in new combinations. New mutations are only rarely advantageous. Without sex any such mutations will be scattered through the population and will have but a scant chance of being combined within one individual. Sex changes all of this. It brings together rapidly all of the good things that mutation offers. There is no need to wait hopefully for a sequence of favorable changes in a single line of self-replicating organisms.

Unfortunately, however, this optimistic argument overlooks one massive problem: the biological cost of sex. Females give birth. Males do not. Hence, although an asexual organism reproduces itself completely, a sexual organism has to share its offspring with its mate—and for a female, unless she can get some balancing recompense like male parental care, this is a poor bargain. From an individualistic perspective, therefore, there appears to be a massive selective force against sexuality, whatever the long-term benefits to the group. And yet, obviously, sex exists and thrives, even. Most paradoxically, there are species which alternate sexu-

George C. Williams John Maynard Smith

ality with asexuality. If sex is such a biological burden, one would expect them to become exclusively asexual, virtually immediately.

In the past two decades, much has been written on the evolution of sex. I want to highlight three men who have been at the forefront of theoretical thought on the subject. The first is the American George C. Williams, author of the vehemently pro–individual selection classic, *Adaptation and Natural Selection.* Williams felt that "the prevalence of sexual reproduction in higher plants and animals is inconsistent with current evolutionary theory," and that hence "there is a kind of crisis at hand in evolutionary biology" (Williams 1975, v). The solutions toward which Williams was drawn were all ones that posited an immediate (huge) advantage that sex must give the individual, in the face of the massive handicap it imposes. Since sex shuffles the genes, there must be some massive benefit to having a range of genotypes in one's offspring—and this clearly must imply that the offspring face a varied environment. With sexual reproduction you are covering your options.

If, alternatively, the environment were constant and entirely to be anticipated, one uniquely best-suited genotype would offer the greatest chance for survival.

One distinctive feature of all of Williams's models is that they apply exclusively to organisms with high fecundity, like trees which have literally millions of seeds. But, even if these are in the majority, they are clearly not the only organisms in the world. Moreover, those organisms in respects most obvious to, certainly most interesting to, ourselves tend to be of (comparatively) low fecundity. What of them? Williams argues that, paradoxical though it may seem, we have no reason to doubt (and much reason to think) that sexuality is maladaptive! "If and when any form of asexual reproduction becomes feasible in higher vertebrates, it completely replaces sexual. So in these forms sexuality is a maladaptive feature, dating from a piscine or even protochordate ancestor, for which they lack the preadaptations for ridding themselves" (Williams 1975, 102–103). Sex may be socially and psychically important to us humans, but if our genes could do otherwise, it would go the way of our tails.

I shall relate Williams's thinking to thoughts on P/progress, but first we must move back to England and to our other two evolutionists. In 1963, John Maynard Smith, student of Haldane, became acutely aware of the need to promote an individual-selectionist approach to evolutionary change. This realization was partly a function of reading Hamilton on hymenopteran sociality, and it was partly a negative reaction to a book then attracting considerable attention, *Animal Dispersion in Relation to Social Behaviour,* by V. C. Wynne-Edwards—whose central thesis was that organisms have group adaptations to regulate population sizes. Thus primed, Maynard Smith was ready at the end of the decade to try the application of the mathematics of game theory to problems of animal behavior. Game theory presupposes that the actions (or non-actions) of a "player" (in this case, animals) are directed toward their own biological benefits. Through this work came a concept which has attracted massive theoretical and empirical attention: an Evolutionary Stable Strategy (ESS), being such that "if all the members of a population adopt it, then no mutant strategy could invade the population under the influence of natural selection" (Maynard Smith 1982, 10).[1]

Like Williams, Maynard Smith (1978) found himself worrying about sex. Unlike Williams, Maynard Smith was by no means convinced that we have at hand all of the materials for an adequate solution. Dividing the problem into two, the *origin* of sex and its *maintenance,* he was convinced that only a strict individualist approach would speak to the first issue. The burden of sex is just too great not to have some balancing compensation. This is not to say that Maynard Smith was entirely convinced by Williams's story of the virtues of variation in the face of environmental diversity, or rather not to say that he was convinced that this is a full and satisfying explanation. Instead, he favored as an explanation that features of sexuality "hitchhike" their way into favor, by being associated with and carried by other favorable genes. These features are not directly of value themselves, but they promote genes (and combinations) which do have higher selective values.

With respect to the maintenance of sex, for all his individualistic yearnings, Maynard Smith allowed that here there might be a role for group selection. The evidence is that asexuality leads to extinction. The question is whether, as the group-selectionist argument supposes, the propensity to extinction of asexuality is great enough to balance the burden of sexuality. If the move to asexuality is uncommon and unlikely, as Maynard Smith thought true, then he believed that the tradeoff is possible. "It is difficult to see how one could test this proposition, but it is not obviously false" (Maynard Smith 1978, 70). Suppose a sexual species is widely scattered, with forms (morphs) adapted to different sub-environments. One member becomes asexual and rapidly takes over the sub-environment which it inhabits. The rest of the species, in the other sub-environments, continue on because they are better adapted to their own sub-environment than the asexual clone. Before long, evolutionarily speaking, the asexual clone goes extinct. The remaining sexual groups survive, and perhaps take over the asexual group's sub-environment. Sex has been explained, through a group-selective process.

My third evolutionist is William D. Hamilton himself. Recognized today as the most innovative thinker in the field since Fisher—a man whose work he read "with great enthusiasm"—he has little time for the options of Williams and Maynard Smith. For him, the mechanisms must be adaptively positive, individual-selectionist through and through.

> The main defect of those early models [of Williams and Maynard Smith] is that they would work for high fecundity organisms but not for low, as admitted by both of them. And since half of the interesting living world is low-fecundity organisms including us, I thought one needed something more and I didn't like to think that their models were just right for fungi and prolific trees and we've got to find some other completely different mechanism that would prop up sex in humans and pandas and

> the like—that would seem to me to be unlikely and unaesthetic and therefore I wanted some modification of their models that would encompass everything at once and that was what I was looking for. (Interview with author, Spring 1991)

As with his explanation of hymenopteran sociality, Hamilton's explanation is simple (an epistemic value he much cherishes). Go back to the notion of an arms race. We have the predator and the prey, the attacker and the defender, in an ongoing battle. Now one, now the other "rearms." The versions discussed so far have involved some sort of cumulative change, and thus the notion of (comparative) progress. But what if the race simply involves staying one step ahead of your rival, perhaps in an endlessly oscillating state? This is the key to Hamilton's explanation of the evolution and maintenance of sex. Sexuality and recombination are forever shuffling the genes, so that new organisms are formed. These are then one step ahead of their rivals in an ongoing race (Hamilton 1980; Hamilton, Axelrod, and Tanese 1990; Seger and Hamilton 1988).

What could the rivals be that set up this relentless pressure, thereby justifying the crippling costs of sexual reproduction? In a word—parasites! Hamilton points out that organisms are forever locked in a battle with microorganisms, these being adapted to prey upon their hosts. It is necessary to raise defenses. Unfortunately, no protective measure can be a once-and-for-all phenomenon. Parasites generally have very rapid generation times and the subsequent potential for rapid evolution—evolution around the defenses. Hosts therefore must themselves forever move and change the defenses, and this is possible thanks to sex: "the hosts' best defense may be one based on genotypic diversity, which, if recombined each generation, can present to the parasites what amounts to a continually moving target" (Seger and Hamilton 1988, 176).

Associated with this view of the evolution of sex is Hamilton's theory about sexual selection. In species where parasite defense is especially crucial, an organism (generally the female) that invests in the production and care of offspring has a strong need to check the quality of would-be mates. Hamilton hypothesizes that the prospective mate advertises its ability to combat parasites (with the implication that its offspring would be similarly capable) by being showy or brightly colored—its appearance in any particular case being a direct function of its health. The males (as it generally is) put themselves on display, and the females can and do choose (Hamilton and Zuk 1982; Hamilton 1990; but see Cox 1989 and Hamilton and Zuk 1989).

In recent years, Hamilton has pushed these views with vigor and with style. Because of the respect with which he is regarded, his thinking has commanded much attention. As yet, however, no hypothesis has won full acceptance. That is not our worry. For our purposes, we have covered enough ground. The time has come to pull back.

Thoughts of Progress?

What do we find when we pull back? What has any of this talk about the evolution of sex to do with the topic of biological progress or to the related issue of cultural Progress, in one of its various forms? At the most obvious and straightforward level, not a great deal. The issue of sex is hardly a matter of raw data. Indeed, a great deal of theory is presupposed even to make it an issue, certainly an issue of some pressing significance. And this theory is hardly one that merely reflects the facts. But, *prima facie,* the issue of progress is totally irrelevant.

Yet, is it? Although there is overlap between our three evolutionists, Maynard Smith for instance admitting to the "great influence" of Williams, they are not close to agreement. There is the adaptive/non-adaptive status of sex in mammals, for one point of difference, and the possible role of group selection in sex maintenance, for another. Since, apparently, the evidence is not definitive, we are led to dig down to uncover submerged values, cultural values even. At that point, we find that Williams, Maynard Smith, and Hamilton take radically different attitudes to the question of progress in biology. Simply because there are differences, it does not follow that they must be relevant. But, the trail is hot!

Williams is evolution's muckraker. In *Adaptation and Natural Selection,* not only did he take after the illicit group-selectionist thinking he saw tainting the field, he directed his wrath against the rampant progressionism which so corrupted evolutionary thought. Julian Huxley was singled out for particular scorn: "I would maintain, however, that there is nothing in the basic structure of the theory of natural selection that would suggest the idea of any kind of cumulative progress. An organism can certainly improve the precision of its adaptation to current circumstances . . . but as perfection is approached the opportunity for further improvement would correspondingly diminish. This is certainly not a process for which the term 'progress' would be at all appropriate" (Williams 1966, 34). Progress of an ongoing kind is simply impossible: "Ultimately you reach a point where, with selection as the editor, there's a limit to what selection can accomplish against recurrent and unfavourable mutation—you run up against this, because the genome gets bigger and bigger" (interview with author, April 1988).

The flip side to this pessimism is an enthusiasm on Williams's part for the writings of T. H. Huxley, especially the rather morbid themes of "evolution and ethics." Williams has co-edited an edition of this very work, in which he argued that Huxley is absolutely correct in his somber view of the organic world. (See Huxley 1989.) Ignoring any progressivist sentiments which might be found in Huxley's writings, Williams seizes on the relativism of adaptation, and even more on the positively unpleasant aspects of the organic world, especially of human nature. Although educated by the Jesuits, he has an almost Jansenist attitude to our species and seems to roll together the non-progressiveness of biological evolution and the non-Progressiveness of human accomplishments. Apparently, military life (at the end of the Second World War) had a significant impact: "Certainly, in the army I witnessed some really miserable behaviour towards each other that colored my view of human nature, and showed me some of its less wholesome aspects" (interview with author, April 1988).

Williams is negative toward P/progress—or, rather, let me put things more carefully: he is negative about a Progress simply modeled on progress, or conversely. He does endorse a kind of Progress, but it is a kind that flies in the face of what most people would call "progress." "I would not equate maladaptive with bad. In fact, more likely the converse. If something is adaptive, I am more inclined to think it evil, or at least suspect. This is certainly a conclusion that people would draw (I hope) from my Huxley book" (letter to author, February 1, 1991). There are similar ideas to be found in a book he has just co-authored on Darwinian implications for medicine (Nesse and Williams 1995). Thanks to the reproductive virtues of *Homo sapiens,* many of the more unpleasant aspects of human existence came courtesy of natural selection. In words he has used elsewhere: "Mother Nature is a wicked old witch" (Williams 1993). Categorize Williams as you will, he is certainly not neutral about P/progress.

What about Maynard Smith? He is no old-fashioned monad-to-man progressionist; but, explicitly, he does believe in biological progress of a kind. There are stages or levels to life, these can be ordered hierarchically, and history had to be a passage from the lowest to the highest. In all, Maynard Smith believes there are eight "levels of complexity, differing in the organization of the genetic material":

1. Replicating molecules
2. Populations of molecules in compartments
3. Prokaryotic cells
4. Eukaryotic cells
5. Multicellular organisms
6. Demes; social groups
7. Species
8. Groups with cultural inheritance

These are rising orders of "units of evolution," the latter of which are entities which have evolved "characteristics that ensure their own survival and reproduction" (Maynard Smith 1988a, 222). All the time, the danger is that (selective) processes at the lower levels will corrode away entities at a higher level. Although there are always tensions between successive levels, units at upper levels can cohere and persist despite destructive forces at lower levels. This does not imply that progress is in any sense an "inevitable" consequence of natural selection. Recently, Maynard Smith has written of this as the "fallacy of progress" (Maynard Smith and Szathmáry 1995, 4). "However, one can recognize in the evolution of life several revolutions in the way in which genetic information is organized" (Maynard Smith 1988a, 229). He also makes much of the significance of the division of labor (making explicit reference to Adam Smith) in life's rise through the hierarchy (Maynard Smith and Szathmáry 1995, 210). It is this, together with new ways of transmitting biological information, which makes possible, in an upward sequential fashion, "the major transitions in evolution." With a familiar endpoint: "Human society is the final level of complexity so far achieved by living organisms" (Maynard Smith 1988a, 229).

This all meshes nicely with Maynard Smith's broader philosophical perspective. Although he laughs at himself now, as a schoolboy Maynard Smith was, like Haldane (1932a), much taken with the futurist novel *The Last and First Men* by Olaf Stapledon, a key theme in which is that humans as we know them today evolve (in one branch) into a form of super-beings (interview with author, Spring 1991). More influential, no doubt, was the fact that—again like Haldane, indeed one gathers in major part because of Haldane—Maynard Smith was drawn to Marxism. Although that enthusiasm has long waned, particularly thanks to the Lysenko affair, he still has a tendency to judge things in a dialectical, hierarchical fashion. Indeed, Maynard Smith is even prepared to say that modern mathematics has now given a theoretical underpinning to at least one of Engel's laws, namely that of quantity into quality (Maynard Smith 1981, 37). Moreover, specifically with respect to our own species, he demands that we "recognize that culture transcends biological constraints—at least as much as it is shaped by those constraints" (p. 77). All of this is seasoned, as it was for Haldane, with a good dose of scientism, in the sense of belief in the Progressive nature of science: "Unfashionable as it may be to say so, we really do have a better grasp of biology today than any generation before us, and if further progress is to be made it will have to start from where we now stand" (p. 11). This goes with an enthusiasm for the philosophy of Karl Popper.

I have spoken already of William Hamilton's warm feelings toward the work of Fisher, and these readily extend to the way that Fisher promoted his Fundamental Theorem as a kind of progress-building inverse of the second law of thermodynamics:

> Q. What about the Fundamental Theorem?
> A. I was quite enthusiastic about it. [Although recognizing limitations] I still find it a very useful idea.
> Q. Fisher on the analogy with thermodynamics?
> A. Yes, I liked that idea, I never understood thermodynamics well enough to know how accurate the analogy is, but I certainly agreed with him that it had the look of a principle that was of equal relatedness and I also agreed with him that it was a principle that to some extent showed that we didn't need to worry about the degradation of everything by the Second Law of Thermodynamics. That there were indeed processes that would work to create more pattern in the universe rather than less.
> Q. So you agree that evolution adds up to something?
> A. Yes . . . I certainly believe that evolution produces some of the most interesting, complex patterns that the universe has.
> Q. Would those complex patterns include humans?
> A. Yes, I think so.
> Q. Do you see evolution as some sort of upward climb?
> A. Yes, I do very much. And, God knows what the end of it is to be. It's hard to say whether we're far advanced or just starting. But, I see it as definitely a progressive, pattern-generating force. I don't see any reason why it has to approach any asymptote. (Interview with author, Spring 1991; following quotes from same interview)

Of course, one might argue that complexity or pattern generation in itself has no essential value; but this would not have been the position of Fisher, nor it seems, is it that of Hamilton.

Q. Worth? Value coming out of evolution?

A. Yes. If one thinks life is worth living one can hardly think otherwise, can one? Otherwise one would commit suicide.

Unlike Fisher, Hamilton has no strong religious faith. But when he talks of his belief in the Progress of science and technology—a Progress which reflects the progress of biology—one catches glimpses of distinctively Fisherian concerns about the inherent dangers of technological Progress for long-term human progress.

Q. Technology?

A. I do have faith in the power of technology. There are some aspects of the use of technology that I'm very nervous of. I'm basically rather anti medical advances because I think they rather help the life and survival that I don't really like the idea of. But there are many other kinds of problems where I think there is value of all kinds, for example. And I also like the idea that technology can help us in pure science to understand experience. There are inventive tools like the computer, which in every way benefit the user, and I imagine there are other instruments which do the same.

Q. Does science advance?

A. Oh yes, definitely!

Q. You link advance in science to advance in technology?

A. Yes, quite a lot. I think computers are one good example. One might say that computer technology has nothing to do with mathematics, but mathematicians are admitting that computers are giving them new problems and giving them new solutions.

Behind the Explanations

Let us go back now to sex and its putative explanations. I do not want to make any silly, overly strong claims. I certainly do not want to argue that the three scientific positions I have reviewed are mere epiphenomena of three biological views on progress, in turn functions of three cultural views on Progress. There is an evidential gap, however, and I would argue that the ways in which it is filled are more than just consistent with the broader (P/progress-related) positions taken by the theorists.

Start with Williams. He does not like the idea of progress in evolution. Indeed, he argues that Progressive successes in the moral realm, however caused, depend on spectacular failures in biological progress. For Williams, the triumph of adaptation is generally to be regretted—a claim which at once ties a strong link between his views on sex and his views on P/progress. His thinking is that sexuality in so-called higher organisms is maladaptive. It would be in their biological interests to get rid of it, but they cannot. This conclusion is surely just what one would expect of a latter-day Jansenist. If one believes that evolution is inherently non-progressive, and that even humans have no special status, the very opposite in fact, the fact that we are saddled with a crippling biological burden is the kind of gloomy cosmic joke one might expect of nature.

Indeed, this adaptive failure is about the only grounds one might have for thinking that sex, in itself, is a morally good thing, rather than something to be fought by our sense of right and wrong. Lest I be accused of philosophical fantasizing, let me requote, in fuller context, from the letter given above: "Do I think mammalian sex is maladaptive? Yes, in the sense I originally claimed. If a viable method of asexual reproduction presented itself it would take over. On the other hand, I would not equate maladaptive with bad. In fact, more likely the converse. If something is adaptive, I am more inclined to think it evil, or at least suspect."

With Maynard Smith, one likewise finds connections between his general thought (including that on P/progress) and his views on sex and its evolution. Maynard Smith admits openly that his youthful Marxist enthusiasms have left their mark on his general thinking on the evolutionary process. Most Darwinians follow the leader, steering away in horror from the thought of one-step, pre-adapted forms that represent a significant break with their predecessors. Maynard Smith, to the contrary, allows that: "I have always had a soft spot for 'hopeful monsters'; new types arising by genetic mutation, strikingly different in some respects from their parents, and taking a first step in the direction of some new adaptation, which could then be perfected by further small changes" (Maynard Smith 1981, 153). Elsewhere, he ties the possibility of saltational change to dialectical thought: "We are now familiar with the idea that gradual changes in the parameters of a dynamic system can, at critical points, lead to sudden and discontinuous changes in system behaviour. It seems certain that gradual changes in genetic constitution can lead to discontinuous changes in phenotype" (p. 136).

In addition, Maynard Smith has an ambivalent attitude toward the evolution of humankind, which reflects his philosophical position. In some respects, he is drawn toward seeing us as part of the biological scene, and indeed he argues for the evolutionary foundations of human incest barriers; but, in other respects, he thinks that we humans transcend mere biology. His reasons for ambivalence are complex and go beyond Marxism, to the fact of his coming to manhood when Hitler was in full power (Maynard Smith 1988b, 46). What is not complex is the fact that the kind of progressive view of evolution he endorses both supports and is supported by his feelings that we humans are special.

Yet, what specifically of the issue of sex? In Maynard Smith's case there is an obvious link, namely between his willingness to think in group-selective terms for the maintenance of sex and his belief that, at one level, species can function as units of evolution. And, indeed, he himself makes the connection explicitly.

> The gene pool can be thought of as carrying the genetic information of a species, rather as the genome does of an individual. When a species splits in two, the daughter species carry most of the ancestral genes, so that "species heredity" is ensured. Species, therefore, are candidates as units of evolution. The snag, of course, is that with-in species, between-individual selection is ubiquitous, and strong enough to prevent species-level adaptations.
>
> Sexual reproduction may constitute an exception to this rule. Theory suggests that sex can confer an advantage on the population by accelerating evolution and by delaying the accumulation of deleterious mutations. The taxonomic data suggest that parthenogenetic varieties are short-lived in evolutionary time. (Maynard Smith 1988a, 228)

As before, my claim is hardly that there is a tight deductive link between the cultural commitment to Progress and the theory of the maintenance of sex. But, as before, I would argue that there is more than mere contingency linking the two. Maynard Smith is no longer an ardent Marxist, and he is certainly no communist. But he is sympathetic to a world view, including an evolutionary world view, which sees an upward P/progress in terms of hierarchy, with the possibility (if not necessity) of more holistic thought at higher levels of the hierarchy. This world view does not necessitate a commitment to group selection, yet it certainly makes it possible, especially as one approaches "higher" organisms like humans—and the admission of such a possibility is no minor act for a man who has been at the forefront of the move to individual selectionism.

Finally, we come to Hamilton. The possible link is easy to make. Like Fisher, we should expect Hamilton to see human evolution and culture in terms of the general theory. This he does, and—as is shown by a discussion of human fashions—there is feedback to his thinking on sexual selection. Remember that males attract females by advertising their health and fitness through their finery:

> Of course underlying these interpretations there is a biological interpretation of similar events and styles in human life. Why did Beau Brummell in Regency England dress up as he did? Was it to find a wife, or to find an "affair"? And, why did the women who admitted themselves attracted to such men dress themselves in equal finery? Many were married: but we may note a possibility that in a time when marriages were often arranged for financial advantage, their mates may have been more rich than attractive or healthy. Of course, there are differences between humans and animals, including in the human case (shades of Fisher!): "the transferable wealth which to my mind overshadows even the rational thought." Nevertheless, "some accounts from long and careful observations of bird sexual and social relations suggest a degree of complexity of motivation and deceit that makes the comparison worthwhile—in both directions." (Hamilton 1990, 343)

We should expect Hamilton, like Fisher, to think that his theory of sex confirms that the end products of selection are in themselves good.

> There were several reasons for not liking [Williams's position on sex].
>
> One is that things as close to us as lizards do sometimes abandon sex, and also it's annoying in ordinary human terms that human reproductive technology is going ahead so fast that soon it will probably be possible for us to abandon sex—I don't like it, and since the sex to be abandoned will likely be the male and being a man myself, I don't like to think we have to be in this position of regarding males as really useless. And so I produced a reason of why we are here and why we should be kept going.
>
> Q. Are you serious about that?
>
> A. I like to put it into my talks as mainly a joke, but I do think that males are the less serious sex in biology—and having seen them abandoned in many very successful insects that I know of, I felt I had to do the best I could. (Interview with author, Spring 1991)

Even more explicitly, in answer to a question about his feelings on Williams's view that human sexuality is maladaptive, Hamilton replied:

> Yes, I do feel uncomfortable with that, and I do feel that I have good reason now—much better than I had 10 years ago—for thinking that the human male, or any male, is a necessary part of a progressive kind of higher evolution such as we have.

Need one say more? Hamilton's general views on P/progress are feeding into his theorizing about sex—and, no doubt, in turn spreading out from that theorizing to support his general philosophy.

Let me sum up. Williams believes in a kind of cultural Progress, which depends on a fight against or outright denial of biological progress. We must battle adaptation and inasmuch as we succeed, we have done well. Sexuality, specifically sexuality in higher organisms like ourselves, is a good and to be cherished precisely because it is an adaptive handicap. Maynard Smith believes in a progressionist hierarchy with humans at the top, a belief which maps his philosophy of Progress. To rise through the hierarchy—a process which is clearly going to be temporal as well as conceptual—one needs more sophisticated mechanisms, sexuality in particular, and to get this one needs new causal mechanisms, group selection in particular. Hamilton thinks individual selection leads to progress and that this echoes or reflects Progress. It is not so much that comparative progress (an arms race) leads directly to absolute progress, but rather that a comparative-progress-type situation leads to sexuality, and then this makes possible the evolution of an absolutely better kind of organism, human males as well as females. In all three cases there is a gap between evidence and theory, a gap filled in three different ways by three different thoughts of Progress.

Thanks primarily to the efforts of the synthetic theorists (on both sides of the Atlantic), we have now a professional discipline of evolutionary studies. Grant that professionalism excludes overt discussions and commitments to progress in science. What we find, indeed, is that there is a new generation of evolutionists for whom P/progress is essentially irrelevant. In the tradition from Weldon down through the ecological geneticists, today's evolutionists work on short-term processes and their mechanisms. Admittedly I am taking just one or two cases as examples for a whole community, but the point is that these scientists cherish the epistemic values of mature science, which they properly think they are producing. What we also find, however, is that there is a continuing belief in Darwinian comparative progress and that this belief is linked to thoughts of social Progress. We find, in addition, that comparative progress has a tendency to edge up into absolute progress, and that similar moves are made in the cultural realm. As might have been expected, however, by the time that the shift has been made to absolute P/progress, the domain has moved from that of professional science into the world of popular culture, as I have been characterizing it.

Because progress has been eliminated from professional evolutionism to protect its status as "professional"—rather than from conviction that P/progress is wrong—we might expect to find also that P/progress, although hidden, stands ready to influence the ways that theorists might fill evidential gaps between data and meaning. And this we do find. I do not mean to argue that there is a generally held position on P/progress. There never has been, and there is not now. Nor is culture simply wagging the tail of professional science. But if our discussion of theorizing on the evolution of sex shows anything, it is that P/progress broadly understood continues to shape the thinking of some of the most professional of evolutionary biologists. The mature science has cultural undertones.

13

Contemporary Debates

In this chapter, I want to build upon the conclusions just drawn at the end of the last chapter. To do so, I shall look at two highly visible areas of evolutionary research—one old and one new. I seek the gaps between theory and evidence, and the ways in which evolutionists fill them.

Paleontological Escalation

Paleontology is still very much the science of the popular domain. Think of the general fascination with dinosaurs or with the search for human origins. Perhaps it is to be expected that there is no one more contemptuous of the field than fellow evolutionists, particularly those of a mathematical turn. Listen to a leading population geneticist on a recent trendy paleontological hypothesis: "I think it's done very little new in any conceptual sense. It's resurrected an extreme version of a historically debunked theory of evolution on the basis of virtually no evidence" (R. Lande, interview with author, January 1989).

Paleontologists are aware of others' regard, or disregard, for their work, and this makes them defensive. Hence, they are going to be pretty cagy about public professions of their personal convictions, most especially about professions of belief in progress. It was pronouncements on such topics that got them such a bad reputation in the first place. Osborn's shadow still chills. Entirely typical in voicing his opinions on the sorts of things to which progress supposedly leads is one of the brightest and best of the younger crop of paleontologists, Jack Sepkoski (a student of Stephen Jay Gould): "I see intelligence as just one of a variety of adaptations among tetrapods for survival. Running fast in a herd while being as dumb as shit, I think, is a very good adaptation for survival" (interview with author, January 1989).

Yet, if we have learned anything by now, it is that we should not take evolutionists entirely at face value, especially when they are being self-consciously value-free. Sepkoski, to continue this one case, has been a major enthusiast for a recent paleontological hypothesis, supposing that there are periodic, mass extinctions and, subsequently, openings for new niches. There are certainly times when this position—particularly when it is cast in terms of a favorite theme of American history, the frontier—seems more than just a neutral, disinterested report on objective reality:

> Mass extinctions have probably been good for the evolving biosphere. I said "good" and I've got to explain why I said "good"—in the sense that they've probably promoted diversity.
>
> Real evolutionary innovations, probably coming in during the rebound of these extinction events, clear out a lot of diversity. Clear out a lot of biomass. We're back into semi-frontier days. Sort of environment where you don't have to be real good to get on, so something very new and different may be able to grab hold of a piece of the ecological pie, and hold it giving rise to new kinds of organisms.
>
> So mass extinctions are good in that sense. They promoted evolutionary innovation. (Interview with author, January 1989)

Immediately Sepkoski qualified his comments, recognizing that while "wholesale death has produced some interesting consequences," it is not particularly good for the concerned individuals.

Whether or not one looks upon the conquest of the West by the White Man to be an especially good thing, something with much to do with Progress, it is hardly now necessary to emphasize that, as with the student of today's organisms, the day-to-day work of the paleontologist has little or nothing to do with progress. There may well be other structuring principles, thoughts of adaptive utility for instance, but these have scant immediate connection with overall improvement. Recently highlighted (by Gould 1989 and others) as a paradigm of good paleontological research are studies of peculiar, soft-bodied creatures from a quarry in Canada (the "Burgess Shale"). Although this work demanded much creative insight, not to mention sheer manual dexterity, it did not in itself demand or involve a direct commitment to upward progression. (See, for instance, Whittington 1975.) In the analyses of the Burgess Shale fossils long-extinct animals are brought to life, but the implications for the cosmic scheme of things are simply left untouched.

That progress is now excluded from the daily life of the paleontologist must be granted. Yet, does this imply that there is no feel at all for progress? Start with the relativistic kind of progress that Darwin inaugurated. Does this appear in paleontology? The answers are mixed. A fairly significant majority position shows a familiar dynamic between stability and change. If you are just asking about diversity, then the feeling seems to be that, allowing for the frequent extinctions, the fossil record indicates a reasonably steady state. Thus, for instance: "At the end of the Devonian or the beginning of the Carboniferous, fishes had come to occupy all available carnivorous niches in the aquatic environment. Thereafter, fishes remained at (declined to) an *apparent* equilibrium in diversity demonstrated as an effective upper limit on diversity (at whatever rank) which they could not/did not exceed" (Thomson 1977, 396, his italics).

However, if you are asking about "improvement" in some sense, there is an equally strong feeling that this does appear in the fossil record (Schopf 1977). Fish "can be seen to have undergone significant morphological 'advancement' (for example from the chondrostean to holostean grade among Actinopterygii, or the 'cladodont' to 'hybodont' grade in Chondrichthyes)" (ibid.). Improvements are found also by those who work on plants. Consider the accompanying figure, analyzing the morphological evolution of a certain group of (Cretaceous) angiosperms, through time. The overall message is that although some species ("morphological entities") retain "primitive" characteristics, the general pattern is one of "advance," where apparently this reflects approach to the "modern condition." And the very language of discussion reveals the value judgment: "The most striking feature of Zone 1 leaves is their 'first rank' leaf architecture: i.e., poor definition of vein orders, irregularity of spacing, angle of departure, course, and branding patterns of secondary and higher-order veins, and incomplete differentiation of blade and petiole, a syndrome of characters originally postulated to be primitive in dicots on the basis of comparative studies of Recent forms" (Doyle 1977, 530).

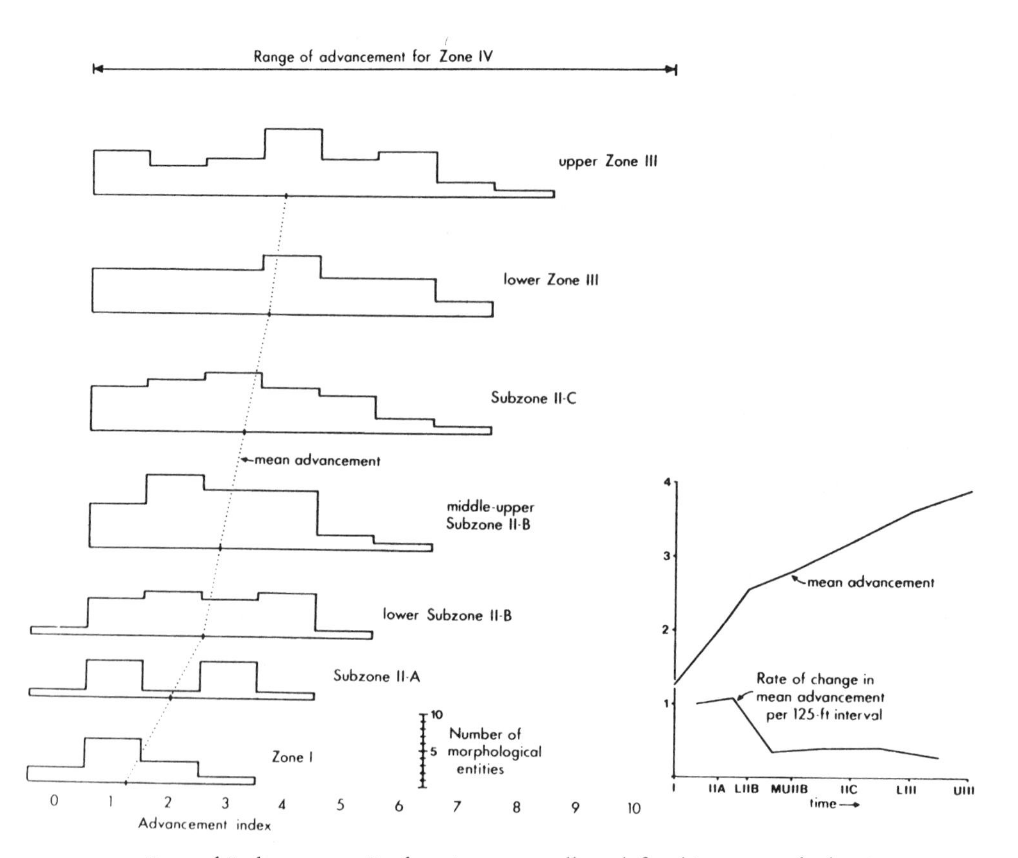

Rate of "advancement" of angiosperm pollen, defined in terms of adaptive improvements (such as shape and size). (From Doyle 1977.)

What about the causal processes supposed for such improvements? The favored explanation is of selection brought about by some kind of competition between organisms. A most detailed treatment comes from Geerat J. Vermeij (1987), who sees selection regularly bringing on patterns of "escalation" in life's history. Explicitly invoking the metaphor of an arms race, he argues that predator-prey interactions lead quantitatively to more organisms evolving weapons of attack or defense, and qualitatively to more improved weapons of attack or defense. Overall: "if selection among individuals predominates over other processes of evolutionary change, and if enemies are the most important agencies of this selection, the incidence and expression of traits that enable individual organisms to cope with their enemies (competitors and predators) should

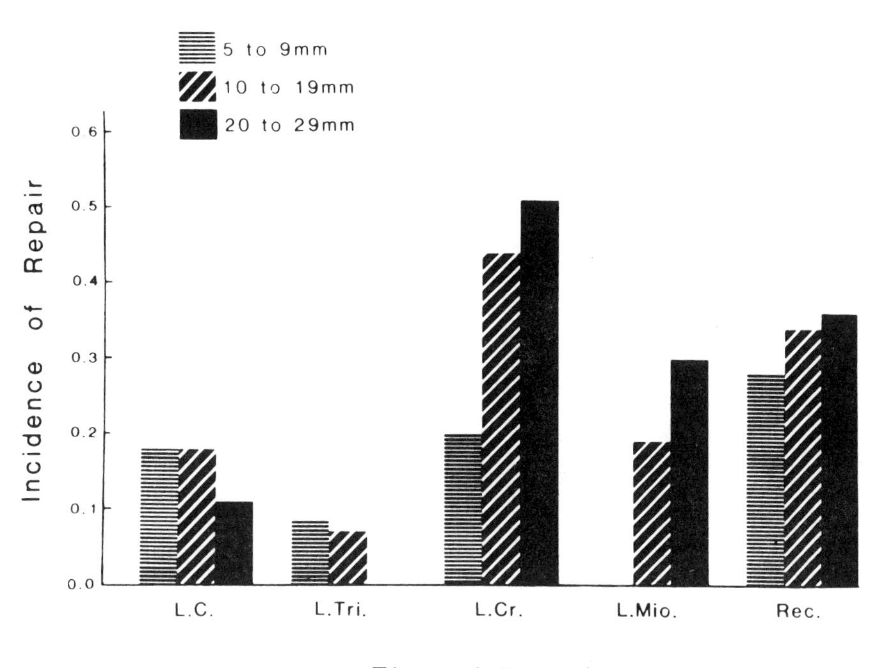

Was there an improvement in the ability of shellfish to repair themselves after attack, and have predators now evolved more efficient methods of attack? This chart suggests that the predators may have improved (hence inflicting damage requiring more repair) or the defenders may have improved (hence being better at repair) or both, and that after time a balance of sorts was reached, as happens in military arms races. (From Vermeij 1987.)

be found to increase within specified habitats over the course of time" (p. 359). Backing this, Vermeij argues that: "On the whole, the evidence from fossils is in accord with the hypothesis of escalation." And he supports this claim by listing a number of features that have improved: metabolic rate, shell-breaking capacity, broken-shell repair ability (!), dental specialization, locomotive ability, and more (p. 359).

Also pertinent is another idea which has received some considerable attention in the past two decades, the so-called Red Queen hypothesis. The brain child of Leigh van Valen (1973), the hypothesis claims that in a sense evolution goes nowhere—or, rather, like the Red Queen in *Alice Through the Looking Glass,* by running fast, one stands still. Less metaphorically, van Valen documents the empirical fact that, within major

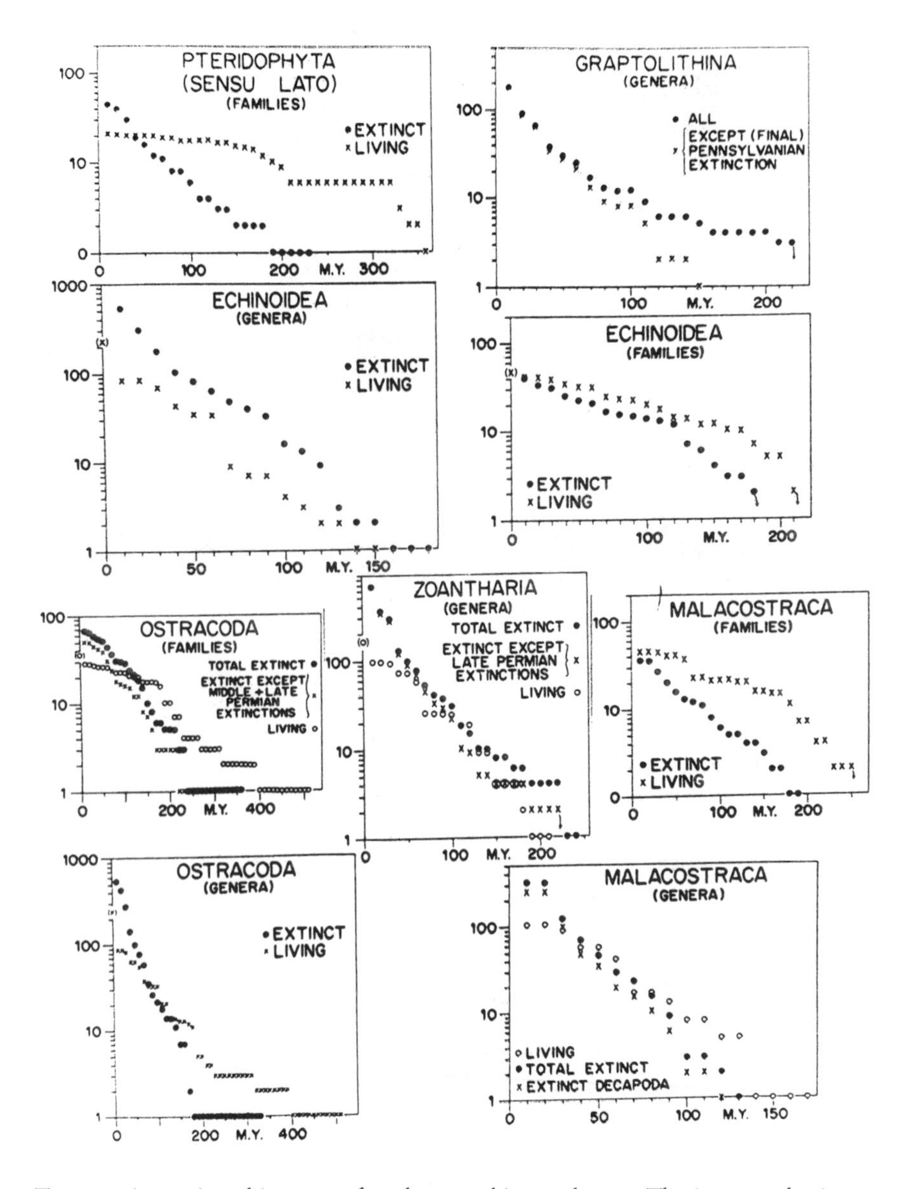

Taxonomic survivorship curves for plants and invertebrates. The (more or less) straight-line graphs (with negative slopes) suggest that for any group the rates of extinction are constant. (From van Valen 1973.)

organic groups, sub-groups seem to go extinct at a regular rate; he concludes that therefore the overall group can keep up its diversity only by producing its new sub-groups at the extinction rate. In other words: "The effective environment of the members of any homogeneous group of organisms deteriorates at a stochastically constant rate" (p. 6).

As an orthodox Darwinian (he was a student of both Simpson and Dobzhansky and a postdoctoral fellow with Maynard Smith), van Valen does not seek a causal explanation beyond those which make selection paramount. Basically he suggests that there are only so many resources available, and as one group succeeds others must fail (in the long run), which leads to a kind of averaging effect. Hypothesize an adaptive landscape in a specified resource space:

> The amount of resources is fixed and can be thought of as an incompressible gel neutrally stable in configuration, supporting the peaks and ridges. If one peak is diminished there must be an equal total increase elsewhere, in one related peak or more uniformly. Similarly, increase in a peak results in an equal decrease elsewhere.
>
> Species occupy this landscape and can be thought of as trying to maximize their share of whatever resource is scarcest relative to its use and availability. (van Valen 1973, 9)

Since there is a fixed amount of evolutionary space, the appearance of new groups must be balanced by the extinction of old groups—and, taking into account all the groups in a space as a whole, we expect the rates to be roughly uniform.

Prima facie none of these hypotheses bode well for evolutionary progress, even of the comparative kind, as epitomized by Vermeij's "escalation." The very crux of van Valen's insight seems to center on the non-directionality of evolution—running fast in order to stand still. Even in his original discussion, however, van Valen allowed the possibility of comparative progressive change within the overall picture, and now he would tie this in with contemporary discussions: "What Vermeij calls 'escalation' occurs—coevolution between interacting groups of one kind or another, so that adaptations at one time tend to be better than adaptations at an earlier time with common challenges" (interview with author, January 1989).

Likewise, the Norwegian population biologist Nils Christian Stenseth, a sometime associate of Maynard Smith, sees the Red Queen hypothesis, arms races, and comparative progress as a package deal:

> [Around 1970] Maynard Smith could not find a good basis for the common-sense feeling that there is a steady evolutionary increase in complexity and improved adaptedness. These difficulties seem to result from ignoring the biotic component of the environment. With the Red Queen view of evolution, such a continued increase in complexity is indeed expected; if other evolving species are the major component of a species' environment, an evolutionary "race for life" seems unavoidable. One evolutionary improvement achieved by one species must necessarily be neutralized by an even better adaptation by the co-existing species if they are to continue to exist. (Stenseth 1985, 67)

We are getting close to notions of absolute progress; but, before we turn to them directly, let us pause and ask about paleontologists' views on Progress. Is the talk of arms races, escalations, and the Red Queen hypothesis linked with parallel views about social and/or intellectual improvement? In Vermeij's case it certainly is, to such an extent that he recognizes that he stands on the verge of slipping from the professional to the popular domain. In the preface to *Escalation and Evolution* he writes: "This book is written for those who, like me, are fascinated with evolution, but I hope it will also be read by historians of human affairs. There are some obvious parallels between the history of life and the cultural history of man—the episodic pace of change, the effects of crises, and the role of population growth, for example—and the fossil record teaches some sobering lessons about the futility of the seemingly unstoppable arms race between nations" (p. xiii). He states explicitly that, with non-scientific readers in mind, he has kept the technicalities to a minimum.

Similar sentiments about social Progress are held by others. "I'm not a complete cultural relativist. I think there has been ethical progress since [the Middle Ages]" (John Sepkoski, interview with author, January 1989). As it happens, although he does believe in social Progress of a kind (certainly scientific Progress), van Valen is loathe to tie together the various P/progressive stands. Others are less reticent. Stenseth, speaking more as a theorist interested in long-term processes than as an evolutionist as such, is forthrightly holistic in his thinking: "I'm a leftish person—I believe in Progress." Moreover: "I see how various parts of my thinking fit together with other parts, including my science" (interview with author, November 22, 1990).

Now, what about absolute progress? What you are generally not going to get from regular paleontologists today are Osborn-like musings about human superiority. You might find some exceptions in related fields, however—the champion of orthogenesis would have felt right at home with the paleoanthropologists: "Can the nuclear family not be viewed as a prodigious adaptation central to the success of early hominids?" (Lovejoy 1981, 348). But, although today's paleontologists may shrink from glorifications of the Norman Rockwell ideal of life—threatened alas by "continually sexually receptive" (p. 294) hussies who wait on the sidelines, ever ready to seduce the providing male from his central biological role as *paterfamilias*—they are certainly open to Darwin-like slides from the comparative to the absolute.

There are hints of "sliding" in Vermeij, who goes so far as to say that: "It is possible, however, that species have improved in their capacity to survive in the *physical* environment. Many of the characteristics associated with competitive and defensive superiority—large body size, high body temperature, parental care of the young, and a tightly sealing exoskeleton, for example—also buffer individuals against short-term fluctuations in temperature and other physical factors" (Vermeij 1987, 42). He thinks we might find one consequence of this kind of improvement to be that today's organisms live longer in an absolute sense than did those of the past.

More generally, Anthony Hallam, quoted at the beginning of this book for his regrets about the (lowly) status of evolutionary studies, epitomizes the tensions of today's workers: worried lest they appear too blatant, yet convinced that the overall picture they are filling in does mean something:

> I'd say that in the crudest way that you can look at the fossil evidence progress is not terribly evident, but I'd immediately qualify that by saying that in certain groups you can see a kind of a progress . . . I do believe in evolutionary progress in a restricted sense. I don't share Gould's view that man is here purely by lucky chance. I think that in particular clades, to use a fashionable term, one can perceive "improvements" in terms of mechanics, in terms of physiology. I think mammals are in many respects cleverer organisms biologically than reptiles, which are in turn cleverer than amphibians . . .
>
> I think the classic story, as amended and modified, is still a story interpretable in terms of progress . . . (Interview with author, Spring 1991)

Even humans are included in the picture, so long as you do not get too teleological.

> I think that man is definitely a success story—but you've got to define your criteria. Consciousness and the ability to analyze the rest of the world and ourselves in the way that we do. The gift of language—all the usual things—is amazing. In those senses, yes, we certainly can be regarded as a pinnacle.

And, expectedly, Hallam subscribes to a belief in human Progress—in democracy and human rights, in concern for the environment, and most especially in the products of the human mind and hand.

> I would say that the clearest, most unequivocal evidence of Progress I can pick up in human history is in technology. There is the most incredible Progress by a variety of criteria.

Plus ça change, plus c'est la même chose.

Punctuated Equilibria Theory

No sooner had Simpson brought the paleontologists within the neo-Darwinian fold than they were chaffing at the bit, anxious to be away again. Take David Raup, professor at Chicago, educated at Harvard, and by his open admission "highly influenced by Mayr and by the people at the American Museum, especially Simpson" (interview with author, January 1989). With such a background, it is hardly surprising that "one of my principle objectives was to apply population genetics explicitly to the fossil record." Unfortunately, harsh reality in the form of incomplete records and vast time scale intervened, and thus: "I've since learned that it simply can't be done." Indeed, it might be that it could not possibly be done. "One could say that normal adaptation, by normal [population-genetical] methods, is just an epiphenomenon and nice, but doesn't help us to explain the big pictures."

Thus liberated, Raup has felt able—compelled—to join others in turning his back on ubiquitous adaptationism: "The standard wisdom is that the mammals did well because the dinosaurs died, and not the other way around." He admits: "It could have still been that mammals were smart little dinks that went around eating dinosaur eggs—I was certainly taught in grade school that it was a competitive replacement." However: "The current view is very strongly the other way around—a rock or something hit the dinosaurs on the head, and the basically inferior mammals were able to move into the niche." (Quotes from interview with author, January 1989.)

Progress and competition in macroevolution

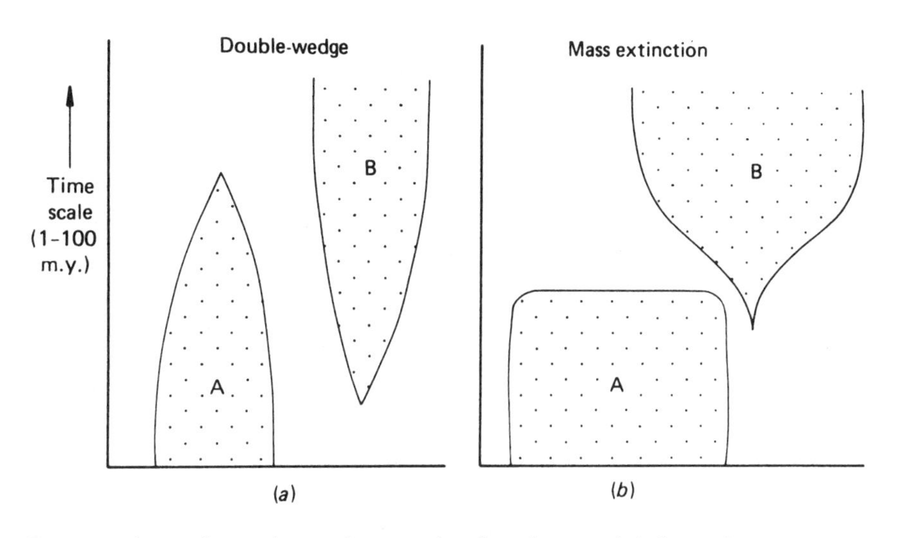

Does one form of organism replace another form because it is better in some sense *(a)*? Or does one form replace another because an extinction (for some independent reason) creates an empty niche for the newer form *(b)*? (From Benton 1987.) Gould and Calloway (1980) refer to situation *(b)* as "ships that pass in the night."

The frontrunner in this critical approach to adaptationism has been Stephen Jay Gould, who along with fellow invertebrate paleontologist Niles Eldredge has promoted the theory of "punctuated equilibria." Turning to this theory, with our eye ultimately on the possible implications for progress, we find that it has had three phases (Ruse 1989). The first was at the beginning of the 1970s, when Eldredge and Gould announced that their profession was being true neither to neo-Darwinism nor to the phenomenology of the fossil record (Eldredge 1971; Eldredge and Gould 1972). Every paleontologist knows that long stretches of identical fossil forms are broken by abrupt changes to new forms. Traditionally, the absence of transitional forms is explained away by the supposedly inadequate fossil record. Eldredge and Gould argued that if you take seriously a hypothesis of Mayr's, the "founder principle"—which supposes that new species generally start from small, isolated, rapidly changing groups—you *expect* that the transitional organisms will be few in number and isolated. Therefore, a jerky fossil record is predicted.

Stephen Jay Gould

Punctuated equilibria *(A)* contrasted with Darwinian gradualism *(B)*. (From Ruse 1982.)

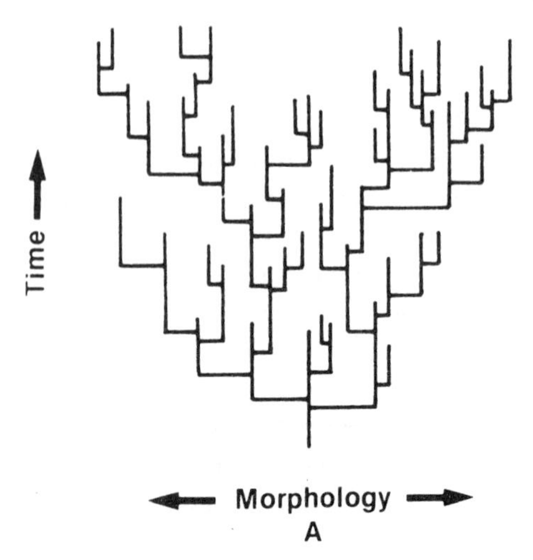

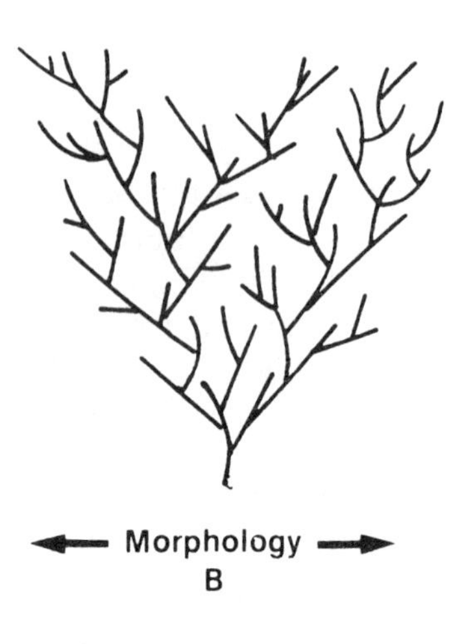

During the 1970s, Gould's thinking especially veered away from this conventional principle. (See Gould and Eldredge 1977 for the mid-point.) Whereas he had begun as an orthodox neo-Darwinian, Gould, like Raup, began to doubt the ubiquity of adaptive advantage. By 1980, the story is fully into the second phase: in a notorious article Gould (1980b) listed some of the key features of the synthetic theory and then declared it "effectively dead"! Rather than the founder principle as a way of achieving rapid change, he began to stress macro-mutations—although he denies that he was ever a full-blown saltationist.

One could be forgiven for thinking that Gould had swung through two right angles, and this was certainly the way most interpreted him. Expectedly, therefore, he triggered considerable hostility from conventional neo-Darwinians, who derided this faith in non-orthodox modes of change and who pointed out that, in far less time than can be recorded in the fossil record, well-known selection pressures could turn a mouse into an elephant (Ruse 1982; Stebbins and Ayala 1981). Perhaps as the result of criticism, Gould's analysis took a sharp turn to a third phase, in which regular Darwinism—that is to say, evolution through selection at the individual level—is incorporated within an "expanded" theory.

The emphasis of this theory is on "hierarchies" (Gould 1982a,b). The forces of change are now seen to operate in a layered fashion. They start with change down at the genetic or molecular level (where drift may be important); then they work at the level of the individual organism with conventional selection; and next, and most significantly, they operate up at the level of species, where the differential survival of groups causes patterns of change that (supposedly) cannot be inferred simply from changes down at the individual level. In thus promoting "species selection," Gould reveals his opposition to "reductionism," apparently a besetting sin of the conventional evolutionist.

We now have much more than a mere expansion of orthodoxy. In Gould's theory there is a distaste for simple adaptationism, at whatever level. Thus, for instance, in a (1984) discussion of the shape of fossil snail *(Cerion)* shells, the aim is to show that only giant and dwarf species have a distinctive "smokestack" form, because certain growth constraints which hold back normal-sized shells are released (and ineffective) at extreme sizes. But, as always with Gould, the immediate aim is not the long-term aim. The latter is an attack on the overwhelming relevance of selectionist arguments.

> Evolution is a balance between internal constraint and external pushing to determine whether or not, and how and when, any particular channel of development will be entered. Natural selection is one prominent mode of pushing, but most engendered consequences of any impulse may be complex, nonadaptive sequelae of rules in growth that define a channel. Most changes must then be prescribed by these channels, not by any particular effect of selection. Natural selection does not always determine the evolution of morphology; often it only pushes organisms down a preset, permitted path. (Gould 1984, 191–192)

We are a long way from the Cain and Sheppard approach to snail biology. The emphasis is on *form* rather than *function.* Gould is an evolutionist who looks first to homologies rather than to adaptations, to Unity of Type rather than to Conditions of Existence. By his own admission, what fascinates him are the underlying groundplans—*Baupläne*—of organisms, rather than the immediate functional utilities of particular characteristics (Gould 1977b, 1982a). It is Oken rather than Cuvier who is his mentor, as he shows incidentally by backing his *Cerion* discussion with a favorable reference to Darlington's friend, the Russian evolutionist Nikolai Vavilov, who was a prime representative of his country's Germanic-style evolutionism (Adams 1980).

Now, what has any of this to do with progress? Quite a lot, one might think, because (continuing for a moment with the continental theme) the love of revolutions, the anti-reductionism, the fondness for form over function—all seem explained by explicit admissions by Gould that he works from a Marxist base: "In the light of this official philosophy, it is not at all surprising that a punctuational view of speciation, much like our own, . . . has long been favored by many Russian paleontologists . . . It may also not be irrelevant to our personal preferences that one of us learned his Marxism, literally at his daddy's knee" (Gould and Eldredge 1977, 43). The case seems to be complete, and given the fact that Marx and Marxists are such explicit Progressionists (and progressionists), one is all set to find progressionism running right through the theory of punctuated equilibria.

Certainly Hallam, who (although no Marxist) inclines to punctuated equilibria theory, makes a direct biology/culture link:

> I was very impressed by somebody making a point about how serious it [the environmental crisis] is. That if you transfer a frog from a bowl of cold water directly into a bowl of hot water, it will just jump out. It knows what's good for it. But if you put a frog in a bowl and you slowly

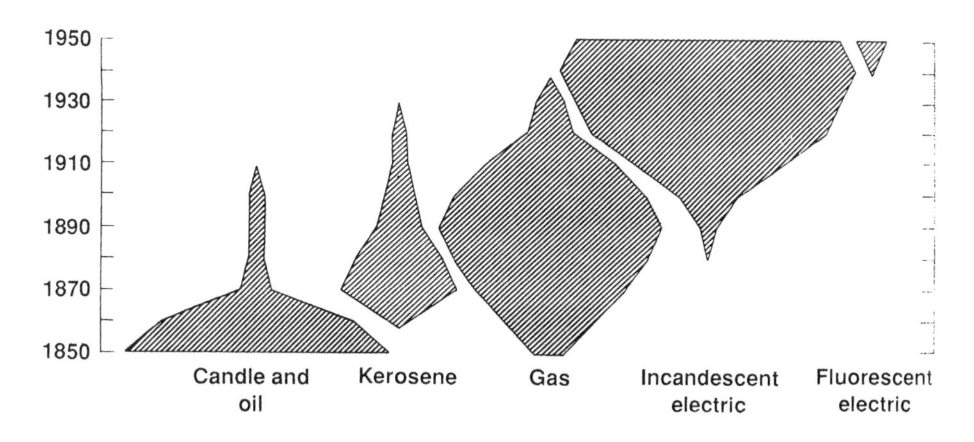

The evolution of lighting, as shown here, displays much the same pattern as organic evolution follows, yet without the causal force of natural selection. (From Gould, Gilinsky, and German 1987.)

heat that water up, it will stay in that bowl until it dies. It can't react to gradual change. And I think that's a good analogy to the human species. It's creeping up on us too gradually. Our whole physiology and mentality is geared up to good crisis response. But not to gradual, inexorable changes. And that's what we have to guard against. (Interview with author, Spring 1991)

In Gould's case one must beware of drawing overly simple connections. The gradualist/revolutionist dichotomy has never been interpreted in fixed ways according to political conviction (Masters 1989). Gould himself has been busily backtracking on any significant relationship between his political beliefs, past or present, and his theory (Gould 1981). Most important, he takes his attack on adaptationism to be in major part a direct attack on biological progress! Not that Gould is against any kind of meaningful interpretation of the fossil record: "You must separate questions of directionality from questions of progress. Certainly I think there are directions in life's history. There's certainly patterns in the history of life. I think there are directional patterns. Time does have an arrow. You can tell which way is up" (interview with author, April 1988). When a new kind of life-form (like mammals) evolves, within early sub-groups there is an initial large increase in diversity before a more gradual falling off and extinction. This pattern contrasts with a more evenly distributed diversity within later sub-groups.

But even here Gould does not want traditional Darwinian interpretations, which explain the bottom-heavy nature of early clades (evolutionary branches) as representing "adaptive radiations." (Adaptive radiation occurs when new life forms with specialized characteristics evolve to take advantage of open ecological opportunity, such as faced the mammals after the extinction of the dinosaurs.) Gould, rather, draws an analogy with similar clade patterns in the world of cultural change (see Gould, Gilinsky, and German 1987, fig. 6, p. 1440), in which natural selection is not at work, and argues:

> Deeper principles of structural organization must regulate and constrain the shape of genealogical systems in time. Natural selection may generate the changes that fit these molds in natural history; some other process may serve the same causal role in archeology. But the structural principles that fashion the molds and set the constraints upon pathways of change may be more abstract, and therefore common to a broad range of disciplines wedded to differing immediate mechanisms.
>
> Our documentation of the bottom-heavy asymmetry of life's clades, for example, may form part of a larger and more general study of innovation. (Gould, Gilinsky, and German 1987, 1441)

For all that he allows directionality, when it comes to biological progress Gould is withering in his contempt. "Why is it that people are always hung up on the notion of progress? I don't think there's a great deal to it, more than that humans are latecomers" (interview with author, April 1988). He speaks of progress as a "noxious, culturally embedded, untestable, nonoperational, intractable idea that must be replaced if we wish to understand the patterns of history" (Gould 1988, 319). There is simply no process guaranteeing that that which we value (especially ourselves) will come out on top.

Here Gould's anti-adaptationism really comes to the fore, as he stresses the purely contingent nature of life's history. Extinctions great and small make a mockery of progress, considered either as diversity or as absolute improvement of form. Life's tape played over and over again would never play the same tune. In his (popular) account of the fossils of the Burgess Shale, Gould stresses how many different forms there were and how easily the sub-set of actual survivors might have drawn on other members of the much larger set of potential survivors.

> Our most precious hope for the history of life, a hope that we would relinquish with greatest reluctance, involves the concepts of progress and predictability . . .

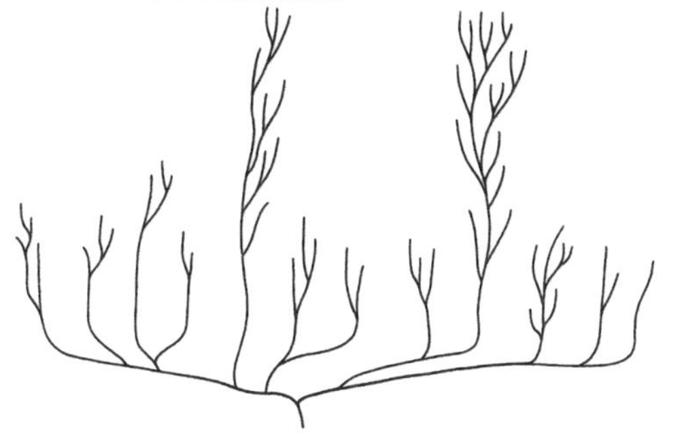

Gould's illustrations of two rival conceptions of life's history. Above, the conventional view of a gradual spread of forms, according to which success is controlled by adaptive function; below, the true picture (as shown by the Burgess Shale) of rapid initial diversification (without much regard to adaptive value) and subsequent random extinction. (From Gould 1989.)

> Burgess disparity and later decimation is a worst-case nightmare for this hope of inevitable order . . . If the human mind is a product of only one [of the existing set of organisms], then we may not be randomly evolved in the sense of coin flipping, but our origin is the product of massive historical contingency, and we would probably never arise again even if life's tape could be replayed a thousand times. (Gould 1989, 233–234)

One thing is certainly clear. We have moved now from disinterested reporting on objective reality to the point where values are being infused. Even those who work directly on the Burgess Shale feel that Gould has no right to assume that their efforts prove the non-directionality of change. Who is Gould to say that selection was unimportant? Who is he to deny that the organisms of that era progressed to something rather special today? "We [humans] may be one off, but damn it all we are the most interesting phenotype around!" (Simon Conway Morris, interview with author, Spring 1990). Indeed, given his objections to biological progress, and given also that Gould happily draws analogies between biological history and cultural history in order to make a point, one might be forgiven for concluding that so far from Marxism (or any related philosophy) being an influence, he is against progress precisely because he is against cultural Progress.

However, paradoxically, the opposite is the case.

> Q. What about Progress?
> A. I have ambiguous feelings. It's so easy to see how it's been misused as part of the larger-scale justification for racism and determinism—any time you needed any general instrument of oppression, the progress doctrine was pretty good because it gave you the power—you must be progressively better and then you tie it in with other things. On the other hand, I think when you get away from Darwinism—which given the nature of the Mendelian mechanism and the way natural selection works is not inherently progressivist—and you move into human culture, which has a Lamarckian mode of inheritance, then you do have a justification for a more linear sort of Progress. I do believe that cultural history in the sense of complexification of technology is Progressive. Now that doesn't mean it's good! The worst thing is to equate Progress with good. It may lead to the destruction of all of us. Cultural Progress may be the arch-destroyer of culture, for all I know. (Interview with author, April 1988)

He adds that he has little time for naysayers of technological Progress: "The easiest way to counter that is to read the grief of any nineteenth-century person who's faced the death of their children. Then anybody tell me that Progress is a bad thing!" And he extends his thinking to science itself. "Sure, I think science is Progressive. Yes, in the sense that there's some physical reality out there and we manage to learn more and more about it." The key connection is that Gould—or rather, let me stress, Gould *now*—sees belief in biological progressionism as a major barrier to cultural Progress, and for this reason he is led to oppose the former strongly. This point comes through most strongly in *The Mismeasure of Man* (1981). In this book, which does not discuss punctuated equilibria but was written at the height of the controversy, Gould argues strenuously that racist doctrines have been founded on theories of progressionism. These restrictive doctrines have stood solidly in the way of admirable aspects of Progress.

Thus the public, and the private, Gould. Yet, there may be more to the story. Indeed, ultimately, there may be a biology/culture, progress/Progress link in Gould's thought. Notice how I have been stressing that it is *now,* today, that Gould opposes biological progress. This was not always so. In 1977, just at the point when the punctuated equilibria controversy was getting into high gear, Gould published his major contribution to evolutionary biology—of greater relevance to the scientific community than punctuated equilibria, to judge by the attention paid to it.[1] This book can with reason be considered one long testament to an adaptively fueled progressionism leading to humankind. *Ontogeny and Phylogeny* is a funny hybrid, part historical account of the embryology/evolution connection and part an argument that a key factor in evolution is changes in the timing and rate of various stages of development. Specifically, Gould argued that selection alters these rates, even though the morphological products developed after such changes may not be tightly adaptive; and as part of the general picture he also maintained that those animals which take the strategy of few offspring but increased parental care (*K*-selected) typically do this by retaining juvenile characteristics in the adult ("neoteny").

Crucial to our interests is that although Gould was intending to address fellow professionals, perhaps because of the atypical nature of the book, he left little doubt (for all that he noted the controversial status of the idea) that he saw *K*-selected organisms as somehow higher than those selected to have lots of offspring but to provide little care (*r*-selected).

> Evolutionary trends toward greater size and complexity form the classical subject matter of "progressive" evolution as it is usually conceived—the slow and gradual fine tuning of morphology under the continuous control of natural selection. These trends display three common features marking them almost inevitably as primary products of *K*-selective regimes:
>
> 1. A primary role for morphology in adaptation—usually leading to increased complexity, improvement in biomechanical design, or at least the continual exaggeration of specialized structures with clear functions.
> 2. A general tendency to increasing size—Cope's rule . . .
> 3. In most cases, a delay in the absolute time of maturation . . . Larger animals with a generally increased level of morphological differentiation almost surely mature later than their much smaller and more generalized ancestors. (Gould 1977b, 34)

Moreover, as a kind of climax to the book, we find that we humans are one of evolution's major success stories:

> I have been trying to deemphasize the traditional arguments of morphology while asserting the importance of life-history strategies. In particu-

> lar, I have linked accelerated development to *r*-selective regimes and identified retarded development as a common trait of *K* strategists . . . I have also tried to link *K* selection to what we generally regard as "progressive" in evolution, while suggesting that *r* selection generally serves as a brake upon such evolutionary change. I regard human evolution as a strong confirmation of these views. (Gould 1977b, 399)

Even the very examples underline the point. The "profound" differences between chimpanzees and humans are reflected in the former's inability to type the *Iliad,* rather than the latter's inability to swing through trees.

Gould has moved on since making this argument, less with respect to the science and more in his refusal to identify any selection-produced state with progress. Note, however, that not only has Gould been prepared to link evolution and progress but that in his realization that a commitment to biological progress (as fueled by selection) is a bar to social Progress, he is hardly denying the special status of humankind. The opposite, in fact. Hence, in exploring Gould's present position, it is worth considering the general direction his thinking, especially his thinking about causes, has taken. Given the very name "punctuated equilibria," it is tempting to suggest that Gould has moved from being a Darwinian to being a Spencerian. His theory does relate directly back to American dynamic equilibrium views of the 1920s and 30s, in that it was proposed as an extension of neontological thought to paleontology and it promoted Mayr's founder principle up front. Right behind this theory lies the whole metaphor of the adaptive landscape, and we know that that was hardly progress-neutral.

These views as such hardly make Gould a progressionist today, especially since he has modified his thinking since his early support for the founder principle. But the past has a nasty way of showing its forgotten (or suppressed) face. For all his protestations, Gould has just edited a highly progressionist popular book on evolution, covering the history of life from the primitive to a stunning bare-breasted Cro-Magnon beauty (Gould 1993). In the same vein, one notes in his work a continental connection, which comes out ever more strongly as Gould pushes down or constrains British adaptationism. Even though he would now deny any British Darwinian kind of progress, his ideas have the flavor of the Germanic idea of upward climb—not Marxist as such, but harkening back to common roots. If only because of the favorable reference to Vavilov, who like all Russian evolutionists was openly progressionist (in a German way), we should entertain this possibility. In the Germanic type

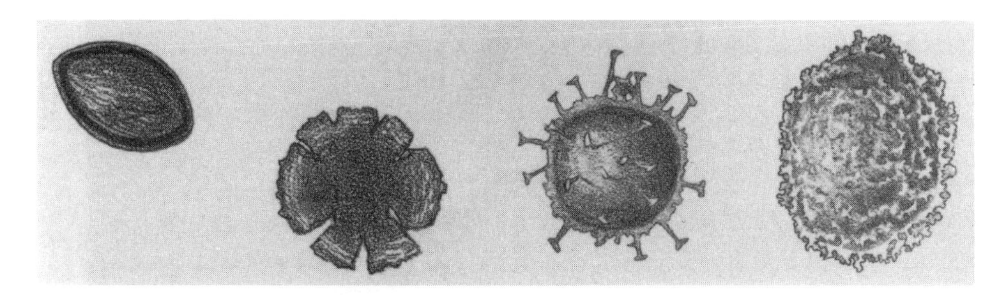

Precambrian acritarchs from the beginning of the first chapter of *The Book of Life,* edited by Stephen Jay Gould.

A Cro-Magnon woman from the end of the final chapter of *The Book of Life.* The important point is not that Gould really believes in a simplistic progression of life, for he surely does not, but that by giving his name to a book such as this, a work for the general reader, he is helping to perpetuate a public perception of evolutionary progress.

of progressionism to which I refer, *hierarchy* plays the crucial role: "If a historical system begins with simple components (as ours presumably did), and if complexity requires hierarchy and the bonding of lower-level individuals into higher entities (with a partial suppression of their independence and an altered status as parts of a larger whole), then a structural ratchet will ordain increasing complexity—'progress' if you will—as hierarchy builds historically" (Vrba and Gould 1986, 226).

Even if we disregard the "ratchet" analogy, which is another link between Gould and his Americans, the case to take seriously a Germanic (as opposed to Darwinian) notion of progress in Gould's thinking is strengthened by other comments in Gould's writings explicitly sympathetic to *Naturphilosophie* (Gould 1982a), not to mention a vision of humankind that puts us above the vulgarly biological: "No one doubts, for example, that the human brain became large for a set of complex reasons related to selection. But, having reached its unprecedented bulk, it could, as a computer of some sophistication, perform in an unimagined range of ways bearing no relation to the selective reasons for initial enlargement. Most of human society may rest on these nonadaptive consequences" (Gould 1982a, 108). At the very least, as Maynard Smith recognizes, excerpts like this one imply a temporal progression mapping the conceptual progression: Humans could not have appeared until lower levels had begun to function. Note also the technological analogy.

I doubt that one can tie up all of the knots neatly. Even if there is a deeper progressionist level to Gould's thought, and you may well remain unconvinced, I am happy to agree that one does not logically have to be a progressionist to think about macro-evolution. Indeed, after being pressed, Gould's co-author, Eldredge, finally blurted out: "To tell you the truth, Michael, I find thinking about the issue of progress like thinking about the issue of God . . . I don't care! I really don't care! I call myself an 'agnostic' but the truth is, I really don't care. And that's about how the idea of progress strikes me" (interview with author, Spring 1992). Gould is different. He does care, if only because he knows that open enthusiasm for progress is unlikely to lift the low-grade status of paleontology, a personal crusade.

Not that he should expect any thanks. Gould is today's most brilliant contributor to popular science. From his Harvard-located base in professional science, he moves across the divide with deceptive ease, contributing a monthly column ("This View of Life") to the popular science magazine *Natural History,* publishing collections of articles as well as single-theme books (for instance, Gould 1977a, 1980a, 1987), and lecturing to audiences of literally thousands. More deliberately than Simpson, even though it was from the older man that he took the title of his column, Gould has produced an extensive body of work, often showing in his popular writings the social and metaphysical currents beneath the smooth surface of his and others' professional science.

Yet, although many admire him, there is resentment of him as well. Perhaps, simple jealousy—or, perhaps, people do not like having their faults, especially if they are real, made public. Tension may arise as fellow professionals see their careful labors torn open before the vulgar: the move to professional respectability is too recent in this field and not yet sufficiently appreciated (especially by other professionals) for public deconstruction from within. Or perhaps some feel an unease (anger?) when Gould shows he does not think the professional/popular divide sacrosanct, as when he uses the popular realm to make points (say in favor of punctuated equilibria) which he clearly intends to reflect back into the professional realm. When people like Dobzhansky and Simpson addressed their messages to a popular audience, they had the authority of pathbreakers; they were doing it, despite their true wishes, because their evolution had started in the popular realm and there simply was not yet a body of professionals around to criticize. Gould should know better. Especially he should know better than to show that his apparently orthodoxly professional, epistemologically legitimate desire to critique progress (as in Gould 1988) is driven by a popularly pronounced, culturally based promotion of a better Progress (as in Gould 1981, 1989).

Whatever the true reason, hackles are raised. In my interviews with evolutionary scientists, I often asked about the effect of Gould's popular writing on his professional reputation. One respondent was particularly blunt: "It crucifies him. I hope he has no idea what people actually say about him. He's not as bad off as Carl Sagan, but he's approaching that." Science is a rough game.[2]

Sociobiology

I turn now to that topic which has caused the greatest recent controversy in evolutionary theory. In the 1970s, a number of evolutionary biologists came to believe that the time had come to frame their work on social behavior into a semi-autonomous sub-discipline. This would have its own identity but be loosely located within the evolutionary—more precisely, within the neo-Darwinian—family. It would be an enterprise with the same logical standing as (say) paleontology or biogeography (Ruse 1979b). To this end, these evolutionists started promoting their brain child, complete with a new name: sociobiology.[3] The focus of this movement was the world's outstanding expert on the social insects, Edward O. Wilson of Harvard University. It was he who (in 1975) produced the "Bible" of the movement, *Sociobiology: The New Synthesis,* this title consciously reflecting the volume of Julian Huxley a generation earlier. What made sociobiology more than just a topic of academic concern was the fact that Wilson and his fellows extended their discussion from the

Edward O. Wilson, "framed" between a picture of Charles Darwin (to the left) and a picture of Herbert Spencer (to the right).

animal world to that of humans. As one might expect, social scientists reared back in pain and fear (Sahlins 1976). Less expected, and more interesting, was the strong negative reaction by some students of the life sciences, including evolutionary biologists.

Here, although I appreciate the fact that they were not the only players, I shall concentrate exclusively on Wilson and his greatest critic, fellow Harvard biology department member, Richard C. Lewontin. Turning at once to *Sociobiology,* we find it divided into three parts. First, in "Social Evolution," comes a discussion of the causal nuts and bolts of the subject. These include the elementary principles of population genetics, supplemented with such items as the Hamiltonian analysis of kin selection. Highlighted as significant is the "multiplier effect"—the major behaviorial consequences that might follow from a minor genetic blip: "A small evolutionary change in the behavior pattern of individuals can be amplified into a major social effect by the expanding upward distribution of the effect into multiple facets of social life" (Wilson 1975a, 11). Interestingly, for all his detailing of the importance of individual selection, Wilson revealed an empathy for a group perspective on the workings of selection. He stands virtually alone among sociobiologists in this feeling,

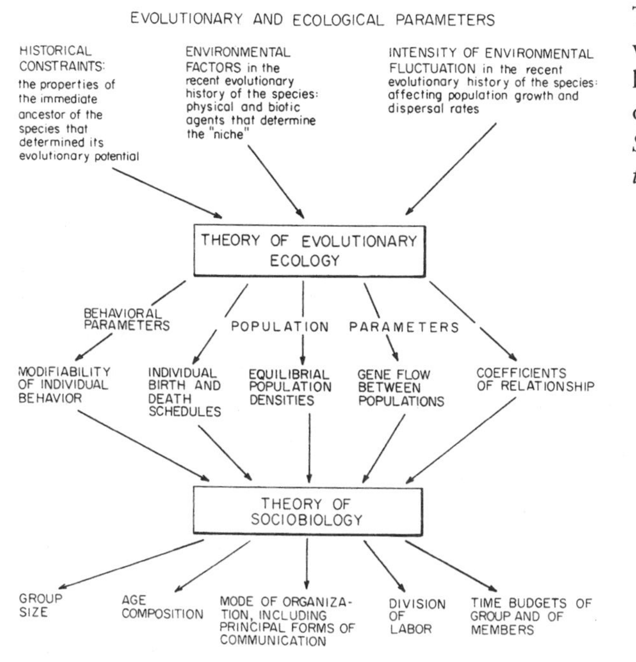

The place of sociobiology with respect to the rest of evolutionary and ecological theory, according to *Sociobiology: The New Synthesis.*

and has indeed been criticized on this score. However, these holistic yearnings were significant and prove important.

Next, in "Social Mechanisms," Wilson reviewed issues to do with communication, aggression, sex, parental care, and the like. Here, he retrod and expanded on much of the ground covered in an earlier book, devoted exclusively to the social insects (Wilson 1971). Finally, in the third and final part of his book, "The Social Species," Wilson ranged across all that we know about the social life of animals, from slime molds to humans. It was the final chapter, "Man: From Sociobiology to Sociology," later expanded into a work exclusively on our species, *On Human Nature,* which caused the controversy. Inviting us to "consider man in the free spirit of natural history, as though we were zoologists from another planet completing a catalog of social species on Earth" (p. 547), Wilson plunged in, following his own prescription. One learns that we are in respects a supremely successful social species, and the reason for our success is really not so hard to see. Our social behavior is a direct function of our distinctive characteristics: large brains, bipedalism, linguistic abilities, and more. Thanks to these things that make humans human, we are supremely skilled at working together in social situations.

Of course, recognizing what makes us work today does not at once explain why we evolved as we did. To this latter end, Wilson (1975a) proposed what he calls an "autocatalytic" model of evolution. Consciously drawing on the example of cybernetic thinking (p. 564), he suggested that a species may change undramatically for a long while under the force of natural selection and then sometimes the pressure builds up, a sort of "threshold" is reached, and an intensive feedback (selection-powered) evolution takes over. In the human case, this happened twice. First, we evolved gradually from the apes—evolved upward, literally, as we became bipedal. About three million years ago, with hands freed to work with tools and so forth, the first threshold was reached and brain and mental evolution took over in a big way, swelling our brain up from ape to human dimensions. Then: "The second, much more rapid phase of acceleration began about 100,000 years ago. It consisted primarily of cultural evolution and must have been mostly phenotypic in nature, building upon the genetic potential in the brain that had accumulated over the previous millions of years" (Wilson 1975a, 565–566).

What sort of being has this mental evolution produced? We are very clever, undoubtedly, but animals nevertheless. Our minds are shaped or constrained by what Wilson was later to label "epigenetic rules," capacities or dispositions which are made actual during development (Lumsden and Wilson 1981). These incline us to think and behave in certain ways rather than others—in certain adaptive ways rather than other non-adaptive ways. Thus, culture today and the humans within reflect that biological heritage which made us what we are. Culture is very flexible; but, using a phrase from *On Human Nature,* "the genes hold culture on a leash" (Wilson 1978, 167).

What hold do the genes have on us? The picture that Wilson sketches is well known. Humans live in groups—are biologically disposed to live in groups—which break down into families. The males are mildly polygynous, that is to say there is a tendency toward one man with multiple mates. Even if this trait is not formally codified into polygynous marriage, at least there is a propensity toward promiscuity. Females, to the contrary, are much more "coy" and discriminating about sexuality—after all, it is they who are left holding the baby. Humans, males particularly, tend toward aggressiveness, although obviously there are checks. Indeed, compared with other mammals, we are in respects quite peaceable. What we do have is a highly developed ethical sense, which makes us feel urges of obligation toward our fellows (Ruse and Wilson 1986). Since, for other obvious biological reasons, we have urges toward selfish ends, whether these be toward food or mates or whatever, our behavior includes much calculation of obligations, debts, perceived justices and injustices real or apparent, and the like. Overall: "Individual behavior, including seemingly altruistic acts bestowed on tribe and nation, are directed, sometimes very circuitously, toward the Darwinian advantage of the solitary human being and his closest relatives" (Wilson 1978, 58–59).

Finally, we have religion. This plays a major role in the life of the Wilsonian human. It gives meaning to existence, it reinforces morality, it encourages the weak and controls the sinner, and above all it helps to bond us into one group or tribe. In the human case, it is in the context of religion that Wilson most inclines toward a group-selection analysis. Humans would rather believe than know. We virtually beg for indoctrination. And there must be a reason for this desire, a reason which Wilson finds in the benefits of a socially cohesive group. Religion exists to make us work together, because directly or indirectly we all thereby benefit.

Is our behavior fixed for eternity? Not necessarily. Thanks to the multiplier effect, significant genetically founded cultural changes in a very short time are possible (Wilson 1975a, 569–572). How short is "short"? Certainly no more than a thousand years, and possibly quite a bit less than this. The genes modify or alter the epigenetic rules, and then these in turn affect culture, perhaps quite dramatically. At this point, feedback occurs, and fairly significant selection pressures are felt back down at the level of the gene. Thus, if some cultural phenomenon proves highly adaptive, very quickly we can expect some form of genetic assimilation, by which biology strengthens the ways we think and behave.

Biology as Religion

Wilson is unique among today's evolutionists in the extent to which he openly and deliberately pushes his progressionism. One only has to look at the chapter headings of *Sociobiology: The New Synthesis* to get the full force of Wilson's commitment to a very familiar notion of upward climb: "the colonial microorganisms and invertebrates"; "the social insects"; "the cold-blooded vertebrates"; "the birds"; "evolutionary trends within the mammals"; "the ungulates and elephants"; "the carnivores"; "the nonhuman primates"; "man: from sociobiology to sociology." The text backs Wilson's progressionist thinking. He sees a general and rather traditional upward ladder; but, more than this, as one might expect from a student of social behavior, Wilson sees progression in the social world also. The reader is left in little doubt that this is a very important product of evolution indeed: "Four groups occupy pinnacles high above the others: the colonial invertebrates, the social insects, the nonhuman mammals, and man. Each has basic qualities of social life unique to itself" (Wilson 1975a, 379).

Apparently there is a paradox in this progression, because although "the sequence just given proceeds from unquestionably more primitive and older forms of life to more advanced and recent ones, the key properties of social existence, including cohesiveness, altruism, and cooperativeness, decline" (ibid.). You need not despair, however, for it turns out that the human peak pushes up above the others, and progress in all of the senses is back on track: "Man has intensified [the] vertebrate traits while adding unique qualities of his own. In so doing he has achieved an extraordinary degree of cooperation with little or no sacrifice of personal survival and reproduction. Exactly how he alone has been able to cross to this fourth pinnacle, reversing the downward trend of social evolution in general, is the culminating mystery of all biology" (Wilson 1975a, 382). It is this mystery that we have seen Wilson addressing in his theory of "autocatalytic" evolution.

For Wilson, progression is a deeply pervasive aspect of the biological world.

> Q. Do you believe in progress?
>
> A. I think there is such a thing as progress in evolution, which stays pretty close to the term as defined in the English language or its equivalent in ordinary parlance. Let me give you an example. The transition from prokaryote to eukaryote is evolutionary progress. It represents evolutionary progress in the sense that it built upon a preexisting order and achieved conspicuously higher degree of sustained complexity. It permitted the opening—or shall we say, achieved the opening—of a whole array of new ecological niches, allowing life to attain much larger size forms, far higher levels (when multicellularity was added) of internal homeostasis, and the conquest of the land and the air. I think we can stay with "progress" as a term to describe that.
>
> Q. Humans?
>
> A. We have to be very careful when we say that progress in the evolutionary sense is in any way to be equated with Progress in the human socio-politico sense. But I think that in the pure less-value laden sense of the word, we would have to see quite a lot of evolution as

> progressive, for the reasons that I've just said. In the sense that it represented important advances in sustained complexity, and expanded life as a whole into modes of existence and parts of the environment that had not been occupied before.
>
> Q. Natural selection?
>
> A. I think it's a clear-cut product of natural selection. (Interview with author, April 1988)

Wilson draws a distinction between "success," meaning "having a long life as a species," and "dominance," meaning "having an impact on the ecosystem." We humans are the all-time dominant species. We bid fair for success also, but only time can tell. We represent "a new order, a new way of doing things, and built upon the others, a sustained new level of complexity that has carried that species and therefore a representative of life that produced it, into new modes of existence and new environmental niches." (See also Wilson 1990.)

Why is Wilson so deeply a committed progressionist? He himself relates his belief to his Darwinism; but, although his belief in natural selection and consequent adaptationism contributes to his progressionism—humans succeed because they can do things better than others—there has to be more than that behind it. At one level we should include Herbert Spencer. Wilson openly sympathizes with Spencer's overall progressive philosophy. At another level, we see in Wilson's writings the Wrightian landscape working at full blast. Note that when people like Wilson talk of adaptive "pinnacles," we have dropped all pretense that the "landscape" might be described as a chopping, changing seascape. And at a third level we should not discount the influence of Wilson's intellectual grandfather, America's previous great expert on the social insects, Harvard faculty member, big chum of L. J. Henderson, and dynamic equilibria enthusiast: William Morton Wheeler (Wheeler 1923, 1927; Evans and Evans 1970). Like his predecessor, Wilson moves back and forth across the human/hymenopteran divide, now looking at sociality in one group, and now in the other, and all in a highly P/progressionist manner.[4]

Moving to the question of cultural Progress and its possible significance, again Wilson provides grist for our mill. The crucial autocatalytic phase of evolution is expressed in metaphors drawn from cybernetics. Remarkably, more broadly, Wilson seems committed to a cultural equivalent of Agassiz's three-fold law of parallelism—cultures follow each other progressively, in a way that matches a progressive scale of extant cultures, reflecting the cultural development of the individual.

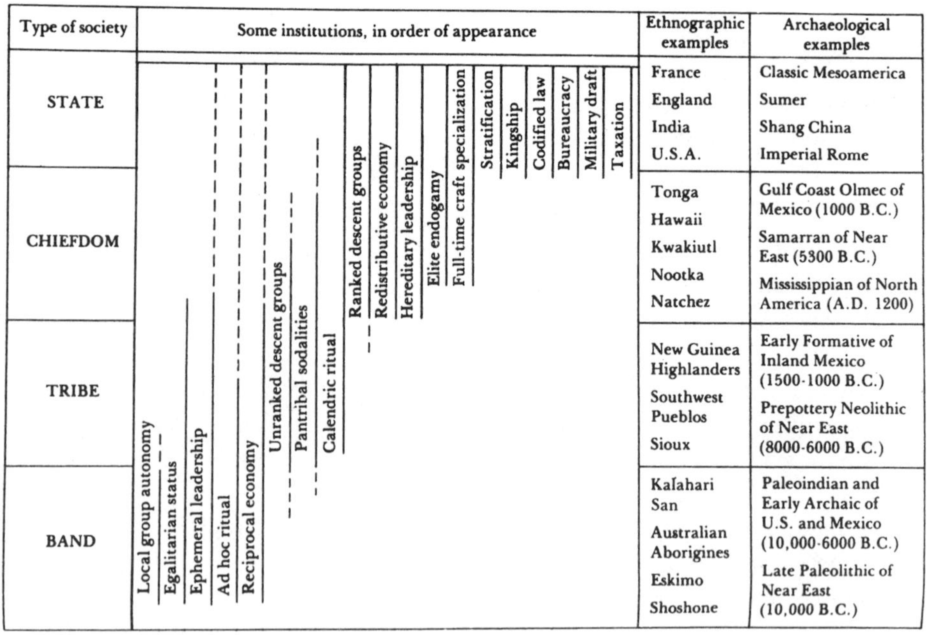

Type of society	Some institutions, in order of appearance	Ethnographic examples	Archaeological examples
STATE	Stratification; Kingship; Codified law; Bureaucracy; Military draft; Taxation	France England India U.S.A.	Classic Mesoamerica Sumer Shang China Imperial Rome
CHIEFDOM	Ranked descent groups; Redistributive economy; Hereditary leadership; Elite endogamy; Full-time craft specialization	Tonga Hawaii Kwakiutl Nootka Natchez	Gulf Coast Olmec of Mexico (1000 B.C.) Samarran of Near East (5300 B.C.) Mississippian of North America (A.D. 1200)
TRIBE	Unranked descent groups; Pantribal sodalities; Calendric ritual	New Guinea Highlanders Southwest Pueblos Sioux	Early Formative of Inland Mexico (1500-1000 B.C.) Prepottery Neolithic of Near East (8000-6000 B.C.)
BAND	Local group autonomy; Egalitarian status; Ephemeral leadership; Ad hoc ritual; Reciprocal economy	Kalahari San Australian Aborigines Eskimo Shoshone	Paleoindian and Early Archaic of U.S. and Mexico (10,000-6000 B.C.) Late Paleolithic of Near East (10,000 B.C.)

A cultural version of the law of parallelism from *On Human Nature.*

In my opinion the key to the emergence of civilization is *hypertrophy,* the extreme growth of pre-existing structures . . .

Hypertrophy can sometimes be witnessed at the beginning. One example in its early stages is the subordination of women in elementary cultures. The !Kung San of the Kalahari Desert do not impose sex roles on their children. Adults treat little girls in apparently the same manner as little boys, which is to say with considerable indulgence and permissiveness. Yet, . . . small average differences still appear. From the beginning the girls stay closer to home and join groups of working adults less frequently. During play, boys are more likely to imitate the men, and girls are more likely to imitate the women. As the children grow up, these differences lead through imperceptible steps to a still stronger difference in adult sex roles . . .

So only a single lifetime is needed to generate the familiar pattern of sexual domination in a culture. (Wilson 1978, 89, 91)

At a personal level, also, Wilson has a strong belief in Progress, which has given him an optimism and faith in science and technology. Crucial here were his childhood experiences, growing up in the South during the Depression and the subsequent Second World War.

> Q. Progress?
> A. Absolutely! I don't believe that someone who worried about progress and dominance and success can claim to be doing this in a culture-free objective model of inquiry. I think the study of evolutionary progress and dominance and success is just dandy! And I would admit that I have a basically optimistic manner . . .
> Q. Culture?
> A. Certainly culture matters. In my case, I've grown up in a culture that is heavily devoted to millenarianism, that is the belief that—I don't believe this—that Christ will soon come and we'll all go to Paradise. Boosterism—the American spirit of entrepreneurship and the belief in the unlimited ability of individuals to rise and prosper. The Protestant quality of belief in the work ethic and the rewards of work.
> In the Roosevelt era—and this was era in which people were desperately poor, Alabama was a Third World country—the country believed in Roosevelt. People believed in this socialism that Roosevelt was projecting. Alabama was the center of some of the most revolutionary nature of socialistic changes ever instigated in this country.
> The Tennessee Valley authority dammed the Tennessee and Tombigbee rivers, it created major dams and hydro-electric power through an area that had previously been without electric power largely, and removed hookworm and malaria and many other diseases through massive public health projects and generally injected new life into the economy of a part of the country which since the Reconstruction and the failure of the Insurrection of 1861–65 had been pretty near rock bottom.
> So this was part of the country which had a lot of despair, but it was also a part which had a real belief in the future. Along came the Second World War and many southerners, being a highly militaristic culture, thought that that was just fine. Military careers were thought of extra and important value in the South. And this was a part of the country that went into the Second World War with the belief that, yes, this was a good war and with a firm belief that we were going to prevail. So there was no impairment of the basic spirit of optimism and faith in the future. At least across large elements of the American middle-class Southern white population. (Interview with author, April 1988; see also Wilson 1994; Wright 1987)

We have a strong case for linking Wilson's beliefs about cultural Progress with his beliefs about biological progress. But there is yet more, the most important factor of all. To understand Wilson, and this includes his progressionism, you must go to the heart of the man, and this heart is religious. He grew up in a religion-intoxicated culture. At the age of fifteen he was "born again" at a revivalist meeting. And although shortly thereafter he went to the University of Alabama, fell among evolutionists, and lost his faith, the emotional impact of his earlier experiences and the bonding with his fellow humans has never left him.

> Q. Religion?
>
> A. Of all the humanists I know, I'm the one who gives the most credence to religion as a psychological and deep sociological phenomenon. I think it's evolved, I think it's deeply genetic, I think that it affects a large part of human behavior, and it's a most powerful social influence that can never be stamped out.
>
> Q. Background?
>
> A. I had some pretty strong religious experiences as an adolescent, being in that fundamentalist culture. Such rites of passage which are a good deal more emotional and deep-reaching and personal than one finds, I suspect, in most religions. Then being converted out of it, my love of science and discovery of Darwinian thinking and all that that entails about the new way of looking at the world, then I was never very satisfied with just a purely secular interpretation of the world and do see the value of the great rewards in identifying your life and your life's efforts with something larger than yourself. But what can a secular humanist search for that is larger than himself? Two things. One is the perception of the human species as an evolving species—the human gene pool—hence the considerable interest I have in sociobiology to interpret behavior, including religious behavior, as a product of the human gene pool. Our sacred genetic heritage . . . And the other religious outreach that I have then is towards biodiversity. What is our greatest heritage? What's worth preserving? What really is awe inspiring, other than the human species itself, this remarkable product of evolution, the totality of biodiversity on earth, which we've scarcely begun to explore, which is virtually sacred—if I can use that term—in our heritage. . .
>
> So a dedication to those conceptions . . . is a reasonably satisfying substitute for the millennium. (Interview with author, April 1988)

Explicitly, Wilson sees his evolutionism as a secular faith. As a myth.

> The evolutionary epic is mythology in the sense that the laws it adduces here and now are believed but can never be definitely proved to form a cause-and-effect continuum from physics to the social sciences, from this

world to all other worlds in the visible universe, and backward through time to the beginning of the universe. (Wilson 1978, 192)

For Wilson, his evolutionism yields the only things of true value. It sustains one's faith and sense of oneness with the only group with full powers of thought and action. And it offers the supreme moral dictates: to preserve and further humankind, and to promote organic diversity (both for its own sake and even more because we humans have evolved in symbiotic relation with nature and without diversity, let alone the utilitarian goods of drugs and foods from unexploited species, we perish).

Biological progressionism is integral to this vision. Because of progression, we humans stand at the top of the peak, with all of the valuable qualities we possess. Because of progression, through our religious spirit we transcend the selfishness of the gene and unite with our fellows. And because of progression, we have our purpose and obligations in life. That which has evolved, that which has evolved to the top, has value, and thus we *ought* to preserve and continue it. This means we have a *moral* obligation to preserve and further the human species and, because we are entwined with the rest of the living world, to promote biodiversity. Wilson frequently speaks with the fervor and rhetoric of the evangelical preachers, urging us to repent of our carelessness toward the Earth before it is too late. To this end, Wilson is much involved in the campaign to preserve the Brazilian rain forest (Wilson and Peter 1988).

Although he is uncomfortable with such an interpretation, there is a strong dispensationalist air to the closing passage of *Sociobiology*. We are warned that, as things go now, we have but a hundred years before we will have total self-knowledge with no myths or religion to shield us from its awful effects. Apparently, through a philosophy of progress we must reengineer ourselves and our world to meet this challenge.

A Rival Vision

The criticisms of Wilson fell essentially into two parts: first there were those of a more scientific ilk, and then second there were more metaphysical or philosophical objections (Ruse 1979b). Together they were taken to be a powerful counter to the illicit ideology embodied in *Sociobiology: The New Synthesis*. Both can be found in the writings of Richard Lewontin, a man who in 1975 (the year of the publication of *Sociobiology*) was rightly regarded as *the* outstanding population geneticist of his time.

Dobzhansky surrounded by friends and students. In the front row, Richard Lewontin is second from the left, Francisco Ayala (a sometime priest) is on the far right; Bruce Wallace stands directly behind Lewontin, and Leigh van Valen (with beard) is at the top right corner.

Dobzhansky's star student, he excelled at theory and had (a decade earlier) pioneered radically new techniques, based on molecular biology, for detecting genetic variation in populations. (See especially Lewontin 1974.)

The scientific criticism was two-fold. First, it was objected that there simply is no hard evidence for Wilson's central claims. Lewontin did not want to argue generally against sociobiology. He was prepared to allow that animal behavior has evolved and that selective mechanisms have been important. He was also prepared to allow that humankind falls in some sense under the evolutionary umbrella: "It is undoubtedly true that human behavior like human anatomy is not impervious to natural selection and that some aspects of human social existence owe their historical manifestations to limitations and initial conditions placed upon them by our evolutionary history" (Lewontin 1977, 29). But this is a far cry from saying that, for instance, we have any hard evidence for the sociobiology of homosexuality—something Wilson (1978) explains through kin selection, making homosexuals on a par with worker ants. Is it the case that homosexuals really do reproduce less than heterosexuals? And, if they do, does their behavior really have any adaptive significance? "We can dispense with the direct evidence for a genetic basis of various human social forms in a single word, 'None' " (Allen et al. 1976, 185).

Second, it was objected that the major aspects of humankind, specifically our culture, flatly contradict any possibility of a significant application of biology to our species. Human cultures rise, flourish, decline, and decay in time spans which hardly allow for a tremor through the gene pool. "In a mere 30 generations, Islam rose from nothing to be the greatest culture of the Western World and then declined again into powerlessness." Biology could not possibly have been significant. Likewise: "How are we to explain, on a genetic basis, the immense cultural differences between present day populations?" (Lewontin 1977, 29).

In *Sociobiology,* the multiplier effect was supposed to account for major cultural differences among human groups. It takes only a small genetic difference to bring about a massive cultural difference. About this, Lewontin has been scathing. The multiplier effect is an unsavory wheeze to avoid any genuine contact with reality. "No evidence is given for the existence of such an effect, nor are we told how it would be measured, quantified or specified" (ibid.). Likewise, later work Wilson did with a young Canadian physicist, Charles Lumsden, trying to put his ideas on a formal basis, was dismissed as shoddy and essentially unscientific (Lumsden and Wilson 1981, 1983; Lewontin 1981). Anyone can get any kind of change, so long as they push the causal factors to an unreasonably high pitch.

In a way, though, the science has driven Lewontin less than the philosophy. He has a range of objections. We learn that the enterprise is unfalsifiable—beyond empirical check. Against every critical onslaught, it steps back or sideways. Then the theorizing employs false metaphors. Ideas appropriate in one field are used illicitly in another, and often then illegitimately moved back to the first field. Slavery in the human world is employed as an uncomfortable metaphor in the ant world, at which point there is at least a hint that its naturalness among the Hymenoptera justifies its naturalness among humans. Sociobiologists are guilty of the sin of reification: giving something a name and then assuming that it must have a reality. Aggression, for instance, is a social notion with a host of meanings. Sociobiologists rip it from context, assume that it must have a direct genetic basis, and treat it as though it were a real unit, like eye color. Along with this practice goes the assumption that all characters can be neatly divided into separate kinds, even though a century of genetics has shown how simplistic a view this really is (Lewontin 1977).

Running through the whole of Lewontin's critique—and here he joins forces explicitly with Gould, with whom he has co-authored a major attack—is a general objection to the ubiquitous adaptationism of the sociobiologists (Gould and Lewontin 1979).[5] People like Wilson assume that every last thing must have a direct adaptive function, produced and controlled in the tightest possible manner by natural selection. But, the critics argue, this is just not true. It is not true of animals and plants. It is not true of human morphology. Hence, it is certainly not true of human behavior. The organic world is a medley of causes and effects, and many of these—constraints, allometry (differential growth), historical accident, drift, and much more—quite prevent a perfectly optimized, design-like functioning of selective forces. Sociobiology is mere "Darwinizing," akin to "harmonizing," "in which facile harmonies are built spontaneously around a theme for the sake of a few moment's enjoyment" (Lewontin 1977, 22).

From what foundation would Lewontin himself have us view evolutionary biology? With fellow Harvard biologist Richard Levins, he has written that "as working scientists in the field of evolutionary genetics and ecology, we have been attempting with some success to guide our own research by a conscious application of Marxist philosophy." There are, we learn, real contradictions in nature, and only through the dialectical method can the biologist grasp them: "For us, contradiction is not only epistemic and political, but ontological in the broadest sense. Contradictions between forces are everywhere in nature, not only in human social institutions. This tradition of dialectics goes back to Engels, who wrote, in *Dialectics of Nature* (1880), that 'to me there could be no question of building the laws of dialectics of nature, but of discovering them in it and evolving them from it' " (Levins and Lewontin 1985, 279).

In his own major work, *The Genetic Basis of Evolutionary Change* (based on the 1969 Jesup lectures and published the year before *Sociobiology*), Lewontin tried deliberately to cast his argument in a dialectical fashion. The central topic is that of the dispute between classical and balance hypotheses, with the Muller/Dobzhansky clash set as a conflict between thesis and antithesis. But there is more than just science at stake:

"Indeed the whole history of the problem of genetic variation is a vivid illustration of the role that deeply embedded ideological assumptions play in determining scientific 'truth' and the direction of scientific inquiry" (Lewontin 1974, 57). Nothing could be resolved until molecular techniques could reveal how much variation there truly is with populations. But then, instead of a simple victory for the balance side, or a happy stable synthesis, we find that the debate goes on. The dialectic continues—pointing, in Lewontin's opinion, to the need for a further, more holistic level of genetic understanding.

Where do humans fit into the dialectical model, and what has any of this to do with sociobiology? In Lewontin's Marxist opinion, the fault of Wilson and others lies in the illicit attempt to "reduce" humankind to a lower level of understanding—they fail to realize that humans introduce a whole new level of existence.

> Consciousness allows people to analyze and make deliberate alterations, so adaptation of environment to organism has become the dominant mode. Beginning with the usual relation, in which slow genetic adaptation to an almost independently changing environment was dominant, the line leading to *Homo sapiens* passed to a stage where conscious activity made adaptation of the environment to the organism's needs an integral part of the biological evolution of the species. As Engels (1880) observed in "The Part Played by Labor in the Transition from Ape to Man," the human hand is as much a product of human labor as it is an instrument of that labor. Finally the human species passed to the stage where adaptation of the environment to the organism has come to be completely dominant, marking off *Homo sapiens* from all other life. (Levins and Lewontin 1985, 69–70, a collection of essays; passages cited originally written by Lewontin)

In the world of human social relations, it is just a "category mistake," to use the language of the philosophers, to think that genetic explanations are adequate or even appropriate. It is for this reason that Wilsonian sociobiology is not simply wrong but positively wrong-headed. It is no wonder that all it offers is a thinly veiled apology for all of the reactionary elements in society.

Has any of this anything to do with progress and/or Progress? One expects Lewontin, as a Marxist, to believe in some kind of social Progress, and in some sense this is indeed the case. "Evolutionary theories of social systems, specifically Marxism and some of its variants are explicitly progressivist and perfectionist" (Levins and Lewontin 1985, 27). (This is from a book dedicated "To Frederick Engels, who got it wrong a lot of the time but who got it right where it counted.") However, Lewontin is certainly no simple-minded biological progressionist, most particularly not one of a Darwinian variety. He argues strongly against the idea that there is any ready measure of organic complexity, or that if there is, it will be of any aid to the progressionist. In any case, Lewontin denies that we can have a proper analogy between biological change and social change: "Marxist historical theory predicts the eventual replacement of one mode of production by another universally, although for long periods different, contradictory modes may coexist" (p. 18).

But, having said this, as in the case of Gould let us not forget that a Darwinian-type progress is not the only kind of biological progress we have ever encountered. There is also a Hegelian-transcendentalist-type of progress, a hierarchy driven by a kind of historical necessity, with humans at the top. This is the kind of progress a Marxist should understand, and there are at least hints of sympathy with something like it in what Lewontin writes. He is committed to a hierarchical view of nature: "As against the reductionist view, which sees wholes as reducible to collections of fundamental parts, we see the various levels of organization as partly autonomous and reciprocally interacting" (p. 288). Additionally, Lewontin seems to see the working of the law of quantity into quality with the production of humankind, and his language used does rather imply that humans are in some sense above, perhaps even better, than other organisms: "For Marxists the evolution of humans from prehumans and the inclusion of human history in natural history presupposed both continuity and discontinuous, qualitative change" (p. 254). The point is, apparently, that as human society "arises" out of animal societies, it "transforms" the adaptations it possesses, thereby making new needs (p. 46). At the same time: "Once the products of human labor become commodities, produced for exchange, they acquire a new set of properties beyond their physical and chemical structure or their utility" (p. 262) Perhaps much of this argument is value-free, but there is more than a hint of Engels, who with Marx did value humans above the rest.

In a sense, therefore, there is a kind of biological progress at the heart of Lewontin's thinking—a kind of progress which goes back (via Marx) to the transcendentalists. In a very real sense, therefore, one can claim that the clash between Wilson and Lewontin is not between a progressionist and a non-progressionist but between people with very different views of biological progress. Indeed, this is absolutely fundamental to their disagreement. The Wilsonian human is deeply and absolutely *centered* in its biological past from which it has risen, whereas the Lewontinian human is unique because it has *transcended* its biological past. For Wilson, our ethic of biodiversity concern stems from our having come up from, but remaining part of, nature. For Lewontin, our ethic of social and political concern stems from our having risen above nature. Rooted in rival progressive visions, with humans at the top, the two interpretations are rival secular religions, one might say.

Finally, let us ask whether the Wilson/Lewontin clash throws any light on the matter of professionalism, which is so crucial in the history of the relationship between progress and evolutionism. We are not disappointed. Lewontin thought Wilson was betraying the standards of good science, and thereby forsaking that goal of promoting excellence in evolutionism—a goal which he and Wilson had shared as younger men. Consider his comments in an interview which he gave after he had penned a very unpleasant review of the first collaborative effort of Wilson with Lumsden.

> One of the reasons my book review of Lumsden and Wilson had a kind of sneering tone is that it is the way I genuinely feel about the project, namely that it is not a serious, intellectual project. Because I have only two possibilities open to me. Either it is a serious intellectual project, and Ed Wilson can't think, or he can think, but it is not a serious project and therefore he is making all the mistakes he can—he does. If it is a really deep serious project, then he simply lowers himself in my opinion as an intellectual. (Segerstrale 1986, 75)

The point is that Wilson is simply not doing the kind of work that one expects of a first-class, professional evolutionist. As is revealed in a television interview about racist views in biology, Lewontin felt that he himself was doing such work: "everything that modern genetics, the kind of genetics that goes on in our lab, for example, tells us is that Darwin was right in the first place. That most of the genetic variation that occurs in the human species and indeed in most species, is between individuals within any group, and rather little of it is between groups" (Segerstrale 1986, 66). The way to interpret these comments is not simply as a case of blowing one's own horn, but rather as an expression of deep-felt convictions about the way one should do science—evolutionary science, in particular.

What about Wilson? He too was professionally trained, with high ideals for his science. He too was sensitive to the need to keep and promote a respectable profile for his work. Indeed, much he has done has been precisely with an eye to pushing evolutionary theory to higher areas of inquiry not previously explored, that it might thereby have a stronger profile. During the sociobiology controversy, Wilson was conscious of the danger of slipping beneath professional criteria. *Sociobiology: The New Synthesis* is a glamorous book, with big pages and lots of pictures, not to mention eye-catching phrases, as with the notorious first chapter heading: "The Morality of the Gene." Not surprisingly, some of the reviews were sharply critical of the popular nature of this work; but, trying to turn the table on his tormentors, Wilson replied that in *Sociobiology* he had not answered certain questions of a more general nature, precisely because it was not an appropriate forum. In less professional places, he had taken up such issues. "In an article published in the *New York Times Magazine* on October 2 [1975], I felt free to go well beyond the science in the book to discuss some of my personal feelings about the implications of sociobiology" (Wilson 1975b, 60; the article Wilson mentioned, "Human Decency Is Animal," was reprinted in Caplan 1978, 267).

The barbs did strike home. For the sequel to *Sociobiology: The New Synthesis,* the Simpsonian ploy of altering the level of discourse, and thus making criticism inappropriate, is activated: "*On Human Nature* is not a work of science; it is a work about science" (Wilson 1978, p. x). Ultimately, however, without judging the specific validity of Lewontin's criticisms, we can conclude that he was right to sense something odd about Wilson's approach to the professional/popular divide. Truly, Wilson sides with the Huxley grandson in rejecting the ground rules of the Huxley grandfather. Uniquely among today's leading evolutionists, Wilson really wants to keep progress up front in his professional science—and gives way only under pressure. For Wilson, any concealment—and there is not much—is done from pure expediency, not conviction.

There is a ready explanation why this should be so. Like Julian Huxley, Wilson believes that one can derive ethical norms from the history of life. Against T. H. Huxley, Wilson denies the fact/value distinction. He believes that there is no ultimate distinction of logical type between material claims and moral claims. If progress be true, it has as much right in science as anywhere else. Hence, it is not that Wilson is unique in being influenced by non-epistemic values. It is just that he is today alone in that he cannot see, other than for prudential reasons, why these values cannot be made quite explicit when he is writing as a professional scientist. Epistemologically, in his heart he sees them as no bar to mature science.

In this, and indeed in all of the preceding historical chapters, I have been painfully aware of the temptation to pick and choose one's characters and topics to suit one's conclusions. In this chapter, particularly, I do not pretend that I have given anything like an adequate scientific survey of evolutionists at work today. Nevertheless, I have looked at some of the big controversies and some of the major players, with much the same conclusion as before. If you are a professional evolutionist working on micro-problems, then there is neither need nor temptation to bother yourself with questions of progress—and, most probably, *qua* professional evolutionist you will not. But those thinking about the broader scope of things will be open to progressionist influences—and, although as professionals they will feel the need to curtail their public enthusiasm, there will be a strong inclination to let such influences come seeping back in. It is hard to keep a good cultural value down!

14

Conclusion

Our long history is finished. The significance of the cultural value of Progress to the history of evolutionism has been confirmed, many times over. Evolutionary thought is the child of Progress, and for its first hundred years was but a pseudo-science, supported and justified by its cultural content. Charles Darwin changed its standing. Thanks to the *Origin,* evolutionary thought became respectable, albeit still only as popular science. There it remained frozen, for nearly another hundred years, courtesy in major part of Darwin's lieutenant, Thomas Henry Huxley. Then the mathematicians gave the field both models and status and, building on these contributions, the evolutionists of the 1930s and 1940s crossed the divide to professional science. But the price was the expulsion of progress, no small sum to people still firmly committed to Progress.

We have surely covered a lot of ground to explain the Janus face of evolutionary thought today: fascinating yet disturbing. For nigh two centuries, evolution functioned as an ideology, as a secular religion, that of Progress—usually against, although sometimes with Providence. Moreover, let there be no mistake that at the popular level, which for most people is the beginning and the end of their acquaintance with evolution, Progress continues to ride high. I have yet to find a museum or a display or a chart or a book which is not overtly progressionist. Go to the Smithsonian in Washington, D.C., and marvel at the huge "Tower of

Humans at the top of the "Tower of Time."

Time," touchingly politically correct in the careful depiction of the evolutionary triumph of a black man, an oriental woman, and an aged white male. Or, go to South Kensington, the home of Richard Owen's natural history branch of the British Museum. From the permanent display on "Man's Place in Evolution" we learn: "When we look at the fossil remains, we find evidence of physical differences. For example, we are taller than the first human beings—the habilines—and we have larger brains

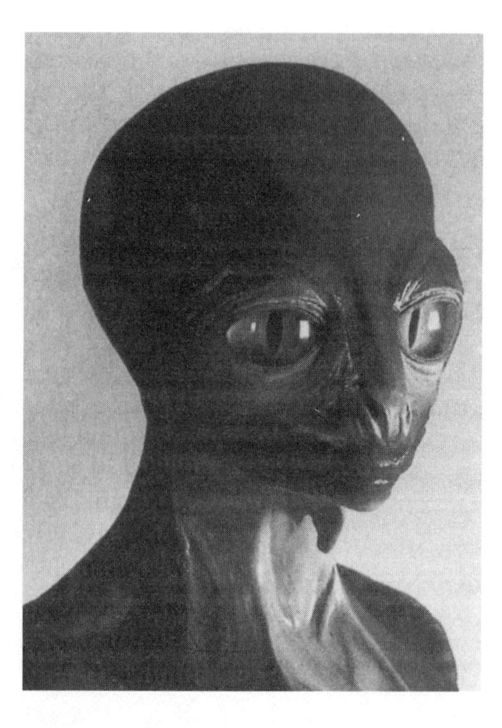

A "dinosauroid"—it is bipedal and has arms, hands, and an opposable thumb. (Model by D. A. Russell and R. Seguin; from Russell 1987, 127.)

than they had. There is also evidence that human behaviour has become much more complicated." These things may be true, but why do we not learn of adaptations we may have lost? Why is it not explained that the earlier adaptations might have done very well in their time? The traveling displays broadcast the same message. Quite incredible was a special exhibit in January 1991, "The Return of the Living Dinosaurs." In the final display the visitor was confronted by a model of how the dinosaurs might have evolved, had they not gone extinct. Herbert Spencer would have loved it. The *Naturphilosophen* would have been beside themselves with joy.[1]

Look at the books that a young enthusiast might encounter. A typical example, drawn from a series for young people (early teens), is *The Young Scientist Book of Evolution: Discoveries and Theories of the Origin of Life* (Cork and Bresler 1985). This has the imprimatur (as consultant) of Mark Ridley, student of Dawkins and in his own right an author of several books on and around evolutionary topics (Ridley 1985, 1986). It really is an excellent book: bright, well-illustrated, historically sensitive, up-to-date (there is something on kin selection), balanced, and

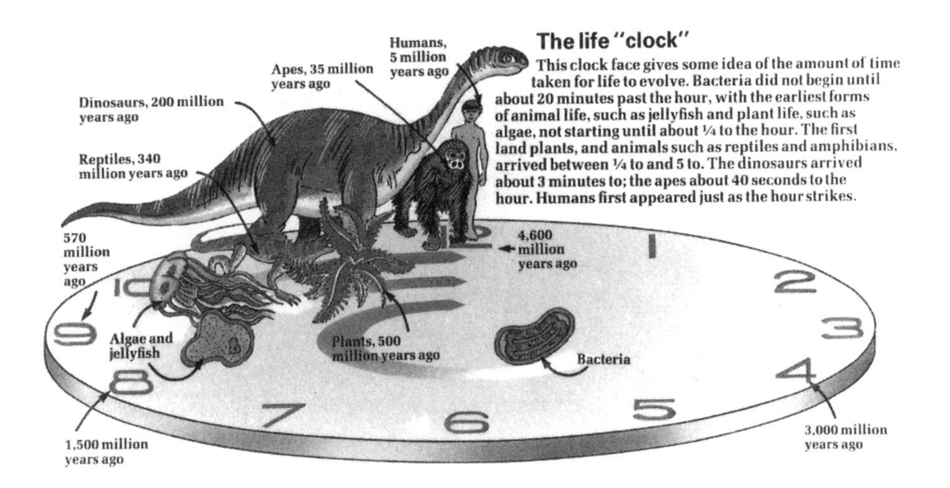

The point of a picture like this is not that it is intended to show or prove progress. It is not. It demonstrates, rather, that progress is an assumption built into people's thinking about evolution. (From Cork and Bresler 1985.)

fair. Creationism is introduced and discussed in a non-sneering manner, although the implication is certainly that it is not a regular scientific theory. It is brief, and a student would probably need a teacher's guidance to understand all of the points being made. But then, what are teachers for, if not to give guidance?[2] The book is also progressionist. The message is unambiguous. It shows itself in the "clock" of life, as the hour hand sweeps around through the bacteria and jellyfish and on to man (note the sex). It shows itself in the time chart of life on earth (p. 21). And it shows itself in the discussion of the fossil record, which takes us again from our friends the jellyfish (p. 22), right along to our grandfathers, *Homo sapiens sapiens* (p. 27). Helpfully also we are given a discussion of our special features (p. 26).

This book is the tip of an iceberg, distinctive in being an excellent book. Where it is not distinctive is in being progressionist, for this is the universal theme, from a recent issue of *Scientific American* on origins to an American Association for the Advancement of Science manifesto on science education, *Science 2061* (AAAS 1989), which promotes an evolutionism tied by the metaphysical thread of progress—until we get to ecology, where equilibrium takes over. But what would one expect when a major consultant to the project, Francisco Ayala, starts off his own textbook on

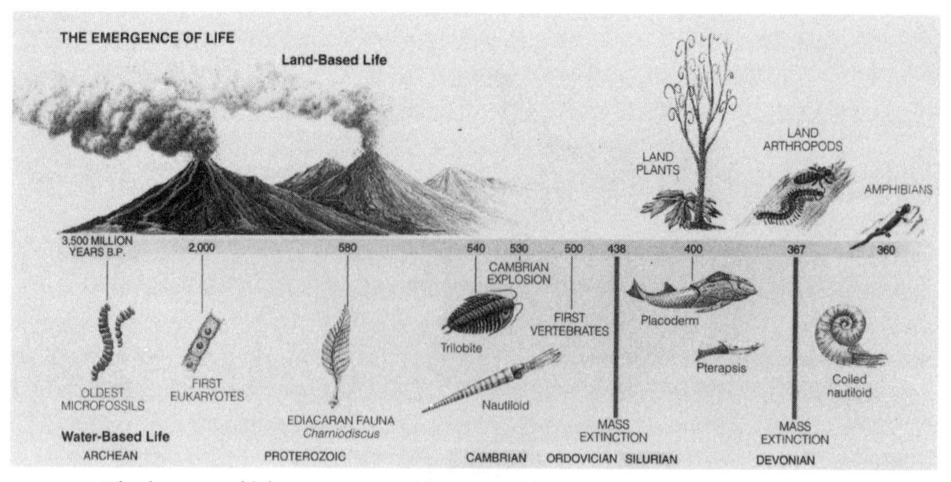

The history of life as envisioned by *Scientific American.* (From Weinberg 1994.)

the subject by saying of evolution that: "*These adaptive changes occasionally give rise to greater complexity of developmental pattern, of physiological reactions, and of interactions between populations and their environment*" (Dobzhansky et al. 1977, 8, italics in original)?

At the popular level, *Progress* and *evolution* are synonyms, with all of the attractions, ambiguities, and animadversions that that implies. Little wonder that the sensitized observer feels, at the least, discomfort. But what of the professional science, the science in the mature vein? Here, too, much has been explained in our trek through the history of the discipline. Most particularly, we now know of the reasons behind its lesser status as a professional science. Not only has evolution functioned as an ideology, as a secular religion, but for many professional biologists that has been its primary role. It has not been a mature (or proto-mature) science, governed by epistemic norms, nor has that necessarily been an end ardently sought. Very belatedly has evolution been brought to professional standing, as a result of steps taken when there were major competitors and detractors within the life sciences, like molecular biology. Moreover, the popular story that progress was expelled under the twin forces of Darwinism (the relativity of natural selection) and Mendelism (the randomness of mutation) is just plain wrong. After Darwin, progress rode high; after Mendel, progress continued to ride high. The expulsion of progress occurred less because the epistemic factors were overwhelming and more simply because its practitioners wanted the

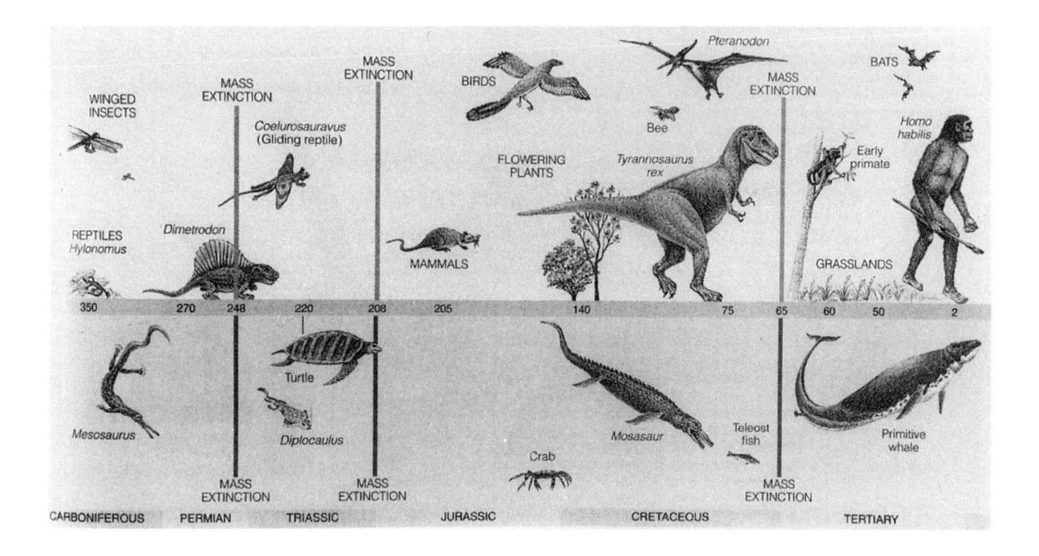

status as professionals. Even today, particularly in areas like paleontology, professional evolution's roots are still deep in popular culture, and that shows.

What then of our subsidiary hypotheses? Begin with the question of P/progress and of evolutionary theory today. In this respect, the most important finding—important also to emphasize, for there will be many eager to misinterpret my history—is that there is no comfort for those who argue that evolutionary theory (in whatever form) necessarily and always is impregnated with progressionism, and that now as from the beginning Progress has been the only (or major) reason for being an evolutionist. Whatever the significance of Progress, and the point of this book is to show that it is great, the history of evolutionary thought has been one of ever greater manifestation of epistemic norms. Think of Darwin and consilience; of Fisher and the other mathematicians determined to achieve coherence within their work and consistency without; of Weldon and of people today (Nicholas Davies and Geoffrey Parker) who prize predictive potential; of Hamilton and the elegance of kin selection and that same urge that continues to drive him. The marks of scientific maturity have grown. Furthermore, an important finding in the last two chapters has been that there truly is a functioning, professional, evolutionary biology *not* permeated by ideas of progress. If that is what you seek, you can find it.

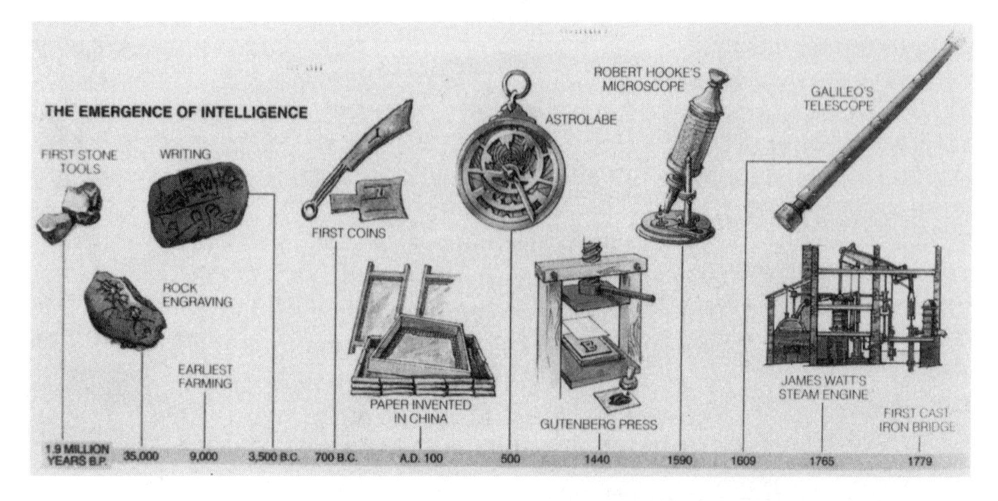

The emergence of intelligence as envisioned by *Scientific American.* Note that this chart appeared in the same article as the chart of the history of life and was intended to complement it. (From Weinberg 1994.)

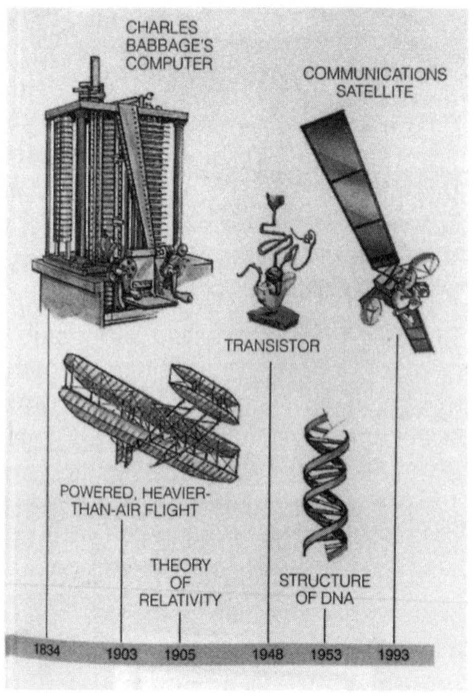

The cover of a recent book on evolutionary theory by Niles Eldredge. Even though the author is himself indifferent to progress, the design has given Wright's landscape a progressionist twist.

Does this mean that such biology is entirely culture-free, and if it is not is this a sign of weakness? My study does not answer the first part of this question, but then I have not taken a stand on the second part. Certainly, one might conclude from the work of a non-progressionist like Geoffrey Parker that other cultural values have influenced him, and may indeed have become part of his science. Whether these values must necessarily persist has gone unanswered, as has the question of whether their elimination would necessarily be followed by an influx of other cultural values. Since this is not a comparative study across science today, I cannot make sweeping judgments about culture and science generally.

One thing this study has shown is that the elimination of the influence of Progress does not seem to necessitate the substitution of another cultural value of the same category. A plausible initial hypothesis points to the fact that Progress is no longer a value cherished by many and suggests that evolutionary theorizing reflects this changed attitude. While it is true that some evolutionists have promoted beliefs about decline and decay, however, this does not seem to have been a general and lasting trend. If anything, today's evolutionists seem rather favorable to Progress, and thoughts of decline are absent—oddities like George Williams excepted. Perhaps "equilibrium" is in the same cultural category as "progress" and plays a like role. But this has hardly been proven, and even less has equilibrium been shown a force which is growing and pushing aside Progress. At the least before answering this suggestion one would need to grapple with the fact that there are other reasons for favoring equilibrium hypotheses, most particularly mathematical tractability. I would point out, nevertheless, that while the direct influence of Progress may be gone, or at least much diminished, it has declined as the result of a kind of meta-cultural value or, if you like, a value within the culture of science. Progress was not expelled simply by the force of the epistemic values. What truly counted was the desperate desire on the part of evolutionists to be taken seriously as professionals, a desire reinforced by such things as the respect everyone has for mathematical virtuosity, as exhibited by the population geneticists. This search for professional respect led to a greater emphasis on work guided by epistemic norms and was aided by downplaying progressionism. Cultural values of one kind led to the diminution of cultural values of another kind.

Switch the other way, and consider the hypothesis that progressionism may well exist in today's top-quality evolutionary thought but that it is of little moment because, *qua* cultural value, it has been effectively neutralized. Today one can define "progress" in epistemic terms, measures of complexity and the like, and by doing so take all of the tension out of the issue. Here, it is useful to hark back to the distinction between *evaluation*—measuring against a standard—and *valuing*—wanting or having an interest in something. Comparative progress is alive and well in today's evolutionary biology, call it "arms race" or "escalation" or the "Red Queen hypothesis." This kind of progress corresponds to evaluation, since it is all a matter of success in a certain set dimension, and is epistemically trouble-free. Absolute progress undoubtedly occurs in some of today's (would-be) professional evolutionary biology. I have shown that. But, at the risk of simply closing debate by fiat, I can only say that no satisfactory epistemic criterion of such progress has yet been given. *Adaptation and Natural Selection* still offers the definitive critique.

COMPLEXITY AND EVOLUTION

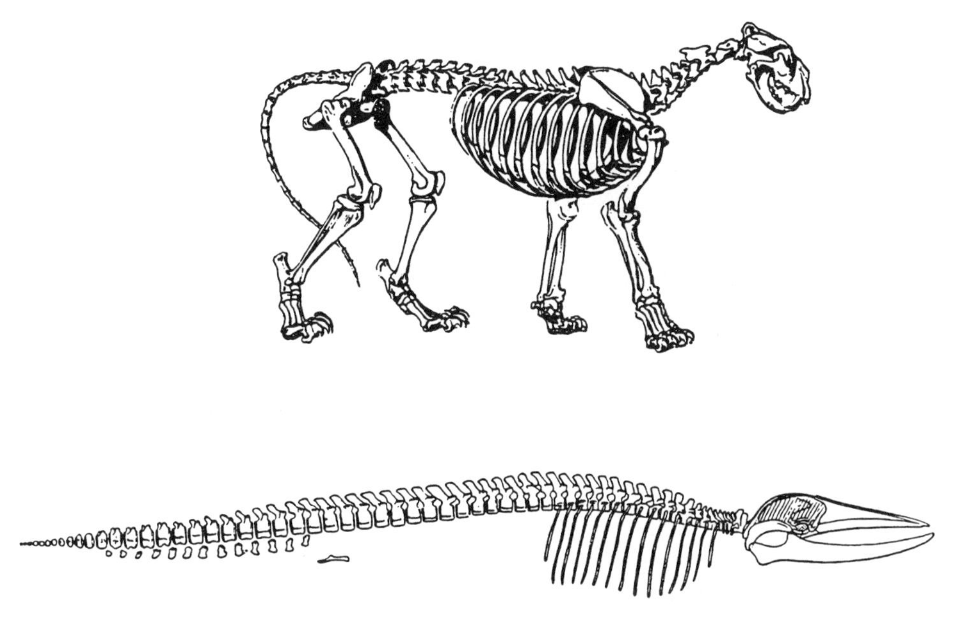

Drawings (not to scale) of the skeletons of a lion and a whale. The former is much more complex than the latter, although from an evolutionary viewpoint they are surely at the same level on the tree of life. The simplicity of the whale is no less a product of natural selection than the complexity of the lion. (From McShea 1991.)

More recent work, for instance on measures of complexity, simply shows what people like Simpson (himself a progressionist) said all along—namely, that there is just no good reason to think that complexity is a necessarily ever-increasing product of the evolutionary process (McShea 1991). The lion backbone is considerably more complex than the whale backbone, but who is to say which is the "higher" organism? What is interesting and surely pertinent is how infrequently (Simpson again excepted) evolutionists bother to discuss and define their notion of "progress." It is just taken as "obvious." It is also interesting that those who are evolution-as-progress boosters today are often more willing to acknowledge the subjectivity of what they are about. Dobzhansky's student, Ayala—"Well, I would say by many definitions, including very biologically meaningful definitions, humans are more progressive than any other organism" (interview with author, May 22, 1988)—is a case in point: " 'progressive' is an evaluative term that demands a subjective commitment to a particular standard of value" (Ayala 1988, 95).

I doubt that these brief comments about epistemic criteria will silence enthusiasts for progress. It will be objected that everyone knows that selection does not necessarily lead to complexity and progress. The point is whether, perhaps through arms races, a ratchet effect kicks in and so, overall, one does get absolute progress. Or it will be argued that, articulated epistemic definitions or not, there clearly was progress of a kind from life's beginnings to the arrival of multicellular organisms. Perhaps this actuality opens the door for the possibility of more progress, including that which happened recently in the primate line. A third counter will be that, complexity having exhausted its plausibility, there may be another epistemic criterion or flag waiting to take its place. Perhaps information theory would be of help here (Dawkins 1992). Or perhaps average taxon age span is what we should be looking at (Raup 1988).

In response, let me simply stress that, ultimately, my real concern is not with the logical possibility or impossibility of an epistemic criterion of progress, or even the chances of such a criterion for a Darwinian. Appearances and arguments to the contrary, perhaps there is an adequate characterization of complexity (or similar feature) and progress. (Kauffman 1993 is one recent writer who thinks there is.) My key point is that progress is not in evolutionary thinking today because of pure epistemic factors, nor relatedly has there been an ever-increasing move from the cultural to the epistemic. Whatever the logic of the case, for the evolutionists of the past, as well as for William Hamilton, Edward O. Wilson, and others today, progress comes from the cultural value of Progress. (See Ruse 1993 for an overview of current scientific thinking on the subject of progress.)

One final major question remains to be answered. Grant the existence today of a non-progressionist evolutionary biology. A major conclusion of this study is that some of the most significant of today's evolutionists are Progressionists, and that because of this we find (absolute) progressionism alive and well in their work. Or perhaps we come across a variant like Gould, who may not be a progressionist but who might plausibly be said to be a non-progressionist precisely because of his Progressionism! The question, to concentrate on the norm, is whether this is a transient situation, or if it is something likely to extend into the future. Are we now witnessing the death throes of progressionism in professional evolutionary biology, or will it be with us for many years, perhaps always? Is the epistemic maturity of evolutionary biology to be ever threatened by the cultural value of Progress? Let me offer three suggestions as to why Progress, with its related progressionism, may not disappear that quickly—suggestions, not logical entailments, for to prove more would be to prove too much.

First, evolutionists believe in and will continue to believe in progress because at work here is a version of what the cosmologists today refer to as the "anthropic" principle: the very possibility of knowledge sets conditions or constraints on the nature of knowledge. "What must the world be like in order that man may know it?" (Barrow and Tipler 1986, 143). Just as you cannot learn to swim without water, so you cannot know without an organ capable of knowledge. This fact in itself starts to set up its own internal logic. In the case of evolution, you cannot have a theory about it without a theorizer, which just so happens to be one of us humans. But we ourselves are part of the evolutionary process—we are not and cannot stand aside. Hence, there is bound to be a tendency to judge the process from our own perspective. We cannot do otherwise. What this means is that we are forced to value things like the evolution of intelligence, since we are the people asking the questions. As Simpson was wont to say: If animals are so damn smart, why aren't they asking questions? Humans are in a kind of Cartesian fix (where denying *cogito* affirms *sum*), inasmuch as our asking about progress presupposes that we have certain features which enable us to ask about progress. And if, as it happens, no-one else can do likewise, these features in themselves take on significance.

Second, science generally, and evolutionary understanding in particular, is seen to be Progressive (Ruse 1986). Even if you the reader have some historical or philosophical theory about science not being Progressive, remember that it is not shared by the average scientist. They believe that they get ever closer to a true understanding of reality through their theories and models. Science teaches us about the world outside and, for all of the false starts, as time goes by we learn more, and we learn in finer detail. This holds true particularly of evolutionary theory. No-one, certainly no scientist, would pretend other than that we do know a great deal more than previously. Science, therefore, is an area—it is *the* area—of human endeavor where hopes of Progress are quite unsullied by social obscenities like the Holocaust or economic disasters like the Depression or political regressions like fascism. Scientists uniquely, and evolutionary theorists particularly, make the one group in society who have reasonable hopes of Progress. At least, they are the group who think they have reasonable hopes of Progress.

The conclusion is drawn at once. Evolutionists take their belief in scientific Progress and transfer it into a belief in organic progress. We have seen this happen, again and again. Moreover, the particular conception of Progress is accordingly transferred. Some are quite explicit about the transfer. For instance, Dawkins (1976) has a whole theory of cultural units, "memes," acting in the same way as genes and resulting in the evolution of knowledge as much as the evolution of organisms. Likewise, a Popperian like Ayala sees a kind of continuous error-rejecting progress in nature, which develops just as scientific knowledge develops—problems arise, solutions are offered, the best is chosen, and then the whole process repeats (Ayala 1974). Here Ayala follows Popper himself (1972), who was quite explicit in linking Progress and progress. And the same is true of those, influenced by other philosophies, who see their distinctive brand of knowledge accumulation reflecting (and being reflected by) their understanding of evolution. Think, for instance, of the people like T. H. Huxley, who linked the jumps in knowledge due to genius with the jumps in change due to favorable macro-variations.

Third, there is simply a huge amount of self-selection among evolutionists. With respect to biological progress, evolutionists past and present are just not an unbiased sample. This was true before Darwin. So long as evolutionism remained part of popular science, people were looking for a new religion (generally deistic-cum-secular), and part of the creed was both Progress and progress. People were simply not interested in a non-progressivist evolution. Even as evolutionary studies were professionalized, the same self-selection was going on. Remember Dobzhansky's candid admission that, from the very first day of his evolutionary activities, his prime concern was with our own species. He wanted to find some way of justifying the ways of God through biology—something he achieved finally through the Christian progressivism of Teilhard de Chardin (1955). The kind of people who become full-time evolutionists tend to do so precisely because they are looking for a meaning to life. They want to find biological progress. And so, they do. Consider today just the example of E. O. Wilson (1994). He lost his Baptist faith and embraced Darwinism instead. For Wilson, evolution without progress would be a hollow shell. Mere science! It is not pure chance that Wilson is a biologist rather than a physicist. Of course, he is an extreme case, but there is a little bit of what drives Wilson in more of us than we care to admit. And perhaps this is not so surprising when you think that nearly all of us come to evolution through the popular realm—it is not as if we get a disinterested introduction to the subject.

My three reasons do not make a logically compelling explanation for why we should not expect progressionism to disappear from evolutionary theory anytime soon. They are not intended to. But I do not expect to see the demise of progressionism, however professional and mature evolutionary studies may become. This conclusion is reinforced by the consideration that, at the purely conceptual level, evolutionary thinking of the most professional or mature level is still pervaded by metaphors sympathetic to progress, such as "tree of life," "adaptive landscape," and "arms race." Removing phrases like these might make for epistemic purity, but doing so would also banish many virtues, like predictive fertility, and surely result in epistemic sterility.

Return to that wonderful museum in Le Jardin des Plantes. You need not ask. The evolution display is gloriously progressive. More than that. The gallery below the main display primes the visitor with a rapid history of technological Progress through the ages and then launches into the story of human ingenuity in animal and plant breeding, analogously from the primitive to the most sophisticated and complex. At the entrance to the garden stands a statue of Lamarck, who is looking through the gate toward the Seine. Erected by public subscription at the beginning of this century, a bronze bas-relief on the plinth depicts the biologist in old age, blind, comforted by his daughter: "*La postérité vous admirera, elle vous vengera, mon père.*" Yes, indeed.

Notes

Introduction

1. For today's professional historian of science, the worst of all possible sins is Whiggishness, using the past to *justify* (or denigrate) the present, usually by showing how a history of failure led up to the truth (or conversely). Agreeing with Nickles (1995), I am not convinced that Whiggishness is so great a fault or so easily avoidable as many assume. But this is not my game here. Nor is a circular analysis of evolution in terms of evolution. Although as an evolutionist I look to the past for answers, my analysis will not be evolutionary—elsewhere I have been very critical of such analyses (Ruse 1986).

2. Many historians and sociologists today, "social constructivists," argue that all science, professional or not, mature or not, contains cultural values, and that these values are the ultimate determinants of content (for instance Latour and Woolgar 1979; Collins 1985; Young 1985; Shapin and Schaffer 1985; Desmond 1989). Although the findings of this study will certainly be pertinent to this claim, my real quarry is the status of evolutionary thought, not some general thesis about science.

3. The Oxford Movement started in the 1830s, when a group of high-church theologians at the university tried to see how far the premises of the Church of England (the "Thirty-Nine Articles") could be accommodated with Roman Catholicism. It culminated in the next decade with many of its members going over to Rome, including its charismatic leader, John Henry (later Cardinal) Newman (Church [1891] 1970). It is to this that Sedgwick refers at the beginning of his letter.

4. Those readers not immediately familiar with the scientific ideas mentioned in this paragraph will find them more fully discussed at later points in this text.

1. Progress and Culture

1. I must note that not everyone agrees that Progress is an idea unknown to the ancient world. Nisbet (1980) argues that the Greeks had grasped and

endorsed it. Fortunately, this is not my controversy and I can go with general opinion.

2. The Birth of Evolutionism

1. A "theist"—traditionally a Jew, a Christian, or a Muslim—is one who believes that God intervenes miraculously in His Creation. Such a person believes, in other words, in a Providential God. A "deist" is one who believes in God as Unmoved Mover. Such a person expects humans to work Progressively, to achieve their own (and God's) ends.

2. One might argue that Lamarck's work in systematics was forcing him away from a uni-directional chain to a branching tree, and this is surely true. But in the light of Augier's non-evolutionary tree, forced on its author because of contempory work in plant systematics (appropriately he cites Lamarck in support!), the exact relationship between Lamarck's evidence and his evolutionism remains loose. More generally, the fact that professional systematists were having to revise their Chain-of-Being thinking, even invoking the tree, irrespective of evolution, hints that the relationship between evolutionism and professional/mature science might not be logically tight.

3. I do not intend to suggest either that, in fact, epistemic norms are entirely culture-relative or that they have a culture-free existence, like Platonic forms. For what it is worth, although this is not part of my argument, I myself am inclined to think that the norms are rooted in our (shared) biology, but that they take on different guises in different societies (Ruse 1986, 1995). Within the scope of our study, I will suggest that collectively the norms are more or less shared but that, at different times and for different reasons, different norms from the group are cherished and promoted. The wave theorists were very proud of their predictive abilities, especially of their ability to deduce *surprising* phenomena: I referred to the same ability as "fertility," in the Introduction. (See Buchwald 1989.)

4. I am not claiming that Oken was not a professional biologist. He was; although it is worth noting that *Isis* was pitched at more of a popular level.

3. The Nineteenth Century: From Cuvier to Owen

1. There is a parallel here with Kant's objection to evolution, and one suspects that there may have been an influence on the German-educated Cuvier.

2. Following convention, I refer to Darwin's own names for and pagination of his notebooks. I use the exemplary recent edition, Barrett et al. (1987).

3. I use the term *transcendentalism* in the usual sense, namely to refer to one standing in the Germanic idealistic tradition. An attraction is that the term (properly) transfers to nineteenth-century American thinkers.

4. I am not saying that Cuvier did produce a biology free of cultural values. In

fact, you know that I do not think he did. The point is that this is the way he needed to portray biology and, to be fair, the way he thought he was portraying biology. It has not escaped my notice that, successful or not, Cuvier was now employing a kind of meta-value, a value of the culture of science—the desirability of culture-free science—to expel cultural values from science.

5. This point goes to show that the category of "pseudo-science," like the category of "science" itself, cannot really be determined without acknowledging subjective factors. Lyell was using Lamarck as a foil to present his own ideas. It suited Lyell, a canny trained lawyer, to present Lamarck as a worthy opponent. One consequence was that many read Lyell and were converted by the soundness of the ideas—to evolution! Another consequence, for us, is that there is something inherently popular about the *Principles of Geology,* notwithstanding its significance in the history of science. But this is true, as we shall learn also, of an even greater book. Lyell's original intention was to write "Conversations on Geology" for the general public, and there was always that air about the book—not the least in the way that it was designed to be a moneymaker (Rudwick 1969).

4. Charles Darwin and Progress

1. "Pangenesis" supposes that there are little particles given off from all body parts and that these collect and form the sex cells. Its Lamarckian implications were a major attraction.

2. Chambers, unlike Darwin, never saw that *the* problem for the professional was that of design. However, note that Darwin, to an extent, was stuck in the 1830s. By the 1850s, homology was becoming as or more important than the teleology of Paley and Cuvier (Richards 1992). In a sense, the *Origin* was old-fashioned when it appeared, even though part of its genius was to redefine fashion. This cobwebby aura is a function of Darwin's ambiguous status. *Qua* professional, he knew he *should* not publish in the 1830s; *qua* non-professional, he did not *have* to, and when he did its old-fashioned quality showed.

3. *The Bridgewater Treatises,* published in the 1830s, were the definitive statements on British natural theology. It was through his contribution on astronomy that Whewell first gained wide public attention. See Whewell (1833) and, more generally, Gillispie (1950). Darwin's comment underlines my earlier point about his old-fashioned approach to teleology. A younger biologist like Huxley would have been horrified at the suggestion that he was writing a *Bridgewater Treatise,* even a secular one.

6. The Professional Biologist

1. The examination papers can be found in the respective university archives.
2. Without implying that Huxley would (or would not) have accepted Pop-

per's philosophy, I do suspect that Huxley would have agreed with Popper's (1974) first assessment of Darwinism as a "metaphysical research programme," a kind of condition for good (that is, mature) science rather than the working science itself.

3. The following is the breakdown of the natural science honours graduates, by subject, in the years 1886–1899, at Oxford University:

	A	B	C	D	E	F	G	Total
86–89	54	12	29	12	1	1	—	109
90–94	89	19	50	6	5	3	—	172
95–99	108	24	72	12	3	9	1	229
Total	251	55	151	30	9	13	1	510

A: Chemistry; B: Physics; C: Physiology; D: Morphology; E: Botany; F: Geology; G: Astronomy. Note the success of chemistry, *the* key industrial science.

8. British Evolutionists and Mendelian Genetics

1. The chromosomes are paired, with species members having (with some exceptions) the same numbers of chromosomes in the same pairings. Corresponding points on paired chromosomes are known as "loci," and more generally across the species one speaks of *the* locus. The gene variants which can occupy any particular gene locus are known as "alleles" or "allelomorphs." If the alleles on both chromosomes of a pair at some locus are identical, the individual (with respect to that locus) is a "homozygote"; if they are different, a "heterozygote." If the whole organism ("phenotype") of a heterozygote is identical to the phenotype of one of the homozygotes of the respective alleles, that allele is said to be "dominant" over the other, which is "recessive."

2. The influence here was the Manchester philosopher/psychologist Samuel Alexander, whose *Space, Time, and Deity* appeared in 1920. Similar philosophical ideas were espoused by the late-nineteenth-century British philosopher W. K. Clifford (1879), and they influenced Karl Pearson in his *Grammar of Science.*

3. Although Goodrich turned to philosophy to articulate his views on the evolution of mind, I do not sense that he turned to philosophy to help justify the kind of science he was doing. Generally, this holds true of other evolutionists in this century, and in major part was surely because philosophy of science became totally oriented toward physics, quite ignoring biology, and so the compliment was returned. There were some exceptions and they were important.

9. Discipline Building in Britain

1. This is from a manifesto for UNESCO, the non-Christian nature of which so irritated a number of important people that Huxley was denied a full term as Director-General.

2. In the early 1970s, in preparation for a conference on the formation of the synthetic theory, Ernst Mayr sent a questionnaire to the main participants. See Mayr and Provine (1980).

10. The Genetics of Populations

1. In *Creative Evolution* there is also something much like the landscape metaphor for evolution; but Gayon (1992, 354) perceptively suggests that the direct influence may have been Haldane and his more artistically gifted assistant, C. H. Waddington, who were toying with landscape-like metaphors in Haldane (1931b). Remember that Wright's major paper (1931) was prepared in the mid-1920s, whereas the short paper (1932) was prepared and then presented at a congress almost immediately. Hence, one looks for recent influences, although given the date of Wright's letter to Fisher (February 3, 1931) with a proto-version of the landscape, the timing may be a bit tight to give Haldane and Waddington full credit.

11. The Synthesis

1. A "taxon" is a particular group, like *Homo sapiens*. Taxa are themselves grouped into "categories." Thus, *Homo sapiens* is a taxon occurring at the species category level. *Homo* comes at the genus category level. Mayr's point was that the morphology cannot be part of the definition of the species *category*, that class of things which we call species *taxa*. We do use morphology to identify members of particular taxa.

2. I have edited and condensed my own questions. Except as shown explicitly, the responses are given verbatim.

3. Was Creationism a threat to discipline building? It is true that, through this period in America, it kept evolution out of the school textbooks (Nelkin 1977), but the building problems were not that different from those in Britain, where Creationism had no influence.

4. One can hardly say that, by this stage of his life, Mayr was very much open to philosophical influence. The interaction was rather one of Mayr justifying that which he already believed. It may be questioned whether "reductionism," the attempt to explain the larger in terms of the smaller, is a purely epistemic value. It is certainly epistemic inasmuch as it promotes other epistemic values, like simplicity, consilience, and predictive fertility. For the molecular biologist, reduction is very much the mark of desirable mature science. For the Hegelian, cherishing the ever-increasing hierarchy, reductionism and its converse surely have a cultural dimension.

5. There is a major analogy here with the "tree of life." That too is not

necessarily a progressionist metaphor, but it lends itself readily to such an interpretation.

12. Professional Evolutionism

1. In fairness to the memory of Haldane, about whose discipline-building skills I have been harsh, I must note that not only has Maynard Smith made major theoretical contributions to evolutionary thought, but more generally he has been at the forefront of those pushing a more mathematically informed approach to the subject (Maynard Smith 1968, 1974, 1982, 1989a). Additionally, in Britain, in the past decades he has been one of the key players in consolidating the field's professional status, using his post of Dean of Biological Sciences at Sussex University to gather a group of brilliant younger workers, writing a superb introduction to evolutionary thought (Maynard Smith 1958 and subsequent editions), as well as politicking successfully to get evolutionists elected to the Royal Society. Although it was Maynard Smith who sought out the older man, he has always stressed his intellectual and personal debt to Haldane. "I often do not know whether an idea is really my own or borrowed from Haldane" (Maynard Smith 1992, 50). He actually called a mini-autobiography "In Haldane's footsteps" (Maynard Smith 1989a).

13. Contemporary Debates

1. For instance, by the end of 1990, the first co-authored paper on punctuated equilibria (Eldredge and Gould 1972) had received 466 citations, whereas in five years less, *Ontogeny and Phylogeny* had received 670 citations. One sees a similar picture if one looks at citation patterns in *Evolution.* (See Ruse 1995 for a detailed discussion, together with raw data drawn from the *Science Citation Index,* of the influence of punctuated equilibria on the biological community.)

2. A revealingly sour review of a collection of essays by Gould complains that he "violates certain rules of etiquette" by refusing to strive "for clarity in the dual sense of expository simplicity and in making oneself transparent so that the empirical world is visible through the text but the peculiarities of the author are invisible" (Slobodkin 1988, 503). Significantly, the writer was the founder of the Department of Ecology and Evolution at the State University of New York at Stony Brook and one of a number of young turks in the United States in the 1960s—a number including the two antagonists to be introduced in the next sections of this chapter—doing what people like Maynard Smith were doing in Britain. They picked up evolutionary studies from the Synthetic Theorists and took them forward in new ways, especially in ways self-consciously physics-like in the use of sophisticated mathematical models and precise (often molecular-based) experiments (Wilson 1994).

3. English evolutionists like Nicholas Davies would certainly fall within this grouping: an honor they have accepted with more or less enthusiasm, some being reasonably happy with the idea and others preferring not to be set aside from other evolutionists quite so self-consciously.

4. Wheeler, a professional, was careful about where he talked of evolution. In 1910, in a classic book on the ants, he said not a word on evolution; and, even in 1918, in an article for the American Philosophical Society, he apologized for introducing the subject.

5. I am sure it was the sociobiological controversy which sensitized Gould to the social dangers of biological progressionism.

14. Conclusion

1. In fairness, some museum curators today are sensitive to evolution's progressivist tendencies. The new (1993) human evolution exhibit at the American Museum of Natural History even tells us: "We humans often think of ourselves as the culmination of a steady history of evolutionary improvement. But this idea is wrong, for evolution is neither goal-oriented nor merely a matter of species gradually improving their adaptation to the environments. Human evolution, like that of all other species, is the product of a unique series of interactions between our ancestors and an environment that fluctuated in a pattern that will never exactly repeat itself. Were a remote ancestor to replace us today, it is extremely improbable that a new *Homo sapiens* would stride the earth several million years hence." But the display itself goes progressively from "Lucy" *(Australopithecus afarensis)* to *Homo sapiens,* and the taxonomic tableau ("Our Place in Nature") gives the same impression. Upstairs, in the Hall of Primates, the old order reigneth, from tree shrew via chimpanzees to "man."

2. If you are in the State of Ohio, apparently, the guidance will be firmly progressionist. According to a survey, among that sub-section of biology teachers who believe in evolution at all, more favor characterizing the process in progressionist terms than by reference to natural selection (Zimmerman 1987).

Bibliography

Manuscripts and Archival Material

I follow the common editorial convention of enclosing words crossed out in a manuscript in angle brackets: <deletion>; additions to a manuscript are enclosed in double angle brackets: <<addition>>. Locations and abbreviations for archival materials are as follows:

Cain Papers — Arthur J. Cain Papers, American Philosophical Society, Philadelphia

Chambers Papers — Robert Chambers Papers, National Library of Scotland, Edinburgh

Dohrn Papers — Anton Dohrn Papers, Naples Biological Station, Naples

Darlington Papers — C. D. Darlington Papers, Bodleian Library, Oxford

Darwin Papers — Charles Darwin Papers, University Library, Cambridge, England

Dobzhansky Papers — Theodosius Dobzhansky Papers, American Philosophical Society, Philadelphia

Evolution Papers — American Philosophical Society, Philadelphia

Ford Papers — E. B. Ford Papers, Bodleian Library, Oxford

Haldane Papers — Haldane [Family] Papers, National Library of Scotland, Edinburgh; J. B. S. Haldane Papers, University College, London

J. Huxley Papers	Julian Huxley Papers, Rice University Library, Houston
Hyatt Papers	Alpheus Hyatt the Second Papers, Syracuse University Library, Syracuse, New York
Lankester Papers	E. Ray Lankester Papers, British Library, London
Osborn Papers	NYHS: H. F. Osborn Papers, New York Historical Society, New York; AMNH: materials collected in both the main library and the Department of Vertebrate Paleontology Library, American Museum of Natural History, New York
Pearson Papers	Karl Pearson Papers, Watson Library, University College, London
Poulton Papers	E. B. Poulton Papers, Hope Library, Oxford
Q. Wright Papers	Quincy Wright Papers, University of Chicago, Chicago
Royal College Papers	Royal College of Surgeons, London
Royal Society Papers	Royal Society, London
Sheppard Papers	Philip Sheppard Papers, American Philosophical Society, Philadelphia
Simpson Papers	G. G. Simpson Papers, American Philosophical Society, Philadelphia
Spencer Papers	Herbert Spencer Papers, University Library, London
T.D.–E.M. Correspondence	Correspondence of Theodosius Dobzhansky and Ernst Mayr, American Philosophical Society, Philadelphia
T. H. Huxley Papers	Thomas Henry Huxley Papers, Imperial College, London
Wallace Papers	Alfred Russel Wallace Papers, British Library, London

Whewell Papers — William Whewell Papers, Wren Library, Trinity College, Cambridge

Wright Papers — Sewall Wright Papers, American Philosophical Society, Philadelphia

Published Works

Abrams, L. 1905. The theory of isolation as applied to plants. *Science* 22: 836–838.

Adams, M. 1980. Sergei Chetverikov, the Kol'tsov Institute, and the evolutionary synthesis. In *The Evolutionary Synthesis: Perspectives on the Unification of Biology,* ed. E. Mayr and W. Provine, 242–278. Cambridge, Mass.: Harvard University Press.

Agassiz, E., ed. 1885. *Louis Agassiz: His Life and Correspondence,* 2 vols. London: Macmillan and Company.

Agassiz, L. 1842. On the success and development of organized beings at the surface of the terrestrial globe, being a discourse delivered at the inauguration of the Academy of Neuchatel. *Edinburgh New Philosophical Journal* 23: 388–399.

——— 1859. *Essay on Classification.* London: Longman, Brown, Green, Longmans, and Roberts and Trubner.

Aird, I., H. H. Bentall, and J. A. Fraser Roberts. 1953. A relationship between cancer of the stomach and the ABO blood groups. *British Medical Journal* 1: 799–801.

Alexander, S. 1920. *Space, Time and Deity,* 2 vols. The Gifford Lectures at Glasgow, 1916–1918. London: Macmillan and Co.

Allen, E., and others [Sociobiology Study Group]. 1976. Sociobiology: A new biological determinism. *BioScience* 26: 182–186.

——— 1977. Sociobiology: A new biological determinism. In *Biology as a Social Weapon,* ed. Sociobiology Study Group of Boston, 131–150. Minneapolis: Burgess.

Allen, G. E. 1978a. *Life Science in the Twentieth Century.* Cambridge: Cambridge University Press.

——— 1978b. *Thomas Hunt Morgan: The Man and His Science.* Princeton, N.J.: Princeton University Press.

Allen, J. A. 1905. The evolution of species through climatic conditions. *Science* 22: 661–668.

——— 1906. "Barriers" and "bionomic barriers," or isolation and non-isolation as bionomic factors. *Science* 23: 310–312.

Almond, G., M. Chodorow, and R. H. Pearce, eds. 1982. *Progress and Its Discontents.* Berkeley: University of California Press.

American Association for the Advancement of Science. 1989. *Science for All Americans: A Project 2061 Report on Literacy Goals in Science, Mathematics and Technology*. Washington, D.C.: AAAS.

Appel, T. A. 1987. *The Cuvier-Geoffroy Debate: French Biology in the Decades Before Darwin*. New York: Oxford University Press.

——— 1992. The making of a federal patron: Biology at NSF, 1950–52. Unpublished manuscript.

Aristotle. 1984a. De Generatione de Animalium. In *The Complete Works of Aristotle*, ed. Jonathan Barnes, 1111–1218. Princeton, N.J.: Princeton University Press.

——— 1984b. De Partibus. In *The Complete Works of Aristotle*, ed. Jonathan Barnes, 1087–1110. Princeton, N.J.: Princeton University Press.

Ashton, R. 1980. *German Idea: Four English Writers and the Reception of German Thought, 1800–1860*. Cambridge: Cambridge University Press.

Augier, A. 1801. *Essai d'une nouvelle classification des végétaux*. Lyon: Bruyset Ainé.

Ayala, F. J. 1974. The concept of biological progress. In *Studies in the Philosophy of Biology*, ed. F. J. Ayala and T. Dobzhansky, 339–354. London: Macmillan.

——— 1982. The evolutionary concept of progress. In *Progress and Its Discontents*, ed. G. Almond, M. Chodorow, and R. H. Pearce, 106–124. Berkeley: University of California Press.

——— 1985. The theory of evolution: Recent successes and challenges. In *Evolution and Creation*, ed. E. McMullin, 59–90. Notre Dame: University of Notre Dame Press.

——— 1988. Can "progress" be defined as a biological concept? In *Evolutionary Progress*, ed. M. Nitecki, 75–96. Chicago: University of Chicago Press.

Bagehot, W. 1868. Physics and politics. *Fortnightly Review* (London), April 1, 8: 518–538.

Bailey, L. H. 1896. *Plant Breeding, Being Five Lectures upon the Amelioration of Domestic Plants*. New York: Macmillan.

——— 1897. *The Survival of the Unlike: A Collection of Evolution Essays Suggested by the Study of Domestic Plants*, 2d ed. New York: Macmillan.

——— 1898a. *The Principles of Agriculture: A Text-book for Schools and Rural Societies*. New York: Macmillan.

——— 1898b. *Sketch of the Evolution of Our Native Fruits*. New York: Macmillan.

——— 1901. *Botany: An Elementary Text for Schools*, 2d ed. New York: Macmillan.

——— 1904. *Plant Breeding: Being Five Lectures upon the Amelioration of Domestic Plants*, 3d ed. New York: Macmillan.

——— 1911a. *The Country Life Movement in the United States.* New York: Macmillan.

——— 1911b. *The Outlook to Nature,* rev. ed. New York: Macmillan.

Baker, J. R. 1976. Julian Sorell Huxley. *Biographical Memoirs of Fellows of the Royal Society* 22: 207–238.

——— 1978. *Julian Huxley, Scientist and World Citizen, 1887–1975.* Paris: UNESCO.

Baker, J. R., and J. B. S. Haldane. 1933. *Biology in Everyday Life.* London: Allen and Unwin.

Barrett, P. H., P. J. Gautrey, S. Herbert, D. Kohn, and S. Smith, eds. 1987. *Charles Darwin's Notebooks, 1836–1844.* Ithaca, N.Y.: Cornell University Press.

Barrow, J. D., and F. J. Tipler. 1986. *The Anthropic Cosmological Principle.* Oxford: Clarendon Press.

Barry, M. 1837. Further observations on the unity of structure in the animal kingdom. *Edinburgh New Philosophical Journal* 22: 345–364.

Barsanti, G. 1989. *La Scala, la Mappa, l'Albero.* Florence: Sansoni Editore.

——— 1992. Buffon et l'image de la nature. In *Buffon 88,* ed. J. Gayon. Paris: Vrin.

Barthelemy-Madaule, M. 1982. *Lamarck the Mythical Precursor: A Study of the Relations between Science and Ideology.* Cambridge, Mass.: MIT Press.

Bates, H. W. 1862. Contributions to an insect fauna of the Amazon Valley. *Transactions of the Linnaean Society of London* 23: 495–566.

——— [1863] 1892. *The Naturalist on the River Amazon.* London: John Murray.

Bateson, W. 1884. The early stages in the development of Balanoglossus. *Quarterly Journal of Microscopical Science,* n.s. 24: 208–36.

——— 1886. The ancestry of the Chordata. *Quarterly Journal of Microscopical Science,* n.s. 26: 535–571.

——— 1894. *Materials for the Study of Variation, Treated with Especial Regard to Discontinuity in the Origin of Species.* London: Macmillan.

——— 1902. *Mendel's Principles of Heredity: A Defence.* Cambridge: Cambridge University Press.

——— 1922. Evolutionary faith and modern doubts. *Science* 55: 1412.

Beatty, J. 1987. Weighing the risks: Stalemate in the classical/balance controversy. *Journal of the History of Biology* 20: 289–320.

——— 1994. The proximate/ultimate distinction in the multiple careers of Ernst Mayr. *Biology and Philosophy* 9(3): 333–356.

Bellamy, E. 1887. *Looking Backward 2000–1887.* New York: Tricknor.

Ben-David, J. 1972. The profession of science and its powers. *Minerva* 10: 362–383.

Bennett, J. H., ed. 1983. *Natural Selection, Heredity, and Eugenics, Including Selected Correspondence of R. A. Fisher with Leonard Darwin and Others.* Oxford: Oxford University Press.

Benson, K. R. 1979. William Keith Brooks (1848–1908): A case study in morphology and the development of American biology. Ph.D. diss., Oregon State University, Corvallis, Ore.

Benton, M. J. 1987. Progress and competition in macroevolution. *Biological Reviews* 62: 305–338.

Berg, L. S. [1926] 1969. *Nomogenesis or Evolution Determined by Law.* Cambridge, Mass.: MIT Press.

Bergson, H. 1911. *Creative Evolution.* London: Macmillan.

Berlin, I. 1993. *The Magus of the North: J. G. Hamann and the Origins of Modern Irrationalism.* London: Murray.

Bernard, H., et al. 1897. Are the Arthopoda a natural group? *Natural Science* 10: 97–117.

Berry, E. W. 1906. Isolation and evolution. *Science* 23: 34.

Bölsche, W. 1909. *Haeckel: His Life and Work.* Trans. J. McCabe. London: Watts and Co.

Bonnet, C. 1745. *Traité d'insectologie, ou observations sur les pucerons,* 2 vols. Paris: Durand.

Bourdier, F. 1969. Geoffroy Saint-Hilaire versus Cuvier: The campaign for paleontological evolution (1825–1838). In *Toward a History of Geology,* ed. C. J. Schneer, 36–61. Cambridge, Mass.: MIT Press.

Bowden, J. W. 1839. The British Association for the Advancement of Science. *British Critic* 25: 1–48.

Bowler, P. 1975. The changing meaning of "evolution." *Journal of the History of Ideas* 36: 95–114.

——— 1976. *Fossils and Progress.* New York: Science History Publications.

——— 1984. *Evolution: The History of the Idea.* Berkeley: University of California Press.

——— 1986. *Theories of Human Evolution.* Baltimore: Johns Hopkins University Press.

——— 1988. *The Non-Darwinian Revolution: Reinterpreting a Historical Myth.* Baltimore: Johns Hopkins University Press.

——— 1990a. *Charles Darwin: The Man and His Influence.* Cambridge, Mass.: Blackwell.

——— 1990b. *The Invention of the Past.* Oxford: Blackwell.

——— 1994. Are the Arthopoda a natural group? An episode in the history of evolutionary biology. *Journal of the History of Biology* 27: 177–213.

Box, J. F. 1978. *R. A. Fisher: The Life of a Scientist.* New York: Wiley.

Brandon, R. N., and R. M. Burian, eds. 1984. *Genes, Organisms, Populations: Controversies Over the Units of Selection.* Cambridge, Mass.: MIT Press.

Braun, A. 1853. Betrachtungen über die Erscheinung der Verjungung in der Natur, insbesondere in der Lebens- und Bildungsgeschichte der Pflanze [The Phenomenon of Rejuvenescence in Nature, Especially in the Life and Development of Plants]. *Ray Society Botanical and Physiological Memoirs* i–xxv, 1–341.

Brett, H. 1931. *Joachim of Flora*. London: Methuen.

Brewster, D. 1838. Review of Comte, *Cours de Philosophie Positive*. *Edinburgh Review* 67: 271–308.

Brooks, W. K. 1909. Biographical memoir of Alpheus Hyatt (1838–1902). *Biographical Memoirs of the National Academy of Sciences* 6: 311–25.

Buchwald, J. 1989. *The Rise of the Wave Theory of Light: Optical Theory and Experiment in the Early Nineteenth Century*. Chicago: University of Chicago Press.

Buffon, G. L. L. 1749–1767. *Histoire naturelle, générale et particulière,* 15 vols. Paris: Imprimerie Royale.

——— 1954. De la nature de l'homme. In *Oeuvres philosophiques de Buffon,* ed. J. Piveteau. Paris: Presses Universitaires de France.

Burchfield, J. D. 1974. Darwin and the dilemma of geological time. *Isis* 65: 300–321.

——— 1975. *Lord Kelvin and the Age of the Earth*. New York: Science History Publications.

Burkhardt, R. W. 1972. The inspiration of Larmarck's belief in evolution. *Journal of the History of Biology* 5: 413–438.

——— 1977. *The Spirit of System: Lamarck and Evolutionary Biology*. Cambridge, Mass.: Harvard University Press.

Burwick, F. 1986. *The Damnation of Newton: Goethe's Color Theory and Romantic Perception*. New York: Walter de Gruyter.

Bury, J. B. [1920] 1924. *The Idea of Progress; An Inquiry into Its Origin and Growth*. London: MacMillan.

Cabanis, P. J. G. 1802. *Rapports du physique et du moral de l'homme*. Paris: Bossange Frères.

Cain, A. J. 1954. *Animal Species and Their Evolution*. London: Hutchinson.

Cain, A. J., and P. M. Sheppard. 1950. Selection in the polymorphic land snail *Cepaea nemoralis*. *Heredity* 4: 275–294.

——— 1952. The effects of natural selection on body colour in the land snail *Cepaea nemoralis*. *Heredity* 6: 217–231.

——— 1954. Natural selection in *Cepaea*. *Genetics* 39: 89–116.

Cain, J. A. 1992. Common problems and cooperative solutions: Organizational activity in evolutionary studies, 1936–1947. *Isis* 84: 1–25.

——— 1994. Ernst Mayr as *community* architect: Launching the Society for the Study of Evolution and the journal *Evolution*. *Biology and Philosophy* 9 (3): 387–428.

Callebaut, W. 1994. *Taking the Naturalistic Turn.* Chicago: University of Chicago Press.

Canning, G., H. Frere, and G. Ellis. 1798. The loves of the triangles. *Anti-Jacobin,* 16 April, 23 April, and 17 May.

Cannon, W. F. 1961. The impact of uniformitarianism: Two letters from John Herschel to Charles Lyell, 1836–1837. *Proceedings of the American Philosophical Society* 105: 301–314.

Caplan, A., ed. 1978. *The Sociobiology Debate.* New York: Harper and Row.

Cardwell, D. S. L. [1957] 1972. *The Organisation of Science in England.* London: Heinemann.

Carlyle, T. [1831] 1896. Characteristics. In *Critical and Miscellaneous Essays of Thomas Carlyle,* ed. H. D. Traill, 1–43. New York: Chapman and Hall.

——— [1834] 1937. *Sartor Resartus: The Life and Opinions of Herr Teufelsdröckh,* ed. C. F. Harrold. New York: Odyssey Press.

Caron, J. A. 1988. "Biology" in the life sciences: A historiographical contribution. *History of Science* 26: 223–268.

Carpenter, W. B. 1839. *Principles of General and Comparative Physiology.* London: Churchill.

Casey, T. L. 1906. Variation versus mutation. *Science* 57: 632.

Castle, W. E. 1911. *Heredity in Relation to Evolution and Animal Breeding.* New York: Appleton.

——— 1916. *Genetics and Eugenics.* Cambridge, Mass.: Harvard University Press.

——— 1917. Piebald rats and multiple factors. *American Naturalist* 51: 370–375.

Castle, W. E., and J. C. Phillips. 1914. *Piebald Rats and Selection: An Experimental Test of the Effectiveness of Selection and of the Theory of Gamete Purity in Mendelian Crosses.* Washington, D.C.: Carnegie Institution of Washington.

Chambers, C. A. 1958. The belief in progress in twentieth century America. *Journal of the History of Ideas* 19: 197–224.

Chambers, R. 1844. *Vestiges of the Natural History of Creation.* London: Churchill.

——— 1845. *Explanations: A Sequel to Vestiges of Creation.* London: Churchill.

——— 1846. *Vestiges of the Natural History of Creation,* 5th ed. London: J. Churchill.

——— 1847. General preface. In *Selected Writings of Robert Chambers,* iii–iv. Edinburgh: W. & R. Chambers.

——— 1853. *Vestiges of the Natural History of Creation,* 10th ed. London: J. Churchill.

Chambers, W. 1872. *Memoir of Robert Chambers, with Autobiographic Reminiscences of William Chambers LLD.* Edinburgh: Chambers.

Church, R. W. [1891] 1970. *The Oxford Movement: Twelve Years, 1833–1845.* Chicago: University of Chicago Press.

Clark, E. V., E. B. Ford, and C. Thomas. 1957. The Fogou of Lower Boscaswell, Cornwall. *Proceedings of the Prehistoric Society* (London) 23: 213–219.

Clark, J. W., and T. M. Hughes, eds. 1890. *Life and Letters of the Reverend Adam Sedgwick.* 2 vols. Cambridge: Cambridge University Press.

Clark, R. W. 1960. *Sir Julian Huxley, F.R.S.* London: Roy.

——— 1968. *J.B.S.: The Life and Work of J. B. S. Haldane.* London: Hodder and Stoughton.

——— 1972. *A Biography of the Nuffield Foundation.* London: Longman.

Clifford, W. K. [1878] 1879. On the nature of things-in-themselves. In *Lectures and Essays,* ed. L. Stephen and F. Pollack, 2, 71–88. London: Macmillan.

——— 1879. *Lectures and Essays,* ed. L. Stephen and F. Pollack, 2 vols. London: Macmillan.

——— 1885. *The Common Sense of the Exact Sciences,* ed. K. Pearson. London: Kegan, Paul.

Clutton-Brock, T. H., ed. 1988. *Reproductive Success: Studies of Individual Variation in Contrasting Breeding Systems.* Chicago: University of Chicago Press.

Clutton-Brock, T. H., and Charles Godfray. 1991. Parental investment. In *Behavioural Ecology: An Evolutionary Approach,* 3d ed., ed. J. R. Krebs and N. B. Davies, 234–262. Boston: Blackwell Scientific Publications.

Clutton-Brock, T. H., F. E. Guinness, and S. D. Albon. 1982. *Red Deer: Behavior and Ecology of Two Sexes.* Chicago: University of Chicago Press.

Cobb, J. A. 1912. Human fertility. *Eugenics Review* 4: 379–382.

Colbert, E. H. 1980. *A Fossil Hunter's Notebook: My Life with Dinosaurs and Other Friends.* New York: Dutton.

Coleman, W. 1964. *Georges Cuvier, Zoologist: A Study in the History of Evolutionary Thought.* Cambridge, Mass.: Harvard University Press.

Collins, H. 1985. *Changing Order.* London: Sage.

Comstock, J. H. 1875. *Notes on Entomology: A Syllabus of a Course of Lectures Delivered at the Cornell University by J. Henry Comstock, Spring Trimester, 1875.* Ithaca, N.Y.: University Press.

——— 1888. *An Introduction to Entomology,* Part I. Ithaca, N.Y.: J. H. Comstock.

——— 1893. Evolution and taxonomy. In *Wilder Quarter Century Book,* ed. J. H. Comstock, 37–114. Ithaca, N.Y.: Comstock Publishing.

——— 1918. *The Wings of Insects: An Exposition of the Uniform Terminology of the Wing-Veins of Insects and a Discussion of the More General Characteristics of the Wings of the Several Orders of Insects.* Ithaca, N.Y.: Comstock Publishing.

——— 1924. *An Introduction to Entomology,* Parts I and II. Ithaca, N.Y.: Comstock Publishing.

——— 1925. Burt Green Wilder. *Science,* n.s. 61 (1586): 531–533.

Comstock, J. H., and A. B. Comstock. 1895. *A Manual for the Study of Insects.* Ithaca, N.Y.: Comstock Publishing.

Comstock, J. H., and V. L. Kellogg. 1895. *The Elements of Insect Anatomy.* Ithaca, N.Y.: Comstock Publishing.

Comstock, J. H., and J. G. Needham. 1898–99. The wings of insects. *American Naturalist* 32: 43–48, 81–89, 231–257, 335–340, 413–424, 561–565, 769–777, 903–911; 33: 117–126, 573–582, 845–860.

Comte, A. [1822] 1975. Plan of the scientific operations necessary for reorganizing society. In *Auguste Comte and Positivism: The Essential Writings,* ed. G. Lenzer, 9–67. New York: Harper and Row.

——— 1830–42. *Cours de Philosophie Positive.* Paris.

Condorcet, M. J. A. N. C. [1795] 1956. *Sketch for a Historical Picture of the Progress of the Human Mind.* New York: Noonday Press.

——— 1988. *Correspondence inedité de Condorcet et de Mme. Suard,* ed. E. Badinter. Paris: Fayard.

Conklin, E. G. 1897. The embryology of *Crepidula. Journal of Morphology* 13: 1–226.

——— 1902. Karyokinesis and cytokinesis in the maturation, fertilization and cleavage of *Crepidula* and other Gastropoda. *Journal of the Academy of Natural Sciences of Philadelphia* 12: 5–121.

——— 1921. *The Direction of Human Evolution.* London: Oxford University Press.

Conry, Y. 1974. *L'introduction du darwinisme en France au XIXème siècle.* Paris: Vrin.

Cook, O. F. 1906a. Factors of species-formation. *Science* 23: 506–507.

——— 1906b. The nature of evolution. *Science* 24: 303–307.

Cook, R. 1974. *The Tree of Life: Symbol of Centre.* London: Thames and Hudson.

Cope, E. D. [1868] 1886. On the origin of genera. *Proceedings of the Academy of Natural Sciences* (Philadelphia), 20: 242–300. Reprinted in Cope 1886, 41–123.

——— [1870] 1886. On the hypothesis of evolution, physical and metaphysical. First published in *Lippincott's Magazine*; reprinted in Cope 1886, 128–172.

——— [1881] 1886. The developmental significance of human physiognomy. Lecture first published in Cope 1886, 281–293.

——— 1883. *The Vertebrata of the Tertiary Formations of the West,* Book I. Washington, D.C.: Government Printing House.

——— 1886. *Origin of the Fittest.* New York: Macmillan.

——— 1896. *The Primary Factors of Organic Evolution.* Chicago: Open Court.

Cork, B., and L. Bresler. 1985. *The Young Scientist Book of Evolution: Discoveries and Theories of the Origin of Life.* London: Usborne.

Corsi, P. 1988. *The Age of Lamarck.* Berkeley: University of California Press.

Corsi, P., and P. J. Weindling. 1985. Darwinism in Germany, France and Italy. In *The Darwinian Heritage,* ed. D. Kohn, 683–730. Princeton: Princeton University Press.

Cott, H. 1940. *Adaptive Colouration in Animals.* London: Methuen.

Coulton, G. G. 1927. *Five Centuries of Religion.* Cambridge: Cambridge University Press.

Cox, F. E. G. 1989. Letter to Editor. *Nature* 341: 289.

Crosland, M. P. 1967. *The Society of Arcueil: A View of French Science at the Time of Napoleon I.* London: Heinemann.

——— 1978. *Gay-Lussac, Scientist and Bourgeois.* Cambridge: Cambridge University Press.

——— 1992. *Science under Control: The French Academy of Sciences, 1795–1914.* Cambridge: Cambridge University Press.

Crunden, R. M. 1982. *Ministers of Reform: The Progressives' Achievement in American Civilization, 1889–1920.* New York: Basic Books.

Cunningham, A., and N. Jardine, eds. 1990. *Romanticism and the Sciences.* Cambridge: Cambridge University Press.

Cuvier, G. 1810. *Rapport historique sur les progrès des sciences naturelles.* Paris.

——— [1813] 1822. *Theory of the Earth,* 4th ed., trans. R. Kerr, ed. Robert Jameson. Edinburgh: William Blackwood.

——— 1817. *Le règne animal distribué d'aprés son organisation, pour servir de base à l'histoire naturelle des animaux et d'introduction à l'anatomie comparée.* Paris.

——— 1825. Nature. In *Dictionnaire des sciences naturelles,* 261–268. Paris. [Unpublished English translation provided by William Coleman.]

——— 1827. Eloges historique de Daubenton, lu le 5 avril 1800. *Recueil des éloges historiques lus dans les séances publiques de l'Institute Royal de France* 1: 37–80.

Darlington, C. D. 1932. *Recent Advances in Cytology,* with foreword by J. B. S. Haldane. London: Churchill.

——— 1939. *The Evolution of Genetic Systems.* Cambridge: Cambridge University Press.

——— 1947a. The genetic component of language. *Heredity* 1: 269–286.

——— 1947b. *Heredity,* news and views. *Nature* 159: 599.

——— 1948. *The Conflict of Science and Society.* London: Watts.

——— 1953. *The Facts of Life.* London: Allen and Unwin.

——— 1959. *Darwin's Place in History.* Oxford: Blackwell.

——— 1969. *The Evolution of Man and Society.* London: Allen and Unwin.

Darlington, C. D., and K. Mather. 1949. *The Elements of Genetics*. London: Allen and Unwin.

Darwin, C. 1839. *Journal of Researches into the Geology and Natural History of the Various Countries Visited by H.M.S. Beagle*. London: Henry Colburn. [Modern editions often titled *The Voyage of the Beagle*.]

——— 1842. *The Structure and Distribution of Coral Reefs*. London: Smith Elder.

——— 1851a. *A Monograph of the Fossil Lepadidae; or, Pedunculated Cirripedes of Great Britain*. London: Palaeontographical Society.

——— 1851b. *A Monograph of the Sub-Class Cirripedia, with Figures of all the Species. The Lepadidae; or Pedunculated Cirripedes*. London: Ray Society.

——— 1854a. *A Monograph of the Fossil Balanidae and Verrucidae of Great Britain*. London: Palaeontographical Society.

——— 1854b. *A Monograph of the Sub-Class Cirripedia, with Figures of all the Species. The Balanidge (or Sessile Cirripedes); the Verrucidae, Etc., Etc., Etc*. London: Ray Society.

——— 1859. *On the Origin of Species*. London: John Murray.

——— 1862. *On the Various Contrivances by which British and Foreign Orchids are Fertilized by Insects, and On the Good Effects of Intercrossing*. London: John Murray.

——— 1863. [Review of H. W. Bates's paper, "Mimetic Butterflies."] *Natural History Review*, pp. 219–224.

——— 1865. *On the Movements and Habits of Climbing Plants*. London: Longman, Green, Longman, Roberts and Green and Williams and Norgate.

——— 1868. *The Variation of Animals and Plants Under Domestication*, 2 vols. London: John Murray.

——— 1871. *The Descent of Man*, 2 vols. London: John Murray.

——— [1873] 1977. On the males and complemental males of certain Cirripedes, and on rudimentary structures. *Nature* 8: 431–432. Reprinted in *The Collected Papers of Charles Darwin*, 2 vols., ed. P. H. Barrett, 2: 177–182. Chicago: University of Chicago Press.

——— 1875. *Insectivorous Plants*. London: John Murray.

——— 1881. *The Formation of Vegetable Mould, Through the Action of Worms, with Observations on their Habits*. London: John Murray.

——— 1958. *The Autobiography of Charles Darwin, 1809–1882*, ed. Nora Barlow. London: Collins.

——— 1959. *The Origin of Species by Charles Darwin: A Variorum Text*, ed. M. Peckham. Philadelphia: University of Pennsylvania Press.

——— 1977. *The Collected Papers of Charles Darwin*, 2 vols., ed. P. Barrett. Chicago: University of Chicago Press.

——— 1985– . *The Correspondence of Charles Darwin* [9 vols. to date]. Cambridge: Cambridge University Press.

Darwin, C., and A. R. Wallace. 1858. On the tendency of species to form varieties; and on the perpetuation of varieties and species by means of selection. *Proceedings of the Linnaean Society, Zoological Journal* 3: 46–62.

——— 1958. *Evolution by Natural Selection*. Cambridge: Cambridge University Press.

Darwin, C., R. Fitzroy, and P. P. King. 1839. *Narrative of the Surveying Voyages of His Majesty's Ships Adventure and Beagle, Between the Years 1826 and 1836, Describing their Examination of the Southern Shores of South America, and the Beagle's Circumnavigation of the Globe*. London: H. Colburn.

Darwin, E. 1789. *The Botanic Garden*. Part II, *The Loves of the Plants*. London: J. Johnson.

——— 1791. *The Botanic Garden*. Part I, *The Economy of Vegetation*. London: J. Johnson.

——— 1794. *Zoonomia; or, The Laws of Organic Life*, vol 1. London: J. Johnson.

——— 1800. *Phytologia*. London: J. Johnson.

——— 1801. *Zoonomia; or, The Laws of Organic Life*, 3d ed., 4 vols. London: J. Johnson.

——— 1803. *The Temple of Nature*. London: J. Johnson.

Darwin, F., ed. 1887. *The Life and Letters of Charles Darwin, Including an Autobiographical Chapter*, 3 vols. London: Murray.

Darwin, F., and A. C. Seward, eds. 1903. *More Letters of Charles Darwin*, 2 vols. London: John Murray.

Daudin, H. 1926. *Cuvier et Lamarck: les classes zoologique et l'idée de série animale, 1790–1830*, 2 vols. Paris: F. Alcan.

Davenport, C. B. 1906. The mutation theory in animal evolution. *Science* 24: 556–558.

Davie, G. E. 1961. *The Democratic Intellect: Scotland and Her Universities in the Nineteenth Century*. Edinburgh: Edinburgh University Press.

Davies, N. B. 1990. Dunnocks: Cooperation and conflict among males and females in a variable mating system. In *Cooperative Breeding in Birds*, ed. P. B. Stacey and W. D. Koenig, 455–485. Cambridge: Cambridge University Press.

——— 1992. *Dunnock Behaviour and Social Evolution*. Oxford: Oxford University Press.

Davies, N. B., and M. de L. Brooke. 1988. Cuckoos versus reed warblers: Adaptations and counter-adaptations. *Animal Behaviour* 36: 262–284.

——— 1989a. An experimental study of co-evolution between the cuckoo, *Cuculus canorus*, and its hosts, I. Host egg discrimination. *Journal of Animal Ecology* 58: 207–224.

——— 1989b. An experimental study of co-evolution between the cuckoo,

Cuculus canorus, and its hosts, II. Host egg markings, chick discrimination and general discussion. *Journal of Animal Ecology* 58: 225–236.

Davies, N. B., M. de L. Brooke, and A. F. G. Bourke. 1989. Cuckoos and parasitic ants: Interspecific brood parasitism as an evolutionary arms race. *TREE* 4 (9): 274–278.

Dawkins, R. 1976. *The Selfish Gene.* Oxford: Oxford University Press.

——— 1986. *The Blind Watchmaker.* London: Longman.

——— 1988. The evolution of evolvability. In *Artificial Life,* ed. C. G. Langton, 201–220. Redwood City, Calif.: Addison-Wesley.

——— 1992. Progress. In *Keywords in Evolutionary Biology,* ed. E. F. Keller and E. Lloyd, 263–272. Cambridge, Mass.: Harvard University Press.

——— 1995. *A River out of Eden.* New York: Basic Books.

Dawkins, R., and J. R. Krebs. 1979. Arms races between and within species. *Proceedings of the Royal Society of London, B* 205: 489–511.

de Beer, G. R., ed. 1938. *Evolution: Essays on Aspects of Evolutionary Biology Presented to Professor E. S. Goodrich on His Seventieth Birthday.* Oxford: Clarendon.

——— 1947. Edwin Stephen Goodrich. *Obituary Notices of the Fellows of the Royal Society* 5: 477–490.

——— 1963. *Charles Darwin: Evolution by Natural Selection.* London: Nelson.

de Sismondi, J. C. L. 1847. *Political Economy and the Philosophy of Government.* London: John Chapman.

Desmond, A. 1979. Designing the dinosaur: Richard Owen's response to Robert Edmond Grant. *Isis* 70: 224–234.

——— 1982. *Archetypes and Ancestors: Paleontology in Victorian Britain, 1850–1875.* London: Blond and Briggs.

——— 1984. Robert E. Grant: The social predicament of a pre-Darwinian transmutationist. *Journal of the History of Biology* 17: 189–223.

——— 1985. Richard Owen's reaction to transmutation in the 1830's. *British Journal for the History of Science* 18: 30–50.

——— 1989. *The Politics of Evolution: Morphology, Medicine and Reform in Radical London.* Chicago: University of Chicago Press.

——— 1994. *Huxley, the Devil's Disciple.* London: Michael Joseph.

Desmond, A., and J. Moore. 1992. *Darwin: The Life of a Tormented Evolutionist.* New York: Warner.

Dexter, R. W. 1954. Three young naturalists afield. *Scientific Monthly* 71 (1): 45–51.

DiGregorio, M. 1984. *T. H. Huxley's Place in Natural Science.* New Haven: Yale University Press.

Dobzhansky, T. 1937. *Genetics and the Origin of Species.* New York: Columbia University Press.

——— [1943] 1981. Temporal changes in the composition of populations of *Drosophila pseudoobscura* in different environments. *Genetics* 28: 162–186. Reprinted in Lewontin et al. 1981, 305–328.

——— 1953. *Genetics and the Origin of Species,* 3rd ed. New York: Columbia University Press.

——— 1954. Critical review of *The Facts of Life. American Journal of Physical Anthropology* 12 (4): 619–623.

——— 1955. A review of some fundamental concepts and problems of population genetics. *Cold Spring Harbor Symposium in Quantitative Biology* 20: 1–15.

——— 1956. *The Biological Basis of Human Freedom.* New York: Columbia University Press.

——— 1962a. *Mankind Evolving.* New Haven: Yale University Press.

——— 1962b. *The Reminiscences of Theodosius Dobzhansky.* New York: Oral History Research Office, Columbia University, New York.

——— 1967. *The Biology of Ultimate Concern.* New York: New American Library.

Dobzhansky, T., F. J. Ayala, G. L. Stebbins, and J. W. Valentine. 1977. *Evolution.* San Francisco: Freeman.

Dobzhansky, T., and H. Levene. 1955. Developmental homeostasis in natural populations of *Drosophila pseudoobscura. Genetics* 40: 797–808.

Dobzhansky, T., and L. S. Penrose. 1954. Critical review of *The Facts of Life. Annals of Human Genetics* 19 (1): 75–77.

Dobzhansky, T., and B. Spassky. [1944] 1981. Manifestation of genetic variants in *Drosophila pseudoobscura* in different environments. *Genetics* 29: 270–290. Reprinted in Lewontin et al. 1981, 371–391.

Dobzhansky, T., and B. Wallace. 1953. The genetics of homeostasis in *Drosophila. Proceedings of the National Academy of Sciences* 39: 162–171.

——— 1959. *Radiation, Genes and Man.* New York: Henry Holt and Company.

Dobzhansky, T., S. Wright, and W. Hovanitz. [1942] 1981. The allelism of lethals in the third chromosome of *Drosophila pseudoobscura. Genetics* 27: 363–394. Reprinted in Lewontin et al. 1981, 243–273.

Doyle, J. A. 1977. Patterns of evolution in early angiosperms. In *Patterns of Evolution as Illustrated by the Fossil Record,* ed. A. Hallam, 501–546. Amsterdam: Elsevier.

Duncan, D., ed. 1908. *Life and Letters of Herbert Spencer.* London: Williams and Norgate.

Dunn, L. C. 1965. *A Short History of Genetics.* New York: McGraw-Hill.

Dupree, A. H. 1959. *Asa Gray, 1810–1888.* Cambridge, Mass.: Harvard University Press.

Eldredge, N. 1971. The allopatric model and phylogeny in paleozoic invertebrates. *Evolution* 25: 156–167.

Eldredge, N., and S. J. Gould. 1972. Punctuated equilibria: An alternative to phyletic gradualism. In *Models in Paleobiology,* ed. T. J. M. Schopf, 82–115. San Francisco: Freeman, Cooper.

Eliot, C. W. 1877. Address. In *Ninth Annual Report of the American Museum of Natural History,* 49–52. New York, New York: Printed for the Museum.

Ellegard, A. 1958. *Darwin and the General Reader.* Göteborg: Göteborgs Universitets Arsskrift.

Engel, A. J. 1983. *From Clergyman to Don: The Rise of the Academic Profession in Nineteenth Century Oxford.* Oxford: Oxford University Press.

Engels, F. [1880] 1940. The part played by labour in the transition from ape to man. In *The Dialectics of Nature,* 279–296. New York: International Publishers.

Evans, M. A., and H. E. Evans. 1970. *William Morton Wheeler, Biologist.* Cambridge, Mass.: MIT Press.

Falk, R. 1961. Are induced mutations in *Drosophila* overdominant? II. Experimental results. *Genetics* 46: 737–757.

Farrar, W. V. 1975. Science and the German university system, 1790–1850. In *The Emergence of Science in Western Europe,* ed. M. P. Crosland, 179–192. New York: Science History Publications.

Filipchenko, A. 1929. *Variation and Methods for Its Study,* 4th ed. [in Russian]. Leningrad: Gosizdat.

Fisch, M., and S. Schaffer, eds. 1991. *William Whewell: A Composite Portrait.* Oxford: Oxford University Press.

Fisher, R. A. [1918a] 1971. The correlation between relatives on the supposition of Mendelian inheritance. *Transactions of the Royal Society of Edinburgh* 52: 399–433. Reprinted in Fisher 1971, 1, 134–168.

——— [1918b] 1971. The causes of human variability. *Eugenics Review* 10: 213–220. Reprinted in Fisher 1971, 1, 169–176.

——— [1922] 1971. On the dominance ratio. *Proceedings of the Royal Society of Edinburgh* 42: 321–341. Reprinted in Fisher 1971, 1, 414–434.

——— [1927] 1972. Objections to mimicry theory; statistical and genetic. *Transactions of the Royal Entomological Society, London* 75: 269–278. Reprinted in Fisher 1972, 2, 164–173.

——— [1928] 1972. The possible modification of the response of the wild type to recurrent mutations. *American Naturalist* 62: 115–126. Reprinted in Fisher 1972, 2, 243–254.

——— 1930. *The Genetical Theory of Natural Selection.* Oxford: Oxford University Press.

——— [1934] 1973. Indeterminism and natural selection. *Philosophy of Science* 1: 99–117. Reprinted in Fisher 1973, 3, 230–248.

——— [1947] 1974. The renaissance of Darwinism. *Listener* 37: 1001. Reprinted in Fisher 1974, 4, 616–620.

——— [1950] 1974. *Creative Aspects of Natural Law: The Eddington Memorial Lecture.* Cambridge: University Press. Reprinted in Fisher 1974, 5, 179–184.

——— 1971–1974. *Collected Papers of R. A. Fisher,* 5 vols., ed. J. H. Bennett. Adelaide: University of Adelaide.

Fisher, R. A., and E. B. Ford. [1947] 1974. The spread of a gene in natural conditions in a colony of the moth *Panaxia dominula. Heredity* 1: 143–174. Reprinted in Fisher 1974, 4, 623–655.

Fisher, R. A., E. B. Ford, and J. Huxley. [1939] 1974. Taste-testing the Anthropoid apes. *Nature* 144: 750. Reprinted in Fisher 1974, 4, 250–252.

Flower, W. H. 1890. Presidential Address. *Report of the Fifty-Ninth Meeting of the British Association for the Advancement of Science, Newcastle Upon Tyne, September 1889,* 1–24. London: John Murray.

Forbes, D. 1952. *The Liberal Anglican Idea of History.* Cambridge: Cambridge University Press.

Ford, E. B. 1931. *Mendelism and Evolution.* London: Methuen.

——— 1942. *Genetics for Medical Students.* London: Methuen.

——— 1953. An above-ground storage pit of the la tène period. *Proceedings of the Prehistoric Society* (London), 19: 121–126.

——— 1964. *Ecological Genetics.* London: Methuen.

——— 1965. *Genetic Polymorphism.* London: Faber and Faber.

Ford, E. B., and J. S. Huxley. 1927. Mendelian genes and rates of development in *Gammarus chevreuxi. British Journal of Experimental Biology* 5: 112–134.

Fox, R. 1984. Science, the university, and the state in nineteenth century France. In *Professions and the French State, 1700–1900,* ed. G. Geison, 66–145. Philadelphia: University of Pennsylvania Press.

Franklin, B., et al. 1970. Report of the commissioners charged by the King to examine animal magnetism. In *Foundations of Hypnosis from Mesmer to Freud,* ed. M. N. Pinterow, 89–124. Springfield, Ill: Charles C. Thomas.

Freeman, R. B. 1977. *The Works of Charles Darwin: An Annotated Bibliography Handlist,* 2nd ed. London: Dawson.

Fye, W. B. 1985. H. Newell Martin—a remarkable career destroyed by neurasthenia and alcoholism. *Journal of the History of Medicine and Allied Sciences* 40: 133–166.

Gayon, J. 1992. *Darwin et l'après Darwin: Une histoire de l'hypothèse de séléction naturelle.* Paris: Kimé.

Geison, G. 1978. *Michael Foster and the Cambridge School of Physiology: The Scientific Enterprise in Late Victorian Society.* Princeton, N.J.: Princeton University Press.

Geoffroy Saint-Hilaire, E. 1818. *Philosophie anatomique.* Paris: Mequignon-Marvis.

——— 1819. Memoires sur l'organisation des insects: Premier memoire. *Journal Complementaire Dictionaire Scientifique Medical* 5: 340–351.

——— 1820. Second memoire: Sur quelques régles fondamentales en philosophie naturelle. *Journal Complementaire Dictionaire Scientifique Medical* 6: 31–35.

——— 1825. Recherches sur l'organisation des Gavials. *Mémoires du Muséum d'Histoire Naturelle* 12: 97–155.

——— 1830. *Principes de philosophie zoologique*. Paris: Pichon et Didier, Rousseau.

——— 1833a. Considerations sur des Ossemens fossiles la plupart inconnus, trouves et observes dans les Bassins de L'Auvergne. *Revue Encyclopedia* 59: 76–95.

——— 1833b. Sur le degré d'influence du monde ambiant pour modifier les formes. *Academie des Sciences Paris Memoires* 12: 63–92.

——— 1835. Loi universelle (attraction de soi pour soi), ou clef applicable à l'interpretation de tous les phénomènes de philosophie naturelle. In *Etudes progressives d'un naturaliste, pendant les années 1834 et 1835*. Paris: Roret, Denain et Delamarre.

——— 1838. *Fragments biographiques precedes d'études sur la vie, les ouvrages et les doctrines de Buffon*. Paris: F. D. Pillot.

George, H. [1879] 1926. *Progress and Poverty*. Garden City, N.Y.: Doubleday, Page.

Ghiselin, M. 1974. A radical solution to the species problem. *Systematic Zoology* 23: 536–544.

Ghiselin, M., and L. Jaffe. 1973. Phylogenetic classification in Darwin's monograph on the subclass Cirrepedia. *Systematic Zoology* 22: 132–140.

Gibbon, C. 1878. *The Life of George Combe*. London: Macmillan.

Giere, R. 1988. *Explaining Science: A Cognitive Approach*. Chicago: University of Chicago Press.

Gillispie, C. C. 1950. *Genesis and Geology*. Cambridge, Mass.: Harvard University Press.

Goethe, J. W. 1887–1919. *Goethes Werke*. Weimar: Weimarer Ausgabe.

——— 1962. Bildung und Umbildung organischer Naturen. In *Schriften zur Anatomie, Zoologie, Physiognomik*. Munich: Taschenbuch Verlag.

Goldschimdt, R. 1940. *The Material Basis of Evolution*. New Haven: Yale University Press.

Goodrich, E. S. 1906. Notes on the development, structure, and origin of the median and paired fins of fish. *Quarterly Journal of Microscopical Science* 50: 333–376.

——— 1912. *The Evolution of Living Organisms*. London: T. C. and E. C. Jack.

——— 1918. On the development of the segments of the head in Scyllium. *Quarterly Journal of Microscopical Science* 63: 1–30.

——— 1922. Some problems in evolution. In *British Association for the Advancement of Science: Report of the Eighty-Ninth Meeting, Edinburgh 1921, September 7–14*, 75–85. London: John Murray.
——— 1924a. *Living Organisms: An Account of Their Origin and Evolution*. Oxford: Oxford University Press.
——— 1924b. The origin of land vertebrates. *Nature* 114: 935–936.
——— 1930. *Studies on the Structure and Development of Vertebrates*. London: Macmillan.
Gotthelf, A., and J. G. Lennox, eds. 1987. *Philosophical Issues in Aristotle's Biology*. Cambridge: Cambridge University Press.
Gould, S. J. 1977a. *Ever Since Darwin*. New York: Norton.
——— 1977b. *Ontogeny and Phylogeny*. Cambridge, Mass.: Belknap Press.
——— 1980a. *The Panda's Thumb*. New York: Norton.
——— 1980b. Is a new and general theory of evolution emerging? *Paleobiology* 6: 119–130.
——— 1981. *The Mismeasure of Man*. New York: Norton.
——— 1982a. Darwinism and the expansion of evolutionary theory. *Science* 216: 380–7.
——— 1982b. The meaning of punctuated equilibrium and its role in validating a hierarchical approach to macroevolution. In *Perspectives on Evolution*, ed. R. Milkman, 83–104. Sunderland, Mass.: Sinauer.
——— 1983. The hardening of the synthesis. In *Dimensions of Darwinism*, ed. M. Grene, 71–93. Cambridge: Cambridge University Press.
——— 1984. Morphological channeling by structural constraint: Convergence in styles of dwarfing and giantism in *Cerion*, with a description of two new fossil species and a report on the discovery of the largest *Cerion*. *Paleobiology* 10: 172–194.
——— 1987. *Time's Arrow, Time's Cycle*. Cambridge, Mass.: Harvard University Press.
——— 1988. On replacing the idea of progress with an operational notion of directionality. In *Evolutionary Progress*, ed. M. H. Nitecki, 319–338. Chicago: University of Chicago Press.
——— 1989. *Wonderful Life: The Burgess Shale and the Nature of History*. New York: Norton.
——— ed. 1993. *The Book of Life*. New York: Viking.
Gould, S. J., and C. B. Calloway. 1980. Clams and brachiopods—ships that pass in the night. *Paleobiology* 6: 383–396.
Gould, S. J., and N. Eldredge. 1977. Punctuated equilibria: The tempo and mode of evolution reconsidered. *Paleobiology* 3: 115–151.
Gould, S. J., and R. C. Lewontin. 1979. The spandrels of San Marco and the Panglossian paradigm: A critique of the adaptationist program. *Proceedings of the Royal Society of London, B: Biological Sciences* 205: 581–598.

Gould, S. J., N. L. Gilinsky, and R. Z. German. 1987. Asymmetry of lineages and the direction of evolutionary time. *Science* 236: 1437–1441.

Grant, R. E. 1814. *Dissertatio Physiologica Inauguralis, de Circuitu Sanguinis in Foetu.* Edinburgh: Ballantyne.

——— 1825. On the structure and nature of the *Spongilla friabilis. Edinburgh Philosophical Journal* 14: 270–284.

——— 1833–34. Lectures on comparative anatomy and animal physiology. *Lancet* 1 and 2, 60 lectures.

——— 1841. *On the Present State of the Medical Profession in England.* London: Renshaw.

——— 1861. *Tabular View of the Preliminary Divisions of the Animal Kingdom.* London: Walton and Maberely.

Gray, A. 1836. Lindley's natural system of botany. *American Journal of Science and Arts* 32: 292.

——— 1846. [Review of] *Explanations: A Sequel to the Vestiges of the Natural History of Creation. North American Review* 42: 465–506.

——— [1859] 1889. Memoir on the botany of Japan, in its relations to that of North America, and of other parts of the Northern Temperate Zone. *Memoirs of the American Academy of Science and Arts,* n.s. 6. Reprinted in part in *Scientific Papers of Asa Gray,* 125–141. Boston and New York: Houghton, Mifflin.

——— [1860a] 1876. [Review of] *The Origin of Species by Means of Natural Selection. American Journal of Science and Arts.* Reprinted in *Darwiniana,* 7–50. New York: Appleton.

——— [1860b] 1876. Natural selection not inconsistent with natural theology. *Atlantic Monthly* 6, 109–116, 229–239, 406–425. Reprinted in *Darwiniana,* 72–145. New York: Appleton.

——— 1879. *Structural Botany,* 6th ed. [*Gray's Botanical Text-Book,* vol. 1.] New York and Chicago: Ivison, Blakeman, Taylor.

——— 1881. *Structural Botany,* 6th ed. London: Macmillan.

Gray, J. L., ed. 1894. *Letters of Asa Gray,* 2 vols. Boston: Houghton, Mifflin.

Green, J. H. 1865. *Spiritual Philosophy: Founded on the Teaching of the Late Samuel Taylor Coleridge,* ed. J. Simon. London.

Greene, J. 1959. *The Death of Adam.* Iowa: Iowa State University Press.

——— 1977. Darwin as a social evolutionist. *Journal of the History of Biology* 10: 1–27.

——— 1990. The interaction of science and world view in Sir Julian Huxley's evolutionary biology. *Journal of the History of Biology* 23: 39–55.

Greg, W. R. 1868. On the failure of "natural selection" in the case of man. *Fraser's Magazine* (September), 78: 353–362.

Gregory, W. K. 1913. Critique of recent work on the morphology of the vertebrate skull, especially in relation to the origin of mammals. *Journal of Morphology* 24 (1): 1–42.

——— 1937. Biographical memoir of Henry Fairfield Osborn, 1857–1935. *National Academy of Sciences of the United States of America Biographical Memoirs,* 51–119.

Gruber, J. W. 1965. Brixham Cave and the antiquity of man. In *Context and Meaning in Cultural Anthropology,* ed. M. E. Spiro, 373–402. New York: Free Press.

Gruber, M. E. 1981. *Darwin on Man,* 2d ed. Chicago: University of Chicago Press.

Gulick, J. T. 1906. Isolation and the evolution of species. *Science* 23: 433–434.

Guppy, H. B. 1906. *Observations of a Naturalist in the Pacific between 1896 and 1899.* London: Macmillan.

Haeckel, E. 1862. *Die Radiolarien (Rhizopoda radiaria): Eine Monographie.* Berlin: G. Reimer.

——— 1866. *Generelle Morphologie der Organismen,* 2 vols. Berlin: Reimer.

——— [1868] 1869. Monographie der Moneren. *Jenaische Zeitschift für Medecin und Naturwissenschaft,* band iv, heft 1. Translated as: Monograph of Monera. *Quarterly Journal of Microscopical Science,* n.s. 9: 27–42, 113–134, 219–232, 327–342.

——— 1876. *The History of Creation: or the Development of the Earth and Its Inhabitants by the Action of Natural Causes,* 2 vols. London: Kegan Paul, Trench.

——— 1896. *The Evolution of Man,* 2 vols. New York: Appleton.

Hahn, R. 1971. *The Anatomy of a Scientific Institution: The Paris Academy of Sciences, 1666–1803.* Berkeley: University of California Press.

——— 1975. Scientific careers in eighteenth century France. In *The Emergence of Science in Western Europe,* ed. M. P. Crosland, 127–138. New York: Science History Publications.

Haldane, J. B. S. 1923. *Daedalus, or Science and the Future.* London: Kegan Paul, Trench, Trubner.

——— 1924a. A mathematical theory of natural and artificial selection. Pt. I. *Transactions of the Cambridge Philosophical Society* 23: 19–41.

——— 1924b. A mathematical theory of natural and artificial selection. Pt. II. The influence of partial self-fertilisation, inbreeding, assortative mating, and selective fertilisation on the composition of Mendelian populations, and on natural selection. *Proceedings of the Cambridge Philosophical Society* 1: 158–163.

——— 1926. A mathematical theory of natural and artificial selection. Pt. III. *Proceedings of the Cambridge Philosophical Society* 23: 363–372.

——— 1927a. A mathematical theory of natural and artificial selection. Pt. IV. *Proceedings of the Cambridge Philosophical Society* 23: 607–615.

——— 1927b. A mathematical theory of natural and artificial selection. Pt. V. *Proceedings of the Cambridge Philosophical Society* 23: 838–844.

——— 1930. A note on Fisher's theory of the origin of dominance and on a correlation between dominance and linkage. *American Naturalist* 64: 87–90.

——— 1931a. A mathematical theory of natural and artificial selection. Pt. VII. Selection intensity as a function of mortality rate. *Proceedings of the Cambridge Philosophical Society* 27: 131–136.

——— 1931b. A mathematical theory of natural and artificial selection. Pt. VIII. Metastable populations. *Proceedings of the Cambridge Philosophical Society* 27: 137–142.

——— 1932a. *The Causes of Evolution.* New York: Cornell University Press.

——— 1932b. *The Inequality of Man and Other Essays.* London: Chatto and Windus.

——— 1934a. Anthropology and human biology. In *Congrès International des Sciences Anthropologiques et Ethnologiques,* 53–64. London: Institut Royal D'Anthropologie.

——— 1934b. *Human Biology and Politics (The Norman Lockyer Lecture).* London: British Science Guild.

——— 1938a. *Heredity and Politics.* London: Allen and Unwin.

——— 1938b. *The Marxist Philosophy and the Sciences.* London: Allen and Unwin.

Haldane, J. B. S., and J. S. Huxley. 1927. *Animal Biology.* Oxford: Clarendon.

Hamilton, W. D. 1964a. The genetical evolution of social behaviour I. *Journal of Theoretical Biology* 7: 1–16.

——— 1964b. The genetical evolution of social behaviour II. *Journal of Theoretical Biology* 7: 17–32.

——— 1980. Sex versus non-sex versus parasite. *Oikos* 35: 282–290.

——— 1990. Mate choice near or far. *American Zoologist* 30 (2): 341–352.

Hamilton, W. D., R. Axelrod, and R. Tanese. 1990. Sexual reproduction as an adaptation to resist parasites. *Proceedings of the National Academy of Sciences, USA* 87 (9): 3566–3573.

Hamilton, W. D., and M. Zuk. 1982. Heritable true fitness and bright birds: A role for parasites. *Science* 218: 384–387.

——— 1989. Letter to Editor. *Nature* 341: 289–290.

Hanen, M. P., M. J. Osler, and R. G. Weyant, eds. 1980. *Science, Pseudo-Science and Society.* Waterloo, Ontario: Wilfred Laurier Press.

Hardy, G. H. 1908. Mendelian proportions in a mixed population. *Science,* n.s. 28: 49–50.

Harrison, J. 1971. Erasmus Darwin's view of evolution. *Journal of the History of Ideas* 32: 247–264.

Harrison, J. A. 1970. *Biosocial Aspects of Sex: Proceedings of the Sixth Annual Symposium of the Eugenics Society, London, September 1969.* Oxford: Blackwell, For the Galton Foundation.

Harvard University. 1893. Report of the Committee on Zoölogy. In *Reports of the Visiting Committees to the Board of Overseers.* Cambridge, Mass.: Harvard University.

Hegel, G. W. F. [1817] 1970. *Philosophy of Nature.* Oxford: Oxford University Press.

——— 1977. *Phenomenology of Spirit,* A. V. Miller. Oxford: Oxford University Press.

Helfand, M. S. 1977. T. H. Huxley's *Evolution and Ethics:* The politics of evolution and the evolution of politics. *Victorian Studies* 20: 159–177.

Hempel, C. G. 1966. *Philosophy of Natural Science.* Englewood Cliffs, N.J.: Prentice-Hall.

Henderson, L. J. 1913. *The Fitness of the Environment.* New York: Macmillan.

——— 1917. *The Order of Nature.* Cambridge, Mass.: Harvard University Press.

Henson, P. M. 1990. Evolution and taxonomy: J. H. Comstock's research school in evolutionary entomology at Cornell University, 1874–1930. Ph.D. diss., University of Maryland.

Herbert, S. 1974. The place of man in the development of Darwin's theory of transmutation: Part 1, July 1837. *Journal of the History of Biology* 7: 217–258.

——— 1977. The place of man in the development of Darwin's theory of transmutation: Part 2. *Journal of the History of Biology* 10: 155–227.

Herrick, C. J. 1946. Progressive evolution. *Science* 104: 469.

Herschel, J. F. W. 1827. Light. In *Encylopaedia Metropolitana,* ed. E. Smedley et al. London: J. Griffin.

——— 1831. *Preliminary Discourse on the Study of Natural Philosophy.* London: Longman, Rees, Orme, Brown, and Green.

——— 1841. Review of Whewell's *History* and *Philosophy. Quarterly Review* 135: 177–238.

Hesse, M., and M. Arbib. 1986. *The Construction of Reality.* Cambridge: Cambridge University Press.

Heuss, T. 1991. *Anton Dohrn: A Life for Science.* Berlin and New York: Springer-Verlag.

His, W. [1881] 1967. On the principles of animal morphology. In *The Interpretation of Animal Form,* ed. W. Coleman, 167–178. New York: Johnson Reprint Corp.

Hodge, C. 1872. *Systematic Theology,* 3 vols. London and Edinburgh: Nelson.

Hodge, M. J. S. 1971. Lamarck's science of living bodies. *British Journal for the History of Science* 5: 323–352.

——— 1992. Biology and philosophy (including ideology): A study of Fisher and Wright. In *The Founders of Evolutionary Genetics,* ed. S. Sarkar, 231–293. Dordrecht: Kluwer Academic Publishers.

Hofstadter, R. 1959. *Social Darwinism in American Thought.* New York: Braziller.

Houghton, W. E. 1957. *The Victorian Frame of Mind.* New Haven: Yale University Press.

Howarth, J. 1987. Science education in late-Victorian Oxford: A curious case of failure? *English Historical Review* 102: 334–371.

Hufbauer, K. 1982. *The Formation of the German Chemical Community, 1720–1795.* Berkeley: University of California Press.

Hull, D., ed. 1973. *Darwin and His Critics.* Cambridge, Mass.: Harvard University Press.

——— 1988. *Science as a Process.* Chicago: University of Chicago Press.

Huxley, J. S. 1912. *The Individual in the Animal Kingdom.* Cambridge: Cambridge University Press.

——— 1923. *Essays of a Biologist.* London: Chatto and Windus.

——— 1924. The negro problem. *The Spectator,* November 29, pp. 821–822.

——— 1936. Natural selection and evolutionary progress. In *British Association for the Advancement of Science, Report of the Annual Meeting, 1936, Blackpool, September 9–16,* 81–100. London: Office of the British Association.

——— 1940. *The New Systematics.* Oxford: Clarendon Press.

——— 1941. *Man Stands Alone.* New York: Harper.

——— 1942. *Evolution: The Modern Synthesis.* London: Allen and Unwin.

——— 1948. *UNESCO: Its Purpose and Its Philosophy.* Washington, D.C.: Public Affairs Press.

——— 1953. *Evolution in Action.* London: Chatto and Windus.

——— 1954a. The evolutionary process. In *Evolution as a Process,* ed. J. Huxley, A. C. Hardy, and E. B. Ford, 1–23. London: Allen and Unwin.

——— 1954b. Scientific humanism, evolution, and human destiny. Los Angeles, Calif.: American Humanist Association.

——— 1955. Morphism and evolution. *Heredity* 9: 1–52.

——— 1957. *New Bottles for New Wine.* London: Chatto and Windus.

——— 1959. Introduction to Teilhard de Chardin's *The Phenomenon of Man,* 11–28. London: Collins.

——— 1964. *Essays of a Humanist.* London: Chatto and Windus.

——— 1970. *Memories.* London: Allen and Unwin.

——— 1973. *Memories II.* London: Allen and Unwin.

Huxley, J. S., A. C. Hardy, and E. B. Ford, eds. 1954. *Evolution as a Process.* London: Allen and Unwin.

Huxley, L., ed. 1900. *The Life and Letters of Thomas Henry Huxley,* 2 vols. London: Macmillan.

Huxley, T. H. [1853] 1898. On the morphology of the Cephalus Mullusca, as illustrated by the anatomy of certain heteropoda and pteropoda collected during the voyage of the HMS Rattlesnake in 1846–50. *Philosophical Transactions of the Royal Society* 143 (1): 29–66. Reprinted in Huxley 1898, 1, 152–193.

——— [1854] 1893. On the educational value of the natural history sciences. Address delivered in St. Martin's Hall. Reprinted in Huxley 1893b, 38–65.

——— [1854] 1903. Review of *Vestiges of the Natural History of Creation*, Tenth Edition. *British and Foreign Medico-Chirurgical Review* 13: 425–439. Reprinted in Huxley 1903, 1–19.

——— [1855] 1898. On certain zoological arguments commonly adduced in favour of the hypothesis of the progressive development of animal life in time. *Royal Institute Proceedings* 2: 82–85. Reprinted in Huxley 1898, 1, 300–304.

——— [1858] 1898. On the theory of the vertebrate skull. *Proceedings of the Royal Society* 9: 381–457. Reprinted in Huxley 1898, 1, 538–606.

——— [1859] 1893. The Darwinian hypothesis. *Times*, December 26. Reprinted in Huxley 1893a, 1–21.

——— [1862] 1894. Geological contemporaneity and persistent types of life: The Anniversary Address to the Geological Society. Reprinted in *Discourses: Biological and Geological*, 272–304. London: Macmillan.

——— 1863. *Evidence as to Man's Place in Nature*. London: Williams and Norgate.

——— [1865] 1893. Emancipation—black and white. *The Reader*, May 20. Reprinted in Huxley 1893b, 66–75.

——— [1867–68] 1898. Remarks upon the Archaeopteryx Lithographica. *Proceedings of the Royal Society of London* 16: 243–248. Reprinted in Huxley 1898, 3, 340–345.

——— [1868] 1898. On the animals which are most nearly intermediate between birds and reptiles. *Geological Magazine* 5: 357–365. Reprinted in Huxley 1898, 3, 303–313.

——— [1870] 1894. Presidential address to the Geological Society: Paleontology and the doctrine of evolution. Reprinted in *Discourses: Biological and Geological*, 340–388. London: Macmillan.

——— [1871] 1893. Administrative nihilism. *Fortnightly Review*. Reprinted in *Methods and Results*, 251–289. London: Macmillan.

——— 1879. *Hume*. London: Macmillan.

——— [1880] 1893. The coming of age of *The Origin of Species*: Royal Institution Lecture. Reprinted in Huxley 1893a, 227–243.

——— [1880] 1898. On the application of the laws of evolution to the arrangement of the Vertebrata and more particularly of the Mammalia. *Proceedings of the Scientific Meetings of the Zoological Society of London*, 649–662. Reprinted in Huxley 1898, 4, 457–472.

——— [1883] 1903. The Rede Lecture. *Nature* 28: 187–189. Reprinted in Huxley 1903, 69–79.

——— 1888. *American Addresses: With a Lecture on the Study of Biology*. New York: D. Appleton and Co.

——— 1893a. *Collected Essays: Darwiniana*. London: Macmillan.

——— 1893b. *Science and Education: Essays*. London: Macmillan.

——— 1893c. *Evolution and Ethics and Other Essays*. London: Macmillan.

——— 1898. *The Scientific Memoirs of Thomas Henry Huxley,* 4 vols., ed. M. Foster and E. R. Lankester. London: Macmillan.

——— 1903. *The Scientific Memoirs of Thomas Henry Huxley, Supplementary Volume,* ed. M. Foster and E. R. Lankester. London: Macmillan.

——— 1989. *Evolution and Ethics with New Essays on Its Victorian and Sociobiological Context,* ed. J. Paradis and G. C. Williams. Princeton: Princeton University Press.

Huxley, T. H., and J. S. Huxley. 1947. *Evolution and Ethics, 1893–1943*. London: Pilot.

Huxley, T. H., and H. N. Martin. 1875. *A Course of Practical Instruction in Elementary Biology*. London: Macmillan.

Hyatt, A. 1889. *Genesis of the Arietidae*. Smithsonian Contributions to Knowledge, 673. Washington, D.C.: Smithsonian Institution.

——— 1893. Phylogeny of an acquired characteristic. *American Naturalist* 27: 867–877.

——— 1894. Phylogeny of an acquired characteristic. *Proceedings of the American Philosophical Society* 32: 349–647.

——— 1897. The influence of woman in the evolution of the human race. *Natural Science: A Monthly Review of Scientific Progress* 11: 89–93.

Hyatt, A., and J. M. Arms. 1890. *Guides for Science-Teaching. No. VIII: Insecta*. Boston: D. C. Heath.

Hyman, S. E. 1962. *The Tangled Bank: Darwin, Marx, Frazer and Freud as Imaginative Writers*. New York: Atheneum.

Iggers, G. G. 1982. The idea of progress in historiography and social thought since the enlightenment. In *Progress and Its Discontents*, G. A. Almond, M. Chodorow, and R. H. Pearce, 41–66. Berkeley: University of California Press.

Inglis, B. 1971. *Poverty and the Industrial Revolution*. London: Hodder and Stoughton.

Jackson, R. T. 1913. Alpheus Hyatt and his principles of research. *American Naturalist* 47: 195–205.

James, W. 1880. *The Principles of Psychology*. New York: Henry Holt.

Jameson, R. 1826. Observations on the nature and importance of geology. *Edinburgh New Philosophical Journal* 1: 293–302.

Jensen, J. V. 1991. *Thomas Henry Huxley: Communicating for Science*. Newark: University of Delaware Press.

Jepsen, G. L., E. Mayr, and G. G. Simpson, eds. 1949. *Genetics, Paleontology and Evolution*. Princeton, N.J.: Princeton University Press.

Jerison, H. J. 1985. Issues in brain evolution. *Oxford Surveys in Evolutionary Biology* 2: 102–134.

Johnson, P. E. 1991. *Darwin on Trial*. Washington, D.C.: Regnery Gateway.

Jordan, D. S. 1905a. Ontogenetic species and other species. *Science* 22: 872–873.

——— 1905b. The origin of species through isolation. *Science* 22: 545–562.

——— 1906. Discontinuous variation and pedigree culture. *Science* 24: 399–400.

Jordanova, L. J. 1976. The natural philosophy of Lamarck in its historical context. Ph.D. diss., Cambridge University.

——— 1984. *Lamarck*. Oxford: Oxford University Press.

Kant, I. [1790] 1951. *Critique of Judgement*, trans. J. H. Bernard. New York: Hafner.

Kauffman, S. A. 1993. *The Origins of Order: Self-Organization and Selection in Evolution*. Oxford: Oxford University Press.

Kavalowski, V. 1974. The "vera causa" principle: An historico-philosophical study of a metatheoretical concept from Newton through Darwin. Ph.D. diss., University of Chicago.

Keohane, N. O. 1982. The enlightenment idea of progress revisited. In *Progress and Its Discontents*, ed. G. A. Almond, M. Chodorow, and R. H. Pearce, 21–40. Berkeley: University of California Press.

Kettlewell, H. B. D. 1955. Selection experiments on industrial melanism in the Lepidoptera. *Heredity* 9: 323–342.

——— 1956. Further selection experiments on industrial melanism in the Lepidoptera. *Heredity* 10: 287–301.

——— 1961. The phenomenon of industrial melanism in the Lepidoptera. *Annual Review of Entomology* 6: 245–262.

——— 1973. *The Evolution of Melanism*. Oxford: Clarendon.

Kielmeyer, C. F. 1938a. Ideen einer Entwicklungs-geschichte der Erde und ihrer Organisation, Schreiben au Windischmann, 1804. In *Gesammelte Schriften*, ed. F. H. Holler, 203–210. Berlin: F. Keiper.

——— 1938b. Uber die Verhaltnisse der organischen Urafte untereinander in der Reihe der verschiedenen Organizationen: Die Gesetze und Folgen dieser Verhaltnisse. In *Gesammelte Schriften*, ed. F. H. Holler. Berlin: F. Keiper.

Kimmelman, B. A. 1987. A Progressive Era discipline: Genetics at American agricultural colleges and experimental stations, 1900–1920. Ph.D. diss., University of Pennsylvania.

King-Hele, D., ed. 1981. *The Letters of Erasmus Darwin*. Cambridge: Cambridge University Press.

Knight, D. 1975. German science in the Romantic period. In *The Emergence of Science in Westerm Europe,* M. P. Crosland, 127–138. New York: Science History Publications.

——— 1986. *The Age of Science: The Scientific World-View in the Nineteenth Century.* Oxford: Blackwell.

——— 1990. Romanticism and the sciences. In *Romanticism and the Sciences,* ed. A. Cunningham and N. Jardine, 13–24. Cambridge: Cambridge University Press.

Kohler, R. E. 1991. *Partners in Science: Foundations and Natural Scientists, 1900–1945.* Chicago: University of Chicago Press.

Kottler, M. J. 1974. Alfred Russel Wallace, the origin of man, and spiritualism. *Isis* 65: 145–192.

Kovalevskii, V. 1873. On the osteology of the Hyopotamide. *Philosophical Transactions of the Royal Society of London* 163 (1): 19–94.

Kuhn, T. 1957. *The Copernican Revolution.* Cambridge, Mass.: Harvard University Press.

——— 1962. *The Structure of Scientific Revolutions.* Chicago: University of Chicago Press.

——— 1977. Objectivity, value, judgment, and theory choice. In *The Essential Tension: Selected Studies in Scientific Tradition and Change,* 320–339. Chicago: University of Chicago Press.

Lamarck, J. B. 1778. *Flore françoise.* Paris: Imprimerie Royale.

——— 1794. *Recherches sur les causes des principaux faits physique* . . . Paris: Maradan.

——— 1796. *Réfutation de la théorie pneumatique ou de la nouvelle doctrine des chimistes moderne* . . . Paris: Agasse.

——— 1800–1810. *Annuaires météorologiques.* Paris: the author (1800–1805); Maillard (1806); Treuttel and Wurtz (1807–1810).

——— 1801a. *Système des animaux sans vertebres* . . . Paris: Deterville.

——— 1801b. Physique terrestre. Unpublished manuscript MS 756. Paris: Muséum National d'Histoire Naturelle.

——— 1802a. *Recherches sur l'organisation des corps vivants,* . . . Paris: Maillard.

——— 1802b. *Hydrogéologie.* Paris: Agasse.

——— 1809. *Philosophie zoologique.* Paris: Dentu.

——— 1815–1822. *Histoire naturelle des animaux sans vertèbres.* Paris: Verdiere.

——— 1820. *Système analytique des connaissances positives de l'homme, restreintes à celles qui proviennent directement ou indirectement de l'observation.* Paris: Belin.

Lane, A. C. 1906. Isolation by choice. *Science* 23: 702.

Lankester, E. R. 1877. Notes on the embryology and classification of the animal kingdom, comprising a revision of speculations relative to the origin and significance of the germ layers. *Quarterly Journal of Microscopical Science* 17: 395–454.

——— 1880. *Degeneration: A Chapter in Darwinism.* London: Macmillan.

——— 1881. Limulus an Arachnid. *Quarterly Journal of Microscopical Science,* n.s. 21: 609–649.

——— 1906. Inaugural address before the British Association. *Science* 24: 225–238.

——— 1923. *Great and Small Things.* London: Methuen.

Laplace, P. S. 1796. *Exposition du système du monde.* Paris: Imprimerie du Cercle-Social.

——— [1814] 1951. *A Philosophical Essay on Probabilities,* 6th French ed., ed. and trans. F. W. Truscott. New York: Dover Publications.

Latour, B. 1987. *Science in Action.* Cambridge, Mass.: Harvard University Press.

Latour, B., and S. Woolgar. 1979. *Laboratory Life: The Construction of Scientific Facts.* Beverly Hills: Sage.

Laudan, L. 1977. *Progress and Its Problems: Towards a Theory of Scientific Growth.* Berkeley: University of California Press.

——— 1981. *Science and Hypothesis.* Dordrecht: D. Reidel.

Laurent, G. 1987. *Paléontologie et evolution en France de 1800 à 1860: Une histoire des idées de Cuvier et Lamarck à Darwin.* Paris: Editions du C.T.H.S.

Law, E. 1820. *Considerations on the Theory of Religion,* new edition, ed. George Henry Law. London.

Leigh, E. G. 1986. Ronald Fisher and the development of evolutionary theory. I. The role of selection. *Oxford Surveys of Evolutionary Biology* 3: 187–223.

——— 1987. Ronald Fisher and the development of evolutionary theory. II. Influences of new variation on evolutionary process. *Oxford Surveys of Evolutionary Biology* 4: 212–263.

Lenoir, T. 1982. *The Strategy of Life: Teleology and Mechanics in Nineteenth Century German Biology.* Dordrecht: Reidel.

Lerner, I. M. 1954. *Genetic Homeostasis.* New York: John Wiley.

Lester, J. 1995. *E. Ray Lankester and the Making of Modern British Biology,* with assistance of P. Bowler. Oxford: British Society for the History of Science.

Levins, R., and R. Lewontin. 1985. *The Dialectical Biologist.* Cambridge, Mass.: Harvard University Press.

Lewis, D. 1983. Cyril Dean Darlington. *Biographical Memoirs of Fellows of the Royal Society* 29: 113–158.

Lewontin, R. C. 1974. *The Genetic Basis of Evolutionary Change.* New York: Columbia University Press.

——— 1977. Sociobiology—a caricature of Darwinism. In *PSA 1976,* ed. F. Suppe and P. Asquith, 22–31. East Lansing, Mich.: Philosophy of Science Association.

——— 1981. Sleight of hand [review of *Genes, Mind and Culture*]. *The Sciences,* July, pp. 23–26.

Lewontin, R. C., J. A. Moore, W. B. Provine, and B. Wallace, eds. 1981. *Dobzhansky's Genetics of Natural Populations I–XLIII.* New York: Columbia University Press.

Lloyd, F. E. 1905. Isolation and the origin of species. *Science* 22: 710–712.

Lovejoy, A. O. [1911] 1959. Kant and evolution. In *Forerunners of Darwin,* ed. B. Glass, O. Temkin, and W. L. Strauss Jr., 173–206. Baltimore: Johns Hopkins University Press.

——— 1936. *The Great Chain of Being.* Cambridge, Mass.: Harvard University Press.

Lovejoy, C. O. 1981. The origin of man. *Science* 211: 341–350.

Lumsden, C. J., and E. O. Wilson. 1981. *Genes, Mind, and Culture.* Cambridge, Mass.: Harvard University Press.

——— 1983. *Promethean Fire: Reflections on the Origin of Mind.* Cambridge, Mass.: Harvard University Press.

Lurie, E. 1960. *Louis Agassiz: A Life in Science.* Chicago: University of Chicago Press.

Lyell, C. 1826. Transactions of the Geological Society of London. vol. i. 2d series. London. 1824. *Quarterly Review* 34: 507–540.

——— 1830–33. *Principles of Geology, Being an Attempt to Explain the Former Changes in the Earth's Surface by Reference to Causes Now in Operation,* 3 vols. London: J. Murray.

——— 1851. Anniversary address of the president. *Quarterly Journal of the Geological Society of London* 7: xxv–lxxvi.

Lyell, K., ed. 1881. *Life, Letters and Journals of Sir Charles Lyell, Bart.,* 2 vols. London: John Murray.

Mackensie, D. 1981. *Statistics in Britain: 1865–1930.* Edinburgh: Edinburgh University Press.

Macleay, W. 1819–21. *Horae Entomologicae.* London: S. Bagster.

Maienschein, J. 1978. Ross Harrison's crucial experiment as a foundation for modern American experimental embryology. Ph.D. diss., Indiana University.

——— ed. 1987. *Defining Biology: Lectures from the 1890s.* Cambridge, Mass.: Harvard University Press.

——— 1991. *Transforming Traditions in American Biology: 1880–1915.* Baltimore: Johns Hopkins University Press.

Malthus, T. R. [1798] 1965. *First Essay on Population,* ed. J. Bonar. New York: A. M. Kelley.

——— [1826] 1914. *An Essay on the Principle of Population,* 6th ed. London: Everyman.

Manier, E. 1978. *The Young Darwin and His Cultural Circle.* Dordrecht: Reidel.

Marchant, J., ed. 1916. *Alfred Russel Wallace: Letters and Reminiscences,* 2 vols. London: Cassell.

Martin, H. N. [1876] 1967. The study and teaching of biology. *Memoirs from the Biological Laboratory of the Johns Hopkins University* 3: 192–204. Reprinted in *The Interpretation of Animal Form,* ed. W. Coleman, 181–191. New York: Johnson Reprint Co.

Marx, K. [1859] 1977. Preface to *A Critique of Political Economy.* In *Karl Marx: Selected Writings,* ed. D. McLellan, 388–392. New York: Oxford University Press.

——— [1867] 1906. Author's preface to the first edition. In *Capital,* trans. S. Moore and E. Aveling, 11–16. New York: Modern Library.

——— [1873] 1906. Author's preface to the second edition. In *Capital,* trans. S. Moore and E. Aveling, 16–26. New York: Modern Library.

Marx, K., and F. Engels. 1965. *Selected Correspondence.* Moscow: Progress.

Masters, R. 1989. *The Nature of Politics.* New Haven: Yale University Press.

Mayer, A. G. 1911. Alpheus Hyatt, 1838–1902. *Popular Science Monthly* 78: 128–146.

Maynard Smith, J. 1958. *The Theory of Evolution.* Harmondsworth, Middlesex: Penguin.

——— 1968. *Mathematical Ideas in Biology.* Cambridge: Cambridge University Press.

——— 1974. *Models in Ecology.* Cambridge: Cambridge University Press.

——— 1978. *The Evolution of Sex.* Cambridge: Cambridge University Press.

——— 1981. Did Darwin get it right? *London Review of Books* 3 (11): 10–11.

——— 1982. *Evolution and the Theory of Games.* Cambridge: Cambridge University Press.

——— 1988a. Evolutionary progress and levels of selection. In *Evolutionary Progress,* ed. M. Nitecki, 219–230. Chicago: University of Chicago Press.

——— 1988b. *Games, Sex and Evolution.* New York: Harvester and Wheatsheaf.

——— 1989a. *Evolutionary Genetics.* Oxford: Oxford University Press.

——— 1989b. In Haldane's footsteps. In *Studying Animal Behaviour: Autobiographies of the Founders,* ed. D. A. Dewsbury, 347–356. Chicago: University of Chicago Press.

——— 1992. J. B. S. Haldane. In *The Founders of Evolutionary Genetics,* ed. S. Sarkar, 37–51. Dordrecht: Kluwer Academic Publishers.

Maynard Smith, J., and E. Szathmáry. 1995. *The Major Transitions in Evolution.* New York: Oxford University Press.

Mayr, E. 1942. *Systematics and the Origin of Species.* New York: Columbia University Press.

——— 1959. Where are we? *Cold Spring Harbor Symposia on Quantitative Biology* 24: 1–14.

——— 1963. *Animal Species and Evolution.* Cambridge, Mass.: Harvard University Press.

——— 1969. *Principles of Systematic Zoology.* New York: McGraw-Hill.

——— 1972. Lamarck revisited. *Journal of the History of Biology* 5: 55–94.

——— 1982. *The Growth of Biological Thought: Diversity, Evolution and Inheritance.* Cambridge, Mass.: Harvard University Press.

——— 1986. Mayr comments on Eldredge's introduction to a republication of *Systematics and the Origin of Species.* Unpublished manuscript.

——— 1988. *Toward a New Philosophy of Biology.* Cambridge, Mass.: Harvard University Press.

——— 1992. Controversies in retrospect. *Oxford Surveys in Evolutionary Biology* 8: 1–34.

Mayr, E., and W. Provine, eds. 1980. *The Evolutionary Synthesis: Perspectives on the Unification of Biology.* Cambridge, Mass.: Harvard University Press.

McClellan, J. E. 1985. *Science Reorganized: Scientific Societies in the Eighteenth Century.* New York: Columbia University Press.

McClelland, C. E. 1980. *State, Society and University in Germany, 1700–1914.* Cambridge: Cambridge University Press.

McKinney, H. L. 1972. *Wallace and Natural Selection.* New Haven: Yale University Press.

McMullin, E. 1983. Values in science. In *PSA 1982,* ed. P. D. Asquith and T. Nickles, 2, 3–28. East Lansing, Mich.: Philosophy of Science Association.

McNeil, M. 1987. *Under the Banner of Science: Erasmus Darwin and His Age.* Manchester: Manchester University Press.

McShea, D. W. 1991. Complexity and evolution: What everybody knows. *Biology and Philosophy* 6: 303–324.

Medawar, P. [1961] 1967. Review of *The Phenomenon of Man. Mind* 70: 99–106. Reprinted in *The Art of the Soluble,* ed. P. Medawar. London: Methuen.

Merriman, C. H. 1906. Is mutation a factor in the evolution of the higher vertebrates? *Science* 23: 241–257.

Merton, R. K. 1973. *The Sociology of Science: Theoretical and Empirical Investigations,* ed. N. W. Storer. Chicago: University of Chicago Press.

Metcalf, M. M. 1906. The influence of the plasticity of organisms upon evolution. *Science* 23: 786–787.

Mill, J. S. [1843] 1974. *A System of Logic Ratiocinative and Inductive,* 2 vols., ed. J. M. Robson. Toronto: University of Toronto Press.

——— 1859. *Dissertations and Discussions: Political, Philosophical, and Historical.* London: John W. Parker and Son.

——— [1869] 1975. The subjugation of women. In *Three Essays.* London: Oxford University Press.

Millhauser, M. 1954. The scriptural geologists: An episode in the history of opinion. *Osiris* 11: 65–86.
——— 1959. *Just Before Darwin: Robert Chambers and "Vestiges."* Middletown, Conn.: Wesleyan University Press.
Milman, H. H. [1829] 1909. *History of the Jews,* 2 vols. London: J. M. Dent.
Milne-Edwards, H. 1827. Organisation. In *Dictionnaire Classique d'Histoire Naturelle,* 332–344.
——— 1834. *Elements de zoologie: Leçons sur l'anatomie, la physiologie, la classification des moeurs des animaux.* Paris: Crochard.
Montague, A., ed. 1984. *Science and Creationism.* Oxford: Oxford University Press.
Moore, B. 1913. *The Origin and Nature of Life.* London: Williams and Norgate.
Moore, J. 1979. *The Post-Darwinian Controversies: A Study of the Protestant Struggle to Come to Terms with Darwin in Great Britain and America, 1870–1900.* Cambridge: Cambridge University Press.
——— 1982. Charles Darwin lies in Westminster Abbey. *Biological Journal of the Linnean Society* 17: 97–113.
Morgan, C. L. 1927. *Emergent Evolution.* London: Williams and Norgate.
Morgan, S. R. 1990. Schelling and the origins of his Naturphilosophie. In *Romanticism and the Sciences,* ed. A. Cunningham and N. Jardine, 25–37. Cambridge: Cambridge University Press.
Morgan, T. H., A. Sturtevant, H. J. Muller, and C. Bridges. 1915. *The Mechanisms of Mendelian Heredity.* New York: Henry Holt.
Morrell, J., and A. Thackray. 1981. *Gentlemen of Science: Early Years of the British Association for the Advancement of Science.* Oxford: Clarendon.
Morse, E. S. 1902. Memorial of Professor Alpheus Hyatt. *Proceedings of the Boston Society of Natural History* 30: 413–433.
Müller, F. 1869. *Facts and Arguments for Darwin,* trans. W. S. Dallas. London: John Murray.
Muller, H. J. 1949. The Darwinian and modern conceptions of natural selection. *Proceedings of the American Philosophical Society* 93: 459–470.
Muller, H. J., and R. Falk. 1961. Are induced mutations in *Drosophila* overdominant? I. Experimental design. *Genetics* 46: 727–735.
Musson, A. E., and E. Robinson. 1960. *Science and Technology in the Industrial Revolution.* Manchester: Manchester University Press.
Myers, G. 1990. *Writing Biology: Texts in the Social Construction of Scientific Knowledge.* Madison, Wis.: University of Wisconsin Press.
Nagel, E. 1961. *The Structure of Science.* London: Routledge and Kegan Paul.
Nelkin, D. 1977. *Science Textbook Controversies and the Politics of Equal Time.* Cambridge, Mass.: MIT Press.
Nesse, R., and G. C. Williams. 1995. *Evolution and Healing: The New Science of Darwinian Medicine.* London: Weidenfeld and Nicolson.

Newman, W. A. 1993. Darwin and cirripedology. In *History of Carcinology,* ed. F. Truesdale, 349–434. Rotterdam: Balkema.

Nichol, J. P. 1837. *Views of the Architecture of the Heavens.* Edinburgh: William Tait.

——— 1847. Preliminary discussion on some points connected with the present position of education in this country. In *The Education of the People,* ed. J. William, xi–lxxx. Glasgow: William Lang.

Nickles, T. 1995. Philosophy of science and history of science. *Osiris* 10: 139–163.

Niebuhr, R. 1940. *Christianity and Power Politics.* New York: Scribners.

Nisbet, R. 1980. *History and the Idea of Progress.* New York: Basic Books.

Nuffield Foundation. 1946. *The First Report: Report of the Trustees for the Three Years Ending 31 March 1946.* Oxford: University Press.

——— 1959. *Fourteenth Report.* Oxford: University Press.

——— 1963. *Eighteenth Report.* Oxford: University Press.

Numbers, R. 1992. *The Creationists.* New York: A. A. Knopf.

Nyhart, L. K. 1986. Morphology and the German university, 1860–1900. Ph.D. diss., University of Pennsylvania.

——— 1995. *Biology Takes Form: Animal Morphology and the German Universities, 1800–1900.* Chicago: University of Chicago Press.

Oakley, K. P. 1964. The Problem of Man's Antiquity. *Bulletin of the British Museum (Natural History), Geological Series,* vol. 9, no. 5.

Oken, L. [1809] 1847. *Elements of Physiophilosophy,* trans. A. Tulk. London: Printed for the Ray Society.

Olby, R. C. 1963. Charles Darwin's manuscript of *Pangenesis. British Journal for the History of Science* 1 (3): 251–263.

——— 1990. Cyril Dean Darlington. In *The Dictionary of Scientific Biography, Supplement II,* Editor-in-Chief F. L. Holmes, 17: 203–209. New York: Charles Scribner's Sons.

Oldroyd, D. R. 1984. How did Darwin arrive at his theory? The secondary literature to 1982. *History of Science* 22: 325–374.

——— 1986. Charles Darwin's theory of evolution: A review of our present understanding. *Biology and Philosophy* 1: 133–168.

Oppenheimer, J. 1978. Anton Dohrn. In *The Dictionary of Scientific Biography, Supplement I,* Editor-in-Chief Charles Coulston Gillispie, 122–125. New York: Charles Scribner's Sons.

Ortmann, A. E. 1906a. Isolation as one of the factors in evolution. *Science* 23: 71–72.

——— 1906b. A case of isolation without "barriers." *Science* 23: 504–506.

——— 1906c. Dr. O. F. Cook's conception of evolution. *Science* 23: 667–669.

——— 1906d. The fallacy of the mutation theory. *Science* 23: 746–748.

——— 1906e. Facts and theories in evolution. *Science* 23: 947–952.
——— 1906f. The mutation theory. *Science* 24: 214–217.
Osborn, H. F. 1896. Ontogenic and phylogenic variation. *Science* 4: 788.
——— 1910. *The Age of Mammals in Europe, Asia and North America.* New York: Macmillan.
——— 1916. *Men of the Old Stone Age: Their Environment, Life, and Art.* London: George Bell.
——— 1917. *The Origin and Evolution of Life on the Theory of Action, Reaction, and Interaction of Energy.* New York: Charles Scribner's Sons.
——— 1924. *Impressions of Great Naturalists.* New York: Charles Scribner's Sons.
——— 1925. *The Earth Speaks to Bryan.* New York: Charles Scribner's Sons.
——— 1927. *Man Rises to Parnassus: Critical Epochs in the Prehistory of Man.* Princeton, N.J.: Princeton University Press.
——— 1929a. *From the Greeks to Darwin: The Development of the Evolution Idea through Twenty-Four Centuries.* New York: Charles Scribner's Sons.
——— 1929b. *The Titanotheres of Ancient Wyoming, Dakota and Nebraska.* Washington, D.C.: U.S. Geological Survey Monograph, 55.
——— 1931. *Cope: Master Naturalist: The Life and Writings of Edward Drinker Cope.* Princeton, N.J.: Princeton University Press.
——— 1934. Aristogenesis: The creative principle in the *Origin of Species. American Naturalist* 68: 193–235.
Ospovat, D. 1981. *The Development of Darwin's Theory: Natural History, Natural Theology, and Natural Selection, 1838–1859.* Cambridge: Cambridge University Press.
Outram, D. 1984. *Georges Cuvier: Vocation, Science and Authority in Post-Revolutionary France.* Manchester: Manchester University Press.
Owen, R. 1841. Report on British fossil reptiles, Pt. II. In *Reports of the British Association for the Advancement of Science 1841,* 60–204. London: John Murray.
——— 1843. *Lectures on the Comparative Anatomy and Physiology of the Invertebrate Animals.* London: Longman, Brown, Green and Longmans.
——— 1846. Report on the archetype and homologies of the vertebrate skeleton. *Report of the Sixteenth Meeting of the British Association for the Advancement of Science,* 169–340. London: John Murray.
——— 1848. *On the Archetype and Homologies of the Vertebrate Skeleton.* London: Voorst.
——— 1849a. *On the Nature of Limbs.* London: Voorst.
——— 1849b. *On Pathenogenesis or the Successive Production of Procreating Individuals from a Single Ovum.* London: Voorst.
——— 1851. Principles of geology by Sir Charles Lyell, etc. *Quarterly Review* 89: 412–451.

——— 1858. Oken, Lorenz. *Encyclopaedia Britannica,* 8th ed., 16, 498–503. Edinburgh.

——— 1860. Darwin on the *Origin of Species. Edinburgh Review* 111: 487–532.

——— 1863. On the *Archaeopteryx* of von Meyer, with a description of the fossil remains of a long-tailed species from the lithographic Stone of Solenhofen. *Philosophical Transactions* 153: 33–47.

——— 1865. Preface. In *Spiritual Philosophy: Founded on the Teaching of the late Samuel Taylor Coleridge,* by J. H. Greene. London: Macmillan.

——— 1992. *The Hunterian Lectures in Comparative Anatomy, May and June 1837,* ed. P. R. Sloan. Chicago: University of Chicago Press.

Owen, R. 1894. *The Life of Richard Owen,* 2 vols. London: Murray.

Page, D. 1867. *Man: Where, Whence, Whither? Being a Glance at Man in his Natural History Relations.* Edinburgh: Edmonston and Douglas.

Paradis, J. 1989. *Evolution and Ethics* in its Victorian context. In *Evolution and Ethics: T. H. Huxley's "Evolution and Ethics" with New Essays on Its Victorian and Sociobiological Context,* ed. J. Paradis and G. C. Williams, 3–55. Princeton, N.J.: Princeton University Press.

Parascandola, J. 1971. Organismic and holistic concepts in the thought of L. J. Henderson. *Journal of the History of Biology* 4: 63–113.

Parker, G. A. 1969. The reproductive behaviour and the nature of sexual selection in *Scatophaga stercovaria* L. (Diptera: Scatophagidae)—III. Apparent intersex individuals and their evolutionary cost to normal searching males. *Transactions of the Royal Entomological Society* (London), 305–323.

——— 1970a. The reproductive behaviour and the nature of sexual selection in *Scatophaga stercovaria* L. (Diptera: Scatophagidae)—I. Diurnal and seasonal changes in population density around the site of mating and position. *Journal of Animal Ecology* 39: 185–204.

——— 1970b. The reproductive behaviour and the nature of sexual selection in *Scatophaga stercovaria* L. (Diptera: Scatophagidae)—II. The fertilization rate and the spatial and temporal relationships of each sex around the site of mating and oviposition. *Journal of Animal Ecology* 39: 205–228.

——— 1970c. Sperm competition and its evolutionary effect on copula duration in the fly *Scatophaga stercovaria. Journal of Insect Physiology* 16: 1301–1328.

——— 1970d. The reproductive behaviour and the nature of sexual selection in *Scatophaga stercovaria* L. (Diptera: Scatophagidae)—IV. Epigamic recognition and competition between males for the possession of females. *Behaviour* 37: 113–139.

——— 1970e. The reproductive behaviour and the nature of sexual selection in *Scatophaga stercovaria* L. (Diptera: Scatophagidae)—V. The female's behaviour at the oviposition site. *Behaviour* 37: 140–168.

——— 1970f. The reproductive behaviour and the nature of sexual selection in *Scatophaga stercovaria* L. (Diptera: Scatophagidae)—VIII. The origin and evolution of the passive phase. *Evolution* 24: 744–788.

——— 1970g. Sperm competition and its evolutionary consequences in the insects. *Biological Reviews* 45: 525–567.

——— 1974a. The reproductive behaviour and the nature of sexual selection in *Scatophaga stercovaria* L. (Diptera: Scatophagidae)—IX. Spatial distribution of fertilization rates and evolution of male search strategy within the reproductive area. *Evolution* 28: 93–108.

——— 1974b. Assessment strategy and the evolution of fighting behaviour. *Journal of Theoretical Biology* 47: 223–244.

Pauly, P. J. 1984. The appearance of academic biology in late nineteenth-century America. *Journal of the History of Biology* 17: 369–397.

——— 1988. Summer resort and scientific discipline: Woods Hole and the structure of American biology, 1882–1925. In *The American Development of Biology*, ed. R. Rainger, K. R. Benson, and J. Maienschein, 121–150. Philadelphia: University of Pennsylvania Press.

Peacocke, A. R. 1986. *God and the New Biology*. London: Dent.

Pearson, E. S. 1936. Karl Pearson: An appreciation of some aspects of his life and work: Part 1. *Biometrika* 28: 193–257.

——— 1937–38. Karl Pearson: An appreciation of some aspects of his life and work: Part 2. *Biometrika* 29: 161–248.

Pearson, K. 1892. *The Grammar of Science*. London: Walter Scott.

——— 1894. Contributions to the mathematical theory of evolution. *Philosophical Transactions* A 185: 71–110.

——— 1900. *The Grammar of Science*, 2d ed. London: Black.

———. 1906. Walter Frank Raphael Weldon, 1860–1906. *Biometrika* 5: 1–51.

Peirce, C. S. 1877. The fixation of belief. *Popular Science Monthly* 12: 1–15.

——— [1893] 1935.. Evolutionary love. *Monist* 3: 176–200. Reprinted in *The Collected Papers of C. S. Peirce*, C. Hartshorne and P. Weiss, 6, 190–215. Cambridge, Mass.: Belknap Press of Harvard University Press.

——— 1955. *Philosophical Writings of Peirce*, ed. J. Buchler. New York: Dover Publications.

Peterson's. 1994. *Peterson's Guide to Graduate Programs in the Biological and Agricultural Sciences*, 28th ed. Princeton: Peterson's.

Pittenger, M. 1993. *American Socialists and Evolutionary Thought, 1870–1920*. Madison, Wis.: University of Wisconsin Press.

Plantinga, A. 1991. When faith and reason clash: Evolution and the Bible. *Christian Scholar's Review* 21 (1): 8–32.

Pollard, S. 1968. *The Idea of Progress: History and Society*. New York: Basic Books.

Popper, K. R. 1959. *The Logic of Scientific Discovery*. London: Hutchinson.

——— 1972. *Objective Knowledge.* Oxford: Oxford University Press.

——— 1974. Darwinism as a metaphysical research programme. In *The Philosophy of Karl Popper,* ed. P. A. Schilpp, 1, 133–143. LaSalle Ill.: Open Court.

Porter, R. 1989. Erasmus Darwin: Doctor of evolution? In *History, Humanity and Evolution: Essays for J. C. Greene,* ed. J. R. Moore, 39–70. Cambridge: Cambridge University Press.

Poulton, E. B. 1884. The structures connected with the ovarian ovum of Marsupialia and Monotremata. *Quarterly Journal of Microscopical Science,* n.s. 24: 118–128.

——— 1890. *The Colours of Animals.* London: Kegan Paul, Trench, Truebner.

——— 1908. *Essays on Evolution, 1889–1907.* Oxford: Oxford University Press.

Price, G. R. 1972. Fisher's "Fundamental Theorem" made clear. *Annals of Human Genetics* 36: 129–140.

Price, R. 1787. *The Evidence for a Future Period of Improvement in the State of Mankind.* London.

Provine, W. 1971. *The Origins of Theoretical Population Genetics.* Chicago: University of Chicago Press.

——— 1981. Origins of the Genetics of Natural Populations series. In *Dobzhansky's Genetics of Natural Populations I–XLIII,* ed. R. C. Lewontin et al. 1981, 1–76. New York: Columbia University Press.

——— 1986. *Sewall Wright and Evolutionary Biology.* Chicago: University of Chicago Press.

Rainger, R. 1991. *An Agenda for Antiquity: Henry Fairfield Osborn and Vertebrate Paleontology at the American Museum of Natural History, 1890–1935.* Tuscaloosa: University of Alabama Press.

Rainger, R., K. R. Benson, and J. Maienschein, eds. 1988. *The American Development of Biology.* Philadelphia: University of Pennsylvania Press.

Raup, D. 1988. Testing the fossil record for evolutionary progress. In *Evolutionary Progress,* ed. M. Nitecki, 293–317. Chicago: University of Chicago Press.

Reddick, J. 1990. The shattered whole: Georg Büchner and Naturphilosophie. In *Romanticism and the Sciences,* ed. A. Cunningham and N. Jardine, 322–340. Cambridge: Cambridge University Press.

Reid, T. [1785] 1863. Essays on the intellectual powers of man. In *Works of Thomas Reid,* 6th ed., ed. W. Hamilton. Edinburgh.

Richards, E. 1987. A question of property rights: Richard Owen's evolutionism reassessed. *British Journal for the History of Science* 20: 129–171.

Richards, R. J. 1987. *Darwin and the Emergence of Evolutionary Theories of Mind and Behavior.* Chicago: University of Chicago Press.

——— 1992. *The Meaning of Evolution: The Morphological Construction and Ideological Reconstruction of Darwin's Theory.* Chicago: University of Chicago Press.

Richmond, M. 1988. Darwin's study of the Cirripedia. In *The Correspondence of Charles Darwin,* ed. F. Burkhardt and S. Smith, 4, 388–409. Cambridge: Cambridge University Press.

Ridley, M. 1985. *The Problems of Evolution.* Oxford: Oxford University Press.

——— 1986. *Evolution and Classification: The Reformation of Cladism.* New York: Longman.

Robertson, C. 1906. Ecological adaptation and ecological selection. *Science* 23: 307–310.

Roger, J. 1989. *Buffon: un philosophe au jardin du roi.* Paris: Fayard.

——— 1993. *Les sciences de la vie dans la pensée française au XVIIIe siècle,* 3d ed. Paris: Éditions Albin Michel.

Romer, A. S. 1949. Time series and trends in animal evolution. In *Genetics, Paleontology and Evolution,* ed. G. L. Jepsen, E. Mayr, and G. G. Simpson, 103–120. Princeton, N.J.: Princeton University Press.

Rudwick, M. J. S. 1963. The foundation of the Geological Society of London: Its scheme for cooperative research and its struggle for independence. *British Journal for the History of Science* 1: 325–355.

——— 1969. The strategy of Lyell's *Principles of Geology. Isis* 61: 5–33.

——— 1974. Darwin and Glen Roy: A "great failure" in scientific method? *Studies in History and Philosophy of Science* 5: 97–185.

——— 1986. *The Great Devonian Controversy.* Chicago: University of Chicago Press.

Rupke, N. A. 1994. *Richard Owen: Victorian Naturalist.* New Haven: Yale University Press.

Ruse, M. 1975a. Darwin's debt to philosophy: An examination of the influence of the philosophical ideas of John F. W. Herschel and William Whewell on the development of Charles Darwin's theory of evolution. *Studies in History and Philosophy of Science* 6: 159–181.

——— 1975b. Charles Darwin's theory of evolution: An analysis. *Journal of the History of Biology* 8: 219–241.

——— 1979a. *The Darwinian Revolution: Science Red in Tooth and Claw.* Chicago: University of Chicago Press.

——— 1979b. *Sociobiology: Sense or Nonsense?* Dordrecht, Holland: Reidel.

——— 1980. Charles Darwin and group selection. *Annals of Science* 37: 615–630.

——— 1982. *Darwinism Defended: A Guide to the Evolution Controversies.* Reading, Mass.: Addison-Wesley.

——— 1984. Is there a limit to our knowledge of evolution? *BioScience* 34 (2): 100–104.

——— 1986. *Taking Darwin Seriously.* Oxford: Blackwell.

——— ed. 1988. *But Is It Science? The Philosophical Question in the Creation/Evolution Controversy.* Buffalo, N.Y.: Prometheus.

——— 1989. *The Darwinian Paradigm: Essays on Its History, Philosophy and Religious Implications*. London: Routledge.

——— 1993. Evolution and progress. *Trends in Ecology and Evolution* 8 (2): 55–59.

——— 1995. *Evolutionary Naturalism: Selected Essays*. London: Routledge.

Ruse, M., and E. O. Wilson. 1986. Moral philosophy as applied science. *Philosophy* 61: 173–192.

Russell, D. A. 1987. Models and paintings of North American dinosaurs. In *Dinosaurs Past and Present*, ed. S. J. Czerkas and E. C. Olson, 115–131. Seattle: Distributed by The University of Washington Press for The Natural History Museum of Los Angeles County.

Russell, E. S. 1916. *Form and Function, a Contribution to the History of Animal Morphology*. London: John Murray.

Russett, C. E. 1966. *The Concept of Equilibrium in American Social Thought*. New Haven: Yale University Press.

——— 1976. *Darwin in America: The Intellectual Response, 1865–1912*. San Francisco: Freeman.

——— 1989. *Sexual Science*. Cambridge, Mass.: Harvard University Press.

Sahlins, M. 1976. *The Use and Abuse of Biology*. Ann Arbor: University of Michigan Press.

Sarkar, S. 1992a. Haldane as biochemist: The Cambridge decade, 1923–1932. In *The Founders of Evolutionary Genetics*, ed. S. Sarkar, 53–81. Dordrecht: Kluwer Academic Publishers.

——— 1992b. Science, philosophy and politics in the work of J. B. S. Haldane, 1922–1937. *Biology and Philosophy* 7 (4): 385–409.

Schaffer, S. 1989. The nebular hypothesis and the science of progress. In *History, Humanity and Evolution: Essays for John C. Greene*, ed. J. R. Moore, 131–164. Cambridge: Cambridge University Press.

Schelling, F. W. J. 1962–67. *Briefe und Dokumente*, 3 vols., ed. Horst Fuhrmans. Bonn: H. Bouvier Verlag.

Scherren, H. 1905. *The Zoological Society of London: A Sketch of Its Foundation and Development and the Story of Its Farm, Museum, Gardens, Menagerie and Library*. London: Cassell.

Schiller, J. 1971. L'échelle des êtres et la série chez Lamarck. In *Colloque international "Lamarck" tenue au Muséum national d'histoire naturelle*, ed. J. Schiller, 87–103. Paris: Blanchard.

Schmidt, H. 1909. *Das biogenetische Grundgesetz Ernst Haeckels und sein Gegner*. Frankfurt: Neuer Frankfurt Verlag.

Schofield, R. E. 1963. *The Lunar Society of Birmingham: A Social History of Provincial Science and Industry in the Eighteenth Century*. Oxford: Clarendon Press.

Schopf, T. J. M. 1977. Patterns of evolution: A summary and discussion. In

Patterns of Evolution as Illustrated by the Fossil Record, ed. A. Hallam, 547–562. Amsterdam: Elsevier.
Schweber, S. 1977. The origin of the *Origin* revisited. *Journal of the History of Biology* 10: 229–316.
——— 1980. Darwin and the political economists: Divergence of character. *Journal of the History of Biology* 13: 195–289.
Secord, J. A. 1989. Behind the veil: Robert Chambers and "Vestiges." In *History, Humanity and Evolution: Essays for John C. Greene,* ed. J. R. Moore, 165–194. Cambridge: Cambridge University Press.
——— 1991. Edinburgh Lamarckians: Robert Jameson and Robert E. Grant. *Journal of the History of Biology* 24: 1–18.
Sedgwick, A. 1831. Address to the Geological Society. *Proceedings of the Geological Society of London* 1: 281–316.
——— 1833. *A Discourse on the Studies of the University.* London: Parker.
——— 1845. Vestiges. *Edinburgh Review* 82: 1–85.
——— 1850. *A Discourse on the Studies at the University of Cambridge,* 5th ed. Cambridge: Cambridge University Press.
Seger, J., and W. D. Hamilton. 1988. Parasites and sex. In *The Evolution of Sex: An Examination of Current Ideas,* ed. R. E. Michod and B. Levin, 176–193. Sunderland, Mass.: Sinauer.
Segerstrale, U. 1986. Colleagues in conflict: An "in vitro" analysis of the sociobiology debate. *Biology and Philosophy* 1: 53–88.
Selzer, J., ed. 1993. *Understanding Scientific Prose.* Madison, Wis.: University of Wisconsin Press.
Shapin, S. 1979. *Homo phrenologicus:* Anthropological perspectives on an historical problem. In *Natural Order,* ed. B. Barnes and S. Shapin, 41–71. London: Sage.
——— 1982. History of science and its social reconstructions. *History of Science* 20: 157–211.
Shapin, S., and S. Schaffer. 1985. *Leviathan and the Air-Pump: Hobbes, Boyle and the Experimental Life.* Princeton, N.J.: Princeton University Press.
Sheets-Johnstone, M. 1982. Why Lamarck did not discover the principle of natural selection. *Journal of the History of Biology* 15: 443–465.
Sheppard, P. M. 1958. *Natural Selection and Heredity.* London: Hutchinson.
Shils, E. 1968. The profession of science. *Advancement of Science* 24: 469–480.
Simpson, G. G. 1928. *A Catalogue of the Mesozoic Mammalia in the Geological Department of the British Museum.* London: British Museum (Natural History).
——— 1944. *Tempo and Mode in Evolution.* New York: Columbia University Press.
——— 1949. *The Meaning of Evolution.* New Haven: Yale University Press.
——— 1951. *Horses.* New York: Oxford University Press.

——— 1953. *The Major Features of Evolution.* New York: Columbia University Press.

——— 1962. *Principles of Animal Taxonomy.* New York: Columbia University Press.

——— 1978. *Concession to the Improbable: An Unconventional Autobiography.* New Haven: Yale University Press.

——— 1987. *Simple Curiosity: Letters from George Gaylord Simpson to His Family, 1921–1970,* ed. L. F. Laport. Berkeley and Los Angeles: University of California Press.

Sinnott, E. W. 1955. *The Biology of the Spirit.* New York: Viking.

Slobodkin, L. B. 1988. [Review of] *An Urchin in the Storm. American Scientist* 76: 503–504.

Smith, A. [1776] 1937. *The Wealth of Nations.* New York: Modern Library.

Smith, R. 1972. Alfred Russel Wallace: Philosophy of nature and man. *British Journal for the History of Science* 6: 177–199.

Smocovitis, V. B. 1988. Botany and the evolutionary synthesis: The life and work of G. Ledyard Stebbins, Jr. Ph.D. diss., Cornell University, Ithaca, N.Y.

——— 1992. Unifying biology: The evolutionary synthesis and evolutionary biology. *Journal of the History of Biology* 25: 1–65.

Sober, E. 1984. *The Nature of Selection.* Cambridge, Mass.: MIT Press.

Spadafora, D. 1990. *The Idea of Progress in Eighteenth Century Britain.* New Haven: Yale University Press.

Spencer, H. 1842. Letter VII. *Nonconformist,* October 19.

——— 1851. *Social Statics; Or the Conditions Essential to Human Happiness Specified and the First of Them Developed.* London: J. Chapman.

——— 1852a. A theory of population, deduced from the general law of animal fertility. *Westminster Review* 1: 468–501.

——— [1852b] 1868. The development hypothesis. *The Leader.* Reprinted in *Essays: Scientific, Political and Speculative,* 1, 377–383. London: Williams and Norgate.

——— 1855. *Principles of Psychology.* London: Longman, Brown, Green, and Longmans.

——— [1857] 1868. Progress: Its law and cause. *Westminster Review* 67: 244–267. Reprinted in *Essays: Scientific, Political and Speculative,* 1, 1–60. London: Williams and Norgate.

——— [1857] 1896. The ultimate laws of physiology. *National Review* 5: 332–355. Reprinted as "Transcendental physiology" in *Essays: Scientific, Political, and Speculative,* 3d ed., 1, 63–107. New York: Appleton.

——— 1904. *Autobiography.* London: Williams and Norgate.

Spengler, O. [1926] 1966. *The Decline of the West.* Authorized translation with notes by Charles Francis Atkinson. New York: Knopf.

Stapledon, W. O. 1930. *Last and First Men: A Story of the Near and Far Future.* London: Methuen.

Stearn, W. 1981. *The Natural History Museum at South Kensington.* London: Heineman.

Stebbins, G. L. 1950. *Variation and Evolution in Plants.* New York: Columbia University Press.

——— 1969. *The Basis of Progressive Evolution.* Chapel Hill: University of North Carolina Press.

——— 1982. *Darwin to DNA, Molecules to Humanity.* San Francisco: W. H. Freeman.

Stebbins, G. L., and F. J. Ayala. 1981. Is a new evolutionary synthesis necessary? *Science* 213: 967–971.

Stebbins, G. L., and Daniel L. Hartl. 1988. Comparative evolution: Latent potential for anagenetic advance. *Proceedings of the National Academy of Sciences* 85: 5141–5145.

Stebbins, R. E. 1974. France. In *The Comparative Reception of Darwinism,* ed. T. F. Glick, 117–167. Austin: University of Texas Press.

Stejneger, L. 1906. C. S. Rafinesque on evolution. *Science* 23: 785–786.

Stenseth, N. C. 1985. Darwinian evolution in ecosystems: The Red Queen view. In *Evolution: Essays in Honour of John Maynard Smith,* ed. P. J. Greenwood, P. H. Harvey, and M. Slatkin, 55–72. Cambridge: Cambridge University Press.

Stewart, D. [1793] 1963. Account of the life and writings of Adam Smith. In *The Works of Adam Smith.* Aalen, Germany: Otto Zeller.

Stimson, D. 1948. *Scientists and Amateurs: A History of the Royal Society.* New York: Schuman.

Sulloway, F. J. 1982a. Darwin and his finches: The evolution of a legend. *Journal of the History of Biology* 15: 1–53.

——— 1982b. Darwin's conversion: The Beagle voyage and its aftermath. *Journal of the History of Biology* 15: 325–396.

——— 1985. Darwin's early intellectual development: An overview of the Beagle voyage (1831–1836). In *The Darwinian Heritage,* ed. D. Kohn, 121–154. Princeton, N.J.: Princeton University Press.

Sumner, J. B. 1816. *A Treatise on the Records of the Creation and on the Moral Attibutes of the Creator . . .* London: Hatchard.

Swainson, W. 1835. *A Treatise on the Geography and Classification of Animals.* London: Longman, Rees, Orme, Brown, Green and Longman.

Swetlitz, M. 1991. Julian Huxley, George Gaylord Simpson and the idea of progress in twentieth century evolutionary biology. Ph.D. diss., University of Chicago.

Taylor, C. 1975. *Hegel.* Cambridge: Cambridge University Press.

Teilhard de Chardin, P. 1955. *Le phénomène humaine.* Paris: Editions de Seuil.

——— 1959. *The Phenomenon of Man*. London: Collins.

Thomson, K. S. 1977. The pattern of diversification among fishes. In *Patterns of Evolution as Illustrated by the Fossil Record,* ed. A. Hallam, 377–404. Amsterdam: Elsevier.

Tindall, G. B., and D. E. Shi. 1989. *America: A Narrative History,* 2d ed. New York: Norton.

Tschetwerikoff, S. S. 1926. On certain features of the evolutionary process from the viewpoint of modern genetics. [In Russian.] *Journal of Experimental Biology* 2: 3–54.

Turgot, A. R. J. [1750] 1895. *The Life and Writings of Turgot,* ed. W. W. Stephens. London: Longmans, Green, and Co.

——— 1895. Discourse at the Sorbonne on the successive advances of the human mind. In *The Life and Writings of Turgot,* ed. W. W. Stephens, 159–173. London: Longmans, Green, and Co.

Turner, J. R. G. 1987. Random genetic drift, R. A. Fisher, and the Oxford School of ecological genetics. In *The Probabilistic Revolution,* ed. L. Krüger, G. Gigerenzer, and M. S. Morgan, 313–354. Cambridge, Mass.: MIT Press.

Turner, R. S. 1971. The growth of professional research in Prussia, 1818 to 1848: Causes and context. *Historical Studies in the Physical Sciences* 3: 137–182.

Van Valen, L. 1973. A new evolutionary law. *Evolutionary Theory* 1: 1–30.

Vaughan, T. W. 1906. The work of Hugo de Vries and its importance in the study of problems of evolution. *Science* 23: 681–691.

Vermeij, G. J. 1987. *Evolution and Escalation: An Ecological History of Life.* Princeton, N.J.: Princeton University Press.

von Baer, K. E. [1828–37] 1853. Über Enwickelungsgeschichte der Thiere [Fragments related to philosophical zoology: Selected from the works of K. E. von Baer]. In *Scientific Memoirs,* ed. and trans. A. Henfry and T. H. Huxley, 195. London: Taylor and Francis.

Vorzimmer, P. 1970. *Charles Darwin: The Years of Controversy.* Philadelphia: Temple University Press.

Vrba, E. S., and S. J. Gould. 1986. The hierarchical expansion of sorting and selection: Sorting and selection cannot be equated. *Paleobiology* 12: 217–228.

Wagar, W. 1972. *Good Tidings: The Belief in Progress from Darwin to Marcuse.* Bloomington: Indiana University Press.

Wallace, A. R. [1855] 1870. On the law which has regulated the introduction of new species. *Annals and Magazine of Natural History* 16: 184–196. Reprinted in Wallace 1870b, 1–25.

——— [1858] 1870. On the tendency of varieties to depart indefinitely from the original type. *Journal of the Proceedings of the Linnean Society, Zoology* 3: 53–62. Reprinted in Wallace 1870b, 26–44.

——— [1864] 1870. The origin of human races and the antiquity of man deduced from the theory of natural selection. *Journal of the Anthropological Society of London* 2: clvii–clxxxvii. Revised and reprinted as "The development of human races under the law of natural selection" in Wallace 1870b, 302–331.

——— [1866] 1870. On the phenomena of variation and geographical distribution as illustrated by the Papilionidae of the Malayan region. *Transactions of the Linnean Society, London* 25: 1–27. Revised and reprinted as "The Malayan Papilionidae, or swallow-tailed butterflies, as illustrative of the theory of natural selection" in Wallace 1870b, 130–200.

——— 1870a. The limits of natural selection as applied to man. In Wallace 1870b, 332–371.

——— 1870b. *Contributions to the Theory of Natural Selection: A Series of Essays.* London: Macmillan.

——— 1876. *The Geographical Distribution of Animals,* 2 vols. London: Macmillan.

——— 1900. *Studies: Scientific and Social,* 2 vols. London: Macmillan.

——— 1903. *Man's Place in the Universe.* London: Chapman and Hall.

——— 1905. *My Life: A Record of Events and Opinions,* 2 vols. London: Chapman and Hall.

——— 1907. *Is Mars Habitable?* London: Macmillan.

Wallace, B. 1958. The average effect of radiation-induced mutations on viability in *Drosophila melanogaster. Evolution* 12: 532–552.

——— 1963. Further data on the overdominance of induced mutations. *Genetics* 48: 633–651.

Waters, C. K., and A. van Helden, eds. 1992. *Julian Huxley: Biologist and Statesman of Science.* Houston: Rice University Press.

Watson, J. 1965. *Molecular Biology of the Gene.* New York: Benjamin.

——— 1968. *The Double Helix.* Signet Books: New York.

Webb, R. K. 1980. *Modern England from the Eighteenth Century to Present,* 2d ed. New York: Harper and Row.

Weinberg, S. 1994. The emergence of life. *Scientific American,* October, pp. 46–49.

Weldon, W. F. R. 1884. On the head-kidney of *Bdellostoma,* with a suggestion as to the origin of the suprarenal bodies. *Quarterly Journal of Microscopical Science* 24: 171–182.

——— 1885. On the suprarenal bodies of vertebrates. *Quarterly Journal of Microscopical Science* 25: 137–150.

——— 1890. The variations occurring in certain decapod Crustacea. I. *Crangon vulgaris. Proceedings of the Royal Society of London* 47: 445–453.

——— 1892. Certain correlated variations in *Crangon vulgaris. Proceedings of the Royal Society* 51: 2–21.

——— 1893. On certain correlated variations in *Carcinus moenas. Proceedings of the Royal Society* 54: 318–329.

——— 1895. Attempt to measure the deathrate due to the selective destruction of *Carcinus moenas* with respect to a particular dimension. *Proceedings of the Royal Society* 57: 360–379.

——— 1898. Presidential Address to the Zoological Section of the British Association. *British Association for the Advancement of Science: Report of the Sixty-Eighth Meeting, Bristol, September 1898,* 887–902. London: John Murray.

Wells, G. A. 1967. Goethe and evolution. *Journal of the History of Ideas* 28: 537–550.

Wells, H. G., J. S. Huxley, and G. P. Wells. 1929–30. *The Science of Life.* London: Amalgamated Press.

Wheeler, W. M. 1910. *Ants: Their Structure, Development and Behavior.* New York: Columbia University Press.

——— 1918. A study of some ant larvae, with a consideration of the origin and meaning of the social habit among insects. *Proceedings of the American Philosophical Society* 57: 293–343.

——— 1923. *Social Life Among the Insects, Being a Series of Lectures Delivered at the Lowell Institute in Boston in March 1922.* New York: Harcourt, Brace.

——— 1927. *Emergent Evolution and the Social.* London: Kegan Paul, Trench, Trubner.

Whewell, W. 1831. [Review of} *Preliminary Discourse . . .* by J. F. W. Herschel . . . *Quarterly Review* 45: 374–407.

——— 1832. [Review of] Charles Lyell's *Principles of Geology. Quarterly Review* 47: 103–132.

——— 1833. *Astronomy and General Physics (Bridgewater Treatise, 3).* London: Pickering.

——— 1837. *The History of the Inductive Sciences,* 3 vols. London: Parker.

——— 1840. *The Philosophy of the Inductive Sciences,* 2 vols. London: Parker.

Whittington, H. B. 1975. The enigmatic animal *Opabinia regalis,* Middle Cambrian Burgess Shale, British Columbia. *Philosophical Transactions of the Royal Society, London* B 271: 1–43.

Williams, G. C. 1966. *Adaptation and Natural Selection.* Princeton, N.J.: Princeton University Press.

——— 1975. *Sex and Evolution.* Princeton, N.J.: Princeton University Press.

——— 1988. Huxley's evolution and ethics in sociobiological perspective. *Zygon* 23: 383–407.

——— 1993. Mother Nature is a wicked old witch. In *Evolutionary Ethics,* ed. M. H. Nitecki and D. V. Nitecki, 217–231. Albany: State University of New York Press.

Williams, W. C. 1971. Robert Chambers. In *The Dictionary of Scientific Biography,* Editor-in-Chief Charles Coulston Gillispie, 3:191–193. New York: Charles Scribner's Sons.

Wilson, E. B. 1896. *The Cell in Development and Inheritance.* New York: Macmillan.

——— 1901. Aims and methods of study of natural history. *Science* 13: 14–23.

Wilson, E. O. 1971. *The Insect Societies.* Cambridge, Mass.: Harvard University Press.

——— 1975a. *Sociobiology: The New Synthesis.* Cambridge, Mass.: Harvard University Press.

——— 1975b. Letter to the editor. *New York Review of Books,* December 11, pp. 60–61.

——— 1978. *On Human Nature.* Cambridge, Mass.: Cambridge University Press.

——— 1990. *Success and Dominance in Ecosystems: The Case of the Social Insects.* Oldendorf/Luhe, Germany: Ecology Institute.

——— 1994. *Naturalist.* Washington, D.C.: Island Books/Shearwater Books.

Wilson, E. O., and F. M. Peter, eds. 1988. *Biodiversity.* Washington, D.C.: National Academy Press.

Wilson, L. 1971. Sir Charles Lyell and the species question. *American Scientist* 59: 43–45.

Winsor, M. P. 1991. *Reading the Shape of Nature: Comparative Zoology at the Agassiz Museum.* Chicago: University of Chicago Press.

Winstanley, D. A. 1935. *Unreformed Cambridge.* Cambridge: Cambridge University Press.

——— 1940. *Early Victorian Cambridge.* Cambridge: Cambridge University Press.

——— 1947. *Later Victorian Cambridge.* Cambridge: Cambridge University Press.

Wittgenstein, L. 1923. *Tractatus Logico-Philosophicus.* London: Routledge and Kegan Paul.

Worthington, W. 1743. *An Essay on the Scheme and Conduct, Procedure and Extent of Man's Redemption.* London: Edward Cave.

Wright, C. 1871. *Darwinism: Being an Examination of Mr. St. George Mivart's "Genesis of Species."* [Reprinted from the *North American Review,* July 1871, with additions.] London: J. Murray.

Wright, R. 1987. *Three Scientists and Their Gods.* New York: Times Books.

Wright, S. 1921. Correlation and causation. *Journal of Agricultural Research* 20: 557–585.

——— [1922] 1986. Coefficients of inbreeding and relationship. *American Naturalist* 56: 330–338. Reprinted in Wright 1986, 13–21.

——— [1923] 1986. Mendelian analysis of the pure breeds of livestock: II. The

Duchess family of shorthorns as bred by Thomas Bates. *Journal of Heredity* 14. Reprinted in Wright 1986, 34–51.

——— [1931] 1986. Evolution in Mendelian populations. *Genetics* 16: 97–159. Reprinted in Wright 1986, 98–160.

——— [1932] 1986. The roles of mutation, inbreeding, crossbreeding and selection in evolution. *Proceedings of the Sixth International Congress of Genetics* 1: 356–366. Reprinted in Wright 1986, 161–171.

——— [1934a] 1986. Physiological and evolutionary theories of dominance. *American Naturalist* 68 (January–February): 25–53. Reprinted in Wright 1986, 173–202.

——— [1934b] 1986. Professor Fisher on the theory of dominance. *American Naturalist* 68 (November–December): 562–565. Reprinted in Wright 1986, 203–206.

——— [1939] 1986. *Statistical genetics in relation to evolution.* Actualités scientifiques et industrielles, 802: Exposés de Biométrie et de la statistique biologique, XIII. Paris, Hermann and Cie. Reprinted in Wright 1986, 283–341.

——— [1948] 1986. Evolution, organic. In *Encyclopaedia Britannica,* 14th ed., revised, 8: 915–929. Reprinted in Wright 1986, 524–538..

——— [1949] 1986. Adaptation and selection. Originally published in Jepson, Simpson, and Mayr 1949, 365–389. Reprinted in Wright 1986, 546–570.

——— 1953. Gene and organism. *American Naturalist* 83 (January): 5–18.

——— 1964. Biology and the philosophy of science. *Monist* 48 (2): 265–290.

——— 1968. *Evolution and the Genetics of Populations: A Treatise,* 4 vols. Chicago: University of Chicago Press.

——— [1978] 1986. The relation of livestock breeding to theories of evolution. *Journal of Animal Science* 46: 1192–1200. Reprinted in Wright 1986, 3–11.

——— 1982. Character change, speciation and the higher taxa. *Evolution* 36 (3): 427–443.

——— 1986. *Evolution: Selected Papers,* ed. W. B. Provine. Chicago: University of Chicago Press.

Wright, S., and T. Dobzhansky. [1946] 1981. Experimental reproduction of some of the changes caused by natural selection in certain populations of *Drosophila pseudoobscura. Genetics* 31: 125–156. Reprinted in Lewontin et al. 1981, 396–426.

Yeo, R. 1993. *Defining Science: William Whewell, Natural Knowledge, and Public Debate in Eary Victorian Britain.* Cambridge: Cambridge University Press.

Young, R. M. 1985. *Darwin's Metaphor: Nature's Place in Victorian Culture.* Cambridge: Cambridge University Press.

Zimmerman, M. 1987. The evolution-creation controversy: Opinions of Ohio high school biology teachers. *Ohio Journal of Science* 87 (4): 115–124.

Credits

Page 51: Modified from G. Barsanti, "Buffon et l'image de la nature," in *Buffon 88*, ed. J. Gayon (Paris: J. Vrin, 1992).

Page 124: Unpublished notes (1828) from the papers of Richard Owen, reproduced by permission of the President and Council of the Royal College of Surgeons of England.

Page 258: Drawing (summer 1857) from the papers of E. D. Cope, Quaker Collection, Haverford College, Haverford, Penn.

Page 272: Redrawn from the papers of H. F. Osborn, American Museum of Natural History, New York.

Page 327: Photo of Julian Huxley by W. Suschitzky from J. R. Baker, *Julian Huxley: Scientist and World Citizen* (Lanham, Md.: UNESCO/UNIPUB, 1978).

Page 339: Painting of E. B. Ford from *Ecological Genetics and Evolution: Essays in Honour of E. B. Ford*, ed. Robert Creed (New York: Appleton-Century-Crofts, 1971).

Pages 369, 371: From S. Wright, "The Roles of Mutation, Inbreeding, Crossbreeding and Selection in Evolution," *Proceedings of the Sixth International Congress of Genetics* 1: 356–366.

Page 374: Redrawn from a letter to R. A. Fisher by S. Wright, American Philosophical Society, Philadelphia, Penn.

Page 390: Photo of Theodosius Dobzhansky from *Evolution*, by T. Dobzhansky, F. J. Ayala, G. L. Stebbins, and J. W. Valentine. Copyright © 1977 by W. H. Freeman and Company. Used with permission.

Page 420: Photo of G. G. Simpson from the American Philosophical Society, Philadelphia, Penn.

Page 421: From G. G. Simpson, *Tempo and Mode in Evolution.* Copyright © 1944 by Columbia University Press. Reprinted with permission of the publisher.

Page 433: Table from G. Ledyard Stebbins, *The Basis of Progressive Evolution.* Copyright © 1969 by the University of North Carolina Press. Used by permission of the publisher.

Page 444: From B. Wallace, "Studies on Irradiated Populations of *D. melanogaster*," *Journal of Genetics* 54 (1956): 2. Reprinted with the permission of Cambridge University Press.

Page 465: Photo of Richard Dawkins by Lisa Lloyd.

Page 488: From J. A. Doyle, "Patterns of Evolution in Early Angiosperms," in *Patterns of Evolution as Illustrated by the Fossil Record*, ed. A. Hallam (New York: Scientific Publishing Co., 1977).

Page 489: From G. J. Vermeij, *Evolution and Escalation*. Copyright © 1987 by Princeton University Press.

Page 490: From L. van Valen, "A New Evolutionary Law," *Evolutionary Theory* (1973): 4.

Page 495: From M. J. Benton, "Progress and Competition in Macroevolution," *Biological Reviews of the Cambridge Philosophical Society* 62 (1987): 305–338. Reprinted with the permission of Cambridge University Press.

Page 496 (top): Photo of Stephen Jay Gould by Paula M. Lerner/Woodfin Camp & Associates.

Page 496 (bottom): Diagrams from Michael Ruse, *Darwinism Defended: A Guide to the Evolution Controversies* (Reading, Mass.: Addison Wesley, 1982).

Page 499: Reprinted with permission from S. J. Gould, N. L. Gilinsky, and R. Z. German, "Asymmetry of Lineages and the Direction of Evolutionary Time," *Science* 236 (1987): 1440. Copyright 1987 American Association for the Advancement of Science.

Page 501: From S. J. Gould, *Wonderful Life* (New York: Norton, 1989).

Page 505 (top): From John Sepkoski, Jr., "Foundations of Life in the Oceans," in *The Book of Life*, ed. S. J. Gould (New York: Viking Books, 1993).

Page 505 (bottom): From Peter Andrews and Christopher Stringer, "The Primates Progress," in *The Book of Life*, ed. S. J. Gould (New York: Viking Books, 1993). Reproduced with permission from Random House.

Page 509: From E. O. Wilson, *Sociobiology: The New Synthesis*. Copyright © 1975 by the President and Fellows of Harvard College. Reprinted by permission of Harvard University Press.

Page 514: From E. O. Wilson, *On Human Nature*. Copyright © 1978 by the President and Fellows of Harvard College. Reprinted by permission of Harvard University Press. Based on Kent V. Flannery, "The Cultural Evolution of Civilizations," *Annual Review of Ecology and Systematics* 3 (1972): 399–426.

Page 518: From R. Lewontin, J. A. Moore, W. B. Provine, and B. Wallace, *Dobzhansky's Genetics of Natural Populations*. Copyright © 1944 by Columbia University Press. Reprinted with permission of the publisher.

Page 527: From "The Tower of Time" (Washington, D.C.: Smithsonian Press).

Page 528: Model by Dale A. Russell and Ron Seguin from Dale A. Russell, "Models, paintings, and the dinosaurs of North America," in *Dinosaurs Past and Present*, vol. 1, ed. Sylvia J. Czerkas and Everett C. Olson (Seattle and London: University of Washington Press, 1987). Reproduced with permission of the Canadian Museum of Nature, Ottawa, Canada.

Page 529: From B. Cork and L. Bresler, *The Young Scientist Book of Evolution*. Copyright © 1991, 1985 Usborne Publishing Ltd.

Pages 530–532: From Steven Weinberg, "The Emergence of Life," *Scientific American*, October 1994, 46–49. Copyright © 1994 by Scientific American, Inc. All rights reserved.

Page 533: From N. Eldredge, *Macro-Evolutionary Dynamics* (New York: McGraw-Hill, 1989). Reproduced with permission of The McGraw-Hill Companies.

Page 535: From D. McShea, "Complexity and Evolution," *Biology and Philosophy* 6 (1991): 315. Reprinted by permission of Kluwer Academic Publishers.

Index